# HANDBOOK ON ALTERNATIVE THEORIES OF INNOVATION

*DEDICATION*

Benoît Godin, born in 1958, passed away on January 5, 2021. He was co-editor of this Handbook and the inspirer of this editorial project. Beyond that, he was a prolific and internationally renowned researcher, known for his original work on the history of measurement statistics in science, technology and innovation; on the history of science and research concepts and models; and on cultures of quantification and metrics. In the last fifteen years of his career, he had turned his attention to the intellectual history of innovation (the idea of innovation and its theology), scrutinizing how the term innovation was used and by which actors, from Greek Antiquity to the present day. In the context of this colossal work, he showed, among other things, that the "superlative" connotation of the term is very contemporary, whereas it had rather been used as an anathema throughout history until the late 1960s or early 1970s.

A creative and prodigious scholar, he was entirely devoted to research and pursued that research with the greatest intellectual rigor and honesty. He leaves an impressive contribution to the study of the history of science and innovation, including the history of the perception of science and innovation in economic theory and society. Perhaps out of modesty, he did not consider himself a historian in his own right, although he made landmark contributions to the history of ideas. He has authored numerous publications, including frequent articles in leading STS journals, as well as important monographs. To name a few: *Innovation Contested: The Idea of Innovation Over the Centuries* (Routledge, 2015); *Models of Innovation: The History of an Idea* (MIT Press, 2017); *Critical Studies of Innovation: Alternative Approaches to the Pro-Innovation Bias* (which we co-edited and which was published by Edward Elgar in 2017); *The Invention of Technological Innovation: Languages, Discourses and Ideology in Historical Perspective* (Edward Elgar, 2019); and his latest book: *The Idea of Technological Innovation: A Brief Alternative History* (Edward Elgar, 2020). He was passionate about the use of words at various times, the transformation of their meaning, their circulation, their reappropriation and the political significance they could have.
This Handbook is dedicated to him.

# Handbook on Alternative Theories of Innovation

*Edited by*

Benoît Godin

*Institut National de la Recherche Scientifique, Montréal, Canada*

Gérald Gaglio

*University Côte d'Azur, France*

Dominique Vinck

*University of Lausanne, Switzerland*

**EE** **Edward Elgar**
PUBLISHING

Cheltenham, UK • Northampton, MA, USA

Published by
Edward Elgar Publishing Limited
The Lypiatts
15 Lansdown Road
Cheltenham
Glos GL50 2JA
UK

Edward Elgar Publishing, Inc.
William Pratt House
9 Dewey Court
Northampton
Massachusetts 01060
USA

Paperback edition 2023

A catalogue record for this book
is available from the British Library

Library of Congress Control Number: 2021944948

This book is available electronically in the **Elgar**online
Sociology, Social Policy and Education subject collection
http://dx.doi.org/10.4337/9781789902303

ISBN 978 1 78990 229 7 (cased)
ISBN 978 1 78990 230 3 (eBook)
ISBN 978 1 0353 1553 6 (paperback)

Printed and bound by CPI Group (UK) Ltd, Croydon, CR0 4YY

# Contents

# Figures

# Contributors

**Fayaz Ahmad Sheikh**, Post-PhD Innovation Researcher at the School of Management, Zhejiang University, Hangzhou, 310058 P.R. China.

**Carolina Bagattolli**, Associate Professor, Department of Economics, Public Policy Postgraduate Program, Federal University of Paraná, Brazil.

**Riza Batista-Navarro**, Lecturer in Text Mining, Department of Computer Science, School of Engineering, University of Manchester, UK.

**Vincent Blok**, Associate Professor, Philosophy Group, Wageningen University, the Netherlands.

**Frank Boons**, Professor of Innovation and Sustainability, Alliance Manchester Business School, University of Manchester, UK.

**Tiago Brandão,** Integrated Researcher, Faculty of Social Sciences and Humanities, Nova University of Lisbon, Portugal.

**Céline Cholez**, Associate Professor, Grenoble Engineering Institute, PACTE Laboratory, Grenoble-Alpes University, France.

**Darryl Cressman**, Assistant Professor in the Philosophy Department in the Faculty of Arts & Social Sciences at Maastricht University, the Netherlands.

**Mónica Edwards-Schachter**, Independent researcher and consultant (social innovation and STEAM education fields). External lecturer, Department of Science Education, Postgraduate Program, University of Burgos, Spain.

**Irwin Feller**, Professor Emeritus, Economics, Pennsylvania State University, United States.

**Gérald Gaglio**, Full Professor of Sociology at the University Côte d'Azur, France.

**Benoît Godin†**, Full Professor, Institut National de la Recherche Scientifique, Montréal, Canada.

**Markus Grillitsch**, Associate Professor, Department of Human Geography and Centre for Innovation Research (CIRCLE), Lund University, Sweden; Inland Norway University of Applied Sciences, Norway.

**Teis Hansen**, Professor, Department of Human Geography and Centre for Innovation Research (CIRCLE), Lund University, Sweden; Department of Food and Resource Economics, University of Copenhagen, Denmark; Department of Technology Management, SINTEF, Trondheim, Norway.

**Alexandra Hausstein**, Senior Researcher, Institute for Technology Futures (ITZ), Karlsruhe Institute of Technology (KIT), Germany.

**Hemant Kumar**, Assistant Professor, Centre for Studies in Science, Technology and Innovation Policy, School of Social Sciences, Central University of Gujarat, Gandhinagar, Gujarat, India.

**Brice Laurent**, Researcher at MINES ParisTech, PSL Research University, CSI – Centre de Sociologie de l'Innovation, France.

**Stine Madsen**, PhD Candidate, Department of Human Geography and CIRCLE – Centre for Innovation Research, Lund University, Sweden.

**Giulio Perani**, Senior Researcher, Department for Statistical Production, Italian National Statistical Institute, Italy.

**Boris Rähme**, Senior Researcher, Center for Religious Studies, Fondazione Bruno Kessler, Trento, Italy.

**Cornelius Schubert**, Associated Professor, Section of Science, Technology, and Policy Studies, Department of Technology, Policy, Society, University of Twente, the Netherlands.

**G.M. Peter Swann**, Emeritus Professor of Industrial Economics, Nottingham University Business School, University of Nottingham, UK.

**Bastien Tavner**, Post-Doc Researcher, Social and Economic Sciences Department, Télécom Paris, Université Paris-Saclay, France.

**Pascale Trompette**, Senior Research Fellow, University of Grenoble Alpes, CNRS, Science Po Grenoble, PACTE, Grenoble, France.

**Ulrich Ufer**, Senior Researcher, Institute for Technology Assessment and Systems Analysis (ITAS), Karlsruhe Institute of Technology (KIT), Germany.

**Harro van Lente**, Full Professor of Science and Technology Studies, Faculty of Arts and Social Sciences, Maastricht University, the Netherlands.

**Dominique Vinck**, Full Professor of Science and Technology Studies, Institute of Social Sciences, University of Lausanne, Switzerland.

**Lucien von Schomberg**, Lecturer in Creativity & Innovation, Greenwich Business School, University of Greenwich, UK.

# Legend of the cover image

There are several theories considered in the Handbook (spinning wheels). Some dominate but not completely. They cover a lot but not everything. Others are more limited, specialized. None of them are square, perfect. They are all a bit twisted, unfinished. There are links between them. Some lead to others. But the whole thing doesn't work without friction either. Sometimes it gets stuck. They all have more or less a core and peripheral elements. The core is never really a hard core, even when it is surrounded by several layers that protect it and then between all these alternative theories, there's debris, stuff without attachments, floating around.

# Introduction to the *Handbook on Alternative Theories of Innovation*

## Benoît Godin, Gérald Gaglio and Dominique Vinck

Innovation has been on scholars' and policy-makers' agendas for over seventy years. From a concept denigrated for 2,500 years, innovation has become an uncontested value. In the twentieth century, a whole literature developed in sociology, economics, policy and management on industrial innovation, organizational innovation, technological innovation, and so on. Theories developed with different perspectives: sociological, psychological, organizational, industrial and economical.

In the last decade or so, alternatives to existing theories exploded, particularly in reaction to the mainstream or dominant theories on technological innovation. On one hand, economic or industrial theories of technological innovation gave rise to new adjectives like architectural innovation, disruptive innovation, transformative innovation and so on. On the other hand, innovation other than industrial innovation and neglected types of innovation were unearthed: dark innovation, innovation through withdrawal, innovation through imitation (Godin and Vinck 2017). Finally, alternatives to industrial innovation emerged: social innovation, responsible innovation, sustainable innovation, inclusive innovation and others. Innovation should be societal, environment and ethical, so it is said.

How do we make sense of this linguistic profusion? Are there new paradigms developing? Many of these concepts have existed for decades, even more, under different names. For example, social innovation as a linguistic expression existed for over a century (Godin 2015). Yet, a new label creates the illusion of progress and provides hope. But is there something else in a new label? Are the alternatives turning things upside down or just refashioning the old? Are the so-called theories in fact theories, or something else? The *Handbook on Alternative Theories of Innovation* offers answers to these questions through a historical, conceptual and critical analysis of the concept of innovation and its usages.

In this Handbook, we call the terms that add an adjective to innovation, like technological innovation and social innovation, X-innovation (Gaglio et al. 2017; Godin 2019). The term technological innovation emerged after the Second World War, with a few occurrences before that date. One of the first such terms coined as an extension to technological innovation is educational innovation, which got into the public discourses in the United States in the 1960s.[1] It was soon followed by public sector innovation (at OECD), then social innovation, then green innovation. What sum up these terms is 'inclusiveness', a key concept of the 1960s (Godin 2020). Innovation is an activity inclusive of every activities, individuals, groups and organizations that, to different degrees, are involved in the innovation process. Innovation is a 'whole'.

The main conceptual framework of this Handbook is conceptual history (Quentin Skinner and John Pocock). We aim to understand the emergence and the context of development of the terms created with innovation, their uses in discourses and theories and the functions they serve. To us, the story of innovation is one of appropriation and contestation. On one hand, people appropriate a word (innovation) for its value-laden quality and, consequently, because of what they can do with it. They extend the word to different phenomena and fields and give to it still more legitimacy. A word with such a polysemy as innovation is a multi-purpose word. It works in the public mind (imaginaries) and among policy-makers. On the other hand, people contest a term (technological innovation) because of its hegemonic connotation. They coin alternative ones (e.g. social innovation) that often become a brand.

As a concept that travels easily between social spheres and between scientific disciplines and as a concept that is programmatic and performative (Godin and Gaglio 2019), innovation remains a relatively unstudied concept. To be sure, studies by the hundreds are published every year on innovation. But this literature studies innovation as a fact or reality, without considering discursive and conceptual issues. What use is made of the concept, for what purposes, by whom, are questions that remain unstudied, with a very few exceptions.

## SCOPE OF THE HANDBOOK

The general objective of the Handbook is the assessment and evaluation of representations of innovation, in the light of new expressed needs emerging from the policy, managerial and social spheres. To this end, and specifically, the Handbook aims to make sense of the alternative theories or approaches of recent years in the light of decades of theories of innovation. The Handbook's purpose is to develop a critical analysis and perspective of the theories developed, not to develop new ones.

The Handbook is a discursive analysis of innovation and of the alternative theories. It addresses three unexplored aspects of innovation and theories of innovation – the historical, the conceptual and the critical. The historical and genealogical aspects concern:

1. The context of emergence of the alternative theories
2. The origin of the alternative theories, evolution and recent developments
3. The goals (explicit and implicit) and rationales of the theories

Regarding the conceptual and discursive aspects, the Handbook contributes:

4. To identify precursor terms
5. To study the semantic field
6. To scrutinize the discourses held in the name of, or uses of, the theories (in the academic community, among practitioners and among policy-makers)

Finally, the Handbook considers the critical aspects of innovation:

7. The 'originality' of the theories: to what extent do the alternatives challenge existing theories?
8. Their normativeness: what role do values and policy play in the new narratives?
9. The politics and ideology of innovation

What do we mean by 'theories'? Theories of innovation fall into two groups, at the least. Some theories are comprehensive theories in terms of scope and issues considered. They explain what innovation is and is not and discuss the constitutive concepts (like innovation, process, system, etc.) with rigor (definition, criteria); they describe how innovation occurs, in terms of processes from generation to diffusion, both in time and space, and consider the actors and institutions involved and their characteristics; they analyze why innovation happens (the factors involved); they identify the effects or consequences of innovation on individuals, organizations and society; and they offer some statistical evidence. Such theories review the state of the art, and they integrate or organize the findings produced to date into a coherent whole.

Whether such broad theories exist in the literature, apart from Everett Rogers' *Diffusion of Innovation* of 1962, is matter of controversy. Perhaps such broad theories of innovation are just impossible. In recent decades, theories of innovation have been replaced by or share the place with a second kind of theories: a voluminous number of conceptual frameworks, approaches, models and empirical surveys, whose contribution to theories is often modest, developed. Theories of innovation are, to many extents, 'restricted' or local rather than comprehensive. Restricted in the sense that they are disciplinary or address only one aspect of innovation, such as the psychological, the organizational or the industrial, and only one dimension: generation or diffusion or consequences. Still others are rather reviews of existing works and ideas collected under the umbrella of a conceptual framework. At the opposite end of the spectrum – the general and macro level – some are theories of change, put under the umbrella of innovation.

In addition, formal theorists are not the whole story. Before (academic) theories, not to mention folk theories, there were theories. For example, practitioners developed thoughts on innovation before academics did and whose representation is essentially that of later scholars (Godin 2019, 2020).

What do we mean by 'alternative' theories? This Handbook studies the theories, or rather the approaches that positioned themselves as a corrective or repair to the mainstream kind of innovation and its study, namely the industrial and technological. It focuses on theories outside the mainstream credo (this credo is constituted of theories of technological change and of 'innovation studies').

What do we mean by 'innovation'? Innovation has two usages. To the moderns, innovation is *novelty*, novelty of any kind: new ideas, things, behaviors and practices. In this sense, innovation is a synonym for novelty and invention. But novelty is only one of the meanings of innovation. As the etymology suggest (in+), innovation is the introduction of something new into practice. To many, innovation excludes some types of novelty: the mental or speculative. Innovation is contrasted to contemplation. Innovation is *action*. What distinguishes innovation from change is purpose, to mainstream theorists at the least. Innovation is designed change, a deliberate initiative. A discovery or an invention becomes an innovation only when it is introduced, applied, adopted, distributed, maybe commercialized, that is, only if it is used and useful: "Research…is an unrealized public investment until the resulting innovations are diffused to and adopted by the intended audience" (Rogers 1962, p. 2).

Mainstream researchers, namely those active in the field of innovation called 'innovation studies' for example, are relatively immune to the new discourses and to new theories, to the extent that are off the beaten track. They are also much reluctant to engage in critical studies (Williams 2019). Given these two considerations, the most relevant scholars on the subject are researchers in search of alternative ways of understanding innovation. As a consequence, many of the contributors to the Handbook are young and outside the main track researchers. These researchers were approached because in the last few years, they have produced original thoughts that point towards a new representation of what innovation is. As the chapter from Mónica Edwards-Schachter documents, the alternative visions or approaches to innovation are voluminous. The Handbook is far from covering them all. We selected some 'popular' ones (e.g. user innovation), others that are precursors (e.g. social innovation) and still others that are more recent or in the making (e.g. frugal innovation). We also considered areas or disciplines that rarely theorized innovation as such (e.g. anthropology).

## STRUCTURE OF THE HANDBOOK

The Handbook is organized in seven parts, corresponding to as many aspects of innovation. First, innovation as a construct or sociopolitical representation. This section considers how innovation, using the concept as such, was considered in the past; how it came to be a virtue in the twentieth century; and what kinds of scenarios for innovation are imagined for the future. Second, the Handbook looks at the theorization of innovation, as it developed over the last century. This section gives a specific attention to the mainstream theories (industrial and managerial), since the alternative approaches generally positioned themselves against these theories. Third and fourth, the Handbook analyzes a sample of alternative theories or approaches developed in the last twenty years or so (societal, environmental and ethical), alternative types of innovation and new public policy instruments. Fifth, authors examine how to support innovation and with which instruments (narratives, measure, and so on). Sixth, the Handbook studies disciplines that remain relatively immune to the discourse on innovation until then – anthropology, psychology, religion and philosophy – and asked why. Seventh and last, we assess critically the theorization of the last few decades.

In the first part (Part I), the Handbook opens with Benoît Godin's contribution on what he calls 'innovation theology', a study of the history and genealogy of the concept of innovation. The chapter documents what the concept of innovation owes to the past, namely religion or theology. Going back to the English Reformation, a key moment when the concept entered our everyday vocabulary, the chapter unearths the concepts that defined innovation at that time, then examines the categories of the then-conception that survive in our modern and hyperbolic concept of innovation.

Harro van Lente turns to conceptions of innovation related to the future. The chapter reviews how recent theories of innovation have highlighted and analyzed the status and the role of imaginaries, or collectively available symbolic meanings and values, in innovation. Harro van Lente introduces a typology of imaginaries depending on whether they provide a complete storyline or only elements of a storyline. This brings a distinction between narratives, graphs, icons and artefacts. How these four types of imaginary figure in studies of innovation is reviewed. Van Lente argues that the notion of 'innovation' itself is an imaginary, a shorthand for progress, superiority and even modernity itself. Likewise, theories of innovation frequently become imaginaries, and, in this way, direct firms, governments and public involvement. To van Lente, imaginaries play an ambiguous role in innovation: they help but also frustrate the search for improvement. Imaginaries are not innocent side-effects of innovation, but vital ingredients, molding the meaning, use and direction of innovations. Without imaginaries there is no innovation.

Following this part of the Handbook dealing with time (past and future), the next part (Part II) takes us to the twentieth century. Benoît Godin delivers a survey of innovation theories developed over the last hundred years – excluding the alternative theories. He documents the psychological, sociological and industrial theories of innovation as a diversity that, over time, came to be dominated by the latter. Irwin Feller's contribution follows, concentrating on the dominant theorization of innovation of the twentieth century: industrial innovation. Feller sets down a baseline set of economic concepts, methodologies and findings against which the explanatory powers of alternative models may be compared and contrasted.

The following part (Part III) of the Handbook considers alternative approaches to innovation, exploring underused dimensions in classical theorization (societal, environmental and ethical). Mónica Edwards-Schachter opens the discussion with a survey or map of innovation forms that have been emerging as a ubiquitous imperative of contemporary society addressing the fast pace of scientific and technological development and pressing global challenges. Innovation has been characterized as 'technological', 'social', 'cultural', 'inclusive', 'environmental', 'open', 'user-centered', 'lean', 'free' and 'responsible', among others. The chapter provides a semantic map of around one hundred innovation types introduced from 1950 to 2019, analyzing relationships between established forms and new incumbents.

The next three chapters look at three major alternatives coined in the last few decades. Cornelius Schubert's concern is 'social innovation'. Social innovation is often framed as an alternative and in opposition to dominant entrepreneurial understandings of technical innovation. In these cases, the term social innovation denotes bottom-up and beneficial processes of social change. The chapter traces the emergence of this specific understanding of social innovation and how it became popular in academic and policy discourse over the last 20 years. Schubert argues that the popularity of social innovations in the policy arena rests on an instrumental notion that strongly resembles the dominant entrepreneurial understandings of technical innovation. This curtails the analytical potential of social innovation for the study of social change.

In another field, Frank Boons and Riza Batista-Navarro look at sustainable innovation. The concept of sustainable innovation is increasingly used to denote research on the relationship between technological change and ameliorating negative ecological and social impacts. It is also one of several conceptual labels that are used to address this topical area. Based on an inclusive definition of sustainable innovation, the authors explore the evolution of academic work in this area. They combine quantitative (i.e. topic modeling) and qualitative (i.e. interpretive) analytical tools to uncover lineages of research on technological change and sustainability, and analyze the distinct language used. The analysis unveils that the field is significantly fragmented, although there are also a few coherent lineages that extend over a long time period.

Then Lucien von Schomberg argues that the emergence of responsible innovation articulates the need for a paradigm shift in the innovation discourse, driving innovation away from mainstream economic interests towards societally desirable outcomes. Von Schomberg reports on three main challenges of responsible innovation – respectively the epistemic, political and conceptual challenge – and discusses to what extent they bring the feasibility of this paradigm shift into question. Von Schomberg argues that the epistemic and political implications of responsible innovation necessarily relate to the widely presupposed concept of technological innovation and commercialized innovation. In this vein, he argues that the discourse of responsible innovation is summoned to the overarching pursuit of technological and economic progress and asks whether, in this light, it is able to realize its ambitions.

The next part (Part IV) turns to other types and conceptualizations of innovation, which also have a potential in terms of alternatives. They are sometimes more specific or overlap with the alternative approaches addressed in the previous section. To begin, Bastien Tavner discusses the rise of 'users' in theories of innovation and policies. To Tavner, User Centred Innovation is developing within the innovation actors as a guiding principle for organizing and conducting innovation projects: (1) the involvement of 'real users', (2) the integration of action-research skills dedicated to users' experience, observation and analysis, and (3) the sharing and discussion of knowledge produced about uses within a multidisciplinary group.

The next chapter studies 'open innovation' as one of those catchwords that has captured the imaginaries of our societies in striving to point out desirable and transformative changes in our political economy. By adopting a historiographical perspective, Tiago Brandão sets out a comprehensive view of the traditions of thought behind the concept of open innovation. The genealogical analysis considers the presentation of several accounts, roaming through the liberal creed of thinkers to the classical management studies, before arriving at the explicit formulation of this conceptual framework by Henry Chesbrough. Brandão concludes that open innovation represents a concept marked by increasing ambiguity that does not escape its entrepreneurial (and management) bias.

Then Darryl Cressman looks at another popular alternative: 'disruptive innovation'. The question of what to make of disruptive innovation is complicated by the fact that it is used frequently and carelessly: it is simultaneously a cohesive management theory, an evocative metaphor and an empty buzzword. In line with the work of intellectual historian Reinhart Koselleck, Cressman traces the history of disruptive innovation from its formal theorization in management theory to its use as a metaphor to describe sociotechnical change.

In a totally different light, G.M. Peter Swann looks back at his notion of 'common innovation' and responds to the critics. By common innovation, Swann means innovation by the common man and woman, by families and by small communities, for themselves. This innovation is associated with the domestic mode of production, a type of innovation that goes back to the Stone Age and that is still widespread today. The chapter considers some of the criticisms that have been leveled at the concept of common innovation.

To continue, Fayaz Ahmad Sheikh and Hemant Kumar invite us to scrutinize grassroots innovation. Producers' innovations are considered as the holy grail for economic development. Yet these innovations have recently come under fire for perpetuating exclusion. Non-dominant innovation models are increasingly being promoted to facilitate inclusion and have been proposed to tackle global concerns such as poverty and climate change. Informal sector innovations have especially captured development scholars' attention, including grassroots innovations (GIs), developed outside the confines of organizational structures and without planned R&D. The chapter lays out the factors spurring GIs' marginalization, namely by deconstructing the elitist portrayal of innovations and investigating the pro-market innovation narrative that has undermined subaltern contributions. By critically synthesizing the extant literature from a multidisciplinary perspective, the chapter highlights the political and economic rationale driving the relative exclusion of GIs.

Céline Cholez and Pascale Trompette add a contribution to such an alternative, with the study of 'frugal innovation'. Since the 2000s, the new approaches in development policies and fighting against poverty turn to private partners (companies and NGOs) for the design of specific and small-scale (even individual) products and services for developing countries contexts. Frugal innovation, targeting underserved people in South countries, aims at articulating high technology solutions, quality principles and best local practices in order 'to do more with less'. The chapter analyzes how the inclusion of end-users became central in the definition of this new design paradigm, often associated with other design concepts as 'grassroots innovation' or 'inclusive innovation'. Cholez and Trompette investigate how the user, with his skills, is considered in the literature that promotes or criticizes frugal innovation. In line with the sociology of innovation, they propose to question this new theory of innovation from how it qualifies and justifies this target, and to what extent its deployment succeeds in impacting people's lives at the base of the pyramid.

The next part (Part V) is about the alternative ways to support innovation, in reframing the instruments. The chapter written by Carolina Bagattolli extends the analysis to international organizations like the OECD. She undertakes a discursive analysis of the 'X-innovation concepts' in innovation narratives present in international organizations, seeking to identify whether they appear as a rupture in the core values of technological innovation. She explains that there are prevailing (i) assertions regarding the potential of these kinds of innovation for solving social and ethical issues, even though with different nuances, (ii) claims – but vague and imprecise – about enlarging participation in science, technology and innovation development, and, furthermore, (iii) these 'x-innovation' terms tend to be presented in ambivalent and ambiguous ways.

To continue, Markus Grillitsch, Teis Hansen and Stine Madsen's focus is the instruments of innovation policy and how they have changed over the last decades. Concentrating on 'transformative innovative policy', they argue that transformative innovation policy is far from new. It has a long history in the literature on innovation. Just to add a few more precursors to Grillitsch, Hansen and Madsen's story: transformative innovation has precursors going back to the 1980s (transformational innovation; Averch 1985, pp. 53–55) and the 2000s (transformative research at the US National Science Foundation). It even closely resembles previous policy concepts such as system innovation. The chapter unfolds the historical and conceptual roots of transformative innovation policy and compares critically the different strands of literature.

Giulio Perani turns to another facet of innovation policy: statistics. The chapter describes the recent experiences of providing a quantification of the efforts by business enterprises to develop and implement innovation. Perani concentrates on the OECD's and the European Union's experiments to define an international standard for business innovation statistics, and critically reviews the results achieved.

The next three chapters (Part VI) deal with innovation theories and alternatives not theorized or newly theorized (appropriated) in the literature. Boris Rähme offers a rich perspective on a neglected topic: innovation in religion studies. The chapter analyses how the vocabulary of innovation has been taken up by religion researchers and scholars to describe transformations in the religious domain, such as the formation of new religious movements and ways in which religion has proved resilient to large-scale modernization processes. Rähme offers a heuristic distinction between three dimensions of the interrelations between religion and innovation: (a) Innovation in religion: how is innovation understood, experienced and practiced within religious traditions and communities? (b) Religion in innovation: how do religious traditions and communities contribute to innovation in various otherwise non-religious contexts? (c) Religion of innovation: has the vocabulary of innovation itself turned into a vehicle for fideistic commitments which are analogous to religious worship? Combining this three-fold distinction with an analysis of the normative presuppositions of innovation-speak, the chapter concludes with a discussion of the idea of a general theory of religion and innovation.

Then Ulrich Ufer and Alexandra Hausstein offer an equally rich analysis of another field: anthropology. In this chapter, they argue that while anthropology has not developed a clear-cut concept of innovation, it plays a considerable role in the conceptual history of innovation, revealed in the two notions of anthropology of innovation as an object of research and of anthropology for innovation as an object of engagement. Anthropologists have shifted emphasis between these two takes on innovation at various instances over the course of the twentieth and early twenty-first centuries, for example by referring to innovation in diffusionist and acculturalist conceptualizations of cultural change, by practical anthropology's engagements for modernization and innovation in developing countries, by anthropology's conceptualizations of innovation as socially embedded, or by cooperation between anthropology and industry.

To complete this section, Vincent Blok philosophically reflects on the concept of innovation. He distinguishes between the process and outcome dimension, and between the ontic and ontological dimensions of innovation. These distinctions lead Blok to propose four characteristics of a philosophical understanding of innovation, with several implications for the object of innovation and its novelty, as well as for the temporality and human involvement in innovation practices. Contrary to the previous two chapters, Blok is not concerned with historical and critical aspects of theories of innovation, but offers a programmatic theorization.

The Handbook concludes with two chapters devoted to conceptual and theoretical elaboration. Brice Laurent discusses three critical approaches to innovation. He presents them as 'styles of critique' that are associated with theories of innovation and comprise descriptive and normative contributions. First, the critique of the ideology of innovation often uses a religious vocabulary to confront 'myths' with established facts, and propose reforms. Innovation, here, is a discourse meant to describe and prescribe, which the critic should evaluate according to the validity of the representations of the world it proposes and the consequences of these representations. Second, innovation can be considered as an engine for critique. In this perspective, critique is part and parcel of innovation, and performed both by concerned groups and by analysts eager to give voice to them. Third, innovation can be understood as situated in heterogeneous regimes that associate it with more or less explicit propositions for social ordering. Critique consists in making these associations visible, and possibly using comparison to de-naturalize them. These three styles of critique are connected to one another, even if they differ according to how they envision the role and the position of the analyst. Confronting them offers ways of opening up new spaces of debate and reflection about innovation.

To go on another aspect of the conceptual dynamics, Gérald Gaglio and Dominique Vinck focus on the contribution of case studies to the theory building. Starting from the analysis of an old concept of 'unintended consequences', they examine its shortcoming and bring forth the idea of collaterality or collateral effects, as a corrective to the negative bias of the former concept. Collaterality is a series of occurrences and ramifications arising from but different from an initial innovation process, and which thereby multiply it. The contribution shows and questions the way to pursue and to renew conceptual production and theoretical shifts.

## NOTE

1.    One may find such a proliferation of terms far earlier, for example in French sociologist Gabriel Tarde's works in the late 1800s. E.g. legislative, judiciary, military, industrial, grammatical, linguistic, theological, religious.

# REFERENCES

Averch, Harvey A. (1985), *A Strategic Analysis of Science & Technology Policy*, Baltimore and London: Johns Hopkins University Press.

Gaglio, Gerald, Benoît Godin and Sebastian Pfotenhauer (2017), *X-Innovation: Re-Inventing Innovation Again and Again*, Project on the Intellectual History of Innovation, INRS: Montreal.

Godin, Benoît (2015), *Innovation Contested: The Idea of Innovation over the Centuries*, London: Routledge.

Godin, Benoît (2019), *Inventing Technological Innovation: Languages, Discourses and Ideology in Historical Perspective*, Cheltenham, UK and Northampton, MA, USA: Edward Elgar Publishing.

Godin, Benoît (2020), *The Idea of Technological Innovation: A Brief Alternative History*, Cheltenham, UK and Northampton, MA, USA: Edward Elgar Publishing.

Godin, Benoît and Gerald Gaglio (2019), 'How does Innovation Sustain "Sustainable Innovation"', in Frank Boons and Andrew McMeekin (eds), *Handbook of Sustainable Innovation*, Cheltenham, UK and Northampton, MA, USA: Edward Elgar Publishing, pp. 27–37.

Godin, Benoît and Dominique Vinck (eds) (2017), *Critical Studies of Innovation: Alternative Approaches to the Pro-innovation Bias*, Cheltenham, UK and Northampton, MA, USA: Edward Elgar Publishing.

Rogers, Everett M. (1962), *Diffusion of Innovations*, New York: Free Press.

Williams, Robin (2019), 'Why Science and Innovation Policy Needs Science and Technology Studies?', in Dagmar Simon, Stefan Kuhlmann, Julia Stamm and Weert Canzler (eds), *Handbook on Science and Public Policy*, Cheltenham, UK and Northampton, MA, USA: Edward Elgar Publishing, pp. 503–522.

# PART I

# VISIONS OF INNOVATION

# 1.  Innovation theology

*Benoît Godin*

In 1967, Robert Charpie, who was President at Union Carbide Electronics and Chairman of the influential US Department of Commerce's report *Technological Innovation: Its Environment and Management* of the same year, claimed that "technological innovation is the driving force behind economic growth.... Technological innovation is truly the cutting edge of the economic growth process and the economic vitality of the Nation is intimately enmeshed with those companies that are successful in the technological innovation process" (Charpie 1967, pp. 357–8). By innovation, Charpie meant "that process by which a new idea is successfully translated into economic impact within our society by providing better products and simultaneously creating new jobs in the manufacturing and application of those products"; by technological innovation he referred to "innovations which flow directly from new technical ideas, inventions, or discoveries and which result in new products or services with essential technical content" (p. 357). Every practitioner of the time (engineers, managers, policy-makers and their advisers and consultants) shared this representation of innovation. Such a view was an article of faith in the second half of the twentieth century. "Do we need to innovate?" asked the OECD a few years later. "Yes because it is one way, perhaps one of the best ways, to react in a rapidly changing society" (OECD 1969, p. 15). As Jack Morton, engineer at Bell Laboratories, put it in 1971: "Innovation is certainly a 'buzz-word' today. Everyone likes the idea; everyone is trying to 'innovate'; and everyone wants to do better at it tomorrow" (Morton 1971, p. 73).

The historiography of innovation, although no writing really deserves that name, attributes the scholarly origin and study of the concept of innovation to the economist Joseph Schumpeter in the 1930s–1940s. This is a legendary tale (Godin 2019, 2020). Schumpeter simply used a concept that was becoming popular, as many others did in the first half of the twentieth century. Moreover, there is no theory of innovation in Schumpeter's works – Schumpeter does not concern himself with explaining the generation of innovation and its diffusion in space and time. Schumpeter's concern is economic development, of which innovation is a means.

Where does the concept come from then and what does it mean? This chapter looks back in time and suggests that religion is a key source of our modern concept of innovation, a not dissimilar thesis to that of Max Weber on Protestantism and the spirit of capitalism (Weber 1930). Innovation is a loaded term that helped enforce the Reformation. To contemporaries, the concept was an inclusive term that covers both religion and politics, giving rise to a secular term for heresy. In the nineteenth century, innovation was re-conceptualized to serve modern society. This was a task to which social reformers including Christians devoted some energies, and they did so in the light of, or in reaction to, the religious discourse.

Religion is not the whole story of course. Technology is a major source of the concepts that define the semantic field of innovation and the discourses it provoked in the twentieth century, through economics and the market ideology. At the same time, the discourse on technological innovation espouses a semantic that has deep roots in history. The aim of this chapter is to document these roots. Fundamental categories from the Reformation and Counter-Reformation's connotation of the concept remain in our modern language. They just changed meaning, from the negative to the positive. Innovation theology is the study of such a genealogy.

## INNOVATION AND THE REFORMATION

The concept of innovation is of Greek origin (Godin 2015). To the ancient Greeks, innovation (*kainotomia*) meant change to the established order. Innovation was political. The concept entered the Latin vocabulary (*innovo*) with a totally different meaning. Innovation is renewing, in line with many terms beginning with "re" (e.g. reformation). The seventeenth century changed this meaning again. Innovation came to be discussed as a deviance or heresy and associated with anything not conforming to orthodoxy. Still again, in the nineteenth century, innovation acquired a new meaning as something totally new and instrumental to political, social and economic reform. Clearly, changes in language and vocabularies indicate changes in values.

The English Reformation is a key moment in this genealogy. It was at that time that innovation entered our everyday vocabulary (English and French). The concept served to support the Reformation. Kings and churches forbade innovation; bishops supported these instructions with sermons, and followers (e.g. controversialists, pamphleteers) developed arguments to this end – normative, legal and cultural.

As an innovation, but not called such at the time, the Reformation's reformers had to develop means to secure the Reformation: political, administrative and legal. To this list one must add language. In proclamations, declarations and statutes, monarchs used the concept of innovation to control the conduct of their subjects. The use of the concept began as an instruction not to innovate. Henry VIII's private correspondence of the 1530s is full of letters to councilors and ambassadors as messengers, instructing them that His Majesty will not "endure" or "tolerate" innovation. In a second step, innovation became a public injunction. In 1548, Edward VI issued *A Proclamation Against Those that Do Innouate*, the first-ever royal injunction denouncing innovation.[1] The proclamation placed innovation in context, constituted an admonition not to innovate (not to change but to respect the new doctrine and discipline of the Church) and imposed punishments on offenders (England and Wales. Sovereign. Edward VI 1548):

> Considering nothing so muche, to tende to the disquieting of his realme, as diversitie of opinions, and varietie of Rites and Ceremonies, concerning Religion and worshippyng of almightie God…; [considering] certain private Curates, Preachers, and other laye men, contrary to their bounden duties of obedience, both rashely attempte of their owne and singulet witte and mynde, in some Parishe Churches not onely to persuade the people, from the olde and customed Rites and Ceremonies, but also bryngeth in newe and strange orders…according to their fantasies…is an evident token of pride and arrogance, so it tendeth bothe to confusion and disorder…: Wherefore his Majestie straightly chargeth and commandeth, that no maner persone, of what estate, order, or degree soever he be, of his private mynde, will or phantasie, do omitte, leave doune, change, alter or innovate any order, Rite or Ceremonie, commonly used and frequented in the Church of Englande … Whosoever shall offende, contrary to this Proclamation, shall incure his highness indignation, and suffer imprisonment, and other grievous punishementes.

From that time onward, the concept of innovation served every cause, political and ecclesiastical, and soon became an accusation. Throughout his reign (1625–1649), King Charles I suffered the accusation of innovating. The Presbyterian Scots and the English Parliament were particularly violent in their words against Charles, who was accused of "popish innovation". It is during this period that the concept became polemical. Everyone (archbishops, bishops, parliamentarians) accused the others (Puritans, Catholics, separatists) of innovation in religion and government.

During the Reformation and afterward, the concept of innovation was used predominantly in the pejorative sense. The very few positive uses that exist are legal and spiritual. For example, Thomas More uses it for renewal of the soul: a return to original or pure soul (before sin). Overall, however, the negative meaning of the concept of innovation, the dominant connotation, continued until late in the nineteenth century.

## A CONCEPT THAT TRAVELS

The religious representation of innovation was highly influential. First, consider science. According to modern standards, the experimental method was certainly a great innovation. However, at the time of the Reformation and after, no one thought of calling this method an innovation. To take just one example, Francis Bacon never used the concept of innovation in the positive sense to name his experimental project. Except for a short essay *On Innovation*, where he discussed the nature of innovation from both the positive and negative sides, Bacon used the concept in the negative sense, mainly in his political and moral writings, and regularly reminded his readers of His/Her Majesties' (Elizabeth, James) injunctions not to innovate in matters of religion (Godin 2016a).

This is also the case with inventors. "Projectors", as inventors were called at the time, never used innovation to name their inventions. The practice continued until the nineteenth century. Inventors had to defend themselves against charges of innovating (upon the tradition or in the name of progress), using the word as such (Godin 2016b).

Political thought shared the same meaning of innovation as did science and the arts. In the seventeenth century, political thinkers never talked of the new idea of republicanism in terms of innovation. The word is too negatively charged for such a purpose. Royalists, however, regularly used the concept to denigrate the republicans. That all innovation in government are "dangerous" was a watchword of the time. Innovation is a "plague" (Godin 2012).

The same holds true for social reform. Until late in the nineteenth century, socialism as "social innovation" was described as subversive to the social order, namely to private property (Sargant 1858; *Encyclopedia Britannica* 1888). Yet it is in the nineteenth century that the representation of innovation began to change. The rehabilitation of innovation went through several routes: cultural (making use of history), linguistic (etymology) and utilitarian or instrumental. Among the instrumental arguments is social reform, as supported by the French socialists (Claude-Henri Saint-Simon, Charles Fourier) and their followers, in France (Victor Considérant) and elsewhere (John Patterson). Religion is an integral part of the discourse of the socialists. As Gareth Stedman Jones has suggested, socialism was to many the "new spiritual power" in post-revolutionary France and elsewhere in the western world (Stedman Jones 2010). The concept of innovation, or rather "social innovation", served this "new Christianity", as Saint-Simon called it (Saint-Simon 1825). Comte's *Cours de philosophie positive* praises Catholicism for the introduction of a system of general education for all, an "immense et heureuse innovation sociale" [great and happy social innovation] (Comte 1841,

p. 366). To Fourier, "Dieu a fait éclore dans nos sociétés des germes d'innovation, bien-faisantes et nuisibles" [God causes to be born in our societies kernels of beneficial or harmful innovation]. Among the good seeds are "innovations domestiques et sociales" [domestic and social innovation] (Fourier 1808, p. 90). Social innovation and the Church's message go hand in hand, as another French socialist put it: "L'évangile, lors même qu'il ne serait pas le livre définitif de la parole divine, sera toujours le guide et le modèle du novateur social" [the gospel, although it is not the definitive book of the divine word, will always be the guide and the model of the social innovator] (Lechevalier 1834, p. 538).

## INNOVATION IN THE TWENTIETH CENTURY

In the twentieth century, innovation became a word of praise. Innovation came to be considered as the source of progress. To be sure, such a discourse began in the decades following the French Revolution. What was previously called "dangerous innovation", like revolution, becomes "happy innovation", a key phrase to Comte. Comte also introduced a contrast that became very popular later on. He contrasts "esprit de conservation" [the spirit of conservation] to "esprit d'innovation" [the spirit of innovation] as two fundamental instincts, and explains social progress as the result of the latter (Comte 1839, pp. 558–9).

Yet a complete rehabilitation of the concept of innovation had to wait until the twentieth century, thanks to or because of technology. The view of change in the seventeenth and eighteenth centuries was eminently conservative. There was no question of progress. Then, after a long period of contestation, a new conception emerged (Godin 2019, 2020). The qualities that had been denounced as social vices emerged as moral virtues. In the name of economic growth, "technological innovation" became instrumental to economic policy, as Charpie claimed. Practitioners started making of innovation a technological affair – before scholars, who only articulated the practitioners' view. "There is little doubt", stated the OECD in one of the first titles on technological innovation ever produced in the Western world, "that if governments succeed in helping to increase the pace of technical innovation, it will facilitate structural changes in the economy, and increase the supply of new and improved products necessary for Member Governments to achieve rapid economic growth and full employment and without inflation" (OECD 1966, p. 8). Religion, or rather a new kind of "religion", remains in the background here. Innovation is the modern belief or new faith:

> Most current social, economic and environmental challenges require creative solutions based on innovation and technological advance. (OECD 2010, p. 30)

> Innovation is our best means of successfully tackling major societal challenges, such as climate change, energy and resources scarcity, health and ageing, which are becoming more urgent by the day. (European Commission 2010, p. 2)

> Innovation is a key driver of productivity, growth and well-being, and plays an important role in helping address core public challenges like health, the environment, food security, education and public sector efficiency. (OECD 2015, p. 11)

In the late 1960s-early 1970s, technological innovation gave rise to a growing literature concerned with organizational strategies and public policies for industrial innovation: in management, economics and research policy. Then starting around 1990, new terms began to appear that argued for a different kind of innovation: social innovation (a term of the nineteenth century), responsible innovation, sustainable innovation, and so on. An adjective rather than an object defines what innovation is. This has to do with the "quality" of innovation: we need a different type of innovation. Innovation must be societal (social), ethical (responsible) and environmental (sustainable).

## INNOVATION THEOLOGY

In *Political Theology*, Carl Schmitt claimed that "All significant concepts of the modern theory of the state are secularized theological concepts" (Schmitt 1922, p. 36).[2] Schmitt's is a controversial thesis that did not remain uncontested. In *lieu et place* of secularization, some prefer to follow Hans Blumenberg and talk of "reoccupation". Some issues – which may have been discussed in religion first – are subsequently dealt with anew (reinvested, reoccupied) in light of a different context (Quesne 2007; Monod 2012; Pankakoski 2013). Above, I talked of "traveling". The concept of innovation traveled from the religious sphere to other social fields that appropriated it for their own ends. At first the concept had essentially the same connotation as the religious one in these fields. In the following centuries, it acquired new meanings.

Innovation theology, I suggest, is the study of the theological genealogy of innovation. Many of our modern concepts owe their origins to religion, or at the least religion contributed significantly to the use and diffusion of such concepts. Such is the case of progress (eschatology) (Lowith 1949), freedom (Troeltsch 1912), liberty (tolerance) (Tierney 1995), liberalism (Siedentop 2014), revolution (resistance) (Kelley 1988, 1995), governmentality (*oeconomia*) (Agamben 2007; Leshem 2016), absolutism (Kantorowicz 1955, 1957; Oakley 1968), sovereignty and constitutionalism (Tierney 1982; Blickle 1992), capitalism (Weber 1930) and economic thought (Todeschini 2008a, 2008b). "Secularization" itself found its first expression in the government of the Church. Innovation is another such concept. Our modern representation of innovation carries survivals of a century-old conception, that of religion.

In order to unearth these survivals, we must study the vocabulary used to talk of innovation at the time of the Reformation. As an accusation above all, innovation was talked of in terms of a series of core or constitutive categories. One is *change*. The Reformation was a period of considerable change. Changing Christianity to Protestantism has been a long process enmeshed in political struggles. This was a time when change, particularly change to religion, law and politics, was prohibited. The vocabulary used to make sense of (acceptable) change was a series of terms beginning with "re": reformation, renovation, restoration, regeneration, renewing (Ladner 1959). All of these terms mean the purification of "infirmities" or what has been "corrupted", "deformed" and "abused". It is a matter of improving what exists, not of changing it. In this context, innovation is the emblematic and contested concept of change. Innovation is changing to the worst: *altering* – a recurrent co-word with innovating – "polluting", "poisoning", "perverting", "prostituting" and "adulterating" things; changing or departing from what is exemplary – customs, laws and government and, particularly in religious matters, Antiquity (Primitive Church and Scripture) – or from the true Protestant religion, by "adding" to or "subtracting" from it, or wholly substituting something else, "breaking" or "overthrowing" it.

What about *novelty*, as the etymology of the word innovation suggests (*novus*). Novelty (something new) itself is a key word of the time, but it is not the issue. Novelty is mere "imagination", in a pejorative sense: "fancy", "fantasy". Novelty is also "men's invention", "device" and "forgery". Again, the etymology must be taken seriously (in+). Innovation is more than mere novelty. Innovation is an *activity* (what we call process today): "*in*troducing" something new into the world, new ideas (doctrine) or activities (worship) into practice. In this sense, innovation is deliberate change. One category that serves this discourse is *liberty*, not in the modern sense (autonomy) but in the sense of excessive and arbitrary liberty (as being opposed to the social order), namely licentiousness. "Private opinion" and "private men" were key terms used at the time to discuss such a liberty. A related category is *design* together with "scheme", and later, "project". The innovator has a purpose, a scheme or design: "overthrowing" the social order; reintroducing "popery" in Protestantism. He is never alone. He creates a whole "sect" that follows him.

What distinguishes innovation from *heresy* (meaning = choice), a key word of the time, is the scope of the liberty or deviance. To be sure, the same vocabulary used against heresy is used to discuss innovation.[3] Innovation too is called heretical ("heretical innovation"). But with the decline of persecution, and of the Inquisition in the Christian world (for political rather than humanist purposes such as individual freedom), innovation came to include more than religious heresy. Innovation is religious *and* political deviance. This distinguishes the use and function of the concept of innovation from heresy in Protestant England. Heresy, as strictly religious, is a word of Papal power. By the time of Charles' reign, the word heresy was used mainly as a polemical. Kings as Sovereign of the Church and controversialists used the concept to refute their opponent (the Catholic) using his own words: using a Popish word against the Papists themselves.

Innovation may be an activity or process and it is discussed as such, with emphasis on the "innovator", but what is feared are the effects of innovation. Innovation is "sudden" and "violent" and, particularly after the French Revolution, is often discussed in terms of "revolt" and what we call *revolution* today (tumult, rebellion, subversion, sedition, division, faction, war, disorder, schism, disobedience), and contrasted to reformation, which is gradual. Innovation is destructive of the social order. This is why innovation is to be feared. The innovator foments a plan to "subvert" things for his own purposes. Innovation may be private as to origin, but it is public with regard to its consequences. Innovation may begin as a small or indifferent thing (adiaphora)[4] but over time it leads to a chain reaction. It creeps imperceptibly, "little by little", into the whole world.

The Reformation's representation of innovation endured for more than three centuries. In turn, the modern representation began to take shape with political and social reformers in the nineteenth century, those who began to re-conceptualize innovation as an instrument for "progress". Until then, the instruments of progress (civilization) were knowledge (including moral knowledge), education (including religious education), law and commerce. The social reformer added "innovation" to this list. However, this step was neither easy nor self-evident. To make his case, the social reformer had to develop arguments against the then-common discourse on innovation, the religious discourse.

The same categories that served to define innovation in the past define innovation today. But the categories of the previous centuries have acquired new meanings. To the moderns, innovation is *novelty*, novelty of any kind: new ideas, things, behaviors and practices. In this sense, innovation is a synonym for novelty. But novelty is only one of the meanings of innovation. To practitioners and some scholars, innovation excludes some types of novelty. Innovation is contrasted to contemplation. Innovation is *action*, as was understood in the past centuries. A discovery or an invention becomes an innovation only when it is introduced, applied, adopted, commercialized, that is, only if it is used and useful: "Research…is an unrealized public investment until the resulting innovations are diffused to and adopted by the intended audience" (Rogers 1962, p. 2). Among writers of the early twentieth century, innovation as action was often discussed in terms of "energy".

The next category in the re-conceptualization of innovation is originality. An innovator initiates something for the "first" time, as was understood at the time of the Reformation. This perspective gave rise to innovation as creativity, a concept that had nothing to do with innovation prior to the twentieth century. Again, it is a matter of re-describing precursor categories previously used in the negative sense: invention, imagination. The most important impact of the concept of creativity in recent decades is, perhaps, a definition of innovation as originality, in the sense of origin: the "first" adoption of a new behavior or practice or the "first" commercialization of a new idea or thing. "The first enterprise to make a given technical change is an innovator. Its action is innovation. Another enterprise making the same technical change later is presumably an *imitator*, and its action, *imitation*" (Schmookler 1966, p. 2).

The most important dimension of innovation is *effects*, again as was understood during the Reformation. The difference between reformation and innovation, that is, between renovating and altering, gave rise to the contrast between "meliorating"/"improvement" and innovation. Early in the twentieth century, economists made a distinction between improvement and invention. An improvement enhances on existing things, while an invention is something entirely new. The same distinction came to define innovation. The more an innovation differs from predecessor innovations, the more it has transformational effects (structural or generic), effects that are called "revolutionary". The effect dimension of innovation explains why scholars study successful innovations and not failures. An invention without effects is not an innovation.

In sum, one observes a shift in the vocabulary over time, from the negative to the positive. "A recurring pattern of Western thought is that ideas originally presented to justify an existing order of things often prove to have revolutionary implications, when they are taken over by critics of the existing order" (Tierney 1982, p. 49). But at the same time, clear survivals of past connotations inhere in our current meaning. Alteration (in the pejorative sense) changed to "difference", "creativity" and "originality"; liberty is discussed in terms of "initiative" and "entrepreneurship"; scheme and design became directed or "planned change", "strategy" and "policy", thus giving a programmatic character to innovation; revolution gave rise to "radical" and "major" innovation as key concepts to the students of innovation.[5] The New Christianism of the nineteenth-century, together with socialism, gave us "social innovation". To many scholars, the term social innovation is situated within a left-wing ideology today, either explicitly or implicitly. Social innovations favor (or should favor, to be so named) the non-institutional, the "alternative" and the "marginal". Furthermore, the "community" and non-profit organizations are favored sources of social innovation and the focus of many studies. Autonomy, liberty, democracy, solidarity and liberation are key words that came into use in theories on social innovation. Social innovation is "democratic, citizen- or community-oriented and user-friendly"; it assigns significance to what is "personalized, small, holistic and sustainable"; its methods are diverse, not restricted to standard science, and include "open innovation, user participation, cafés, ethnography, action research", and so on (Mulgan 2007).

Despite the lingering survivals of past meanings, two entirely new perspectives define innovation today. One is the future. *Philosophes* of the twentieth century added a future perspective to innovation.[6] Innovation became a programmatic concept. Scholars and others sanctified innovation and made of it a way to modern salvation. Previous uses of the concept of innovation were made in the present form (accusing someone of innovating) or the past form (the "Happy Innovations" of previous ages, as examples or models). Furthermore, there was no theory of innovation, in contrast to the field of political thought, where theories have abounded for centuries. As Brian Tierney put it on popular sovereignty: "We are in a world of *mentalités*, of taken-for-granted presuppositions rooted in the corporate life of the Middle Ages, that the canonists never explained because it never occurred to them that they needed any explanation" (Tierney 1982, p. 21). Using a future point of view, a modern philosophe imagines and prescribes a new political (economic) order. Innovation creates new means for a better future (the re-enchantment of the world).

As moderns, we are accustomed to talking of the Reformation as an innovation. The Reformation is one of those moments that, together with the Renaissance and the French Revolution, changed the world. Indeed, the word innovation began to be used at that time. But it was the critics that discussed the Reformation in terms of innovation. The reformers considered the Reformation as an improvement, a renewing, not an innovation. It was a matter of *re*dressing, namely eliminating corruption, hence a vocabulary of "reforming", "restoring", "correcting" and "amending".[7] The Reformation was imitation not innovation. It is past-oriented. Innovation is altering the past. In contrast, a central feature of the modern concept of innovation is a future orientation. Innovation is an activity or process conducted in the present, but whose purpose is imagining or constructing the future. This is a totally new meaning.

The second perspective is the market (Godin 2019). Beginning after World War II, innovation acquired a meaning related to industry. Innovation is the commercialization of inventions or goods embodying knowledge (technology) and thus contributes to economic growth. In a matter of only a few decades, this representation became the spontaneous representation of innovation. Increasingly, this representation is contested. Today, people expect innovation to be societal, environmental and ethical, rather than strictly economic.

## CONCLUSION

Innovation is a religion in the sense that it has become a modern belief and faith. But innovation theology is more than that, as this chapter documented. Innovation theology suggests that the modern representation of innovation developed out of the religious one, shifting this representation to an entirely opposite and positive representation, reinvesting or reoccupying it with new meanings. "Il est au fond logique", claims Jean-Claude Monod, "qu'une 'force' dominante de la culture à une époque donnée ne disparaisse pas sans laisser quelques traces profondes dans les pensées ou dans les institutions ultérieures" (Monod 2012, p. 12).[8] We have simply turned a negative concept into a superlative one.

Innovation theology is not the end of the story. The modern concept of innovation may be a secularized concept, but the religious concept is the result of a "theologization" of a political concept. I may have stressed the religious origin here, but the concept previously had a political (and negative) connotation, originating in classical Greece. It traveled from politics to religion, first as a positive concept, in the spiritual sense (innovo: renewing the soul), then back to the negative (during the Reformation).[9]

In addition to constituting a contribution to the debate on secularization (or rather desacralization), innovation theology is also a contribution to intellectual and conceptual history. Intellectual historians have largely ignored religious ideas (Kelley 1988; Coffey 2009). With a focus on political ideas and their sources in political and legal thought, intellectual historians have sidestepped religious sources – they have also neglected innovation as a political concept. Innovation theology corrects this trend. Innovation theology studies one of the contributions of religion to political, social and economic ideas, in this case innovation. Innovation theology suggests a new area of research: unearthing the historical (religious) assumptions and categories that define innovation today.

Finally, innovation theology is also a contribution to the critical study of innovation. Scholars who study innovation have espoused a pro-innovation bias, or at least sympathy, toward innovation, with few reflective thoughts on either their object of study or their conceptual constructs. The semantic morphology of the field bears witness to this situation. With a vocabulary focusing on the market and economic competitiveness, and increasingly on progress of any kind, innovation has become a panacea. Innovation theology opens the black box of the concept, studies its origin and development, and explains its moral valuation, thus allowing one to question the place of innovation among the many possible actions on society, for example, maintaining the status quo, imitation, resistance, withdrawal.

## NOTES

1. Edward's reign was ruled by a Regency Council (under his uncle Edward Seymour, Duke of Somerset, then John Dudley, Duke of Northumberland) because Edward never reached his majority.
2. One may find the same idea expressed before Schmitt. "Politiks both civil and Ecclesiastical", claimed George Lawson in his *Politia sacra et civilis* (1689), "belong unto theology, and are but a branch of the same" (cited in Tierney 1982, p. 99). Three hundred and fifty years before Lawson, Marsillius of Padua argued, in *Defensor pacis* (1324), that the provision of natural law could be found in divine law (as well as human law) (Tierney, 2014, p. 124).
3. For some official texts on heresy, see Peters (1980); Hughes and Larkin (1964, pp. 57–60, 181–8).
4. In Reformation England, adiaphora was a major argument used by the conformists. Certain practices, ceremonies and rites, qualified as Romish by the reformers, were permitted because they were indifferent, like kneeling at communion, wearing ecclesiastical vestments (cap, surplice), ringing bells, lighting candles at Mass, observing saint's days.
5. To be sure, the word revolution remains in the modern vocabulary – "revolutionary" innovation is contrasted to "minor", "incremental" or "gradual" innovation – but as an attribute of innovation rather than as an effect, although the semantic reference is to effects (an innovation is revolutionary because of its important effects on society).
6. I include in this category all of the writers who produce thoughts on innovation: scholars, engineers, managers and policy-makers. At the time of the Enlightenment, the French called such men *philosophes* (as distinct from professional philosophers). The *philosophes* included men of letters, men of science, statesmen, government administrators, journalists and scholars.
7. With the Act of Supremacy of 1534, Henry VIII became the "only Supreme Head in earth of the Church of England" with "full power and authority to visit, repress, redress, reform, order, correct, restrain and amend all such errors, heresies, abuses, offences, contempts, and enormities" of the Papacy for the "conservation of the peace, unity and tranquility of this realm". Elizabeth I's Act of Supremacy of 1559 instituted visitations "for reformation, order and correction" of "all manner of errors, heresies, schisms, abuses, offences, contempts and enormities" in the Church.
8. "It is inherently logical that a dominant 'force' of culture at a given point in time does not vanish without leaving some profound traces on thought and on subsequent institutions".

9.   Both Ernst Kantorowicz and Brian Tierney documented such a dialectical process, the first in the case of the concept of "king's two bodies" and the second on constitutionalism: there has been "a transference of definition from one sphere to another, from theology to law,…just as, vice versa, in the early centuries of the Christian era the imperial political terminology and the imperial ceremonial had been adapted to the needs of the Church" (Kantorowicz 1957, p. 19); "the typical process that occurred was the assimilation of a text of Roman private law into Church law, its adaptation and transmutation there to a principle of constitutional law, and then its reabsorption into the sphere of secular government in this new form" (Tierney 1982, p. 25).

# REFERENCES

Agamben, Giorgio (2007), *The Kingdom and the Glory: For a Theological genealogy of Economy and Government*, Stanford, CA: Stanford University Press [2011].

Blickle, Peter (1992), *Communal Reformation: The Quest for Salvation in Sixteenth-Century Germany*, Studies in Central European Histories, Volume 1, Leiden: Brill.

Charpie, Robert A. (1967), 'Technological Innovation and Economic Growth, in National Academy of Science', in *Applied Science and Technological Progress*, report to the US House of Representatives, Washington: USGPO, pp. 357–64.

Coffey, John (2009), 'Quentin Skinner and the Religious Dimension of Early Modern Political Thought', in Alister Chapman, John Coffey and Brad S. Gregory (eds), *Seeing Things Their Way: Intellectual History and the Return of Religion*, Notre Dame, IN: University of Notre Dame, pp. 46–74.

Comte, Auguste (1839), *Cours de philosophie positive*, Volume 4, Paris: Bachelier.

Comte, Auguste (1841), *Cours de philosophie positive*, Volume 5, Second Edition, Paris: Ballière et Fils [1864].

*Encyclopedia Britannica* (1888), *Communism*, T. Thomas and S. Baynes (eds), Volume 6, Third Edition, New York: Henry G. Allen, pp. 211–19.

England and Wales. Sovereign. Edward VI (1548), *A proclamation against those that doeth innouate, alter or leaue doune any rite or ceremonie in the Church, of their priuate aucthoritie: and against them which preacheth without licence*, set furth the .vj. daie of Februarij, in the seconde yere of the Kynges Maiesties most gracious reigne, Excusum Londini: In aedibus Richardi Graftoni regij impressoris. Cum priuilegio ad imprimendum solum.

European Commission, (2010), *Communication from the Commission to the European Parliament, the Council, the European Economic and Social Committee and the Committee of the Regions*, Europe 2020 Flagship Initiative Innovation Union, COM (2010) 546, Brussels.

Fourier, Charles (1808), *Théorie des quatre mouvements et des destinées générales*, Chicoutimi, Quebec: Les presses du réel [1998].

Godin, Benoît (2012), 'The Politics of Innovation: the Controversy on Republicanism in Seventeenth-Century England', *Redescriptions*, 16, 77–105.

Godin, Benoît (2015), *Innovation Contested: The Idea of Innovation over the Centuries*, London: Routledge.

Godin, Benoît (2016a), 'Representation of Innovation in Seventeenth-Century England: A View from Natural Philosophy', *Contributions to the History of Concepts*, 11 (2), 24–42.

Godin, Benoît (2016b), 'Technological Innovation: On the Emergence and Development of an Inclusive Concept', *Technology and Culture*, 57 (3), 527–56.

Godin, Benoît (2019), *The Invention of Technological Innovation: Language, Discourse and Ideology in Historical Perspective*, Cheltenham, UK and Northampton, MA, USA: Edward Elgar Publishing.

Godin, Benoît (2020), *The Idea of Technological Innovation: A Brief Alternative History*, Cheltenham, UK and Northampton, MA, USA: Edward Elgar Publishing.

Hughes, Paul L. and James P. Larkin (1964), *Tudor Royal Proclamations*, Volume I, New Haven and London: Yale University Press.

Kantorowicz, Ernst H. (1955), 'Mysteries of State: An Absolutist Concept and its Late Medieval Origins', *Harvard Theological Review*, 48 (1), 65–91.

Kantorowicz, Ernst H. (1957), *The King's Two Bodies: A Study in Medieval Political Theology*, Princeton, NJ: Princeton University Press [2016].

Kelley, Donald R. (1988), 'Ideas of Resistance before Elizabeth', in Heather Dubrow and Richard Strier (eds), *The Historical Renaissance: New Essays on Tudor and Stuart Literature and Culture*, Chicago, IL: University of Chicago Press, pp. 48–76.

Kelley, Donald R. (1995), 'Kingship and Resistance', in Noel B. Reynolds and W. Cole Durham (eds), *Religious Liberty in Western Thought*, Michigan and Cambridge: William B. Eerdmans Publishing, pp. 235–68.

Ladner, Gerhart B. (1959), *The Idea of Reform: Its Impact on Christian Thought and Action in the Age of the Fathers*, Cambridge, MA: Harvard University Press.

Lechevalier, Jules (1834), 'Des paroles d'un croyant', *Revue du progrès social*, 1 (5), 518–38.

Leshem, Dotan (2016), *The Origins of Neoliberalism: Modeling the Economy from Jesus to Foucault*, New York: Columbia University Press.

Lowith, Karl (1949), *Meaning in History: The Theological Implications of the Philosophy of History*, Chicago, IL: University of Chicago Press.

Monod, Jean-Claude (2012), *La querelle de la sécularisation, de Hegel à Blumenberg*, Paris: Librairie philosophique J. Vrin.

Morton, Jack A. (1971), *Organizing for Innovation: A Systems Approach to Technical Management*, New York: McGraw-Hill.

Mulgan, Geoffrey (2007), *Social Innovation: What It Is, Why It Matters and How It Can Be Accelerated*, Working Paper, Center for Social Entrepreneurship, Oxford: Said Business School.

Oakley, Francis (1968), 'Jacobean Political Theology: The Absolute and Ordinary Powers of Kings', *Journal of the History of Ideas*, 29 (3), 323–46.

OECD (1966), *Government and Technical Innovation*, Paris: OECD.

OECD (1969), *The Management of Innovation in Education*, Center for Educational Research and Innovation (CERI), Paris: OECD.

OECD (2010), *Innovation and the Development Agenda*, Paris: OECD.

OECD (2015), *The Innovation Imperative: Contributing to Productivity, Growth and Well-Being*, Paris: OECD.

Pankakoski, Timo (2013), 'Reoccupying Secularization: Schmitt and Koselleck on Blumenberg's Challenge', *History and Theory*, 52 (May), 214–45.

Peters, Edward (ed.) (1980), *Heresy and Authority in Medieval Europe*, Philadelphia, PA: University of Pennsylvania Press.

Quesne, Philippe (2007), 'Réinvestissement: une nouvelle histoire?', in Michael Foessel, Jean-François Kervegan and Myriam Revault d'Allones (eds), *Modernité et sécularisation : Hans Blumenberg, Karl Lowitz, Carl Schmitt, Leo Strauss*, Paris: CNRS Editions, pp. 93–105.

Rogers, Everett M. (1962), *Diffusion of Innovations*, New York: Free Press.

Saint-Simon, Henri (1825), *Nouveau Christianisme: Dialogues entre un conservateur et un novateur*, Paris: Rossange Père.

Sargant, William L. (1858), *Social Innovators and Their Schemes*, London: Smith, Elder and Co.

Schmitt, Carl (1922), *Political Theology: Four Chapters on the Concept of Sovereignty*, Chicago, IL: University of Chicago Press [1985].

Schmookler, Jacob (1966), *Invention and Economic Growth*, Cambridge, MA: Harvard University Press.

Siedentop, Larry (2014), *Inventing the Individual: The Origins of Western Liberalism*, Cambridge, MA: Harvard University Press.

Stedman Jones, Gareth (2010), 'Religion and the Origin of Socialism', in Ira Katznelson and Gareth Stedman Jones (eds), *Religion and the Political Imagination*, Cambridge: Cambridge University Press, pp. 171–89.

Tierney, Brian (1982), *Religion, Law and the Growth of Constitutional Thought, 1150–1650*, Cambridge: Cambridge University Press.

Tierney, Brian (1995), 'Freedom and the Medieval Church', in Noel B. Reynolds and W. Cole Durham (eds), *Religious Liberty in Western Thought*, Michigan and Cambridge: William B. Eerdmans Publishing, pp. 64–100.

Tierney, Brian (2014), *Liberty & Law: The Idea of Permissive Natural Law, 1100–1800*, Washington: The Catholic University of America Press.

Todeschini, Giacomo (2008a), *Richesse franciscaine*, Paris: Verdier.
Todeschini, Giacomo (2008b), 'Theological Roots of the Medieval/Modern Merchants' Self-Representation', in Margaret C. Jacob and Catherine Secretan (eds), *The Self-Perception of Early Modern "Capitalists"*, New York: Palgrave, pp. 17–46.
Troeltsch, Ernst (1912), *Protestantism and Progress: A Historical Study of the Relation of Protestantism to the Modern World*, New York: G.P. Putnam.
Weber, Max (1930), *The Protestant Ethic and the Spirit of Capitalism*, London: Allen and Unwin.

# 2. Imaginaries of innovation

*Harro van Lente*

Innovations are intimately connected to ideas, values and images of what is better, healthier, richer. Typically, novel technologies are accompanied with rosy images about their usefulness, benefits and desirability at large. Medical technologies, for instance, promise to improve health and well-being, while manufacturing technologies promise to enhance competitiveness and energy technologies promise to bring convenience. Likewise, plans for new technologies may be accompanied by negative notions of threat, danger or injustice, and then it can stop or change the course of developments. Nuclear technologies would bring central command and erode democracy, genetic technologies are said to corrupt natural orders, social media may render us credulous. Such promises, warnings, ideals and narratives – in short, imaginaries – can relate to particular technical products such as an electric car, or to generic technologies such as genetic testing. In either case, imaginaries offer particular notions of what problems are at stake, what solutions could help and who should be involved. That is, imaginaries are not just accompanying innovations, but frame them as well.

This chapter reviews how recent theories of innovation have highlighted and analysed the status and the role of imaginaries. For this purpose, I define imaginaries as *collectively available symbolic meanings and values*. In theories of innovation, the role of imaginaries is taken up widely. A recent example is the work of Levenda (2019) and colleagues, who studied the viability of regional projects of energy. They argue that novel directions in energy typically appear as global projects, at the expense of local ownership. This particular framing, or what they call sociotechnical imaginary, after Jasanoff and Kim (2009), undermines the possibilities of renewable energy. Their argument resonates with Bruno Latour's (2018) recent plea for taking locality seriously in order to address the climate crisis. Another example is how evolutionary theories of technical change argue that collective ideas about successful routes are needed for decision-making processes. The argument is that the inherent uncertainty of innovation does not allow rational optimization, hence routines and heuristics are required to make choices nonetheless (Dosi et al. 1988). The sociology of expectations, a third example, studies how representations of the future structure innovation processes. When particular notions about the future are collectively shared, they will interfere with the very development they reflect on, leading to additional legitimation, guidance and coordination of the innovative activities (van Lente 2012). Many other examples can be found, too, in the work on so-called 'sociotechnical imaginaries' that govern policy discourses (Jasanoff and Kim 2009), or in National Innovation System approaches (Edquist 2004).

This chapter looks at a variety of imaginaries as well the various ways in which they are part of recent innovation theories. The review in this chapter is organized as follows. First, I will investigate the concept of imaginary by tracing some roots in social theory which pave the way for the mechanism of performativity, and the sociology of expectations. After all, narratives of progress, improvement and efficiency resonate with long-standing cultural notions, values and identities. A next step is to delineate a typology of imaginaries, distinguishing narratives, pictures, graphs and symbols; these four will be discussed subsequently in different sections.

Then I will discuss how 'innovation' itself appears as an imaginary. The notion of innovation has become an imaginary itself, a shorthand for improvements, for progress, for superiority; a shorthand even for the project of modernity itself. In this way, 'innovation' captures the attention and becomes part of a symbolic universe in which actors have to operate. In a final section I reverse the gaze: instead of asking how imaginaries appear in theories of innovation, the question is how theories of innovation themselves become imaginaries, directing firms, governments or public involvement in innovation. The chapter concludes with a reflection on the ambiguous role of imaginaries in innovation: how they help but also frustrate the search for improvement. The message is that imaginaries are not innocent side-effects, but vital ingredients, moulding the meaning, use and direction of innovations. Without imaginaries, no innovation.

## IMAGINARIES IN SOCIAL THEORY

At this point is it appropriate to clarify the notion of 'imaginary', which I have defined as collectively available symbolic meanings and values. In fact, this characterization goes back to the classic sociological insight formulated by William Thomas and Dorothy Swaine Thomas that "if men define situations as real, they are real in their consequences" (Thomas and Thomas 1928, pp. 571–2). This so-called Thomas theorem stipulates that ideas or representations of social activities are intrinsic parts of these developments themselves, and that this is a fundamental sociological phenomenon. This claim has been elaborated in various ways, including the famous self-fulfilling prophecies of Robert Merton. He goes some steps further with claiming that social events or situations may be *produced* by the ideas we have about those events or situations. His example is about a bank in trouble, only because rumours circulated that the bank was in trouble, prompting clients to run to the counter and withdraw their money. This indeed brought the bank into trouble; it would happen to any bank in this situation. While before the rumours started spreading the bank was in normal business, the collective withdrawal of savings brought the bank severe difficulties, thus fulfilling the originally false rumours. This perverse effect of initially false ideas becoming true when circulating and acted upon, has become known as a 'self-fulfilling prophecy'. As Merton (1948, pp. 195–6), who coined the phrase, argued:

> The self-fullfilling prophecy is, in the beginning, a *false* definition of the situation evoking a new behavior which makes the originally false conception comes *true*. The specious validity of the self-fulfilling prophecy perpetuates a reign of error. For the prophet will cite the actual course of events as proof that he was right from the very beginning [...] Such are the perversities of social logic.

In innovation studies, the famous example of a self-fulfilling prophecy is Moore's Law, which refers to the prediction of Gordon Moore in 1965 that the number of components on an integrated circuit, a chip, would double during the next years. This prediction was subsequently used in the semiconductor industry to set targets and to coordinate efforts, and thus became 'true' (Mody 2016).

The key notion that ideas about reality shape reality, is the cornerstone of symbolic interactionism, which understand social reality as being forged by shared meanings and symbolic worlds, creating a context for further social action (Berger and Luckmann 1966). People live both in natural as well as symbolic environments. Following on this, Cornelius

Castoriadis (1975) used the term 'imaginary' to study society in his book *The Imaginary Institution of Society*. Later, John Thompson (1982, 1984) highlighted the concept of 'social imaginary' and defined it as "the creative and symbolic dimension of the social world, the dimension through which human beings create their ways of living together and their ways of representing their collective life" (Thomson 1984, p. 6).

A related strand of scholarly work on the status and force of imaginaries, is how discourses on complex social issues, such as environmental problems or organizational change, are shaped by frames and storylines (Boje 2008). Maarten Hajer (1995, p. 56) defines storylines as "the generative sort of narrative that allows actors to draw upon various discursive categories to give meaning to specific physical or social phenomena". According to his study, storylines perform a number of roles. First, they help to reduce the complexity of an issue or a problem and thereby also help agreement to be reached. Second, they can become ways to invoke a complex argument amongst stakeholders without recourse to a larger debate. Finally, a storyline provides a narrative that can link different expertise together. An example is how 'sustainability' works as a generative narrative in policy discourse, forging coalitions and pointing to viable directions.

In a more reflexive tone, the philosopher Charles Taylor (2004) used the notion of 'social imaginaries' in his study of the transition in Western culture from the hierarchically organized society to an egalitarian, horizontally organized society. The transition is reflected and effected by what he called social imaginaries. Gradually, the pre-modern social imaginaries were replaced by novel ones, he argued, including the rise of a public sphere, the emergence of a public opinion, the spread of a market economy as an independent sphere, and the ideal of self-government of citizens.

## IMAGINARIES IN INNOVATION: A TYPOLOGY

In theories of innovation imaginaries come in different forms and can be grouped in various ways. As will be discussed below, Neo-Schumpeterian theories point to iconic examples of past success, the 'exemplars', with their heuristic value for current attempts. The sociology of expectations, on the other hand, stresses the omnipresent representations of the future. Other studies point to so-called 'sociotechnical imaginaries', that is to say the persistent and coherent ways in which national projects are framed and executed. The set of imaginaries thus contain iconic examples, future projections, stable discursive constellations and master narratives.

While imaginaries in innovation are utterly diverse, what they share is a reference to the imagined. For the purpose of this chapter I suggest a typology based on the following two aspects of this reference to the imagined. First, imaginaries may add a connection to the imagined by offering a full storyline, with a plot, a script and protagonists; or, alternatively, by offering elements of a storyline, like an iconic example, which can be used in other narratives. Second, these offerings may appear discursively as texts, or materially, as objects. These two basic distinctions can be taken to design a typology, see Table 2.1.

In the next sections, these four basic types will be discussed subsequently. *Texts* and narratives have a content, a message; they are a basic expedient of storytelling, including a plot, a script and protagonists. Texts and narratives provide a history, a background for the present and an outline for future priorities. They typically circulate in organizations (Boje 2008) but also in society at large (Garud et al. 2014).

*Table 2.1*          *Typology of imaginaries*

|  | Full storyline | Element of a storyline |
|---|---|---|
| Text | *texts and narrative* | *symbols and icons* |
|  | with plot, script, protagonist | shared representation of desirable directions and values |
| Object | *graphs and curves* | *artefacts and image* |
|  | connecting past, present and future (by extrapolation) | staging an artefact in a story of desired or likely futures |

The storyline of promises and challenges can also be contained in *graphs and curves*. Think about the already mentioned graph of exponential increase of performance of micro-chips, known as Moore's Law, which structured the semiconductor industry for decades. In general, linear or exponential graphs bring a story about how the world is changing, how the past, present and future are connected, and what this implies for action.

Imaginaries do not always offer complete storylines. They can be more modest and just provide an ingredient, like a *symbol* or an icon, that can be included in several storylines. Think about the symbols of modernity, 'freedom', 'speed' or 'mobility', which offer powerful ingredients for storylines that guide innovations. Another example is the iconic and omnipresent picture of the blue earth in a black universe, which signifies the vulnerability of 'spaceship earth', as well as the ingenuity of mankind to make such pictures from space (Crook 2018).

It is also possible that an *artefact* plays this symbolic role, think about how cars symbolize freedom or masculinity, or 'the computer', bringing along imaginations of control and domination. In general models and prototypes have this symbolic function. They are not constructed to be sold and to fulfil a functional need; they are produced as a test and as a signal to a future market and to competitors: look what they are capable of, in principle. Such artefacts are not just material proofs of what is technically feasible, but also inform the audience on how the manufacturer sees the future. It informs the audience with a message about the future stakes and adds urgency to find a position.

## TEXTS AND NARRATIVES

Texts bring a message, have a content, display a storyline. Texts imply a narrative, with a beginning, an end, a goal; with movements, actions and actors. Texts build on and draw from other texts, which in their turn refer to other texts, etcetera. These networks of signification, or semiotics, render the content intelligible and significant. Given that innovations are affairs dealing with multiple actors, alignment between actions and decisions is needed, forged by a sense of direction. Many studies of innovation have indicated how texts accomplish this, as they bring along a fictive world, with actors and their efforts (Callon et al. 2009).

The archetypical narrative has, according to Joseph Campbell's (1968) classical study, the structure of a hero's adventure. The adventure typically is a journey, with a number of stages. The hero is called to action with a noble task, he (or she) has to be persuaded to perform this and to leave the everyday world. A series of trials lies ahead, with adversaries and allies, with battles and sieges and finally the hero returns, transformed, in the ordinary world. This fundamental storyline of promises ahead and challenges to be addressed is not only prominent in age-old myths and Hollywood screenplays, but also appears in 'innovation journeys' (Van de Ven 2017).

The 'sociology of expectations' starts from the observation that expectations, or imaginaries about the future, abound in innovation (Konrad et al. 2017). Such expectations are more than claims about the future that eventually will be false or true; more importantly they also shape what is going on in the present. Expectations thus truly matter in sociotechnical change; the sociology of expectations has explored the way expectations not only represent, but also propel research fields and innovation trajectories. In the first place, expectations legitimize decisions: investments in new technologies, by governments, research institutes or firms, can be defended by referring to their promised future gains. Expectations also guide innovation processes; given that future gains are always uncertain, it is difficult to calculate the best direction. Hence, what is seen as 'promising' will be used a shortcut for a calculation of the optimal direction. Likewise, expectations can also be a yardstick to measure actual performance: are the improvements following the promises or do they fall short? Finally, expectations may also coordinate research and development efforts, when researchers and firms position themselves in the expected development. When, for instance, brain research is the promised field, researchers, universities and even countries will decide what domain of research they will occupy (imaging, modelling, experimenting), based on what they see others doing, or what they expect others will be doing (van Lente 2012). Actors are part of strategic games, similar to prisoners' dilemmas or coordination games, which shape the emergence and development of the field. This performative aspect of expectations – they do not just reflect but shape some reality – is key in the sociology of expectations.

Expectations typically are connected and appear as building blocks for more encompassing narratives. When such building blocks appear jointly regularly and predictably, they may aggregate into 'visions' or 'Leitbilder', and can thus be defined as more or less coherent package of expectations, in a recognized narrative (Kuusi and Meyer 2002; Sovacool et al. 2019). Examples are visions about 'the hydrogen economy', or 'the internet of things'. Such overarching 'umbrella' imaginaries offer a fertile ground for more specific promises and other kinds of symbolic forms associated with innovation.

Expectations are forceful – and reinforced – due to their circulation amongst engineers, firms, government, who operate in a 'sea of expectations' (van Lente 2012). When expectations are readily available, they allow justifying particular research directions, coordinating efforts or shaping research agendas. This condition – imaginations of the future that circulate and are available – leads to particular dynamics, such as the pressure to fulfil the promise once it is widely accepted: the promise-requirement cycle (van Lente and Rip 1998).

Actors may pick up different roles in these strategic games. So-called 'enactors' are committed to a particular technological promise and provide argument in favour of this option, while other ('selectors') seek a technological solution; they are indifferent to particular options but need to choose between competing options. Such groupings exchange and assess expectations at conferences or trade journals, or 'arenas of expectations', as Sjoerd Bakker (2011) and colleagues have labelled them, following the work of Garud and Ahlstrom (1997). In their study of the Hydrogen Program at the Department of Energy in the US, they also investigated how the DoE assessed the credibility of expectations about hydrogen as an energy carrier (Bakker et al. 2012a, 2012b). This, they note, depends on three achievements in texts: historical progress towards the current level of technological performance; a path to even higher levels of performance; an end target, mapped to societal needs.

Of course, what is promised will not necessarily come true. In fact, what eventually results from innovation activities will always be different from what was expected: it will be less, or more, or just different. Such deviations do not speak for themselves; it is to be negotiated how to evaluate expectations that are not met. Some will take it as evidence that the whole project should be reconsidered and redirected, see for example the case of telecare, where failing expectations prompted a revision of the imaginaries guiding the project (Pols 2012). Others, however, will see failing expectations as a proof that the project requires even bigger investments: the expectations are seen as 'right', yet the efforts were not enough. The classical example is nuclear fusion, which during the last 50 years consistently promises to be 'ready' within 50 years. Here, failed expectations have led to ever more investments (Braams and Stott 2002).

In short, innovation is surrounded by the imaginaries that stipulate ideas, plans and future solutions that merit attention, seek credibility and forge investments. It is about what will or should happen, in some near or not so near future; it is about things that are not 'yet' there. In this sense, innovation is truly about the imaginary.

## GRAPHS AND CURVES

A storyline can also be contained in a *graph* or a curve. Think about the graph of exponential increase of performance of micro-chips, known as Moore's Law, mentioned above. Firms and researchers in the industry took the graph as a yardstick to measure their efforts and to decide on further investments. In every round, the design and production of new micro-chips became more expensive, more or less in the same exponential vein. Hence, only a few global players could sustain the race around the powerful graph. In this way, the graph structured the semi-conductor industry for decades (Mody 2016). Other well-known graphs in innovation are the so-called lifecycles, which depict 'embryonic', 'growth', 'maturity' and 'decline' stages, as if innovations resemble organic existences. Such imaginaries are powerful: when an innovation is depicted as mature or declining, it is less interesting for further investments. Here, the metaphors of lifecycles are often self-fulfilling, too.

Graphs or curves may also capture and carry a societal debate around technologies. A good example is the graph of global warming that showed a gradual lowering of global temperatures during the last two millennia followed by a rapid increase in recent decades. While it was contested and continuously refined, this 'hockey stick' curve epitomized the debate around fossil fuels and the need to invest in renewable energy sources (Mann et al. 1998; Mann 2018). Imaginaries may become the carriers of a whole debate.

Arguably the most powerful curve of innovation stems from the theory of Everett Rogers (2003) on the diffusion of innovations, which was first published in 1962. Drawing from studies on mass communication he outlined a theory of the gradual spreading, or diffusion, of novelty. The famous S-curve, based on the logistic function, describes a slow start, an accelerated growth and a final saturation. The S-curve tells that while spreading is slow in the beginning, it will reach saturation in the long run. His theory also brings categories of adopters, including innovators, early adopters, the early and late majority and laggards. The imaginary of the S-curve comes with the suggestion that novelties cannot be stopped and that the only important question regarding innovation is the speed of spreading. The concomitant grouping of adopters also has normative consequences: while it is good to be seen as an early adopter, one should avoid being a 'laggard', although some take pride in resisting trends and being old-fashioned.

In general, graphs and curves in innovation, when widely circulated, bring a story about how the world is changing, about how the past, present and future are connected. They inform what is at stake and they implicitly set an agenda of relevant questions and options. In this way, they bring an urgency to prepare better for such changes and to design a robust strategy.

## SYMBOLS AND ICONS

Innovations are informed and shaped by cultural notions of what is real or desirable – and, in due course, they reshuffle these notions as well. An important theme in innovation studies is the emergence of a culturally shared idea around an innovation. A famous example is the study of Trevor Pinch and Wiebe Bijker (1984) of the history of the bicycle at the end of the nineteenth century. They stress that the idea of what a bicycle *is* and what it *should do* was open and fluid during the first decades. There was no common interpretation of the novel two-wheel contraptions that was offered. Instead, an amazing range of different models were presented, reflecting different solutions for different problems. The popular model with a very large front wheel and a small rear wheel, for instance, was for young male users a means to expose their skills and courage, while elderly users saw it as an unsafe machine. This shows, what Pinch and Bijker call 'interpretative flexibility': what an artefact is and does is not just given, but interpreted differently. They also note a convergence of meanings at some point in the development, a 'closure'. In the broader social negotiations about what a bicycle is and should do, a particular interpretation emerged of what a bicycle should look like, what criteria it should meet and which uses are adequate. This dominant interpretation became the imaginary of what a 'real' bicycle is. After the closure designs tended to look alike, reflecting the image of an iconic bicycle: two equally sized wheels with rubber tyres, with pedals in between, connected to the rear wheel by a chain. The development of an innovation thus also included the genesis of an icon.

James Utterback (1994) studied the rise of iconic examples in more industries, such as television, car manufacturing, typewriters and plain glass manufacturing. In each case he found a typical pattern of two waves. Before the rise of an iconic example, what he labelled a 'dominant design', a wave of product innovation occurs, with an increasing number of (small) firms launching an increasing array of models to compete in an emerging market. The models differ significantly, highlighting various aspects of what important features for users could be. Some early typewriters, for instance, are lightweight to enhance movability; others are elegantly designed to adorn desks or are very robust to allow intensive usage. For two decades the models were all very different, embodying compromises; they all competed, as it were, for what a 'real' typewriter could be. At some point a 'dominant design' became successful, which did not excel in all the different aspects, such as speed, weight, costs, design, versatility, but provided sufficient performance for many user groups. It was not a 'maximizer' but a 'satisficer'. The dominant design in typewriters was the Underwood Nr. 5, introduced in 1899, two decades after the introduction of the first typewriters. This became the iconic typewriter, what we now see as a 'real' typewriter. After the introduction of the dominant design, new offerings resembled each other and radically different models became rare. The rate of product innovation dropped and as the competition shifted to costs and quality, process innovations increased. So, like in the case of bicycles, the development of typewriters not only brought novel artefacts and novel industries, it also produced an icon, which, in turn shaped the further developments.

Culturally embedded symbols in innovation may also appear as broader sets of values, ideas and goals. Studies of so-called 'sociotechnical imaginaries' analyse how such encompassing notions shape innovations. Jasanoff and Kim (2009) use and defend the notion in a comparison of the development and employment of nuclear power in the United States and South Korea. The notion of 'sociotechnical imaginary' is to explain long-term cross-national variation in science and technology policy. It is defined as "collectively imagined forms of social life and social order reflected in the design and fulfillment of nation-specific scientific and/or technological projects" (Jasanoff and Kim 2009, p. 120). Where the imaginary in the US was about the risk of runaway accidents and catastrophic environmental damage, the imaginary in South Korea was about the risk that nuclear developments would halt and that the nation would depend on foreign powers. Accordingly, the urgency derived from the two imaginaries are different, too. While the policy focus in the US was on containing radiation release, the focus in South Korea was on building a sufficient domestic capacity for power generation. Jasanoff and Kim also stress that the sociotechnical imaginaries are intimately related to and informed by legal systems, cultural preferences and knowledge infrastructures. Pfotenhauer and Jasanoff (2017) extended the concept of sociotechnical imaginaries beyond state policies to investigate localized cultural differences. They study how the successful and seemingly generic MIT model – first developed in Boston – has been organized differently in the UK, Singapore and Portugal. It shows that the governance, the logics and languages of hopes and dreams are very different: the three MIT models have been adapted to societal aspirations. Despite their common outlook, the political particularities also challenge the notion of 'development', that is the tendency to understand local differences in terms of more or less 'developed'. Sociotechnical imaginaries thus symbolize imagined forms of social life implied in national or regional technological projects, which characterize a particular political culture regarding technological innovation.

## ARTEFACTS AND IMAGES

Technological artefacts themselves may figure as imaginary, too. A case in point are prototypes, which are produced to test a particular design or set-up, as well as to show to competitors and potential markets what the manufacturer sees as a direction for the future. Sjoerd Bakker and colleagues (2012a, 2012b), for instance, analysed how prototypes of cars with hydrogen as an energy carrier convey messages of the future. Each prototype contains design choices, for instance concerning the drivetrain or storage of the fuel: as gas or as liquid. In this way, a design of prototypes can be read as a collection of choices, as a statement about a future world and how things matter in the future. This is maybe easier to discern in historical examples, when the viewer no longer automatically aligns with the imagination of the future: old designs convey futures of the past.

Artefacts, like prototypes, thus may function as signposts for further innovations. In the evolutionary theories of technology, the Douglas DC-3 aircraft model is the prime example of an artefact that symbolizes which way to go. It was developed in the 1930s, it had a metal skin and piston engines on low wings, and it provided an exemplary model, or an 'exemplar', for the Douglas company and its competitors. The Douglas DC-3 was an imaginary of a successful aircraft and this imaginary guided subsequent innovations in the aircraft industry for decades. The example figures in a seminal article "In search of useful theory of innovation" by

Richard Nelson and Sidney Winter (1977), where they outlined a new, evolutionary direction for innovation theories. They started with the question why particular sectors have different growth rates. In their search of a new theory they go against the mainstream neo-classical assumptions of perfect rationality and they problematize the possibility of maximization. Instead, they suggest an evolutionary approach to study innovation, with variation and selection. In this view innovations are the outcomes of, on the one hand, the generation of novelty by firms ('variation') and, on the other hand, the fate of novelty in 'selection environments', which includes markets, regulatory arrangements and networks of firms, in supply chains or competing clusters. Firms, the producers of variations, are guided by their 'routines', which constitute a collective memory, a repertoire of skills and a set of orientations. In this way, routines are an equivalent of a genetic profile of a firm, which can be more or less successful in a selection environment. Nelson and Winter and evolutionary economics in general, argue that research and development is an open-ended adventure with unclear directions, unclear gains and unclear costs. While decisions about innovation cannot be calculated or maximized, they still have to be taken. Instead of calculating what the best option is – which is impossible – firms follow 'search routines' or 'heuristics', that is, rules of thumb that promise but cannot guarantee success. Given that 'routines' to set out new directions and projects are embedded in firms, the variations that firms produce are not just random but follow a pattern. Nelson and Winter label the regularity in the sequence of variations as a 'technological regime'; Giovanni Dosi and colleagues (1988) later relabelled the regularity as a 'technological paradigm'. The strength of the pattern may even result in a situation where other options are not seen as viable anymore, a so-called 'lock-in' (Berkhout 2002).

Unlike in biological evolution, variation in evolutionary economics is not blind (Rosenberg 1983). The routines and the heuristics, symbolized by the 'exemplar', inform and guide researchers and engineers in their search for novelty. Instead of blindly generating variations, firms are informed by beliefs about how selection will occur and what will be successful. Developers of novel aircraft believed that when they followed the basic characteristics of the DC-3, their variation would have more chances in a harsh selection environment. Selection, also unlike biological evolution, is not independent from variation: those who have a stake in a particular variation will seek to influence the selection environment. Car manufacturers, for instance, try to delay or to speed up particular innovations (Bladh 2019); nuclear power companies seek to ensure favourable regulation and governmental investments (Jasanoff and Kim 2009); big firms try to modify the market with huge marketing efforts (Soete 2013).

## INNOVATION AS IMAGINARY

While imaginaries are important ingredients of innovation processes, as texts, graphs, symbols or artefacts, innovation is already an imaginary in itself. Since the Industrial Revolution innovations have generally been embraced as the fruits of ingenuity and the sources of progress (Freeman and Louçã 2001). As Pfotenhauer and Jasanoff (2017, p.784) argue, "[i]nnovation has become a go-to answer, a panacea that carries the promise of curing socioeconomic ailments almost irrespective of what these ailments are or how they have arisen". In this way, the modernist belief in rationality and progress has landed on innovation, in particular in the second half of the twentieth century. Helga Nowotny (2006) even argues that 'innovation' has become a modern equivalent of 'progress'.

A key source here is the famous, or infamous, 'linear model', which holds that science first brings insights, to be developed into practical applications, which subsequently diffuse through adoption and ultimately bring progress, in terms of health, safety and well-being, but above all in terms of economic growth. Godin (2006, 2014) concluded that the linear model is an invention in itself, and that its emergence is related to a particular historical period, after WWII, when the State became a patron of science and innovation. In the same move, with the rise of the welfare state, the State became the organizer of economic growth (Cohen and Todd 2018). These two novel responsibilities merged in the linear model. Godin (2006) also argues that the concomitant distinction between fundamental and applied research resonates with the age-old opposition between intellectual work (*theoria*), seen as higher, and practical work (*praxis*), seen as lower. Hence, the linear model reinforces the notion that pure fundamental research is superior over applied research and comes essentially first, also temporally.

According to Joly (2019), the linear model of innovation makes two other distinctions, too. Apart from a particular order of activities – the sequence of basic research, applied research and products bringing societal benefit – it also postulates a task division between constituencies. According to the 'master narrative' of the linear model, science and technology are the sources of innovation, with firms and academic laboratories as the key actors. In a further development, other societal actors are more prominent as 'adopters' and 'users' of the innovations. Activities and decisions in this master narrative are framed by the language of 'competitiveness' and economic growth; the overall value to assess innovation processes is economic welfare. A clear example of the master narrative at work is the 2000 Lisbon Agenda of the EU, which states the objective that 3 per cent of GDP should be spent on research in order to keep Europe healthy and competitive. In mainstream studies of innovation, that is 'innovation studies', the linear model has been criticized, including the putative task division and the particular sequence of activities (Fagerberg and Verspagen 2009). Instead, scholars now stress the interaction between societal parties and the iterations of activities. For instance, scientific knowledge often is produced in response to practical problems of development, so basic research may appear later than the application.

The linear model has been the dominant imaginary of innovation during the last half century, but has been complemented by other models. The critical turn in the 1970s challenged the authority of experts on innovation and called for the rights of non-experts, like citizens and stakeholders, to have a say about priorities, directions and criteria. According to Joly (2019), three alternative models (or 'moral economies') challenge the dominant imaginary of the linear model. First, the model of 'users innovation', which points to the creativity of users: they should not be seen as passive recipients of innovation ('adopters') but are innovative sources themselves. The idea is that when the role of users is better recognized and exploited, firms and society will get better innovations. Second, the model of 'distributed innovation' challenges the assumption that innovation has a point source at fundamental research by stressing the shared origin of innovations. The key example is open source software (OSS), which is developed and refined by networks of engineers and users. The third alternative model, 'social innovation', points to the dynamics of ongoing societal change, with key roles for social entrepreneurs, local communities and knowledge brokers. Innovation, then, is to be understood as part of social transformations, such as reduction of inequality or the fight against poverty. Whether the list of alternatives to the linear model is exhausted by these three remains to be seen. The important point is that while the linear model may be challenged, innovation itself continues to figure as an imaginary in politics and society at large.

# THEORIES AS IMAGINARIES

At this point, it is also fitting to appreciate *theories* of innovation as imaginaries. Theories of innovation are phrased and used by scholars at universities and research institutes, and circulate widely amongst policymakers and firms. Theories in general introduce a vocabulary and a perspective to make sense of the realities societal actors have to operate in (Weick 1995). Theories bring distinctions, guidance and solace; they invite ways to act in the world. In this way, theories are performative: they do not just describe the world, but they also shape the world accordingly (Callon 1998; MacKenzie 2006).

Traditionally, innovation has not been a central topic in social theories, although many traces can be found. Adam Smith, for instance, stressed the importance of technology in the Wealth of Nations, as it facilitates the division of labour, while Karl Marx analysed the disruptive powers of technology, as a means to extract value from labour. Mainstream neo-classical economic models tend to consider innovation as an external factor, as a force that somehow developed outside the realm of economic decisions, but which had, of course, consequences for those decisions. It did not need further explanation or study. Evolutionary theories, as discussed above, were developed in the 1980s to depart from this tradition and to put innovation at centre stage (Dosi et al. 1988). These alternative approaches were also labelled as 'neo-Schumpeterian', as they are inspired by the legacy of Joseph Schumpeter, who studied the dynamics of economic change in the 1930s and coined the idea of 'creative destruction': innovation brings novelty and new opportunities, but also destroys established industries. Innovation, in this view, cannot just be a linear increase of productivity and employment, but will bring painful twists and unexpected corners. In such a view, innovation is a factor that requires full attention, as it can be a friend or a foe. The key notion of 'creative destruction' is now embraced in policy making and management studies. The theories of Schumpeter have influenced governments in their search for innovation. Schumpeter struggled with the question whether sources of innovation are to be found in rebellious young firms daring to challenge the status quo, or, on the contrary in big, powerful firms that have the means to explore and invest in long-term developments. These two claims, known as Schumpeter Mark I and Schumpeter Mark II respectively, are more than scholarly hypotheses. They linger as imaginaries of success amongst innovation policy and entrepreneurial universities: Mark I supports the establishment of science parks and incubators, fed by venture capital and fresh graduates, while Mark II invites strong ties with big firms to test and support novel ideas.

The theory of 'national systems of innovation' (Edquist 2004) provides another example of a theory of innovation as an imaginary suggesting where to go and what matters. In this theory, innovation can only thrive when countries provide the right conditions in terms of education, infrastructure, legislation and capital systems. As a consequence, national governments have become attentive to how they score in terms of such conditions (Kim 2018). This alertness is further propelled by comparative studies by the OECD and others, which often result in rankings in, for instance, R&D investments – a proxy for success. Rankings, in their turn, always provide reasons for concern and incentives for action. When a country has a lower position than the highest – and most are – efforts are needed to prevent the country from further lagging behind; when a country is so exceptional to occupy a number one spot, efforts are needed to secure this position, since competitors are alert and active. Aspirations to have a 'good' national innovation system thus lead to comparison and competition between countries, and to innovation races. The rankings and the races are further enhanced when it is believed that others see them as key for success. The spread of theories of innovation thus does not automatically only lead to insight and judgements, but may also result in herd behaviour and short-sightedness.

## CONCLUSION

This chapter studied how imaginaries, defined as collectively available symbolic meanings and values, figure in recent theories of innovation. I argued that imaginaries in innovation appear as forceful narratives, as compelling graphs, as guiding symbols and as exemplary artefacts. Also the notion of innovation *itself* surfaces as imaginary, as an expression of modernity and progress. In the same vein, *theories* of innovation may transform into imaginary, as they promise success for firms and national governments and suggest where to go. In all these ways, imaginaries frame, invite and facilitate negotiations and interactions.

The overview of this chapter shows that innovation and imaginaries are intimately connected and deeply entangled. Theories of innovation reflect on this in different ways. In evolutionary economics, for instance, imaginaries like the DC-3 aircraft as exemplary achievement ('exemplar') appear as guiding devices for decision making and are bound up in the routines and heuristics that structure the innovation process. The sociology of expectations highlights how technological promises tend to be performative: instead of claims about the future to be refuted or not, they change the world and set actors in motion. The strand of research on sociotechnical imaginaries shows how imaginaries frame the entanglement of innovation with policy making.

Traditionally imaginaries have been seen as side-effects of the 'real' dynamics of innovation, relating only to marketing while leaving the 'real' development in academic research and industrial development, unaffected. Recent theories of innovation, however, stress that imaginaries are an intrinsic part of any innovation, from the first stages of exploration, to the later stages of extensive usage. Imaginaries, as the overview shows, are helpful in many ways. Imaginaries point to viable directions, help actors to make decisions, affect the credibility of innovation projects, and afford sense making in public deliberations on innovation at large. Imaginaries, therefore, are crucial for the shape, the speed and the direction of innovation.

The overview also showed that when imaginaries are more than just side-effects, they cannot be innocent either. When innovation appears as a default promising route, as is implied, for instance, in the linear model of innovation, in Moore's Law, or in the idea of National Systems of Innovation, the room for manoeuvre has decreased. In those cases, it is more difficult to deviate from the paths taken, it is even more difficult to imagine alternative paths. In more extreme cases, global races amongst firms and governments may start, including metaphors of not missing boats, or trains that cannot be stopped. In those cases, parties will base their choices on what they believe the choices of others will be and the whole dynamic may become self-propelling. When every industry follows Moore's Law, it is dangerous not to follow it; when every country seeks to improve its National System of Innovation, it is tempting to follow suit. In these cases, the force of imaginaries leads to a decrease of choice. So while imaginaries in general afford novel directions, they may also generate straightjackets and reduce creativity. The famous saying "imagination rules the world", often attributed to Napoleon Bonaparte, thus is valid in two ways: imaginaries may help to propel innovation, but may also hinder finding better directions.

# REFERENCES

Bakker, Sjoerd, Harro van Lente and Marius Meeus (2011), 'Arenas of expectations for hydrogen technologies', *Technological Forecasting and Social Change*, 78 (1), 152–62.

Bakker, Sjoerd, Harro van Lente and Marius Meeus (2012a), 'Dominance in the prototyping phase – the case of hydrogen passenger cars', *Research Policy*, 41, 871–83.

Bakker, Sjoerd, Harro van Lente and Marius Meeus (2012b), 'Credible expectations – the US Department of Energy's Hydrogen Program as enactor and selector of hydrogen technologies', *Technological Forecasting and Social Change*, 79 (6), 1059–71.

Berger, Peter L. and Thomas Luckmann (1966), *The Social Construction of Reality: A Treatise in the Sociology of Knowledge*, New York: Doubleday.

Berkhout, Frans (2002), 'Technological regimes, path dependency and the environment', *Global Environmental Change*, 12, 1–4.

Bladh, Mats (2019), 'Origin of car enthusiasm and alternative paths in history', *Environmental Innovation and Societal Transitions*, 32, 153–68.

Boje, David M. (2008), *Storytelling Organizations*, London: Sage.

Braams, Cornelis M. and Peter E. Stott (2002), *Nuclear Fusion: Half a Century of Magnetic Confinement Fusion Research*, London: IOP Publishing.

Callon, Michel (ed.) (1998), *The Laws of the Markets*, Oxford: Blackwell.

Callon, Michel, Pierre Lascoumes and Yannick Barthe (2009), *Acting in an Uncertain World: An Essay on Technical Democracy*, Cambridge, MA: MIT Press.

Campbell, Joseph (1968), *The Hero with a Thousand Faces*, Princeton: Princeton University Press.

Castoriadis, Cornelius (1975), *The Imaginary Institution of Society*, Cambridge, MA: Polity Press.

Cohen, Daniel and Jane M. Todd (2018), *The Infinite Desire for Growth*, Princeton: Princeton University Press.

Crook, Tony (2018), 'Earthrise 50+: Apollo 8, Mead, Gore and Gaia', *Anthropology Today*, 34 (6), 7–10.

Dosi, Giovanni, Christopher Freeman, Richard Nelson, Gerald Silverberg and Luc Soete (eds) (1988), *Technical Change and Economic Theory*, London, New York: Pinter.

Edquist, Charles (2004), 'Systems of innovation – a critical review of the state of the art', in Jan Fagerberg, David Mowery and Richard Nelson (eds), *Handbook of Innovation*, Oxford: Oxford University Press, pp. 181–208.

Fagerberg, Jan and Bart Verspagen (2009), 'Innovation studies: The emerging structure of a new scientific field', *Research Policy*, 38 (2), 218–33.

Freeman, Christopher and Francisco Louçã (2001), *As Times Goes By. From the Industrial Revolutions to the Information Revolution*, Oxford: Oxford University Press.

Garud, Raghu and David Ahlstrom (1997), 'Technology assessment: A socio-cognitive perspective', *Journal of Engineering and Technology Management*, 14 (1), 25–48.

Garud, Raghu, Henri A. Schildt and Theresa K. Lant (2014), 'Entrepreneurial storytelling, future expectations, and the paradox of legitimacy', *Organization Science*, 25 (5),1479–92.

Godin, Benoît (2006), 'The linear model of innovation: The historical construction of an analytical framework', *Science, Technology & Human Values*, 31 (6), 639–67.

Godin, Benoît (2014), '"Innovation studies": Staking the claim for a new disciplinary "tribe"', *Minerva*, 52 (4), 489–95.

Hajer, Maarten A. (1995), *The Politics of Environmental Discourse: Ecological Modernization and the Policy Process*, Oxford: Oxford University Press.

Jasanoff, Sheila and Sang Hyun Kim (2009), 'Containing the atom: Sociotechnical imaginaries and nuclear power in the United States and South Korea', *Minerva*, 47 (2), 119–46.

Joly, Pierre-Benoit (2019), 'Reimagining innovation', in Sébastien Lechevalier (ed.), *Innovation Beyond Technology. Creative Economy*, Singapore: Springer, pp. 25–46.

Kim, Eun-Sung (2018), 'Sociotechnical imaginaries and the globalization of converging technology policy: Technological developmentalism in South Korea', *Science as Culture*, 27 (2), 175–97.

Konrad, Kornelia, Harro van Lente, Christopher Groves and Cynthia Selin (2017), 'Performing and governing the future in science and technology', in Ulrike Felt, Rayvon Fouche, Clarke A. Miller and Laurel Smith-Doerr (eds), *The Handbook of Science and Technology Studies* (4th edition), Cambridge, MA: MIT Press, pp. 465–93.

Kuusi, Osmo and Martin Meyer (2002), 'Technological generalizations and leitbilder – the anticipation of technological opportunities', *Technological Forecasting and Social Change*, 69 (6), 625–39.

Latour, Bruno (2018), *Down to Earth: Politics in the New Climatic Regime*, Cambridge: Polity Press.

Levenda, Anthony M., Jennifer Richter, Thaddeus Miller and Erik Fisher (2019), 'Regional sociotechnical imaginaries and the governance of energy innovations', *Futures*, 109, 181–91.

MacKenzie, Donald (2006), *An Engine, Not a Camera. How Financial Models Shape Markets*, Cambridge, MA: MIT Press.

Mann, Michael E. (2018), 'Keynote Address: The hockey stick and the climate wars: Dispatches from the front lines', *Science*, 41 (2), 48–9.

Mann, Michael E., Raymond S. Bradley and Malcolm. K. Hughes (1998), 'Global-scale temperature patterns and climate forcing over the past six centuries', *Nature*, 392, 779–87.

Merton, Robert K. (1948), 'The self-fulfilling prophecy', *The Antioch Review*, 8 (2), 193–210.

Mody, Cyrus M.M. (2016), *The Long Arm of Moore's Law: Microelectronics and American Science*, Cambridge: MIT Press.

Nelson, Richard R. and Sidney G. Winter (1977), 'In search of useful theory of innovation', *Research Policy*, 6 (1), 36–76.

Nowotny, Helga (2006), 'The quest for innovation and cultures of technology', in Helga Nowotny (ed.), *Cultures of Technology And the Quest for Innovation*, New York: Berghahn Books, pp. 1–26.

Pfotenhauer, Sebastian and Sheila Jasanoff (2017), 'Panacea or diagnosis? Imaginaries of innovation and the "MIT model" in three political cultures', *Social Studies of Science*, 47 (6), 783–810.

Pinch, Trevor J. and Wiebe E. Bijker (1984), 'The social construction of facts and artefacts: Or how the sociology of science and the sociology of technology might benefit each other', *Social Studies of Science*, 14 (3), 399–441.

Pols, Jeannette (2012), *Care at Distance: On the Closeness of Technology*, Amsterdam: Amsterdam University Press.

Rogers, Everett M. (2003), *Diffusion of Innovations* (5th ed.), New York: Free Press.

Rosenberg, Nathan (1983), *Inside the Black Box: Technology and Economics*, Cambridge: Cambridge University Press.

Soete, Luc (2013), 'Is innovation always good?' in Jan Fagerberg, Ben R. Martin and Esben S. Andersen (eds), *Innovation Studies – Evolution and Future Challenges*, Oxford: Oxford University Press, pp. 134–44.

Sovacool, Benjamin K., Johannes Kester, Lance Noel and Gerardo Zarazua de Rubens (2019), 'Contested visions and sociotechnical expectations of electric mobility and vehicle-to-grid innovation in five Nordic countries', *Environmental Innovation and Societal Transitions*, 31, 170–83.

Taylor, Charles (2004), *Modern Social Imaginaries*, Durham, NC: Duke University Press.

Thomas, William I. and Dorothy S. Thomas (1928), *The Child in America: Behavior Problems and Programs*, New York: Knopf.

Thompson, John B. (1982), 'Ideology and the social imaginary: An appraisal of Castoriadis and Lefort', *Theory and Society*, 11 (5), 659–81.

Thompson, John B. (1984), *Studies in the Theory of Ideology*, Cambridge: Polity Press.

Utterback, John M (1994), *Mastering the Dynamics of Innovation*, Boston, MA: Harvard Business School Press.

Van de Ven, Andrew H. (2017), 'The innovation journey: You can't control it, but you can learn to maneuver it', *Innovation*, 19 (1), 39–42.

van Lente, Harro (2012), 'Navigating foresight in a sea of expectations: Lessons from the sociology of expectations', *Technology Analysis & Strategic Management*, 24 (8), 769–82.

van Lente, Harro and Arie Rip (1998), 'Expectations in technological developments: An example of prospective structures to be filled in by agency', in Cornelis Disco and Barend J.R. van der Meulen (eds), *Getting New Technologies Together. Studies in Making Sociotechnical Order*, Berlin: Walter De Gruyter, pp. 203–31.

Weick, Karl E. (1995), *Sensemaking in Organizations*, Thousand Oaks, CA: Sage.

# PART II

# THEORIZING INNOVATION IN THE TWENTIETH CENTURY: THE FOUNDATIONS

# 3. Theories of innovation

*Benoît Godin*

Theories of innovation are over one hundred years old, and theoretical thoughts on the subject have multiplied in recent decades. One reads, again and again, that theories of innovation come from economics, particularly from Joseph Schumpeter, as if economics were the whole story. "The term 'innovation'", states historian of technology John Staudenmaier, "appears to have originated in a tradition of economic analysis" (Staudenmaier, 1985: 56). Staudenmaier is not alone in this belief. "The founding framework of innovation" is Schumpeter, states Norbert Alter (Alter, 2000: 8). This is a common attribution of innovation's origin or ancestry in "innovation studies". A dominant and economically oriented field is responsible for such a historiography. However, theories are far more diverse than the standard historical perspective would lead us to believe. Schumpeter was only one among many theorists who studied innovation over the twentieth century. Scholars later resurrected Schumpeter to give legitimacy to a specific framework. As Michael Freeden suggests: "Ideologies are constantly engaged in reconstructing their own history" (Freeden, 1996: 239). Since innovation has become an ideology today, the statement fits the field of innovation studies perfectly well.

The purpose of this chapter is to provide a historical perspective on theories of innovation, and a sense of their broad range. Over the twentieth century, scholars have produced a diversity of theories that cover the psychological, sociological, organizational and economic dimensions of society. Yet, at the same time, this diversity came to be dominated by a few theories, or a particular kind of theory: industrial and technological innovation. What are these theories? How do they explain innovation? With what conceptual apparatus?

This chapter examines a dozen of theories produced from 1890 to the early 1980s, as the foundation of the theorization of innovation. True, theoretical thoughts proliferated in the 1960s and after, and it would be nonsense to limit theories of innovation to those I study here. "There can never be", claimed Louis Tornatzky and his colleagues from the US National Science Foundation in a 1983 review of the literature, "an 'Authorized version' of innovation literature. The field is too complex and too full of value and perspective to make such an exercise practical even if it were desirable" (Tornatzky et al., 1983: x). It is the thesis of this chapter that the theories studied here created the foundations. The frameworks, semantics and discourses of theories of innovation have their roots here.

In this chapter, I limit the analysis to a study of the concepts used to theorize innovation. To do this, I present the main theorists of the twentieth century in chronological order. Without being a strictly linear story, a succession over time is manifest: from sociological to psychological theories, then to managerial and economic theories.

## GABRIEL TARDE

One can find a few conceptual thoughts on innovation before the twentieth century (Godin, 2015). But the first theory of innovation comes just a decade before the twentieth century.

The French sociologist, criminologist and social psychologist Gabriel Tarde (1843–1904), a prominent critic of Emile Durkheim's sociology, produced the first theory of innovation ever written. *Les lois de l'imitation* is a study of what we today call diffusion of innovation, and what Tarde called "imitation". Tarde's was a work of pure sociology: explaining the laws of, and predicting, social transformations. To Tarde, science is the study of regularity and universality. In the social sciences, this takes the form of a study of imitation. Imitation is "à la science sociale ce que la formation d'une nouvelle espèce végétale ou animale est à la biologie" (to social science what the emergence of a new plant or animal species is to biology) (Tarde, 1890: 68, 71). Among the social mechanisms or factors of change is innovation.

According to Tarde, innovation is inventions (practical) and discoveries (theoretical), or "initiatives" in social organization, language, religion, politics, law, industry and useful art (p. 62). Two concepts serve Tarde's theory: invention and imitation. Inventions are the driving force of society (Tarde, 1902), but the diffusion of inventions matters even more. "Une innovation non imitée, est comme n'existant pas socialement" (An innovation not imitated is as though non-existent socially) (Tarde, 1890: 208). Invention is not progress until it becomes an innovation. Innovations are discoveries and inventions applied or used, that is, imitated. "Il y a évolution sociale quand une invention se répand tranquillement par imitation" (There is social evolution when an invention diffuses gradually by imitation) (p. 243).

To Tarde, "les lois de l'invention appartiennent essentiellement à la logique individuelle; les lois de l'imitation en partie à la logique sociale" (the laws of invention belong essentially to individual logic; the laws of imitation in part to social logic) (p. 434), a distinction reproduced again and again in the following decades. "L'invention…n'est pas à mes yeux un fait purement social dans sa source: elle naît…du génie individuel" (Invention…is not to my mind a purely social act in its source: it comes from…individual genius) (p. 54). The study of invention (its origin) is the affair of archeology and the study of imitation (its diffusion) that of statistics (p. 163). Since Tarde, 'invention imitated' (diffused or used) has remained one of the main meanings of innovation.

## BOX 3.1   GABRIEL TARDE

- Invention is a psychological or mental faculty of the individual that combines or assembles existing ideas, ideas originating from other individuals (the social world), a compound of existing or older elements in a new, more complex entity (Tarde, 1890: 105). "Une idée nouvelle est une combinaison d'idées anciennes, apparues en des lieux distincts et souvent fort distants" [a new idea is a combination of old ideas, appearing in distinct places, often considerably far away] (Tarde, 1893: 267).
- Imitation (diffusion) is the essence of social change. Inventions are made (often in parallel) then diffused among individuals.
- This process follows statistical laws. A discovery is first made then it propagates by contagion, according to three phases (S-shaped curve): discovery (slow progress), diffusion (rapid progress) and decline.
- The causes of diffusion are logical (utility, want) and extra-logical (accidental, environmental). There exist two processes of evolution: substitution (of one invention by another) and accumulation (of inventions).

One must turn to others of Tarde's works like *La logique sociale* (1893) to understand the context within which his theory of innovation developed. In that work, Tarde contrasted statics ("logique statique") and dynamics ("logique dynamique"), as Auguste Comte had before him. *La logique sociale* develops an evolutionary view of society, which has conflict as its basic principle – competition ("duels") between new inventions – ("les contradictions…sont l'âme du progrès") (contradictions…are the soul of progress) (Tarde, 1893: 250), and whose resolution is neither predictable nor a "strict unilinear" evolution (pp. 255, 261), but rather a process of 'historical' accumulation (pp. 294–7). Tarde applied the theory to language, religion, sentiments, political economy and arts.

## ABBOT USHER

The next theory of innovation appeared 30 years after Tarde's. In 1929, economic historian Abbott Usher (1883–1965) from Harvard University published *A History of Mechanical Inventions*, from classical to modern times (Usher, 1929). The chapter titled 'The Process of Mechanical Invention', includes a theory of "technical innovation" ("technological" is also used, concurrently with technical or technological change, development, progress and advance). Usher's theory of innovation is essentially psychological. To Usher, innovation is an *act* of creativity and imagination, or insight, as he put it. Innovation is a "mental process" of "combining" – a key term to Tarde – elements of experience into new ones: the establishing of "new organic relations among ideas, or among material agents, or in patterns of behavior", as Usher put it in the second edition (Usher, 1954: 21). Usher positions his theory as an alternative to the mystical accounts of innovation (p. 83):

> The completion of the analysis of the processes of innovation will necessarily be the work of the psychologists. Only advances in the understanding of psychology can give us the kind of knowledge about physiologic and psychological processes that will enable us to exclude the appeal to "subconscious" and "unconscious" processes. Only more complete psychological analysis can furnish full awareness of the complexity of the act of insight.

Usher positioned his work in contrast to "explanation[s] of economic changes in the simple [passive] facts about climate, materials, or soil exhaustion" (p. 2). He begins his analysis with the fact that technical innovation is an integral part of economic history, namely the "dynamic process of history". To Usher, humans actively transform their environment through the technological use of resources. "The analysis of the processes of social evolution in their entirety requires study of processes of invention and processes of diffusion by imitation", particularly discontinuous changes (p. 18). But how?

To Usher, innovation is the cumulative synthesis of individual items over time. This process is not the affair of men of genius. It is a mental activity, certainly, but a pervasive mental activity. Everyone innovates: the scientist, the inventor, the technician, the manager, the artist (p. 10). People innovate by combining preexisting elements into new syntheses, new patterns, or new configurations into a whole (p. 11).

In the 1954 edition of the book, Usher extends his framework to three chapters. Here, Usher makes very few uses of the term innovation. He shifts to "social change" – a then fashionable term – and "novelty", discussed in specific chapters respectively. The context of the theory is the process of invention and diffusion in changing societies. Usher contests the theory of evolution by stages, as he had early in his scholarly career, and brings in the concept of a "sequence of events", from ideas to action (events are defined as "purposes, concepts, patterns of behavior, and explicit action", Usher, 1954: 27). To Usher, action has a "conceptual basis": "The elements of social change appear first as new ideas and concepts. As the new ideas are diffused and assimilated they affect patterns of behavior and political and social actions" (p. 23). Understood as such, social change is not linear but multilinear. Different systems of events give rise to different social systems or cultures.

It is in the 1954 edition that Usher brings in his famous sequence to explain novelty or innovation. Usher offers a four-stage theory of the process of innovation as follows:

- Perception of a problem (or want or need)
- Setting the stage (rudimentary or incomplete pattern)
- Insight (or solution: achievement of a new configuration)
- Critical revision and development (reduction to practice).

Usher's sequence is not the first one developed to explain the innovation process. Yet the psychological sequence is certainly one of the firsts of its kind, joining those of philosophers and psychologists on creativity. The sequence benefited from the insights of Gestalt psychology or, as some called it in the 1930s – in the pejorative sense – the Gestalt "movement". Sequences like these proliferated in the following decades.

## JOSEPH SCHUMPETER

Whether economist Joseph Schumpeter's (1883–1950) main works, *The Theory of Economic Development* (1934) and *Business Cycles* (1939), constitute a theory of innovation is debatable, although innovation holds a central role in them as a factor of economic life: "innovation is the outstanding fact in the economic history of capitalist society" (Schumpeter, 1939: 86). In fact, Schumpeter does not dwell on explaining the generation of innovation – few did that at the time – or its diffusion, as Tarde had.

Like Tarde and Usher before him, Schumpeter placed change at the center of his thinking. He wrote that he dislikes the words statics and dynamics, but he "[kept] to the distinction" nevertheless (Schumpeter, 1934: lxiii). "Static analysis is not only unable to predict the consequences of discontinuous change in the traditional way of doing things; it can neither explain the occurrence of such productive revolutions nor the phenomena which accompany them.... It is just this occurrence of 'revolutionary' change that is our problem, the problem of economic development" (Schumpeter, 1934: 62–3). From that time on, technological innovation had to be discontinuous or revolutionary to be named as such.

What is innovation? To Schumpeter, innovation is "any 'doing things differently' in the realm of economic life", with change in methods of production being one of five types (p. 84), perhaps the major one. Innovation is "the setting up of a new production function" (technological change), namely changes in production according to changes in factors of production, "ways in which quantity of product [output] varies if quantity of factors [input] varies" (p. 87). Innovation is defined via the concept of combination (creativity). Like Tarde and Usher, combination is a key concept to Schumpeter. "This concept plays a considerable part in our analysis. The process of production is combination of productive forces" (Schumpeter, 1939: 14–15). To Schumpeter, combination is "directed towards something different and signifies doing something differently from other conduct" or "innovation". It presupposes a specific kind of "aptitude" (Schumpeter, 1939: 81, footnote).

In Schumpeter's hands, combination was given a new meaning: after the sociological (Gabriel Tarde) and psychological uses (Abbott Usher), he proposed an economic one. Combination serves three definitions. First, it defines the entrepreneur, as it is contrasted to routine and directed labor (Schumpeter, 1934: 78). The entrepreneur combines resources or factors of production (p. 132, 137; Schumpeter, 1949: 228). The concept also serves to define development: "the carrying out of new combinations" (Schumpeter, 1934: 66). Finally, combination defines innovation. Innovation "consists in the carrying out of a New Combination" of factors of production (Schumpeter, 1939: 88).

Schumpeter gave technological innovation a more central place in explaining economic development than any other economists had. But he did not theorize about the process of technological innovation. There is the innovator (the entrepreneur) and there are the followers or copiers. Technological innovation is a given. Schumpeter did not explain how innovations come about; neither did he study the factors and conditions that lead to innovation. The generation and diffusion of technological innovations are not studied.

To Schumpeter, innovation is distinct from discovery and invention. "As long as they are not carried into practice, inventions are economically irrelevant" (Schumpeter, 1934: 88). Again and again, Schumpeter repeated the distinction. One is external to economics (invention), the other internal (innovation). Schumpeter is not the originator of this distinction – to Gabriel Tarde, an innovation not imitated is as though non-existent socially – but he is much cited for it.

## RUPERT MACLAURIN

The economist from MIT William Rupert Maclaurin (1907–1959) was a follower of Schumpeter's ideas. He espoused many of his suggestions, particularly that on the importance of history to economics, and corresponded with him on several occasions. But he went further than Schumpeter. At the same time, and from then on, pure theory changed to "applied" theory. With one exception (Homer Barnett, discussed below), one of scholarship's aims was to contribute to the generation and diffusion of innovation and policy, not just to understand it.

Maclaurin's *Invention & Innovation in the Radio Industry* (1949) is a "historical account" of technological innovation, or "technological change" as he called it alternatively, "from fundamental scientific research to its practical applications in new or improved products and techniques" (p. xiv). Maclaurin was both contributing to the context of the time and trying to change it. As Secretary of one of the committees responsible for Vannevar Bush's report *Science: The Endless Frontier* of 1945, Maclaurin believed that fundamental research was the key factor in economic development: it is "essential to make science penetrate every aspect of industrial life" (Maclaurin, 1949: xiv). Hence the primacy given to fundamental research in his conceptual framework. However, fundamental research is an insufficient condition for technological innovation. "Advances in science are not automatically translated into advances in the practical arts" (Maclaurin, 1949: xiii). Between fundamental research and its applications, there is a "continuum" or "sequence" of activities, with "stages" (applied research, engineering development, production engineering).

Maclaurin's conceptual framework is what in the late 1960s came to be called the "linear model of innovation". It gave research and development (R&D) a central place in the process of innovation, R&D also serving as a proxy for the measurement of innovation. Critics of the linear model of innovation usually stress that fundamental research is not the starting point of technological innovation, or that technological innovation is not a linear sequence beginning with fundamental research. Yet what is missing in the criticisms is Maclaurin's historical perspective, taking the economic historian seriously. Maclaurin was broadening the discourse of the time, on (basic) research leading automatically to technological innovation. Using the radio industry as a case study, he illustrated "the steps which are required to bring a new scientific concept from the theoretical stage to a successful commercial product" (Maclaurin, 1946: 426). Technological innovation is not only the affair of scientists: "The innovator as an individual takes his place with the pure scientist and the inventor as a key figure in material progress" (Maclaurin, 1953: 105). Innovation includes activities other than basic research as necessary stages.

Maclaurin's work is essentially historical. He had no immediate or direct impact on studies of innovation. His work was soon eclipsed by econometrics. Yet Maclaurin initiated the first research program on technological innovation and, from an intellectual point of view, he represents one of the many ways of studying innovation existing in the literature. Maclaurin's linear model remains much discussed and criticized, although without reference to Maclaurin. Perhaps Maclaurin died too young to get the time to defend his ideas. In fact, Maclaurin committed suicide at the age of 52.


## HOMER BARNETT

We now turn back to a type of theory first introduced by Abbott Usher, a psychological theory of innovation. Homer Barnett's *Innovation: The Basis of Cultural Change* was the first comprehensive theory of innovation since Tarde's. Based on several years of fieldwork on cultural change in ethnic and religious communities initiated in 1938, Barnett (1906–1985), a Professor of Anthropology at the University of Oregon, wanted "to formulate a general theory of the nature of innovation and to analyze the conditions for, and the immediate social consequences of, the appearance of novel ideas" (Barnett, 1953: 1). Such a theory "should be a formulation of the rules of thinking as they pertain to the origins of new combinations of thought" (p. 16).

To Barnett, innovation is a psychological process determined by personal as well as social factors and one that has social consequences, a process from mental creation to social acceptance/rejection and diffusion. Barnett's concept of "process" refers to the mechanisms and conditions of innovation, to the how and why, not to the what (description). Barnett concentrated on the mental process of generation, a process not unlike that of economist Usher's "configurations".

According to Barnett, innovation is a process of reorganization and combination: the "linkage or fusion of two or more elements" in a new way (Barnett, 1953: 181). Barnett's theory is entirely founded on the concept of combination, a concept central to Tarde, Usher, Schumpeter and others. "Every innovation is a combination of ideas" and a theory of innovation "should be a formulation of the rules of thinking as they pertain to the origins of new combinations of thought" (p. 16).

Barnett's theory was a reaction to the anthropological controversy on invention *versus* diffusion, whether cultural change results from invention or from the diffusion of invention; whether man is essentially inventive or imitative (Godin, 2017). The invention-diffusion controversy gave rise to influential frameworks combining invention and diffusion as sequential stages of the innovation process. Barnett concluded that everyone innovates, although to different degrees. To Barnett, the imitator is an innovator: he does something new "instead of doing what he is accustomed to do" (Barnett, 1961: 34). Barnett claimed that imitation "necessarily produces a modification"; "assimilation and copying are innovative" (Barnett, 1953: 49–54, 330–2). Tarde believed similarly.

Barnett's theory was also a reaction to the dominance of technology in discourses on invention. Tarde produced a broad meaning of innovation, as Schumpeter identified five economic types of innovation. But technological innovation quickly became the emblematic case of innovation. "Little attention is given" in this book, wrote Barnett, to "the landmarks of technological progress that are usually dealt with in histories of invention" (Barnett, 1953: 2). To Barnett, invention is one and only one kind of innovation. "The real challenge for a theory of innovation lies in the realm of behavior, belief and concept" or "nonmaterial creations" (p. 3).

To Barnett, innovation is "Any thought, behavior, or thing that is new because it is qualitatively different from existing forms" (p. 7). Again like Tarde, Barnett stressed diffusion as the ultimate factor for the progress of societies. The question of "origins" is "only a small part of the problem of cultural change, and from some points of view the least important part…. A study of cultural change takes us beyond the appearance of a new idea into the consideration of its acceptance and rejection" (p. 291). "Acceptance" is one thing (individual) and "diffusion" another (social) (p. 292).

## BOX 3.2   HOMER BARNETT

- All cultural changes are initiated by individuals – cultural conditions influence variations in individuals' innovative possibilities or potential.
- Innovation is driven by wants (a term Barnett prefers to need, which refers to groups rather than individuals, and which is presumably too normative). These wants are either involuntary ("self wants") or directed voluntarily to change ("desire for change").
- The basic process of innovation is configuring preexisting elements into a new whole or recombining existing configurations, either by identification (equivalences), substitution (similarities) or discrimination (differences).
- An important aspect of cultural change is acceptance, or imitation. It requires advocates. The factors that contribute to acceptance are personal and social characteristics of the advocate of change, novelty characteristics and novelty values.
- There are four categories of acceptors and rejecters, based on "lag" in acceptance: dissident, indifferent, disaffected, resentful.

## CHARLES CARTER AND BRUCE WILLIAM

In 1952, the Science and Industry Committee of the British Association for the Advancement of Science (BAAS), an organization founded in 1831, contracted a study on "the problems of speeding up in industry the application of the results of scientific research" to two of its members: Charles Carter (1919–2002) (Chairman) from Queen's University of Belfast and Bruce Williams (1919–2010) (Secretary) from University College of North Staffordshire. The background to the study was the debate on time lags in the application of science in industry. Britain is slow to apply the results of science, as it was said. The lag issue is a major impetus to the representation of innovation as application (of invention).

The study was conducted between 1952 and 1956, and led to a series of three books: *Industry and Technical Progress* (1957), *Investment in Innovation* (1958) and *Science in Industry* (1959). These are certainly among the first books ever written on industry and technological innovation, understood as "science applied", in the sense of the use or application of science in industry (a variant of or a declination on the concept of applied science). Innovation is "the act of bringing [an idea or invention] into practical use" (Carter and Williams, 1957: 15).[1]

Carter and Williams' *Industry and Technical Progress* (1957) is based on a survey of 246 firms and 109 universities, public agencies and departments, and research and trade associations. It studies the factors and conditions of technological innovation according to the "circuit" or "stages" of application of knowledge in industry, from basic research to development to production. The factors studied as leading to technological innovation are: personnel (scientists and engineers), size of organization, management, financial resources, market and environment. To Carter and Williams, eliminating the "gap" between invention and innovation is a matter of "industrial growth", a matter also of "survival" in a competitive world. Innovation is the latest but most important stage of an "investment decision", namely "the action of bringing the new idea into practical use" (p. 76).

*Investment in Innovation* (1958), a subsidiary report to *Industry and Technical Progress*, concentrates on firms' "investment decision" to innovate as a "desire to survive" in a competitive market. The report studies 204 product, process and method innovations, looks at their sources, and offers a classification of innovations as "passive" (responding to a market pressure) or "active" (deliberate searching for new markets). Such a classification led to many others in the following decade – such as the OECD study on *Gaps in Technology* (1968–1970) and economist Chris Freeman's (1921–2010) categorization of firm strategies as offensive, defensive, imitative, dependent, traditional and opportunist (Freeman, 1974).

The third study, *Science in Industry* (1959), extends the previous two analyses to public policy. "We doubt if it can be said that a Government policy on the application of science really exists…The present pattern of Government intervention in favour of technical progress has been formed haphazard from a number of pieces shaped by past history" (Carter and Williams, 1959: 103). The existing aid is dispersed in many Departments but "a general 'Government scientific policy' is something more than the mere addition of the policies of separate Departments" (p. 100). What is an innovation policy or "strategy"? According to Carter and Williams, innovation policy is concerned with a large range of government measures, such as public support to R&D, education, taxation, competition and trade. This is perhaps the first consideration of policy in discussions about technological innovation.

The Carter and Williams studies are quite original and were influential. They paved the way for the study of technological innovation, a kind of study imitated by many others in the following decades. The two dimensions of innovation studied – firm and policy – gave rise to two key features of subsequent discourses on innovation. The consideration of policy in particular is quite unique for the time. Carter and Williams considered innovation policy in broad terms, as it would be the case in the next decades. The focus on a time lag and gap in the use of science in industry also gave rise to a passionate public debate on technological gaps in the 1960s, and the Carter and Williams taxonomy on technical progressiveness later led to a much-cherished indicator: high technology.

## TOM BURNS/GEORGE STALKER AND JERALD HAGE

The systematic study of innovation in organizations (including educational) preceded that on industrial (technology) innovation. Beginning in the early 1950s, Tom Burns (1913–2001), a sociologist at the University of Edinburgh, conducted a series of studies on how firms cope with and adapt to a changing environment. The studies led to *The Management of Innovation*, co-authored with psychologist George Macpherson Stalker from the same university (Burns and Stalker, 1961).

The context of the study is changes in markets and technologies. To Burns and Stalker, changes in organization are due to market change (demand) and technological change (supply), particularly the latter: "How management systems changed in accordance with changes in the technical [technological advance] and commercial tasks [mass production] of the firm, especially the substantial changes in the rate of technical advance" (p. 4); What is "The connexion between progress in material technology and the emergence of new forms of social organization" (p. 19)? From the surveys they conducted, the authors offer two ideal types of organization or management, one adapted to a bureaucratic environment, and the other to a changing organizational environment, "mechanistic system" (a formal organization, bureaucratic, with programmed decision-making) and "organic system" (an informal organization with non-programmed decision-making), the first being appropriate for stable conditions and the second required for conditions of change (pp. 119–22).

Curiously, to the authors, change in organization is not called innovation. The idea of application of a "different kind of management system" under conditions of "novelty and unfamiliarity in both market situation and technical information" (p. vii) is called changes in management (organizational methods) or "institutional change" (new tasks, new people, new departments, new kinds of resources). "By change we mean the appearance of novelties: i.e.: new scientific discoveries or technical inventions, and requirements for products of a kind not previously available or demanded" (p. 96). To the authors, innovation, not defined explicitly (a common practice in the following decades), is new products *and* industrial processes ("adoption" of new technical knowledge "in order to derive new products") (Burns and Stalker, 1961: 249).[2]

Burns and Stalker have had a certain impact among a few researchers who revisited the typology of organization forms in the 1970s and proposed more refined models. Jerald Hage is one of these researchers. Together with Assistant Professor Michael Aiken, sociologist Hage (b. 1932) conducted research to test what he called an "axiomatic" theory of innovation. This led to *Social Change in Complex Organizations* in 1970.

Hage is concerned with explaining why organizations change or do not change: the kinds of change, the causes of changes and the pattern of change. Fundamentally, organizations change because their environment changes. It is a matter of survival. "The major cause of changes in organizational style lies in the environmental elements. When there are rapid changes in its environment, the organization is forced to adapt by altering the rate of program change" (Hage and Aiken, 1970: p. xiv). Hage introduces a series of variables that explain change (and stability) in organizations, emphasizing the need for a system and a temporal perspective: complexity, centralization, formalization, stratification, production, efficiency and job satisfaction.

To Hage, innovation is the means to cope with change. Innovation is defined as program change, namely the addition of new products and services in an organization. When these products and services are first adopted in an organization, they represent an innovation (pp. 13–14). Hage stresses that a new program need not necessarily be new to the world. It suffices for it to be new to an organization adopting it for it to be called an innovation.

To make sense of organizational change, Hage develops a typology of organizations as to whether they are innovative or imitative – dynamic and static (pp. 61–91) – not unlike the typology of Burns and Stalker (organic and mechanical). Hage is concerned with a sociological approach to organizational change, defined as "an alteration in the arrangement of organizational parts" (p. 28), and which he contrasts to the psychological and management approaches (pp. 11–13, 117–39). Yet Hage constantly stresses the role of individuals in explaining organization performance: jobs and "social position types" (kills, powers, rules and rewards). Organizations are not disembodied entities. They exist because individuals created them. Like Tarde, Barnett and Rogers, Hage also studied innovation in a broad sense, not confining himself to technological innovation.

## EVERETT ROGERS

Rogers' theory of innovation is a theory of diffusion of innovation, which borrows several concepts from Barnett.[3] Barnett's theory was much cited in the years after it appeared and subsequently, and his definition of innovation was frequently used by others. Yet Barnett's theory was eclipsed by sociologist Everett Rogers. As with Tarde and Barnett, to Rogers adoption is individual and diffusion is social:

> The adoption process is the mental process [of learning] through which an individual passes from first hearing about an innovation to final adoption…. Diffusion is the process by which an innovation spreads…from its source of invention or creation to its ultimate users or adopters…. The diffusion occurs among persons while adoption is an individual matter. (Rogers, 1962: 76)

The inventor "creates" ("unites by combining"), the innovator "adopts" (pp. 195–6). Rogers offered a definition of innovation as broad as Barnett's: innovation is "an idea, practice or object that is perceived as new by an individual" (p. 13) "or other unit of adoption" (Rogers, 1983: 11). But in the end, Rogers concentrated on technological innovations, as economic historian Abbott Usher suggested: "Most, but not all, innovations discussed here are technological innovations" (Rogers, 1962: 13), because "The adoption of a new idea almost always entails the sale of a new product" (p. 261).[4] Despite Barnett's caution, the pendulum shifted back to technology.

Rogers' *Diffusion of Innovations* reacted to a context, as Barnett had: the pro-innovation bias. If one believes Rogers, by the early 1960s, innovation had become a superlative concept. To Rogers, the then-current studies make the assumption that innovation is always good: "Researchers have implicitly assumed that to adopt innovations is desirable behavior [rational] and to reject innovations is less desirable [irrational]" (p. 142). In contrast, Rogers' theory examined rejection ("discontinuance", pp. 88–95) as well as adoption of innovations, as Tarde's and Barnett's had. Rogers' subjective definition of innovation is witness to this thought. An innovation need not necessarily be a world first or original, but it is new to a user. "It really matters little as far as behavior is concerned whether or not an idea is objectively new as measured by the amount of time elapsed since its first use or discovery. It is the new ness of the idea to the individual that determines his reaction to it" (Rogers, 1962: 11).

Rogers' objectives were both theoretical and practical, like Maclaurin's. Rogers' purpose was to develop a theory of innovation, and he developed one whose breadth is still unsurpassed today. He synthesized the research done on the diffusion of innovation over the previous decades, in every discipline, a task that no scholar had attempted yet. Rogers wanted his research to contribute to reducing the "time lag" between innovation and its acceptance or widespread diffusion, as every one of his colleagues from agricultural sociology had wanted.

Like Maclaurin, one of Rogers' overall purposes was to broaden the discourse of the time on science and scientific research. "Research…is an unrealized public investment until the resulting innovations are diffused to and adopted by the intended audience….Research alone is not enough to solve most problems; the research results must be diffused and adopted before their advantage can be realized" (Rogers, 1962: 2–3). Here, Rogers went one step further than Maclaurin and included adoption and diffusion as stages in the process of innovation:

Innovation → diffusion → adoption (adoption = awareness → interest → evaluation → trial → adoption)

---

## BOX 3.3   EVERETT ROGERS

- The adoption process is composed of five stages, in sequence: awareness, interest, evaluation, trial and adoption (or rejection).
- The (subjective) characteristics of innovation are five: relative advantage, compatibility (with values, experience and ideas), complexity, divisibility, communicability. These characteristics influence the rate of adoption.
- Adopter categories, according to "innovativeness" (time of adoption) are: innovators (first 2.5%), early adopters, early majority, late majority, laggards, non-adopters. The categories are based on an S-shaped curve.
- Importance of opinion leaders and of "change agents" ("a professional person who attempts to influence adoption decisions", p. 254) in adoption.

---

## JACK MORTON

Engineer at Bell Laboratories, Jack Morton (1913–71) was part of thoughts that changed the way innovation was understood in the 1960s and in subsequent decades. Morton had no chance to see the impact of his thoughts. He was assassinated, going out of a bar.

Morton brought in a definition of innovation as a "system". In a book from 1971, titled *Organizing for Innovation: A Systems Approach to Technical Management*, Morton defined a system as:

> an integrated assembly of specialized parts acting together for a common purpose … a group of entities, each having a specialized, essential function. Each is dependent for its system effectiveness

upon its coupling to the system's other parts and the external world ... *Parts, couplings*, and *purpose* are the three characteristics which define every system. (Morton, 1971: 12–13)

Innovation is:

teamwork between science, engineering, and industry ... But our understanding of the innovation process is still incomplete and not widely diffused ... What I hope for is that our understanding of technological innovation will be broadened to include the *totality* [my italics] of human acts by which new ideas are conceived, developed, and introduced. (pp. 2–3)

As Morton put it, under different phrases in preceding papers of the 1960s (pp. 3–4):

Innovation is not just one single act. It is not just a new understanding or the discovery of a new phenomenon, not just a flash of creative invention, not just the development of a new product or manufacturing process; nor is it simply the creation of new capital and consumer markets. Rather, innovation involves related creative activity *in all* [Morton's italics] these areas. It is a connected process in which many and sufficient creative acts, from research through service, couple together in an integrated way for a common goal ... By themselves R&D are not enough to yield new social benefits. They, along with capital resources, must be effectively coupled to manufacturing, marketing, sales, and service. When we couple all these activities together, we have the connected specialized elements of a total [my italics] innovation process.

Morton's definition of innovation as a *total* process, including the phrasing itself, was reproduced regularly in the years following its appearance. Morton was part of a network of practitioners, including policy-makers, which claimed that "The essential virtue in the systems approach to innovation lies in the parts of the process and their linkages with one another, *not in the sequence* [my italics] in which such linkages are performed" (p. 19).

Morton's view is still a stage- or linear model, but with feedbacks, an interactive model as some call it, and is not really dissimilar to Ronald Kline's much cited model (Kline, 1985; Kline and Rosenberg, 1986), produced fifteen years later. Yet, Morton combines this view with that of a system or organizations and institutions, including the external environment, with multiple feedbacks: firm, funding institutions, regulatory agencies, government departments and the world. Morton's exemplar (model) is "the Bell System, whose 'flow charts of the innovation process' evolved from organizational separation of entities and activities to coupling between 1925 and post-World War II". Bell Labs provides "a good case history for the application of the systems approach to the total innovation process" (p. 34). The task of the manager is to make this system work as a unified whole. "It is relatively straightforward, though difficult, to acquire and develop high levels of creative specialization, but it is a much more subtle and complex task to couple them together for the overall purpose of the system" (p. 62).

## GERALD ZALTMAN

In the 1960s, scholars began to study innovation in marketing. Theodore Levitt, Gerald Zaltman and Thomas Robertson are exemplary authors. These writers started to apply the concept of innovation to consumers as purchasers of new products or innovators, and to business strategy, but ended up, or at least Robertson and Zaltman did, with a general theory of innovation. Zaltman constructed his theory along lines similar to those of Rogers: assessing the literature, analyzing the semantics of innovation, studying the characteristics of and resistances to innovation, and theorizing about the innovation process.

Zaltman (b. 1938), a sociologist and Professor of Marketing at Northwestern University, whose book *Innovations and Organizations* Rogers qualified as "influential", contributed greatly to reorienting studies of innovation toward organizations. Zaltman espoused Barnett's and Rogers' definition of innovation as being relative to the context and the time: "any idea, practice or material artifact perceived to be new by the relevant unit of adoption" (Zaltman et al., 1973: 10). Like Tarde, Barnett and Rogers again, he also understood diffusion as a social phenomenon:

> The adoption of innovations refers to the processes whereby an innovation comes to be the most acceptable alternative available at that time…. The diffusion process is the process whereby an innovation is disseminated and accepted among individuals or other adopting units. Adoption occurs at a micro level, whereas diffusion occurs at a macro level. (Zaltman and Stiff, 1973: 417)

Two concepts are central to Zaltman's view. One is (planned) social change, on which he wrote several monographs, as "the alteration in the structure and functioning of a social system" (Zaltman et al., 1973: 158). To Zaltman, the source and outcome of social change is innovation. The second concept is "performance gap", a concept he borrowed from Anthony Downs (b. 1930), an American economist specializing in public policy and public administration, and, again, a concept not very different from William Ogburn's lag. "Performance gaps are discrepancies between what the organization could do by virtue of a goal-related opportunity in its environment and what it actually does in terms of exploiting that opportunity" (p. 2). "The impetus to innovation arises when organizational decision-makers perceive that the organization's present course of action is unsatisfactory" (p. 55). A gap produces a search for alternative courses of action.

On one dimension, Zaltman produced a theory of innovation at a higher level of generality than that produced so far in either sociology or in studies of organizational change. "Most diffusion theorists generally terminate their analysis at the stage of initiation", claimed Zaltman et al. (1973: 3) Consequently, Zaltman produced a "model" of innovation that makes a place for implementation, including discontinuance.

## BOX 3.4   GERALD ZALTMAN

- The innovation process is a sequence (with feedback) composed of two stages with sub-stages: initiation (knowledge awareness, formation of attitudes, decision); implementation (initial implementation, continued-sustained implementation).
- Innovation types or categories are programmed/unprogrammed (slack and distress), instrumental/ultimate, radical/routine.
- Attributes or characteristics of innovation are costs, returns, efficiency, risk and uncertainty, communicability, compatibility, complexity, scientific status, perceived advantage, origin, terminality, reversibility, commitment, interpersonal relationships, publicness, gatekeepers, susceptibility to successive modification, gateway capacity.
- Characteristics of organizations affecting innovation are complexity, formalization, centralization, interpersonal relations, conflict.

• Resistances to innovation are organizational and individual (psychological).

## CHRIS FREEMAN

We now turn to a time – the 1970s – when innovation became an end in itself. This has a lot to do with the studies of innovation becoming a field on its own ("innovation studies"). It also has to do with policy considerations and the national level (innovation beyond the individual and organizational). Tarde and Schumpeter developed theories of social and economic change broadly defined. Innovation was a means to change. Thereafter, innovation became an abstract or a conceptual object.

The history of theorization on industrial innovation is yet to be written. In the rest of this chapter, I limit the analysis to three key scholars. One year after Zaltman, British economist Chris Freeman (1921–2010), a consultant to the OECD and UNESCO in the 1960s and founder of the Science Policy Research Unit (SPRU) in 1966, produced *The Economics of Industrial Innovation*. By that time, the word innovation had become a buzzword. But things were different in economic theory…or were they? According to Freeman, innovation is "an essential condition of economic progress and a critical element in the competitive struggle of enterprises and of nation-states. [But] few [economists] have stopped to examine it". Innovation remains "outside the framework of economic models". It is treated as a "residual" or exogenous variable, and "the social process of innovation" remains a "black-box" to economists (Freeman, 1974: 15–17, 27).

Like previous scholars, Freeman was reacting to a disciplinary context. Freeman espoused a different perspective of innovation as 'muddling through', "rather than the ordered, rational calculations beloved of neo-classical theory" (p. 40). Freeman's interest was the generation, or rather the generators, of innovation, not the use of given inventions (industrial processes). Freeman brought what he called a "balance in coverage of process and product innovations" (p. 37). This was a fruitful 'innovation'. The focus on products led to examining firms as suppliers of technological inventions rather than as users or adopters, in contrast to what writers on technological change had studied: how firms invent new products, what are the conditions for success and the difficulties encountered in introducing technological inventions to the market, is there an optimal size for innovating firms, what strategies are available to the firm, and so on.

While Barnett and Rogers studied individuals as innovators, Freeman looked at firms, a type of analysis first conducted by Charles Carter and Bruce Williams in the late 1950s: he examined the factors that lead to success and failure in technological innovation, the size of firm most conducive to technological innovation, the difficulties of decision-making given the inherent uncertainty and risk of technological innovation, and the strategies available to firms for coping with this uncertainty. Freeman also stressed the role of innovation policy. Indeed, by that time, governments had embraced and started promoting innovation as an instrument of economic policy.

Freeman paved the way for a productive specialty that spawned a voluminous number of articles and books. But Freeman's theorization is far more fragmented than Barnett's or Rogers' or Zaltman's. Like Carter and Williams' works, *The Economics of Industrial Innovation* does not constitute a theory. Freeman's work is a survey or a collection of what is new regarding the then-emerging perspectives on industrial innovation in economics, management and policy – like a handbook – a kind of research program.

Freeman's interest in technological innovation (as product innovation) provided the seed for defining technological innovation as commercialization among scholars: a firm bringing a new product to the market for the first time. To Freeman, technological innovation is "the first *commercial* [Freeman's italics] application or production of a new process or product" (p. 166). At the heart of the phenomenon are an R&D "system", within which are firms where the organized or "professional" activities of R&D are conducted. But commercialization is the key word in Freeman's definition of innovation. He wanted to study the "economic aspects" of the process of innovation: costs, patents, size of firms, decision-making, marketing and time lags. This does not mean that "technical, psychological and other aspects of innovation are unimportant. A more integrated theory of innovation is desirable, but it is beyond the scope of this short book" (p. 39).

The second keyword in Freeman's definition is "first". Innovation is creativity in the sense of originality or priority (first occurrence). Imitation (copying) – a positive concept to the anthropologists of the early twentieth century – is relegated to the idea of lag (and laggard). With this meaning of innovation as first commercialization, we observe a second shift in the concept of creativity applied to economics (the first shift being the combination of factors of production, as Schumpeter had understood it). In Freeman's writings combination also developed a managerial and policy meaning as the combination of organizational activities leading to innovation. To be sure, Freeman does makes analogies with Abbott Usher's "Gestalt" theory of an "imaginative process of 'matching' ideas". "All theories of discovery and creativity stress the concept of imaginative association or combination of ideas", states Freeman: "coupling first takes place in the minds of imaginative people" (Freeman, 1974: 111–12). Yet Freeman does not study this aspect. He expands the theory of the mind to "the whole of the experimental development work and the introduction of the new product" – "linking and coordinating different sections, departments and individuals", "communication within the firm and between the firm and its prospective customer" (p. 112) – and the entrepreneur: "the crucial contribution of the entrepreneur is to *link* the novel ideas and the market" (p. 110).

The context from which Freeman's thoughts emerged was public policy. In the mid-1960s, governments embraced technological innovation as an instrument of economic policy, and published what were among the first titles on "technological innovation". Innovation is good not only for an individual, a social group or an organization, but also for a nation as a whole. There is a need to support the innovators, namely the firms, for their contribution to a country's competitiveness. But how? Academics, together with policy-makers and consultants, had an answer: "a more explicit policy for science and technical innovation is increasingly necessary" (Freeman, 1974: 31).

## EDWIN MANSFIELD

Edwin Mansfield (1930–1997) is a scholar from mainstream economics whose work preceded Freeman's by a few years. As Associate Professor of Economics at the Carnegie Institute of Technology, Mansfield started working on industrial innovation in the late 1950s–early 1960s with support from the RAND Corporation, the Ford Foundation and the Cowles Commission, and contracts from the US National Science Foundation (NSF). The NSF was interested in technological innovation as an output presumably arising from basic research and, over the years, commissioned many studies to determine to what extent this is true, and what links exist between science and economic growth.

Mansfield is perhaps the only economist to consider seriously the influence of sociology on the economic study of technological innovation. In his contribution to the US government report *National Commission on Technology: Automation and Economic Progress*, Mansfield made an analogy between the concepts of leaders and followers from rural sociology and the factors involved in the adoption of innovation in firms (Mansfield, 1966: 124–5). Sociologists stress attitudes, values, social status, abilities, group memberships and farm business characteristics as factors contributing to the early adoption of innovation by individuals. Similarly, based on his studies from the early 1960s, Mansfield identifies economic factors to explain firms' first adoption of innovation, factors such as firm size, rate of growth, profits and management.

After several years of work, Mansfield brought together the studies he had conducted so far, revised them, and in 1968 published *Industrial Research and Technological Innovation*. The book is like Freeman's: a collection of empirical studies of several dimensions of technological innovation, namely industrial R&D (expenses, portfolio and returns), market structure (size of firms), timing of innovation (lags between invention and innovation), and diffusion. To Mansfield, the process of technical change is "the manner in which new processes and products are conceived, developed, commercialized, and accepted" (Mansfield, 1968a: xv). Mansfield's concepts are organized around a three stages triad: invention (R&D) → innovation (first use) → diffusion (imitation as he called it, and widespread use).

Change is once again a major motive for Mansfield's study of technological innovation. The change that interests Mansfield is endogenous change in the market or in firms, rather than contextual change: how industries and firms cope with technology as a factor of economic growth. Mansfield's vocabulary of innovation is like Freeman's: "An invention, when applied for the first time, is called an innovation.... [It] is a key stage in the process leading to the full evaluation and utilization of an invention" (Mansfield, 1968b: 83). A central concept to Mansfield is technological change – also called change in technology – of which innovation is one stage. To Mansfield, technological change is defined according to the "production function", a tool Mansfield uses for measurement, and which Schumpeter made central to his theory. "Technological change results in a change in the production function" (p. 15). "The production function shows, for a given level of technology, the maximum output [new products] rate which can be obtained from given amounts of inputs" (methods of production: capital and labor) (Mansfield, 1968b: 13).

*Industrial Research and Technological Innovation* was one of a series of four books Mansfield produced between 1968 and 1971. In 1968, Mansfield also published *The Economics of Technological Change*, a companion to *Industrial Research and Technological Innovation*, which used the existing literature in addition to Mansfield's own work, adding new topics (or stages to the triad invention-innovation-diffusion): consequences (unemployment) and policy (patents, antitrust policy, government funding of R&D, procurement, higher education).

Mansfield's books are syntheses ("overview" as he called it in 1968b: ix) of existing findings, above all Mansfield's own studies.[5] To be sure, Mansfield's books are (linearly) organized according to a theoretical framework of the "entire" or "total" innovation process by stages, as he called it (Mansfield et al., 1971: 111, 221). But a framework is not a theory. We must turn to Richard Nelson for such a theory.

# RICHARD NELSON

In the same year that Freeman published the second edition of *The Economics of Industrial Innovation* (1982), economist Richard Nelson (b. 1930), together with Sydney Winter, produced *An Evolutionary Theory of Economic Change*. Nelson started his career as a (neoclassical) economist at the RAND Corporation, then moved on to Carnegie Mellon University and Yale University. The book *An Evolutionary Theory* develops a dynamic theory of economic change using mathematical models. Nelson's evolutionary theory is evolutionary in the strict biological sense. It makes abundant use of biological metaphors: inheritance, mutation, variation, selection, adaptation, survival. Evolution here is not limited to the evolution of early anthropologists' and sociologists' thoughts, which focused on gradualism and cumulativeness. Uncertainty (decision-making under uncertainty) is the key concept here.

"In a way, our approach", states Nelson, "represents a formalization of ideas that long have been present in economics, and indeed were dominant before contemporary formalism took over" (Nelson and Winter, 1982: 164). Yet Nelson offers what he calls a "fundamental reconstruction" (pp. 4, 6), an "alternative view" (p. 21), a "major shift" (p. 23), a "radical appraisal" (p. 5). In page after page, Nelson positions his theory in contrast to economic orthodoxy, which he describes as a "narrow set of criteria that are conventionally used as a cheap and simple test" (p. 6), namely the presumptions of maximization and equilibrium, and the mathematics of the production function.

Nelson's theory rests on "three basic concepts" or stages of the process of economic change: routine, search and selection. In fact, Nelson's concepts are a reformulation of the stages of the invention–innovation–diffusion triad within a microeconomic perspective. This was made clear with the first appearance of the theory in 1977: "In an accounting sense we view productivity growth as explained within our proposed theoretical structure in terms of first, the *generation* [my italics] of new technologies, and second, changes in weights associated with the *use* [my italics] of existing technologies" (Nelson and Winter, 1977: 48).

The first basic concept of Nelson's theory is *routine*. Organizations are "relatively rigid": they are "typically much better at the tasks of self-maintenance in a constant environment than they are at major change, and much better at changing in the direction of 'more of the same' than they are at other kinds of change" (Nelson and Winter, 1982: 9–10). Organizations work on the basis of routines, not innovation. Routines are a "set of ways of doing" (p. 400) or

> dispositions and strategic heuristics that shape the approach of a firm to the nonroutine problems it faces.... A firm's routines define a list of functions that determine (perhaps stochastically) what a firm does as a function of various external variables (principally market conditions) and internal state variables (for example, the firm's prevailing stock of machinery, or the average profit rate it has earned in recent periods. (pp. 15–16)

But organizations face changes and must adapt to competition. Thus, the second concept or stage: *search* (generation). Anomalies in routines give rise to problem-solving activities or search ("a firm's activities aimed at improving on its current technology" or "coming up with different new techniques", p. 210; "evaluation of current routines and…their modification", p. 400). Innovation becomes a heuristic (and stochastic) strategy to cope with change: "search routines stochastically generate mutations" (innovation) (p. 400). Nelson includes both invention and imitation in innovation, one of the few economists to do so. These are two "acts of innovation": "the company that produces [invents] the first commercial" product and the "first use" of the product (pp. 263–4).

The third basic concept or stage is *selection* (adoption and diffusion). Some technologies succeed better than others in being selected throughout the economy in response to market and nonmarket factors ("the better of the responses tend to be used more widely", p. 400). "Given a flow of new innovations, the selection environment…determines the way in which relative use of different technologies changes over time" (p. 263). Adapting Schumpeter's vocabulary of invention *versus* innovation, Nelson's concept of "spread" (diffusion) resulting from selection includes both "use in the first who introduce" the innovation, and imitation or "widespread adoption" by other firms (pp. 263–5).

Like every theory of industrial innovation, Nelson's study of spread (diffusion) is restricted to direct users (in the present case, firms), not society at large. What about innovation? Nelson explicitly focuses on the "narrower range of issues", as the "term [is] commonly understood" (p. 278), namely technological innovation or "technical change" as "the implementation of a design for a new product, or of a new way to produce a product" (p. 197).[6] At the more general level, innovation is defined as a "change in routines" (p. 128) and as a "new combination of exiting routines" (p. 130).

## CONCLUSION

This chapter studied over one hundred years of theories of innovation, trying to explain the concepts used. Over the last century, several key concepts were imagined to define what innovation is: combination (creativity), diffusion, application, process, system, technology and commercialization.

The theorists I have surveyed, both academics and practitioners, set the foundations for theories of innovation. They invented the paradigms of the field, they developed the representation of innovation and the concepts used over the twentieth century, and they alerted policy-makers to the role of technological innovation in social and economic development. The following six points summarize the nature and scope of the theories and the representation of innovation within the theories:

1. Theories of innovation are many and diverse: psychological, sociological, organizational and industrial.
2. Theories are both pure and applied, but over time shifted to applied and policy-oriented.
3. Theories cover individuals as well as organizations, but over the years came to be dominated by industrial innovation (firms), together with considerations of national level.
4. Over time, the dominant theories are market-oriented.
5. The theories consider both the generation and the diffusion of innovation, although the latter is rarely studied empirically.
6. The meaning of innovation is both broad (any novelty) and restricted (technology), but over time shifted toward the latter.

Given the diversity of theories produced over the twentieth century, the interesting question is: Why has technological innovation become dominant? There are many reasons for this, one of which is the assumed pivotal role of technology in economic and social progress. "The industrial firm", claimed Keith Pavitt in a review of the literature produced for the OECD in

1971, "is the main agent of technological innovation. It transforms scientific and technological knowledge into new or better goods and services" (Pavitt and Wald, 1971: 29). Another reason is methodological, as Abbot Usher, Everett Rogers and others put it: the ease of measurement. A third reason is public policy, as an instrument and a legitimizing force.

The theories of technological innovation participate in society's market ideology. True, sociologists began studying innovation as an instrument of modernization, broadly defined. But economic issues were not central to these theorists. In contrast, to economists, the development of society occurs through economic progress. The main agent is the innovative firm. These thoughts would not have had the impact they had if there had not been a consensus among governments. Governments, their consultants and international organizations were among the first to produce titles on technological innovation and to engage scholars as consultants. Before scholars, with a few exceptions, governments espoused a conception of technological innovation as the commercialization of inventions. In this context, scholars got a hearing because their theories addressed policy issues directly. Sociologists, economists and scholars from management schools and public policy started studying innovation and technological innovation as an explicit contribution to policy.

## NOTES

1.  I talk here of science applied rather than applied science because, first, applied science is one step or stage only in the "circuit" of the innovation process (basic research, applied research, development). It "involves the movement of an idea from its first discernible beginnings, often far from any possibility of application, to its successful commercial use" (Carter and Williams, 1957: 4). This is what came to be called the "linear model of innovation". Second, again and again Carter and Williams stress that "technical progress does not always require the application of new *scientific* knowledge". It may build on existing knowledge as well, which is "adapted, developed, tried out in new circumstances" (Carter and Williams, 1959: 3). Innovation is a job of "development".
2.  For an explicit definition, see Burns (1956: 147, footnote 1): "an invention becomes an innovation when it is carried into commercial application".
3.  Change agents (Barnett's advocates of change), characteristics of innovations to novelty characteristics and values, and categories of adopters to categories of acceptors and rejecters – defined in both Barnett and Rogers according to time of adoption or "lag".
4.  "The process of innovation can be studied more conveniently in the mechanical field than in most of the conceptual fields, because mechanical apparatus can be traced more accurately and more concretely than religious, ethical, and philosophical concepts" (Usher, 1954: 2). Usher was not alone in thinking this way. As Schumpeter put it: "Economics is the most quantitative of all science. Of *all* the sciences, not just of the social sciences. This obviously makes it much easier to define the economic expression of novelty in an exact manner" (Schumpeter, 1932: 114).
5.  In 1965, Mansfield published an article in which he states explicitly that he summarizes the findings of his own studies ("consider a number of important questions") rather than "integrate[s] the results into a single, all-inclusive theoretical structure" (Mansfield, 1965: 137).
6.  The 1977 paper defines innovation as 1. A "portmanteau to cover the wide range of variegated processes by which man's technologies evolve over time" (Nelson and Winter, 1977: 37); 2. "almost any nontrivial change in product or process, if there has been no prior experience" (p. 48).

## REFERENCES

Alter, Norbert (2000), *L'innovation ordinaire*, Paris: Presses universitaires de France.
Barnett, Homer (1953), *Innovation: The Basis of Cultural Change*, London: McGraw-Hill.

Barnett, Homer G. (1961), The Innovative Process, in *Alfred L. Kroeber: A Memorial*, The Kroeber Anthropological Society Papers, 25: 25–42.

Burns, Tom (1956), The Social Character of Technology, *Impact of Science on Society*, 7 (3): 147–65.

Burns, Tom and George M. Stalker (1961), *The Management of Innovation*, London: Tavistock Publications [1996].

Carter, Charles F. and Bruce R. Williams (1957), *Industry and Technical Progress: Factors Governing the Speed of Application of Science*, London: Oxford University Press.

Carter, Charles F. and Bruce R. Williams (1958), *Investment in Innovation*, London: Oxford University Press.

Carter, Charles F. and Bruce R. Williams (1959), *Science in Industry: Policy for Progress*, London: Oxford University Press.

Freeden, Michael (1996), *Ideologies and Political Theory: A Conceptual Approach*, Oxford: Clarendon Press.

Freeman, Chris (1974), *The Economics of Industrial Innovation*, Middlesex: Penguin Books.

Godin, Benoît (2015), *Innovation Contested: The Idea of Innovation Over the Centuries*, London: Routledge.

Godin, Benoît (2017), *Models of Innovation: The History of an Idea*, Boston: MIT Press.

Hage, Jerald and Michael Aiken (1970), *Social Change in Complex Organizations*, New York: Random House.

Kline, Stephen J. (1985), Innovation is not a Linear Process, *Research Management*, July–August: 36–45.

Kline, Stephen J. and Nathan Rosenberg (1986), An Overview of Innovation, in Rachel Laudan and Nathan Rosenberg (eds.), *The Positive Sum Strategy: Harnessing Technology for Economic Growth*, Washington, National Academies Press: 275–305.

Maclaurin, W. Rupert (1946), Investing in Science for the Future, *Technology Review*, 48 (7): 423–54.

Maclaurin, W. Rupert (1949), *Invention and Innovation in the Radio Industry*, New York: Macmillan.

Maclaurin, W. Rupert (1953), The Sequence from Invention to Innovation and its Relation to Economic Growth, *Quarterly Journal of Economics*, 67 (1): 97–111.

Mansfield Edwin (1965), Rates of Return from Industrial Research and Development, *The American Economic Review*, 55 (1&2): 310–22.

Mansfield, Edwin (1966), Technological Change: Measurement, Determinants and Diffusion, in *National Commission on Technology, Automation and Economic Progress*, Volume 2 (Appendices), Washington: USGPO: 94-132.

Mansfield, Edwin (1968a), *Industrial Research and Technological Innovation: An Econometric Analysis*, New York: Norton.

Mansfield, Edwin (1968b), *The Economics of Technological Change*, New York: Norton.

Mansfield, Edwin, John Rapoport, Jerome Schnee, Samuel Wagner and Michael Hamburger (1971), *Research and Innovation in the Modern Corporation*, New York: Norton.

Morton, Jack A. (1971), *Organizing for Innovation: A Systems Approach to Technical Management*, New York: McGraw-Hill.

Nelson, Richard R. and Sidney G. Winter (1977), In Search of a Useful Theory of Innovation, *Research Policy* 6 (1): 36–76.

Nelson, Richard R. and Sidney G. Winter (1982), *An Evolutionary Theory of Economic Change*, Cambridge, MA: The Belknap Press.

Pavitt, Keith and Salomon Wald (1971), *The Conditions for Success in Technological Innovation*, Paris: OECD.

Rogers, Everett M. (1962), *Diffusion of Innovations*, New York: Free Press [1983].

Schumpeter, Joseph A. (1932), Development, in Becker, Marcus C. and Thorbjorn Knudsen (eds.), *Journal of Economic Literature*, 43 (1), 2005: 106–20.

Schumpeter, Joseph A. (1934), *The Theory of Economic Development: An Inquiry into Profits, Capital, Credit, Interest and the Business Cycle*, New Brunswick: Transaction Publishers [2007].

Schumpeter, Joseph A. (1939), *Business Cycles: A Theoretical, Historical and Statistical Analysis of the Capitalist Process*, New York: McGraw-Hill [2005].

Schumpeter, Joseph A. (1949), Economic Theory and Entrepreneurial History, in Research Center in Entrepreneurial History, *Change and the Entrepreneur*, Harvard University Press: 63–84. Reprinted in Joseph A. Schumpeter, *Essays on Entrepreneurs, Innovations, Business Cycles and the Evolution of Capitalism*, R. Vernon (ed.), New Brunswick (NJ): Transaction Books [1989]: 253–71.

Staudenmaier, John (1985), *Technology's Storytellers: Reweaving the Human Fabric*, Cambridge, MA: MIT Press.

Tarde, Gabriel (1890), *Les lois de l'imitation*, Paris: Les empêcheurs de penser en rond [2001].

Tarde, Gabriel (1893), *La logique sociale*, Paris: Les empêcheurs de penser en rond [1999].

Tarde, Gabriel (1902), L'invention, moteur de l'évolution sociale, *Revue international de sociologie*, 10 (7): 562–74.

Tornatzky, Louis G., John D. Eveland, Miles G. Boylan, William A. Hetzner, Elmima J. Johnson, David Roitman and Janet Schneider, (1983), *The Process of Technological Innovation: Reviewing the Literature*, Washington: National Science Foundation.

Usher, Abbot Payson (1929), *A History of Mechanical Inventions*, New York: McGraw-Hill [1957].

Usher Abbot Payson (1954), *A History of Mechanical Inventions*, Boston: Harvard University Press [1962].

Zaltman, Gerald, Robert Duncan and Jonny Holbek (1973), *Innovations and Organizations*, New York: John Wiley.

Zaltman, Gerald and Ronald Stiff (1973), Theories of Diffusion, in Scott Ward and Thomas Robertson (eds.), *Consumer Behavior: Theoretical Sources*, Englewood Cliffs: Prentice Hall: 417–68.

# 4. Economic approaches to industrial technological innovation

*Irwin Feller*

## PROLOGUE

Framing this Handbook in terms of alternative theories of innovations poses two baseline questions. First, alternatives to what? Second, do the several proffered alternative theories approximate the economist's concept of substitutes, of either/or choices, or are they more properly viewed as complements that variously augment, refine, balance, or bound concepts and precepts contained within one or more other approaches?

Experience in numerous boundary-crossing forums such as this Handbook (e.g., Radnor, Feller and Rogers 1978; Feller 2017) strongly suggests that the answer – explicit or implicit – to the first question is economics. Given this prior assumption, while mindful of several other extensive reviews of the economics of technological innovation (Kelly and Kranzberg 1978; Rosenberg 1982a; Tornatzky and Fleischer 1990; Dosi 1991; Stoneman 1995; Cohen 1995; Hall 2005; Fagerberg, Mowery and Nelson 2005; Fagerberg, Martin and Andersen 2013; Godin 2019), this chapter seeks to advance progress towards critical studies of innovation, as called for by Godin and Vinck (2017, p. 321), by laying down a baseline set of economic concepts, methodologies and findings against which the explanatory powers of alternative models may be compared and contrasted.

Selected for coverage are the following five topics: (1) determinants of the choice of production techniques by firms; (2) determinants of the rate and direction of inventive activity; (3) determinants of the adoption of new production technologies; (4) the impact(s) of technological innovation on the wealth and growth of nations; and (5) the impacts of technological innovation on the distribution of income.

The cluster has the virtue of spanning microeconomic analysis of production and distribution and growth theory while also highlighting a methodological continuum from comparative statics (everything else being equal) to studies of long-term historical processes. Additionally, the cluster both identifies core concepts shared across the five topics but also the absence of any single model of technological innovation. Thus, for example, although the production function concept is central to most theoretical and empirical work on the relationships among technological innovation, economic growth and income distribution it appears but little in analyses of the diffusion of innovations or the relationship between scientific and technological change.

There are obvious disadvantages to this approach, though. First, by focusing on what has been termed the positive and descriptive components of economic thinking and research, it slights the discipline's normative and policy strands. Second, constraints on the length of this chapter perforce entail procrustean choices, as lopped off are intellectually and policy-relevant vital attached limbs, such as science/research policy and innovation studies (Martin 2013). Space constraints also lead to the omission of identifying salient points of connection between economics and the alternative models thereby likely if unduly accentuating the substitute rather than complementary relationships between them.

## INTRODUCTION

Writing in 1959, Jewkes, Sawers and Stillerman observed that "Future historians of economic thought will doubtless find it remarkable that so little systematic attention was given in the first half of the century to the causes and the consequences of industrial innovation" (1959, p. 3). Beginning at approximately the same time, catalyzed by a 1960 National Bureau of Economic Research conference that resulted in the volume, *Rate and Direction of Inventive Activity: Economic and Social Factors* (National Bureau of Economic Research 1962), interest by economists in technological innovation has surged, generating in subsequent decades a voluminous and diverse corpus of theoretical, empirical, historical and policy oriented literatures.

As indicators of this surge, economics was elevated from "other" to "major" status between the 1962, first and 2005, fifth edition of Rogers' *Diffusion of Innovations*, and indeed stands out comparatively among allied social science disciplines for the continuing intellectual vibrancy exhibited in the expanded and novel ways the subject continues to be treated (Feller 2018). Another marker of increased and sustained intellectual interest is the appearance of specialized journals devoted to the study of technological innovation, such as *Research Policy* launched in 1971 and *Economics of Innovation and New Technology* begun in 1990. Additionally, articles related to the topic, broadly defined, now routinely appear in mainstream journals in economics. Manifestly nurturing this increased research activity has been the heightened assessment by international organizations, such as the OECD, as well as national and subnational governments that "innovation" and "innovativeness" are key determinants of economic growth and international (and interregional) competitiveness, an association in good part a product of economic research.

## ECONOMIC APPROACHES

The chapter begins with an outline of the economist's concept of the "abstract" production function, the framing concept that "above all" (Godin 2017, p. 69) has historically shaped how economists have launched inquiries into technological innovation.[1] As both a set of abstract properties and as a template for empirical research, the concept is fundamental to an understanding of how economists address several of the above topics (Rosenberg 1982a). The concept also underpins construction and computation of numerous widely used, politically salient, government statistical readings of an economy's health, such as total factor productivity and output per unit of labor. Starting with this concept also provides a baseline to track the evolution of post-1960s models of technological innovation because most in one way or another begin from the proposition, as expressed by Nelson and Winter, that, "The implicit process characterization of the 'production function' models would appear to be not only rudimentary, but fundamentally misleading" (Nelson and Winter 1977, p. 46).

In the neoclassical formulation of production and distribution, technological innovation represents an alternation in the set of (a) production possibilities available to produce goods and services, and (b) changes in the characteristics of the goods and services available to consumers. Much of the theoretical and empirical analysis cited in the initial decades under review here in fact relates to (a). The formulation highlights that output can be produced by alternative production methods while simultaneously denoting the maximum output that can be produced with a given quantity of inputs. This set of alternatives is taken as given. As expressed by Brown, the existing state of technology, as it were, specifies the "rules" that determine the relationships between inputs and outputs "at any given time" (Brown 1966, p. 12).

The actual choice of technique from this array in turn depends on the relative prices of the inputs. Cost-minimization/output maximization occurs when the ratio of input prices is equal to the ratio of input marginal products. Starting from such a given equilibrium, changes in relative factor prices require the firm to replace the initial technique by one whose input combination newly results in cost minimization. As phrased by Salter, "The best-practice technique at each date is the approach technique having regard to both economic and technical factors" (Salter 1969, p. 23).[2]

This statement highlights three elements that cut across the economics literature on techno-logical innovation, albeit not without disagreements. First, the distinction between technical change, a shift, or movement, from one production technique to another from among the available options, and technological change, the introduction of new options into the set of available production alternatives. Second, the background proposition of an existing inventory of alternative techniques that can be employed to produce goods and services, albeit not as efficiently as the one(s) in use, unless, that is, input prices change. Third, the interplay between changes in new knowledge and changes in factor prices in affecting changes in production techniques. At issue here, indeed the common core question of much of the economic history literature on the development and diffusion of production technologies, is the complexities of parsing the contribution of each to observed shifts in which production technologies in use accord with best practice criteria.

The limitations of this framing concept to studying technological innovation have long been readily apparent to economists, albeit deemed "...not too serious a deficiency in an economic *theory*" (Jones 1976, p. 179). First, it applies primarily to process innovations: alternative methods of producing goods (and services), product characteristics remaining unchanged. This limitation surfaces prominently in parsing the relationship between and among measures of technological innovation, economic growth and standards of living.

Second, although "technology" is an explicit component of the production function, techno-logical innovation in fact is an exogenously determined, or black box, variable. As well stated by Nelson at the start-up phase of the period under review, approaching the subject of inven-tive activity "with the traditional tools of economics" (Nelson 1962, p. 8) does not account for the rate and direction of inventive or innovative behaviors or the processes by which industrial innovations, indeed innovations of any type, diffuse through an economy, or society.

Third, viewed through the prism of production function analysis, technological change is frequently conflated with productivity change, especially in policy and journalistic settings. The two concepts are related to one another but in fact are different. By reducing the quantity of inputs required to produce a given quantity of output, technological change can be a direct contributor to productivity change. But the linkage between the two concepts is neither neces-sary nor sufficient; indeed even the sign of the relationship may not always be positive.

A classic example is learning by doing, or the Horndahl effect, in which production costs decline (and productivity increases) over time as a result of increased experience in employing a given production technique (Arrow 1962). Conversely, technological change can reduce levels of productivity if either investments in the capital and labor necessary to introduce an innovation prove economically misguided, or if a prolonged gestation period involving improvements and investments in complementary production and human capital infrastructures is required in order for the new technologies to positively impact production efficiency. According to David, this later dynamic accounts for lagged contribution of investments in computer technology to productivity increases in the 1970s (David 1990).

## SOURCES OF INVENTION

Several different hypotheses and bodies of empirical findings concerning the determinants of the sources and direction of technological innovation exist within the canonical literature. Among them are induced factor bias, path dependency, variants of linear and non-linear models linking, or not, scientific change and technological innovation, the relative roles of scientific change and market demand influences, and user initiation. This overview can convey only the flavor of these models.

The induced factor-bias hypothesis, one of the earliest models on the determinants of the characteristics of technological innovation, emphasized changes in the relative price of factors of production. As stated by Hicks in the 1932 edition of *Theory of Wages*, "…changes in the relative prices of factors of production is itself a spur to invention, and to invention of a particular kind – directed to economizing the use of a factor which has been become relatively expensive" (Hicks 1932, p. 124). Relatedly, in a well known study, American and British Technology in the Nineteenth Century, Habakkuk attributed differences in the factor intensities of U.S. and U.K. manufacturing to the "dearness" of American labor and its inelastic supply "providing an incentive to devise new labour-saving methods…" (Habakkuk 1962, p. 63).

Plausible as is this hypothesis, it "being quite natural to expect that prices will influence the allocation of resources to different research projects" (Binswanger 1978, p. 23), it has been challenged on the grounds that entrepreneurs were "interested in reducing costs in total, not particular costs such as labor costs or capital costs" (Salter 1969, p. 43; Rosenberg 1969). Given this hypothesized omnibus objective, it followed that whenever an input's cost rose, any advance that reduced total costs would be welcome, it being "irrelevant" whether the savings were more heavily weighted towards one factor or another. Moreover, the strict induced factor-bias hypothesis omitted numerous other relevant influences that could have affected the direction and characteristics of technological innovation, including possible differential research and development costs of moving in one factor saving trajectory or another.

An extensive, now largely archival theoretical literature (Binswanger 1978), focuses on factor-bias propositions. These propositions, however, do contain elements of continuing and contemporary relevance. First, of critical importance to contextualizing the factor bias of induced innovation literature is understanding that it was primarily directed at advancing a model of the dynamics of income distribution, not technological innovation. Its constituent concepts – neutral/non-neutral technological change; elasticity of substitution – have contemporary relevance, being core building blocks in important recent studies on income distribution, such as Piketty's *Capital* (2014) and Atkinson's *Inequality* (2015).

Second, in the longer term course of an economy's technological trajectories, factor bias has been found to exist, at least in selected economic sectors. It is why it exists that becomes interesting and important. Noting differences in the historical factor-intensity trajectories between 1870 and 1970 in the land-intensive, labor extensive, biological science characteristics of Japanese agriculture set against the land-extensive, labor intensive mechanization characteristics of U.S. agriculture, Ruttan (1978, 2001, esp. Chapter 4) has argued that these differences arose not only from technological adjustments made in response to episodic changes in relative prices but also from what he terms induced institutional innovation. By this he meant the (induced) development of institutions (e.g., publicly supported research laboratories; university based academic programs) that selectively nurtured scientific and technological knowledge, human capital expertise, and technology transfer and diffusion mechanisms more in the direction of economizing on one factor than another. These actions were systemic and self-reinforcing: each technical advance tied to a nation's factor endowments created new technical possibilities or encountered new technical bottlenecks, over time further directing the allocation of inventive activity along existing lines, tending though to widen the distance between it and alternative technological paradigms. Historically, changes in production and inventive paradigms tended to be incremental, save for major economic shocks, for example, spikes in energy prices associated with the 1970s spike in crude oil prices, or disruptive scientific advances, for example, the biotechnology revolution in agriculture.

Third, limited as the price-induced factor-bias hypothesis may be as a model of processes of technological innovation, it does catalyze the question, what influences motivate firms, or inventors to search in one direction rather than others for means of improving existing technologies or developing totally new ones. Especially apposite here is Rosenberg's concept of focusing devices and inducement mechanisms in which, (a) the interdependence of the components of complex technologies lead firms to "attack the most restrictive constraint", as, for example, more powerful automobile engines creating the need for improved braking systems, or (b) disruptions in the supply of key inputs leading to increased attention to new technologies that economize on existing supplies and/or furnish substitute inputs (Rosenberg 1969, 1972).

Two other theories of the sources of inventive activity are nominally similar but too easily conflated. One is the linear, or pipeline, model of technological innovation; in simple form, the sequencing from basic research to applied research to any number of end points (Godin 2017). Setting aside the various formulations and critiques of this model, for quite some time a strawman artifact in the academic literature, the assignment in the model of precedence and causative force to science over technology continues to receive wide acceptance, especially in national science and technology policy forums. Here economists, scientists and historians are united in their criticisms (Rosenberg 1976; Brooks 1994; Keller 1984). Noting that "Of course, science and technology have interacted at many points...", Basalla has contended, "Nevertheless, technology is not the servant of science" (Basalla 1988, p. 27). Rosenberg has made this point even more forcefully, arguing that (a) technology is not the "mere application of prior scientific knowledge", but rather a quasi-autonomous body of knowledge about how things can be made ("even when one cannot explain exactly why)"; (b) that technological advances based on this body of knowledge have generated performance characteristics "that demanded a scientific explanation"; and (c) developed the techniques of observation, testing and measurement that undergird scientific progress (Rosenberg 1982b, pp. 141ff).

The second is the relative influence of scientific advances and market influences on the direction of inventive activity. As posited by Schmookler, the question is whether inventions are "... mainly knowledge induced or demand-induced? In the parlance of economics, are they primarily the outgrowth of changes in the conditions of their supply or do they largely reflect changes in the demand for them? (Schmookler 1966, p. 12; also Mowery and Rosenberg 1979). His findings emphasized the importance of a "technical problem or opportunity conceived by the inventor largely in economic terms" as the proximate stimuli to reported inventions (Schmookler 1966, p. 66). Conversely, based on chronologies of important inventions in four industries, "... *in no single instance is a scientific discovery specified as the factor initiating an important invention in any of these four industries*" (Schmookler 1966, p. 67; italics in original).

Contrasted to this analytical strand are system-level findings that innovation increasingly relies on scientific knowledge (Fleming et al. 2019) which catalyze not only specific technologies but also new industrial clusters, for example the role of the Cohen-Boyer discovery (and patent) in the rise of a multi-sector biotechnology industry. More broadly, in recent years, research by economists and scientometricans (Narin, Hamilton and Olivastro 1997; Sorenson and Fleming 2004) have likewise documented the contribution of scientific research, as measured by publications in the peer reviewed scientific literatures, to patents, a proximate predicator of commercial relevance. In 2018, for example, "almost quarter (23%) of USPTO patents in 2016 cited S&E articles, with almost 300 000 S&E articles cited", according to the National Science Board's 2018 Science and Engineering report (NSB 2018, Table 8-6).

Debate on relative influence can be found in an extensive literature that variously emphasizes the necessary, originating role of scientific advance, as in the archetypical linear model; the origins and pathways of technological innovations independent of scientific advance; the catalytic role of technological advance in creating and shaping scientific research; and the complex and iterative interactions between scientific advance and technological change. This diverse array of alternative hypotheses makes tracing separate influences akin to isolating the starting as well as ending point in an Escher etching.

Mowery and Rosenberg's observation midway through the period under review echoes Marshall's classic resolution in his *Principles of Economics* of the debate between Ricardian adherents of the role of costs of production and Austrian proponents of marginal utility analysis that positing either theory as the single most determinant of market based prices was akin to inquiring about which blade of a scissors was responsible for cutting paper. They write, "Successful technological innovation is a process of simultaneous coupling at the technological and economic levels – of drawing on the present state of technological knowledge and projecting it in a direction that brings about a coupling with some substantial category of consumer needs and desires" (1989, p. 8). A recent bibliometric distillation of the relevant literature arrived at much the same conclusion, reporting "...convergence on the mutual importance of the two sources. While science and technology provide the trajectories of innovation, demand is a crucial component in order to direct the trajectory towards the right economic venues" (Di Stefano, Gambardella and Verona 2012, p. 1291).

Additional more formal treatments of the "economics of science" that detail the impacts of costs incentives, human capital, intellectual property rights regimes, and more upon the choice and direction of inventive activity (extending well beyond considerations of Weinberg's classic formulation of criteria for scientific choice) likewise effectively treat supply and demand influences as elements in a set of simultaneous equations (Diamond 1996; Dasgupta and David 1994; Stephan 2012).

Other factors further complicate efforts to tease out the influence of science/technology push factors relative to those occurring on the demand side. The very formulation of the question implicitly subsumes a discrete separation between the entities that generate the knowledge upon which technological change is dependent, for example capital good firms, public sector laboratories; universities and the final user, be it a firm or an individual consumer. This need not be so. Von Hippel's early findings (1976) about the proclivity of researchers to develop (invent) new instruments to further their scientific inquiries, with these new technologies subsequently becoming commercial products highlighted the role of users as sources of invention. His more recent work, *Democratizing Innovation* (2005) highlights further the increased multifaceted ability of users to innovate for themselves, that is to develop exactly what they want rather than rely on manufacturer-centric innovations.

## DIFFUSION OF INNOVATIONS

This section is much abbreviated both because of the availability of several other surveys, as cited above, including my own recently authored monograph (Feller 2018).

A prefatory note though about the contested acceptance of economic approaches to the diffusion of innovations associated with the debates between economists and rural sociologists that accompanied Griliches's 1957 econometric study of the diffusion of hybrid corn. This model presented profitability as the key determinant of the availability of hybrid seed across U.S. states as well as the rate of adoption of hybrid seed by farmers within each state. This explanation contrasted with the importance attached to adopter characteristics and interpersonal interaction and communication patterns previously offered by rural sociologists, leading to a disciplinary based tempest. Largely now an historical footnote, debates about the explanatory power of parallel, occasionally lightly intersecting but also at times competing conceptual approaches to analyses of the diffusion of innovations continues. Thus, while few economists today would disagree with Rosenberg's 1972 observation, that the "...diffusion of (technological) innovations is an essentially economic phenomenon, the timing of which can be largely explained by expected profits, is by now well established..." (Rosenberg 1972, p. 6), it is not assumed here that the statement has universal concurrence.

The diffusion of technological innovations is a complex, non-linear process. In contrast to Schumpeter's theory, or more precisely, metaphor, of creative destruction, which at times has been taken to connote rapid and total displacement of existing production techniques by a newer one, the process is neither necessarily rapid nor total. Mansfield's (1961) pioneer study of the diffusion of industrial innovations for example found that it took 20 years from the date of commercial introduction for all firms in his study to initially adopt a set of (capital-intensive) technologies. Intra-firm diffusion, the process by which firms substituted the innovation for sizeable portions of their pre-existing production capacity took even longer. Numerous historical accounts moreover recount the survival of previously installed techniques for extended periods of time in the face of the introduction of newer, eventually dominant innovations. Mak and Walton (1973) have documented the continued use of flatboats well after the introduction of steamboats, while sailing vessels continued to be employed for conveying long distance high bulk, low value commodities long after the appearance of trans-Atlantic steamships.

Numerous factors have been cited for this phenomenon. These include the juxtaposition of the technical, and thereby, economic limitations of the new technology to competitively match the suite of uses to which existing technologies are deployed, alongside the presence for the defender technology(ies) of a stream of technical improvements that narrow or eliminate the performance gaps between new and old. Accounts of continuing streams of performance-enhancing adjuncts to existing technologies that help explain why they are not necessarily summarily discarded are staples in histories of technologies (Feller 1966; Nasbeth and Ray 1974; Sahal 1981; Mokyr 1990).

Additionally, as Pavitt has observed, "But the new does not always turn out to better than the old" (2005, p. 107). Du Pont's corfam and RCA's videodisc are two examples of high-tech innovations developed and marketed by R&D intensive firms that were market failures (Graham 1986). Kunkle's (1995) account of GE's failure to successfully market its electron microscope with electrostatic lenses in competition with RCA's electromagnetic lens version likewise highlights that the race may not always be to the "technologically" swiftest. Rather, in this case the critical factor in achieving market dominance was held to be RCA's more aggressive and astute efforts to establish early, close connections with influential early adopters.

Technological innovations moreover are not plug-in modules that can be readily substituted for earlier techniques. Rather they may exist within a larger socio-economic–technological framework, such that the benefits of the new technology may not be realizable without substantial changes in the larger setting. If these changes are not forthcoming, prove too costly, or require changes in supporting infrastructure systems – changes that require the actions of others, and are not forthcoming – than seemingly "inferior" technologies may become "locked-in", even if superior ones are available. The ubiquity of the QWERTY keyboard first on typewriters and today on laptops and iPhones, and the dominance of light-water as contrasted with gas-cooled nuclear reactors are two such frequently cited examples (David 1985; Arthur 1989).

A rapidly appearing stream of innovations moreover may indeed impede adoption. Rosenberg (1976) has pointed out that in the case of capital-intensive innovations, expectations of rapid, disruptive technological change can impede the adoption "of new technologies for fear of being "leap-frogged". Much the same may be said of the pace and sequencing of new high-tech consumer products, as has been occurring with iPhones, where the choice is not only whether to adopt or not but when (or which model).

Within the broad consensus among economists about the determinants of the diffusion of industrial innovations, two generic approaches dominate. The first, is essentially econometric in nature. The generally accepted proposition emerging from this line of research is that

> the rate of diffusion of an innovation depends on the average profitability of the innovation, the variation among firms in the profitability of the innovation, the size of the investment required to introduce the innovation, the numbers of firms in the industry, their average size, the inequality in their sizes, and the amount they spend on research and development. (Mansfield et al. 1982, p. 7)

The second approach tends to focus on the diffusion of specific technologies. It too conceptualizes the diffusion process as driven by expectations of profit and thereby dependent on factors relating to supply and demand determinants. In contrast though to the econometric approach, it delves deeper into the technical specifications of the innovation that impact on input requirements, production costs and product characteristics. Moreover, it tends to consider these factors over time, noting for example improvements in original equipment and intermediate goods firms that lower production costs and otherwise increase performance capabilities, allowing the technique to sequentially compete with (and displace) existing technologies. The approach, as in David's (1966) study of the diffusion of the mechanical reaper, also tends to consider both quantitative and qualitative changes in demand for the goods and services produced with the technology.

What is not found though in the economics literature on the diffusion of innovations is a "pro-innovation bias" either at the micro-level with respect to viewing specific technological innovations as inherently economically or societally beneficial such that rapid and extensive adoption is warranted, or at the macro-level, with connotations that promoting more innovation or more innovators are necessarily optimal uses of a society's resources (Soete 2013; Wisnioski, Hintz and Kleine 2019). Soete, for example, has observed, that in contrast to the "surprising" tenor of the business and policy literature that contend that innovation is good for you, the "most common feature of innovation studies" is their attention to innovation failure rather than success (Soete 2013, p. 134). Indeed, he goes further to speak of "destructive creation", that is, innovations that benefit a few "at the expense of many with, as a result, the opposite pattern of a long-term reduction in overall welfare or productivity growth" (loc. cit).

## ECONOMIC GROWTH

In its encomiums to the role of basic research in fostering technological innovation, Vannevar Bush's 1945 report *Science – The Endless Frontier* is widely credited with having provided both the rationale and the metaphor to catalyze the post-World War II U.S. bipartisan political shift that opened the spigots to Federal government support of non-mission oriented basic research.[3] Economists too in no small part played a part in providing the empirical evidence that documented the relationship between technological change and economic well-being.

Economic growth here is defined as intensive growth, that is increases in average per capita incomes, or in colloquial terms, on standards of living. By way of contrast, extensive growth relates to increases in national total output, made possible say by increases in the size of the labor force, expanded tilled acreage, or improved production practices that accommodate (and indeed in a Malthusian world, foster) population increases, the sum of which however are not associated with sustainable increases in per capita standards of living (Jones 1988).

Interest in the distinctions between intensive and extensive economic growth arises from findings that sustained increases in average per capita incomes are a relatively recent historical occurrence (Kuznets 1966). To focus only on the U.S. experience, Gordon has observed, "There was virtually no economic growth for millennia until 1770, only slow growth in the transition century before 1870, remarkably rapid growth in the century ending in 1970, and slower growth since then" (Gordon 2016, p. 2). Rephrased, these empirical findings pose the question: "…how did expansionary growth change to income growth" (Jones 1988, p. 30).

The interest of economists in technological innovation has historically stemmed from their efforts to answer this question (Abramovitz 1989). Intensifying these efforts were the retardation ("crisis") in productivity growth rates in the 1970s and 1980s (Baily and Chakrabarti 1988; Baumol, Blackman and Wolff 1989); the demise of stalwart manufacturing industries – steel; autos; consumer electronics – and criticisms that the big science/mission orientation of U.S. science and technology policy was poorly shaped to meet the diffusion oriented policies of international competitors, especially in emerging "high-tech" industries (Ergas 1987). This interest encompassed the development of new endogenous growth models that emphasized the role of technological innovation as a determinant of economic growth, extensive empirical work directed at conceptualizing and measuring the sources of economic growth, and articulation of rationales for and subsequent evaluation of an emerging portfolio of public sector pro-innovation science and technology policy initiatives (e.g., Advanced Technology

Program; Small Business Innovation Program; Bayh-Dole Act) (Shapira and Kuhlmann 2003; Mowery et al. 2004; Ruegg and Feller 2003).

A convenient starting point to this heightened interest coincident with the Jewkes and Sawyers reference to the seeming disinterest of economists in the subject of industrial techno-logical innovation are the classic studies of Abramovitz (1956) and Solow (1957). Separately and independently, their studies called into question long-standing propositions that the growth of the U.S. economy since the late nineteenth century was primarily a function of capital deepening and capital widening, that is in increases in the capital–labor ratio (mecha-nization), which served to increase output per unit of labor. Instead they each concluded that most of the increase in output per man-hour was attributable to increases in productivity.

The authors, however, used different phrasing in presenting their conclusions, a difference that reverberates through contemporary discourse on the relationships between technological innovation and economic growth. In accounting for the doubling of gross output per man-hour between 1909 and 1949, Solow gave major credit – 87 percent – to what he termed "technical change", a phrasing rich in connotative associations with technological innovation, but more accurately the residual, or unexplained variance in output per man-hour, the dependent varia-ble. Abramovitz, whose study covered the period 1870–1950, was more diffident in using the concept of productivity increase to account for the large otherwise unexplained residual. He wrote:

> Since we know little about the causes of productivity increase, the indicated importance of this element may be taken to be some sort of measure of our ignorance about the causes of economic growth in the United States and some sort of indication of where we need to concentrate our attention. (Abramovitz 1956, p. 11)

These two articles launched a "growth accounting"/sources of economic growth suite of studies covering various time periods in which researchers used increasingly refined measures of the capital and labor inputs, took account of other sources of growth, such as changes in the sectoral composition of output, and modified the specification of the aggregate production function to advance decomposed estimates of the contribution of various sources to estimated total growth in output (Kendrick 1961; Denison 1974; Jorgenson, Gollop and Griliches 1987). The net effect of these studies was to clarify that the 87.5 percent share attributed by Solow to technical change was in fact not a direct measure of technological innovations but rather a portmanteau term for a cluster of (interacting) variables whose importance remained to be accurately measured, much less formally modeled.

Interest in productivity continues to be a staple of economic research, in large part fueled by earlier noted and still present concerns about secular retardation in rates of productivity increase (Williamson 1991; Gordon 2016, pp. 522–31). A summary assessment is that the links between technological innovation and productivity growth, and in turn to economic growth are best seen as complex, iterative and at best long term.

New theories of the relationship between technological innovation and economic growth, loosely characterized as neo-Schumpeterian or evolutionary and endogenous, overlapped with the above stream of empirical work (Verspagen 2005; Walsh 2006). Focusing here only on the latter, in contrast to the essentially neoclassical production function model employed in Solow-like growth models, Romer (1990, 1994) proposed what has come to be known as an endogenous theory of economic growth. This theory incorporates the accumulating set of post-1960 propositions relating to the microeconomics of technological innovation –

a resource-intensive activity undertaken by profit-seeking firms operating in a spectrum of competitive market settings – into a formal growth model. Endogenous growth models resemble Solow-like models in viewing technological change – "improvements in the instructions for mixing raw materials – …as the heart of economic growth" (Romer 1990, p. 872) but differ from them in treating technological innovation as arising "in large part because of the intentional actions taken by people who response to market incentives (thus being endogenous rather than exogenous)" (ibid).

Central to the Romer model, its "most fundamental premise", is the difference it posits between the production of "new instructions" and that of other raw inputs, such as capital and labor. For the former, according to Romer, "Once the cost of creating a new set of instructions has been incurred, the instructions can be used over and over again at no additional cost. Developing new and better instructions is equivalent to incurring a fixed cost. This property is taken to be the defining characteristic of technology" (ibid). Development of new instructions, for example, promoting technological change, thus takes on the form of public good, being non-rivalrous and non-excludable, and thereby likely to be undersupplied via competitive market economies.

Through its emphasis on the public goods characteristics of advances in general purpose or platform technologies, endogenous growth models provided a theoretical grounding to the efforts of policy makers seeking during the late 1970s and early 1980s to ameliorate signs of secular retardation while overcoming political resistance to any initiatives redolent of an industrial policy. Without in any way contending that any direct linkage existed between the advent of endogenous growth models and the emergence of pro-technological innovation policies such as the 1980 Omnibus Trade and Competitive Act and more substantially those of the Clinton Administration years of the 1990s, there yet remains something to be said for Keynes's observation about the power of the ideas of academic scribblers upon those in power.

Above as well as beneath these essentially quantitative and conceptual models of relationships between and among economic growth, technological change, total factor productivity and per capita incomes resides the quotidian matter of standards of living for vast numbers of the population. Viewed at this level, economic growth and its handmaiden, technological innovation, has changed daily life beyond recognition. As tellingly noted as early as 1967 by Nelson, Peck and Kalachek, the emphasis placed by economists on technical change (or innovation) is "not only because of its role as a catalyst, but also because it endows economic growth with much of its capacity for satisfying human wants" (1967, p. 19). Technical change can take the form of improved, more efficient means of producing existing goods – textiles, corn – benefitting consumers mainly in the form of lower cost goods, which frees up income for other uses. But its larger effects – indeed those that correspond more generally to the manner in which it is popularly construed – is to "create the possibility of producing substantially new or improved consumer goods", thereby permitting the "satisfaction of wants which had not been satisfied before" (loc. cit.).

The import of these qualitative assessments is heightened by considering not the oft-cited closing triplet of humankind's plight voiced by Hobbes that life is "…nasty, brutish, and short", but the statement's opening lament that it is also "solitary" and "poor". Few works by an economist better document the contributions of technological innovation to overcoming each of these baneful life conditions than Gordon's (2016) magisterial account, *The Rise and Fall of American Growth* since 1870. The nature of these enhancements is suggested by the book's section headings such as, "The American Home: From Dark and Isolated to Bright and Networked" and "Nasty, Brutish and Short: Illness and Early Death". The following sentence distills the tenor of the narrative: "Manual outdoor jobs were replaced by work in air-conditioned environments, housework was increasingly performed by electric appliances, darkness was replaced by light, and isolation was replaced not just by travel, but also by color images bringing the world into the living room" (p. 1).[4] (Indeed the narrative brings a shock of recognition to someone of an age to recall ice boxes, 78 rpm records, the introduction of frozen orange juice, valise size, "portable" tape recorders and radio programs. It is striking to be reminded of how recent is the appearance of the technological consumables and systems that constitute ordinary life!)

That there is no such thing as a free lunch is a maxim oft cited by economists. The same holds true in their analyses of the welfare impacts of economic growth, including that portion attributable to technological change. Alongside this emphasis within the economics literature on the beneficial aspects of technological innovation on standards of living exists one that highlights the deleterious economic, societal and political effects induced by a technology-driven "growth mania" (Mishan 1971). The concept of externalities or spillover effects, positive or negative, arising from the behaviors of producers and consumers interacting in market economies that underlie the clean air/clean water, toxic substance/hazardous waste environmental regulations of the 1960s and 1970s, is contained in Marshall's *Principles of Economics* (first edition, 1891), and the theory of social cost is the central thesis in Pigou's *Economics of Welfare* (first edition in 1920).


## INCOME INEQUALITY

Returning to the chronological starting period of this overview to trace the evolution of economic thought, consider the following statement by Galbraith: "...Few things are more evident in modern social theory than the decline of interest in inequality as an economic issue" (Galbraith 1958 (1998 edition), p. 69). His explanation for this decline was its displacement by the belief that economic growth – the rising tide lifting all boats; the bigger pie that makes each slice, however relatively small, still larger – "has eliminated the more acute tensions associated with inequality. And it has become evident to conservatives and liberals alike that increasing aggregate output is an alternative to redistribution or even to the reduction of inequality" (Galbraith 1958, p. 80).

Here again the world has turned upside down. Few topics currently attract as much scholarly attention as well as being the cynosure of political debates as secular long trends across most highly developed capitalistic nations towards a decline in labor shares in national income, resulting in increased income inequality (Karabarbounis and Neiman 2013).[5] Indeed, as Piketty notes, "It is of interest to everyone, and that is a good thing" (2014, p. 2).[6]

Numerous economic, societal and political factors have been cited as contributing to these trends (Atkinson 2015; esp. Chapter 3). Among them are the decline in the bargaining power of unions, globalized supply chains, the advent of neoliberal public policies that have frayed income maintenance and social welfare nets while throwing a spanner into escalator programs – education, health, worker training – supportive of upwards economic and social mobility, and a substantial lessening of the degree of progressivity in national tax policies. Added to these factors are the asymmetries implicit both in Griliches-type models of market shaped diffusion processes, as in Cozzens's (2010) account of the shaping of public sector diffusion agencies in developing countries that further advantage existing haves (or early adopters) over other income cohorts in reaping the gains (or avoiding the costs) of technological innovations. And not to be overlooked, indeed recently cited as a "primary reason for the rise of inequality" is "The rise of economists", in policy making quarters, who have been charged with bringing with them a penchant for market forces and a fatalistic acceptance of the workings of market forces (Applebaum 2019).

Of immediate relevance here though are the relationships between recent trends in income inequality and the previous overview of neoclassical economic models of technological innovation. A useful starting point is J.S. Mill's oft-cited dictum that the production of wealth was not an arbitrary thing but had necessary conditions tied to the properties of matter, or rather, in language akin to the role of technology in specification of a production function, "... the amount of knowledge of those properties possessed at a particular place and time". Unlike the laws of production, however, "those of Distribution are partly of human institution: since the manner in which wealth is distributed in any given society, depends on the statues or usages therein obtaining" (J.S. Mill 1848 (1866 edition), p. 41). Or, to put this proposition in contemporary terms, as argued most recently by Atkinson, "It is my belief that the rise in inequality can in many cases be traced directly or indirectly of changes in the balance of power. If that is correct, then measures to reduce inequality can be successful only if countervailing power is brought to bear" (Atkinson 2015, pp. 82–3).

It is consideration of the "laws of production" though, specifically those that relate to technological change that fits within the bounds of this chapter. Of relevance here are less well-known concerns about structural displacement of occupations or regional economies associated with major technological innovations – fracking's effect on coal mining, for example – than an accumulating body of evidence that the characteristics of recent technological innovations writ large across industrial and service sectors have significantly increased the returns to capital relative to those of labor, resulting in an increased share of total outcome accruing to owners of capital relative to those dependent on labor income. The underlying determinants appear to be a combination of the substitution of capital for labor as a result of changes in relative prices, and the (changing?) magnitude of the elasticity of substitution, a measure of the ease with which inputs can be substituted for one another.[7]

Here two theoretical propositions about how the properties of abstract production functions condition how technological change affects the determination of relative factor shares enter the analysis. First, in the case of non-neutral technological innovation, "a factor-saving innovation, ceteris paribus reduces the relative share of income of that factor in all cases". Second, if one factor increases in supply more rapidly than another, and "if the elasticity of substitution is less than unity, then the relative share of the first factor decreases" (Brown 1966, p. 181). In the opposite direction, if the elasticity of substitution exceeds unity, then the relative share of the first factor increases. The relationship of this measure to income distribution dynamics is as follows: "if one factor increases in supply more rapidly than another, and if the elasticity of substitution is less than one, then the relative share of the first factor decreases" (Brown, loc. cit.). If however, the elasticity of substitution is greater than one, then its relative share increases.

There is a complex web of relationships built about these propositions that economists are currently trying to unravel in their empirical research. Most studies of the value of the elasticity of substitution place it at less than 1; accordingly, a rise in capital per worker should produce a decline, not a rise in capital's share. But if technological change is labor-saving, then the rise in the capital–labor ratio would tend to increase capital's relative share. Recent research suggests that this is what has been happening. According to Arpaia, Perez and Pichelmann, most of the declining pattern in labor shares in nine EU15 Member States is governed by "... capital-augmenting technical progress and the assumption of capital-skill complementarity, i.e., the fact that capital equipment is complementary to skilled labour but highly substitutive to unskilled labour" (2009, p. 37). This latter dynamic is what Piketty claims to have occurred between 1970 and 2010, accounting in part, albeit only in part, of the observed increased in capital's share in income in many developed economies (p. 221). Since capital income is more highly concentrated than labor income, this effect contributes to observed increases in income inequality among individuals.

Multiple factors lay beneath these trends. Briefly noting recent findings, they include the displacement via technological change of labor by capital, for example, automation, at a rate higher than such change creates new tasks for labor (reinstatement), especially in manufacturing (Acemoglu and Restrepo 2019); the "skill bias" of technical change over recent decades, creating premiums for those with the requisite skills but exerting downward pressure on returns to unskilled labor (Acemogulu 2002); the destabilizing effects of past periods of higher private rates of return of capital relative to the growth rates of income and output, such that "capital reproduces itself faster than output increases" (Piketty 2014, p. 571); and income-based biases in the manner in which the benefits and costs of new technologies diffuse through a society.

Although the microeconomic, scientific and technological elements that underlie these trends are as yet imperfectly understood, their combined effects on income inequality are seen as worrisome. According to Piketty, no self-correcting mechanism exists to retard the trend towards increases in capital's share, given "the many uses of capital in a diversified advanced economy in which the elasticity of substitution is greater than one" (2014, p. 221). Similarly, Arpaia, Perez and Pichelmann write:

> Our simulations point to the forceful implications of the nature of technological progress given unchanged institutional settings. Not only has the labour share fallen over the past three decades, but it may decline further in the future as a result of capital accumulation and an increasing share of skilled labour in total employment. (2009, p. 37)

## CODA: BUILDING BRIDGES

To return to the underlying if implicit framing of this Handbook, to the extent that alternatives are predicated as "substitutes" rather than as "complements" to economic approaches, claims on behalf of alternative approaches in places appear to conflate the "positive" and "empirical" cast of the economic approaches to technological innovation outlined above with orthogonal treatments about governance regimes, anticipatory developments, normative ends and policy objectives. Additionally, in places they limit their critiques to the historical and indeed contemporary emphasis in much of the economics literature on the effects of innovation on the performance of firms, industries, and national growth rates, slighting the intellectual ferment found in the amorphous but nevertheless substantive activity surrounding calls for a new "innovation studies" research agenda that essentially addresses the issues posed in the following chapters of this Handbook (Fagerberg, Martin and Andersen 2013; cf. Godin 2019).

Sipping rather than drinking thus runs the risk of overlooking the diversity and fullness of the policy analytical and normative exegesis authored by economists. Among the notable but by no means isolated examples can be cited the writings of public intellectuals such as Paul Krugman and Jeffrey Sachs and the Nobel prize awards to M. Kremer, A. Banerjee, and E. Duflo for their experimental approaches to the alleviation of poverty and illiteracy in developing nations – being notable but by means isolated examples. Relatedly, the ubiquity with which economic indicators on economic growth and productivity are cited in government documents, congressional testimony, political rhetoric and the popular press has untowardly served to delimit the economist's spectral bandwidth of research interests and normative horizons for non-economists. Thus at times critical assessments of the impacts of technological innovation conflate critiques of economic models per se with those directed at the workings of price-directed market economies.

The case for finding a (substitute) alternative to economic approaches thus may relate less to its putative disregard for the concerns articulated in other approaches to innovation than to the embeddedness of a substantial but by no means total share of its corpus of work in market-driven economic systems (Pfotenhauer and Juhl 2017). High conceptual, empirical and policy oriented salience may justifiably be accorded to approaches that assign primacy to understanding to societal consequences of technological innovation – the nuanced but telling distinction that Winner (1986) introduced between impacts and side effects – but they do not per se negate the importance of employing the economist's analytical and empirical toolkit to gain an understanding of processes of technological innovation – past, present, and future.

## NOTES

1.  "The abstractness of the production function concept is precisely its source of value: it enables economists to analyze a wide variety of problems, for example the determination of relative income shares, the factors affecting economic growth, and the nature of technological employment" (Brown 1966, p. 11; also Ferguson 1969).
2.  By best practice is meant the technique that corresponds to the cost-minimization/output maximization conditions outlined above. At any point in time though, the set of production techniques in use across an industry and indeed even within a firm typically comprise a spectrum, ranging from best practice to worst practice (technically, techniques where variable production costs are just covered by product prices. Average practice then is the weighted share of output produced by each technique in use (Salter 1969).
3.  For an expanded interpretation of Bush's report that includes justifications for technology development, see Leyden and Menter (2018).
4.  As an offset to this panegyric, see Cowan's (1983) *More Work for Mother* analysis of the equivocal benefits accruing to women from the introduction of electrical household appliances.
5.  Between 1975 and 2012, 42 of the 49 countries covered in the K&N study exhibited downward trends in labor's share of national income. For the U.S., as phrased by Lepore (2018, p. 755), citing Piketty's data, the numbers documenting these trends are "staggering": These trends and concerns about them connect to but are analytically and programmatically distinct from those related to "poverty", and "inequality of opportunity" as addressed say in the Johnson Administration's War on Poverty. Recent trends moreover mark a sharp reversal of those found in earlier periods. Thus between 1929 and 1959 the shares of national income received by the top one, five and twenty percent declined substantially, while that received by the bottom 60 percent, pre-tax, increased from by almost one-third (from 26 to 32 percent) (Kuznets 1966, p. 211).
6.  The two trends are causally related because labor's share, unequally distributed among income size cohorts as it is, is far less unequally distributed than is capital's share. As a benchmark, in the U.S., classified among developed nations as evincing "high inequality", the top 10 percent of labor income recipients received 35 percent of total labor income in 2010 while the bottom 50 percent received 25 percent. By way of contrast, for total capital income, the top 10 percent received 70

percent of total income, while the bottom 50 percent received 5 percent. Summed across sources of income then, the top 10 percent received 50 percent of all income (the top 1 percent received 20 percent), while the bottom 50 percent received 20 percent (Piketty 2014, pp. 247–9).

7.   The elasticity of substitution, a pure number, is defined as the "proportional change in the input ratio attributable to a proportional change in the marginal rate of technical substitution" (Ferguson 1969, p. 41).

# REFERENCES

Abramovitiz, Moses (1956), 'Resource and Output Trends in the United States Since 1870', *American Economic Review Papers and Proceedings*, May 1956, 5–23.

Abramovitiz, Moses (1989), 'Thinking about Growth', in *Thinking about Growth*, Cambridge: Cambridge University Press, pp. 3–79.

Acemoglu, Daron (2002), 'Directed Technical Change', *Review of Economic Studies*, 69, 781–809.

Acemoglu, Daron and Pascual Restrepo (2019), 'Automation and New Task: How Technology Displaces and Reinstates Labor', *Journal of Economic Perspectives*, 33, 3–30.

Applebaum, Binyamin (2019), 'Blame Economists for the Mess We're In', *New York Times*, August 25, 2019, Sunday Review, 9.

Arpaia, Alfonso, Esther Perez and Karl Pichelmann (2009), 'Understanding Labour Income Share Dynamics in Europe', *Economic Papers*, No. 379, 1–51.

Arrow, Kenneth (1962), 'The Economic Implications of Learning by Doing', *Review of Economic Studies*, 29, 155–73.

Arthur, W. Brian (1989), 'Competing Technologies, Increasing Returns, and Lock-In by Historical Events', *Economic Journal*, 99, 116–31.

Atkinson, Anthony (2015), *Inequality*, Cambridge, MA: Harvard University Press.

Baily, Martin and Alok Chakrabarti (1988), *Innovation and the Productivity Crisis*, Washington, DC: Brookings Institution.

Basalla, George (1988), *The Evolution of Technology*, New York: Cambridge University Press.

Baumol, William, Sue Blackman and Edward Wolff (1989), *Productivity and American Leadership: The Long View*, Cambridge, MA: MIT Press.

Binswanger, Hans (1978), 'Induced Technical Change: Evolution of Thought', in Hans Binswanger and Vernon Ruttan (eds), *Induced Innovation*, Baltimore, MD: Johns Hopkins University Press, pp. 13–43.

Brooks, Harvey (1994), 'The Relationship between Science and Technology', *Research Policy*, 23, 477–86.

Brown, Murray (1966), *On the Theory and Measurement of Technological Change*, Cambridge, UK: Cambridge University Press.

Bush, Vannevar (1945), 'Science – The Endless Frontier', Washington, DC: Office of Scientific Research and Development, Report submitted to the President of the United States, July 5, 1945.

Cohen, Wesley (1995), 'Empirical Studies of Innovative Activity', in Paul Stoneman (ed.), *Handbook of the Economics of Innovation and Technical Change*, Oxford, UK: Blackwell, pp. 182–264.

Cowan, Ruth S. (1983), *More Work for Mother*, New York: Basic Books.

Cozzens, Suzan (2010), 'Innovation and Inequality', in Stefan Kuhlmann, Philip Shapira and Ruut Smits (eds), *The Co-Evolution of Innovation Policy: Innovation Policy Dynamics, Systems, and Governance*, Cheltenham, UK and Northampton, MA, USA: Edward Elgar Publishing, pp. 363–85.

Dasgupta, Partha and Paul David (1994), 'Toward a New Economics of Science', *Research Policy*, 23, 487–521.

David, Paul (1966), 'The Mechanization of Reaping in the Ante-bellum Midwest', in Henry Rosovsky (ed.), *Industrialization in Two Systems: Essays in Honor of Alexander Gerschenkron*, New York: John Wiley & Sons, pp. 3–39.

David, Paul (1985), 'Clio and the Economics of QWERTY', *American Economic Review*, 75, 233–90.

David, Paul (1990), 'The Dynamo and the Computer: An Historical Perspective on the Modern Productivity Paradox', *American Economic Review*, 80, 355–61.

Denison, Edward (1974), *Accounting for Sources of United States Economic Growth 1929–1969*, Washington, DC: Brookings Institution.

Diamond, Arthur M. (1996), 'The Economics of Science', *Knowledge and Policy*, 9, 6–49.

Di Stefano, Giada, Alfonso Gambardella and Gianmario Verona (2012), 'Technology Push and Demand Perspectives in Innovation Studies: Current Findings and Future Directions', *Research Policy*, 41, 1283–95.

Dosi, Giovani (1991), 'The Research on Innovation Diffusion: An Assessment', in Nebojsa Nakicenovic and Arnulf Grubler (eds), *Diffusion of Technologies and Social Behavior*, Berlin: Springer, pp. 179–208.

Ergas, H. (1987), 'The Importance of Technology Policy', in Partha Dasgupta and Paul Stoneman (eds), *Economic Policy and Technological Performance*, New York: Cambridge University Press, pp. 51–96.

Fagerberg, Jan, David Mowery and Richard Nelson (eds) (2005), *The Oxford Handbook of Innovation*, Oxford: Oxford University Press.

Fagerberg, Jan, Ben Martin and Esben Andersen (eds) (2013), *Innovation Studies*, Oxford, UK: Oxford University Press.

Feller, Irwin (1966), 'The Draper Loom in New England Textiles, 1894–1914: A Study of the Diffusion of a Technology', *Journal of Economic History*, 26, 320–42.

Feller, Irwin (2017), 'Assessing the Societal Impacts of Publicly Funded Research', *Journal of Technology Transfer*. https://doi.org/10.1007/s10961-017-9602-z

Feller, Irwin (2018), 'The Interweaving of Diffusion Research and American Science and Technology Policy', *Annals of Science and Technology Policy*, 2 (3), 200–306.

Ferguson, Charles E. (1969), *The Neoclassical Theory of Production & Distribution*, Cambridge, UK: Cambridge University Press.

Fleming, Lee, Hilary Greene, Guan-Cheng Li, Matt Marx and Dennis Yao (2019), 'Government-funded Research Increasingly Fuels Innovation', *Science*, June 21, 2019, 364, 1139–41.

Galbraith, John K. (1958; Fortieth Anniversary Edition 1998), *The Affluent Society*, Boston, MA: Houghton Mifflin Company.

Godin, Benoît (2017), *Models of Innovation*, Cambridge, MA: MIT Press.

Godin, Benoît (2019), *The Invention of Technological Innovation: Innovation Language, Discourse and Ideology in Historical Perspective*, Cheltenham, UK and Northampton, MA, USA: Edward Elgar Publishing.

Godin, Benoît and Dominique Vinck (2017), 'Conclusion: Towards Critical Studies of Innovation', in Benoît Godin and Dominique Vinck (eds), *Critical Studies of Innovation*, Cheltenham, UK and Northampton, MA, USA: Edward Elgar Publishing, pp. 319–22.

Gordon, Robert (2016), *The Rise and Fall of American Growth*, Princeton, NJ: Princeton University Press.

Graham, Margaret (1986), *RCA & the Video Disc*, Cambridge, UK: Cambridge University Press.

Griliches, Zvi (1957), 'Hybrid Corn: An Exploration in the Economics of Technological Change', *Econometrica*, 25, 500–22.

Habakkuk, Hrothgar J. (1962), *American and British Technology in the Nineteenth Century*, Cambridge, UK: Cambridge University Press.

Hall, Bronwyn (2005), 'Innovation and Diffusion', in Jan Fagerberg, David Mowery and Richard Nelson (eds) (2005), *The Oxford Handbook of Innovation*, Oxford University Press, pp. 459–84.

Hicks, John R. (1932), *Theory of Wages*, London, UK: Macmillan & Company.

Jewkes, John, David Sawers and Richard Stillerman (1959), *Sources of Invention*, New York: St. Martins Press.

Jones, Eric L. (1988), *Growth Recurring*, Oxford: Clarendon Press.

Jones, Hywel (1976), *An Introduction to Modern Theories of Economic Growth*, New York: McGraw-Hill.

Jorgenson, Dale, Frank Gollop and Zvi Griliches (1987), *Productivity and U.S. Economic Growth*, Cambridge, MA: Harvard University Press.

Karabarbounis, Loukas and Brent Neiman (2013), 'The Global Decline of the Labor Share', *Quarterly Journal of Economics*, 129, 61–103.

Keller, Alexander (1984), 'Has Science Created Technology?', *Minerva*, 22, 160–82.

Kelly, Patrick and Melvin Kranzberg (eds) (1978), *Technological Innovation: A Critical Review of Current Knowledge*, San Francisco, CA: San Francisco Press.

Kendrick, John (1961), *Productivity Trends in the United States*, Princeton, NJ: Princeton University Press.

Kunkle, Gregory (1995), 'Technology in the Seamless Web: "Success" and "Failure" in the History of the Electron Microscope', *Technology and Culture*, 36, 80–103.

Kuznets, Simon (1966), *Modern Economic Growth*, New Haven, CT: Yale University Press.

Lepore, Jill (2018), *These Truths*, New York: W.W. Norton & Company.

Leyden, Dennis P. and Matthias Menter (2018), 'The Legacy and Promise of Vannevar Bush: Rethinking the Model of Innovation and the Role of Public Policy', *Economics of Innovation and New Technology*, 27 (3), 225–42.

Mak, James and Gary Walton (1973), 'The Persistence of Old Technologies: The Case of Flatboats', *Journal of Economic History*, 33, 444–51.

Mansfield, Edwin (1961), 'Technical Change and the Rate of Imitation', *Econometrica*, 29, 741–66.

Mansfield, Edwin, Anthony Romeo, Mark Schwartz, David Teece, Samuel Wagner and Peter Brach (1982), *Technology Transfer, Productivity, and Economic Policy*, New York: W.W. Norton & Company.

Marshall, Alfred E. (1920), *Principles of Economics*, 8th edition, London: Macmillan & Company.

Martin, Ben (2013), 'Innovation Studies: An Emerging Agenda', in Jan Fagerberg, Ben Martin and Ebsen Andersen (eds), *Innovation Studies*, Oxford, UK: Oxford University Press, pp. 168–86.

Mill, John S. (1866), *Principles of Political Economy*, Volume 1, 5th edition, New York: D. Appleton and Company.

Mishan, Edward J. (1971), *Technology & Growth*, New York: Praeger Publishers.

Mokyr, Joel (1990), *The Lever of Riches*, New York: Oxford University Press.

Mowery, David and Nathan Rosenberg (1979), 'The Influence of Market Demand upon Innovation: A Critical Review of Some Empirical Studies', *Research Policy*, 8, 102–53.

Mowery, David and Nathan Rosenberg (1989), *Technology and the Pursuit of Economic Growth*, Cambridge, UK: Cambridge University Press.

Mowery, David, Richard Nelson, Bhaven Sampat and Arvids Ziedonis (2004), *Ivory Tower and Industrial Innovation*, Stanford, CA: Stanford Business Books.

Narin, Francis, Kimberly S. Hamilton and Dominic Olivastro (1997), 'The Increasing Linkage between US Technology and Public Science', *Research Policy*, 26, 317–30.

Nasbeth, Lars and George Ray (eds) (1974), *The Diffusion of New Industrial Processes*, Cambridge, UK: Cambridge University Press.

National Bureau of Economic Research (1962), *Rate and Direction of Inventive Activity: Economic and Social Factors*, Princeton, NJ: Princeton University Press.

National Science Board (NSB) (2018), *Science & Engineering Indicators 2018*, Arlington, VA: National Science Board.

Nelson, Richard (1962), 'Introduction', in *The Rate and Direction of Inventive Activity: Economic and Social Factors*, National Bureau of Economic Research, Special Studies #13, Princeton, New Jersey: Princeton University Press, pp. 3–16.

Nelson, Richard, Merton Peck and Edward Kalachek (1967), *Technology, Economic Growth and Public Policy*, Washington, DC: Brookings Institution.

Nelson, Richard and Sidney Winter (1977), 'In Search of Useful Theory of Innovation', *Research Policy*, 6, 36–76.

Pavitt, Keitt (2005), 'Innovation Process', in Jan Fagerberg, David Mowery and Richard Nelson (eds), *The Oxford Handbook of Innovation*, Oxford University Press, pp. 86–114.

Pfotenhauer, Sebastian and Joakim Juhl (2017), 'Innovation and the Political State: Beyond the Myth of Technologies and Markets', in Benoît Godin and Dominique Vinck (eds), *Critical Studies of Innovation*, Cheltenham, UK and Northampton, MA, USA: Edward Elgar Publishing, pp. 68–93.

Pigou, Arthur Cecil (1920), *Economics of Welfare*, Macmillan, London.

Piketty, Thomas (2014), *Capital*, Cambridge, MA: Harvard University Press.

Radnor, Michael, Irwin Feller and Everett Rogers (1978), *The Diffusion of Innovations: An Assessment*, Report to the National Science Foundation, Grant #PRA76-80388, Evanston, IL: Northwestern University.

Rogers, Everett (1962), *Diffusion of Innovations*, New York: Free Press [2005, 5th edition].

Romer, Paul (1990), 'Endogenous Technological Change', *Journal of Political Economy*, 98 (5), Part 2, S71–S102.

Romer, Paul (1994), 'The Origins of Endogenous Growth', *Journal of Economic Perspectives*, 8, 3–22.

Rosenberg, Nathan (1969), 'The Direction of Technological Change: Inducement Mechanisms and Focusing Devices', *Economic Development and Cultural Change*, 18, 1–24.

Rosenberg, Nathan (1972), 'Factors Affecting the Diffusion of Technology', *Explorations in Economic History*, 10, 3–31.

Rosenberg, Nathan (1982a), 'The Historiography of Technical Progress', in Nathan Rosenberg (ed.), *Inside the Black Box*, Cambridge, UK: Cambridge University Press, pp. 3–33.

Rosenberg, Nathan (1982b), 'How Exogenous is Science', in Nathan Rosenberg (ed.), *Inside the Black Box*, Cambridge, UK: Cambridge University Press, pp. 141–59.

Rosenberg, Nathan (1976), 'On Technological Expectations', *Economic Journal*, 86, 523–35.

Ruegg, Rosalie and Irwin Feller (2003), *A Toolkit for Evaluating Public R&D Investment*, Report to the National Institute of Standards and Technology, NIST GCR-03-857, Washington, DC.

Ruttan, Vernon (1978), 'Induced Institutional Change', in Hans Binswanger and Vernon Ruttan (eds), *Induced Innovation*, Baltimore, MD: Johns Hopkins University Press, pp. 327–57.

Ruttan, Vernon (2001), *Technology, Growth, and Development*, Oxford, UK: Oxford University Press.

Sahal, Devendra (1981), *Patterns of Technological Innovation*, Reading, MA: Addison-Wesley Publishing Company.

Salter, Wilfred E.G. (1969), *Productivity and Technical Change*, 2nd edition, Cambridge, UK: Cambridge University Press.

Schmookler, Jacob (1966), *Invention and Economic Growth*, Cambridge, MA: Harvard University Press.

Shapira, Philip and Stefan Kuhlmann (eds) (2003), *Learning from Science and Technology Policy Evaluation*, Cheltenham, UK and Northampton, MA, USA: Edward Elgar Publishing.

Soete, Luc (2013), 'Is Innovation Always Good', in Jan Fagerberg, Ben Martin and Ebsen Andersen (eds), *Innovation Studies*, Oxford, UK: Oxford University Press, pp. 134–44.

Solow, Robert (1957), 'Technical Changes and the Aggregate Production Function', *Review of Economics and Statistics*, 39 (3), 312–20.

Sorenson, Olav and Lee Fleming (2004), 'Science and the Diffusion of Knowledge', *Research Policy*, 33, 1615–34.

Stephan, Paula (2012), *How Economics Shapes Science*, Cambridge, MA: Harvard University Press.

Stoneman, Paul (ed.) (1995), *Handbook of the Economics of Innovation and Technological Change*, Oxford, UK: Blackwell.

Tornatzky, Louis and Mitchell Fleischer (1990), *The Process of Technological Innovation*, Lexington, MA: Lexington Books.

Verspagen, Bart (2005), 'Innovation and Economic Growth', in Jan Fagerberg, David Mowery and Richard Nelson (eds), *The Oxford Handbook of Innovation*, Oxford: Oxford University Press, pp. 487–513.

von Hippel, Eric (1976), 'The Dominant Role of Users in the Scientific Instrument Innovation Process', *Research Policy*, 5, 212–39.

von Hippel, Eric (2005), *Democratizing Innovation*, Cambridge, MA. MIT Press.

Walsh, D. (2006), Knowledge and the Wealth of Nations, New York: W.W. Norton & Company.

Williamson, Jeffrey (1991), 'Productivity and American Leadership: A Review Article', *Journal of Economic Literature*, 29, 51–68.

Winner, Langdon (1986), *The Whale and The Reactor*, Chicago, IL: University of Chicago Press.

Wisnioski, Matthew, Eric Hintz and Marie Stettler Kleine (2019), *Does America Need More Innovators?*, Cambridge, MA: MIT Press.

# PART III

# ALTERNATIVE APPROACHES TO INNOVATION

# 5.  Mapping innovation diversity

*Mónica Edwards-Schachter*

## INTRODUCTION

Innovation constitutes a ubiquitous imperative of contemporary society at all levels, from individuals and organizations to cities, regions, and nations, and the entire world. The word pervades the social fabric as a "holy grail" (Mashelkar and Prahalad 2010) facing the chaotic, turbulent, and rapidly changing environment that is considered the "new normal," the VUCA[1] term adopted by practitioners that stands for Volatility, Uncertainty, Complexity and Ambiguity (Bennett and Lemoine 2014). Innovation has become the main promise to solving pressing global challenges and achieving a sustainable future (Depledge et al. 2010; Kuhlmann and Rip 2018) is a core "policy imperative" in strengthening the governance of the digitalization, the fourth industrial revolution and sustainability agendas (Smits and Kuhlman 2004; OECD 2011, 2015; OECD/Eurostat 2018; Schwab 2017). Schot and Steinmueller (2018a) and Mazzucato (2018), among others, highlight the role of Science, Technology, and Innovation (STI) policy as an overarching "game-changer" via addressing socially relevant "missions," aligned with the 2030 Sustainable Development Goals Agenda. Accordingly, the task of properly defining, monitoring, and evaluating innovation is even more crucial than before. As pointed out by Dodgson et al. (2015), innovation "is an essential means by which organizations survive and thrive. As the result, innovation must be managed, but before it can be managed it needs to be understood."

But, what do we currently mean by innovation, and to what extent is the realm of innovation being measured? Today innovation looks as an umbrella term involving a myriad of "forms" and "types," in many cases described as "buzz words" or "quasi-concepts" by practitioners, academics, and policy-makers. Innovation is assumed to be technological but also social, cultural, inclusive, green, open, user-driven, frugal, responsible, and so on. The emergence of X-innovations[2] (Gaglio et al. 2019) reveals an increasing "babelization" (Linton 2009; Edwards-Schachter 2016) that challenges not only our understanding about the meaning and purposes of innovation but the consolidation of the young "science of innovation" as a specialty (Godin 2012b; Fagerberg et al. 2013; Godin and Vinck 2017; Martin 2019).

A report published by NESTA points out the "gap between the practice, the theory, the measurement (and subsequently policies) of innovation" that "can produce a misleading view of national innovation performance" (NESTA 2006, p. 17). Miles and Green (2008) refer to many forms of hidden innovations in the cultural sector. Marsili and Salter (2006) talk about the "dark matter of innovation" and, more recently, Martin (2016) uses the same expression – "dark innovation" – to name innovations that have been ignored or are essentially invisible in terms of conventional indicators. Various authors suggest that neglected innovations come from the informal sector of the economy as well as many forms of social innovation, user-innovation, and public sector innovation, among others (Gershuny 1982; Miles and Green 2008; Charmes et al. 2016; Martin 2016). Fagerberg et al. (2013, p. 11) maintain that "the way we conceptualize, define, operationalize and analyze" innovation "is rooted in the past, leaving us with less ability to deal with other less visible forms of innovation." Martin (2016, p. 434) adds that "the challenge for the next generation of IS researchers is to conceptualize, define, and propose improved methods to measure, analyze and understand dark innovation."

This chapter aims to obtain albeit in, brief form, a broad picture "on the nature of this beast called innovation, a concept much used, and abused, nowadays" (Godin 2019, p. 178) by analyzing the proliferation of innovation forms over the last decades. The principal objective is to explore the evolution of the concept, identifying underpinning mechanisms and theoretical cornerstones that may explain this huge variety. A starting point is an assumption that any classification of innovations depends on the theoretical lenses and perspectives adopted. Thus, understanding the elusive concept of innovation spans different theoretical approaches and constructs aimed to explain innovation dynamics and what is understood by the "nature of innovation" (Horn 2005; Godin 2012a, 2012b, 2015; Martin 2012; Salter and Alexy 2014; Edwards-Schachter 2018).

## THE CHANGING NATURE OF INNOVATION

According to the mainstream of the so-called Science Policy and Innovation Studies (SPIS) "nature of innovation" refers to a comprehensive view of the phenomena, considering dimensions, drivers, characteristics, sources, determinants, and effects of innovation, to innovation as a process and an outcome and relationships with technological and socio-technical change (Dosi 1988; Freeman 1991; García and Calantone 2002; Salter and Alexy 2014). Such aspects are described (more or less explicitly) by taxonomies from a dominant interpretation of innovation as technological innovation (Knight 1967; Linton 2009; García and Calantone 2002; Oke 2007; Gault 2018; Hopp et al. 2018). Several policy reports (e.g., OECD 1991; OECD/Eurostat 2018; Prahalad et al. 2009) include explicit references to the evolving nature of innovation, always with the leading role of technological innovation. Management literature has given much attention to the "spiritual nature of innovation" linked to the culture of organizations, which is "the fundamental driver of value in business" and the generation of commercial and business innovations (Buckler 1997, p. 44). Nevertheless, a holistic view of the nature of innovation seems to greatly exceed reductionist scopes on technological innovation. As Pérez (2013, p. 101) affirms, innovation is "an object of study that is constantly being transformed by the very nature of innovation and by its capacity to go beyond technology to modify organizations, institutions, behavior, and ideas," being innovation "a truly evolutionary process in need of dynamic theories." The document *Fostering Innovation to Address Social Challenges* (OECD 2011) points out the limitations of traditional concepts and systems to understand current innovation landscapes and the need to set clear and agreed definitions and "a new framework to better understand the changing nature of innovation and the multiplicity of economic, social and technical drivers" (p. 14). Today innovation researchers can no longer avoid necessary attention to various neglected issues concerning basic questions: who innovates, how and where, innovation processes within hidden sectors of the economy, impact and type of value created, and the unintended desirable and undesirable consequences of innovation (Edwards-Schachter et al. 2012; Godin and Vinck 2017). This study addresses the burgeoning narratives attending to a discernment of innovation forms and types (Martin 2012; Fagerberg et al. 2013; Godin 2015, 2019; Hutter et al. 2015; Edwards-Schachter 2018). A question is posed on whether it would be possible to develop a univocal definition of innovation, overcoming scattered multi-disciplinary approaches on the nature of technological, social, and cultural innovations (Aoyama and Izushi 2003; Nicholls and Murdock 2012; Edwards-Schachter and Wallace 2017).

## METHODOLOGY

Given the objective to obtain a broad picture of innovation forms, a search was performed using databases (Google Scholar, ISI Web of Knowledge, and SCOPUS) from 1950 to 2019 with specific keywords such as "innovation type," "innovation form" and similar terms (e.g., "innovation taxonomy," "innovation category," "innovation typology," "nature of innovation"). The keywords included in the title, abstract, and/or full text were combined using the Boolean operators "AND" and "OR" in the following form: "innovation taxonomy" AND/OR "innovation type." The snowballing technique was also applied, in particular through reviews conducted on specific innovation forms in gray literature (technical and policy reports; e.g., green innovation, responsible innovation, symbiotic innovation, etc.). The purpose was twofold: to distillate most seminal contributions in "analyst constructed typologies" (Patton 1990), including reviews, highly cited publications and the latest ones that represent future trends. The study considered SPIS research (e.g., top journals[3]) but "social sciences" as a general category, the principal limitation being the fragmentation of SPIS literature and that the search only covered documents in English.

## OVERVIEWING TECHNOLOGICAL INNOVATION TYPOLOGIES AND LATEST MODIFICATIONS IN THE OSLO MANUAL

Classical taxonomies of technological innovation come from the economic and managerial strands of innovation studies, focused on the evolution in the production and application of scientific and technological knowledge and R&D activities by large and SMEs firms (see e.g., reviews by Subramanian and Nilakanta 1996; García and Calantone 2002; Coccia 2006; Oke 2007; Linton 2009; Rowley et al. 2011). The term technical innovation is used in the early 1960s (more than technological innovation) associated with manufacturing and industrial innovations and their contribution to technical change (Myers and Marquis 1969; Freeman 1974). Most cited types, as in the example of Knight (1967, p. 482) were:

> (i) Product or service innovation, concerned with the organization's new product or service offerings, (ii) Production-process innovation, referring to the changes to organizational operations and production driven by technological advancements, (iii) Organizational structure innovation, concerned with the organization's authority relations, communication systems, or formal reward systems, and (iv) People innovation, relating to changes to the people (staff) within an organization, including changes in the organizational culture.

The vast majority of documents refer to the seminal ideas and categorization proposed in the *Theory of Economic Development* by Schumpeter (1934),[4] who is considered "the father" of innovation studies.[5] His contribution is cited by the "Frascati Family" Manuals, principally the successive editions of the Oslo Manual in 1992, 1997, 2005, and 2018 as well as the Manual for the Standardization of Technological Innovation Indicators in Latin America or Bogota

*Table 5.1      Definitions of innovation in the Oslo Manual (2005, 2018)*

| Oslo Manual (OECD/Eurostat 2005, p. 46) | Oslo Manual (OECD/Eurostat 2018, p. 20) |
|---|---|
| "An innovation is the implementation of **a new or significantly improved product (good or service), or process**, a new marketing method, or a new organizational method in business practices, workplace organization or external relations" (p. 46) "A common feature of an innovation is that it must have been implemented," i.e. "introduced on the market" (p. 47) Four types of innovation: product, process, organizational and marketing | An innovation is **a new or improved product or process (or combination thereof)** that differs significantly from the **unit**'s previous products or processes and that has been made available to potential users (product) or brought into use by the unit (process). The generic term "unit" describes the actor responsible for innovations and **refers to any institutional unit[7] in any sector, including households and their individual members.** Two main types of innovation: product innovations and business process innovations. Product innovations are divided into two main types (goods and services), while business process innovations are divided into six broad types (1. production of goods or services, 2. Distribution and logistics, 3. Marketing and sales, 4. Information and communication systems, 5. Administration and management, 6. Product and business process development) |

Manual (RICYT 2000), taking part of conceptual debates on the measurement of science, technology and innovation (STI) since the first half of the twentieth century (Godin 2015). Both the first and second editions of the Oslo Manual use the technological product and process (TPP) definition of innovation, centered on the technological development of new products and new production techniques by firms. The edition of 2005 recognizes the importance of innovation in less R&D-intensive industries, such as services in low-technology manufacturing, and spreads out the definition to include organizational and marketing innovations, considered non-technological innovations. The progressive incorporation of the services sector from 1997 onwards acknowledges the shift from a technology-based economy created by industrial production to a service-oriented society that regards knowledge as a central resource (Hipp and Grupp 2005; Windrum and Koch 2008). New innovation forms since the edition of 2005 (Table 5.1) have been introduced by the latest update in 2018.[6]

Gault (2012) proposed to replace the expression "introduced on the market" to "made available to potential users" in order to include the households sector as well as the business sector, opening up the possibility of firms innovating by making free products available to potential users, in addition to introducing products to the market. This was introduced in the Oslo Manual (2018), broadening the scope from business innovation in firms (private sector) to all business enterprises (public and private), also government sector, non-profits organizations (NPO), and households. Such modifications "fully reflect changing models of innovation, including those relating to open innovation, global value chains and global innovation networks" (p. 30). Additionally, it states that "product innovation must be made available to potential users, but this does not require the innovation to generate sales" (Oslo Manual 2018, p. 71) and explicitly introduces the measurement of both "economic and social outcomes of business innovation" (p. 222). Another example is the introduction of the concept of co-innovation[8] or "coupled open innovation" (pp. 133–4) that in turn contemplates different users' roles, user-centered, user-driven, or user as collaborators (Gault 2012).

Although the prevalent focus lies on commercialization and business innovation, public sector innovations are incorporated on the basis of previous taxonomies proposed by Windrum and Koch (2008, p. 5). This typology comprises six types: (1) Services innovation, (2) Service delivery innovation, (3) Administrative or organizational innovation, (4) Conceptual innovation, (5) Policy innovation and (6) Systemic innovation. Conceptual innovation is "the development of new world views that challenge assumptions that underpin existing service products, processes and organizational forms" (the "minimalist state" or minimalist govern-ment is an example of a radical conceptual innovation). Policy innovation can be incremental innovation based on policy learning by government and radical innovation sparked by concep-tual innovation. Systemic innovation refers to change in interactions between organizations and knowledge bases. The first three categories align with product, process and organizational innovations in the Oslo Manual (2018), the remaining three classifications being considered examples of "restricted innovation" (Gault 2018). The term is proposed as a solution to the "problematic cases" of innovation forms that raise questions about how a broader view of innovation could be accommodated (he mentions consumer innovation, social innovation, inclusive innovation, etc.). In those cases, restricted innovation concerns specific subsets of firms that satisfy the Oslo Manual definition, for example, an organization that innovates by providing affordable access of quality goods and services for the excluded population where it is necessary to measure the "social outcomes" and social impact over the short and long term.

Facing the dilemma of the large variation in innovation forms, Borrás and Edquist (2019) agree with the imperative "to restrict" innovation. They acknowledge that in recent years there has been an "inflation" in the use of the innovation concept in a great number of notions. However, Borrás and Edquist affirm:

> We believe that it is not useful to regard them as a certain class of innovation, just as we do not want to consider scientific progress (like Albert Einstein's theory of relativity), institutional change (like changes in constitutional law), or new cultural achievements (a new film) to be innovations. We prefer to see some of these "open innovations," "social innovations," "employee-driven innovations," and so on, as ordinary product and process innovations that have consequences for solving social problems and for mitigating social change. Hence, social innovations are a matter of objectives and directionality (and consequences) of pursuing innovation processes and innovation policies. (p. 28)

Nonetheless, as we will see in the next section, the notion of innovation is increasingly con-tested and claims for a more comprehensive description of its social, cultural, and political nature (Godin and Vinck 2017). New incumbent innovations are influencing the establishment of next innovation policies, being part of a reflective ongoing "movement" in academic and international policy forums around the concept of innovation (Edwards-Schachter 2018; Mazzucato 2018; Kuhlmann and Rip 2018; Schot and Steinmueller 2018a,b; EC 2017, 2019; Gaglio et al. 2019). The Oslo Manual (2018) mentions a future agenda oriented to a more general definition of innovation and its measurement beyond business.

On the other hand, discrepancies between academics and policy-makers also exist with regard to the many adjectives and traditional typologies developed about the intensity or degree of "innovativeness" as well as the significance of technical and/or technological change (Durand 1992; García and Calantone 2002; Linton 2009). Debates on regulations and what is radical, disruptive, or incremental in emerging convergent technologies and the so-called convergent innovation (Bainbridge and Roco 2006), symbiotic innovation (Thomas and Wind 2013), and co-innovation (Lee et al. 2012), which have given rise to a new type of innovation: Responsible Innovation (Sutcliffe 2011; Owen et al. 2013).

Table 5.2 summarizes the abundant adjectives describing technological innovation types and related typologies. It is worth stressing that adjectives such as radical, incremental, and so on appear in ample literature of non-technological and X-innovations, for example, open

*Table 5.2*    *Terminology associated to innovation types centered in technological innovation attending, e.g., the degree of novelty and/or change introduced*

| Adjectives naming innovation types with focus on technological innovation | Examples of typologies |
| --- | --- |
| Adjacent | radical/revolutionary, incremental (Schumpeter 1934) |
| Administrative | systemic, important/major, minor, incremental, unrecorded (Freeman 1974) |
| Ancillary | niche creation, architectural, regular, revolutionary (Abernathy and Clark |
| Architectural | 1985) |
| Associative | autonomous, administrative, systemic (Teece 1984) |
| Autonomous | ancillary, technological, administrative (Damanpour 1987) |
| Breakthrough | incremental innovation, radical innovation, new technology systems, |
| Catalytic | changes in techno-economic paradigms (Freeman and Pérez 1988) |
| Continuous | incremental, modular, architectural, radical (Henderson and Clark 1990) |
| Convergent | incremental, evolutionary market, evolutionary technical, radical (Moriarty |
| Core | and Kosnik 1990) |
| Cumulative | low, moderate, high innovativeness (Kleinschmidt and Cooper 1991) |
| Discontinuous | incremental, new generation, radically new (Wheelwright and Clark 1992) |
| Disruptive | incremental, architectural, fusion, breakthrough (Tidd 1995) |
| Evolutionary | sustaining, disruptive (Christensen 1997) |
| Fundamental | incremental, market breakthrough, technological breakthrough, radical |
| Fusion | (Chandy and Tellis 2000) |
| Generational | radical, really new and incremental (García and Calantone 2002) |
| High | continuous, discontinuous (Boer and Gertsen 2003) |
| Hyper | product, process, position and paradigm innovation (Bessant 2005) |
| Important | administrative, architectural, technical, fundamental, minor, continuous, |
| Incremental | discontinuous, normal, routine, incremental, enabling, disruptive, |
| Low | sustaining, revolutionary, process, product, generational, evolutionary |
| Major/Fundamental | (Linton 2009) |
| Minor | adjacent, core and transformational (Tuff and Nagji 2012) |
| Moderate | |
| Modular | |
| New generation | |
| Niche | |
| Normal | |
| Paradigm | |
| Position | |
| Process | |
| Product | |
| Radical | |
| Regular Innovation | |
| Revolutionary | |
| Routine | |
| Sustaining | |
| Systemic | |
| Systematic | |
| Transformational | |
| Unrecorded | |

*Note:*    See list of authors in Appendix 1. The list considered reviews of taxonomies by Coccia (2006), Oke (2007), García and Calantone (2002) and Linton (2009, see Table 1, p. 730).

radical innovation, radical and incremental green innovations, radical social innovation, and so on.

Degree of change (technical/technological) has been extensively discussed through the introduction of technological trajectories (Nelson and Winter 1982), new technological systems and the concept of "constellations" of innovations, which includes numerous radical and incremental innovations in both products and processes. The expression "technological paradigm" and "change of techno-economic paradigm (technological revolution)" embrace continuous changes and discontinuities in technological innovations as well the direction of advance within a technological paradigm (Dosi 1988; Pérez 2010; Pérez and Murray 2018). Most papers address evolutionary theories based on Schumpeter to characterize incremental versus radical/breakthrough innovation, distinguishing between "continuous" and "discontinuous" technological evolution in analyzing the product-market competition. Classifications differ in management literature with a focus on product development, competition, and market (García and Calantone 2002; Coccia 2006). The taxonomy of Abernathy and Clark (1985) introduces the notion of transilience to examine the creation/destruction of firms' skills/competences and resources linkages in four innovation types (regular, architectural, niche creation or revolutionary). In *The Innovator's Dilemma* Christensen (1997) describes the notion of disruptive technologies as a source of disruptive innovations, which can be or not radical. Different typologies have been proposed based on the different "forces" from which discontinuities may originate and the analysis of innovation diffusion (García and Calantone 2002; Oke 2007; Linton 2009). Hopp et al. (2018) review 40 years (from 1975 to 2016) of academic literature on innovation types, concluding that the field of research has not only grown exponentially but also became increasingly diverse in terms of both its contributing academic disciplines and the number of distinct topics studied, being "clearly multidisciplinary" (Hopp et al. 2018, p. 462).

## BEYOND THE TIP OF THE ICEBERG: OLD TYPES, RE-EMERGENCE OF INNOVATION FORMS, AND NEW ENTRANTS

A fact is a proliferation of competing for epistemic and academic communities as well as the creation of varied journals[9] representing the demand of a "semantic extension" in the conceptualization of innovation (Gaglio et al. 2019). Table 5.3 shows around one hundred innovation forms in alphabetic order (Appendix 2 lists representative authors). Some innovations are more recent than others or more or less related to technological innovation, being most of them dated in the last two decades. Various authors argue that social innovation predates technological innovation and re-emerged in evolved forms in the last decades (Godin 2012a, 2015; Edwards-Schachter and Wallace 2017; Gaglio et al. 2019). A similar trend is observed in recent literature discussing the roots and place of cultural innovation (O'Brien and Shennan 2010; Wijngaarden et al. 2019).

Several types stem to the classical problem of distinguishing between different attributes[10] of innovation (Downs and Mohr 1976), that is, the sector where the innovation activity and action come from (individuals or organizations from the private, public, and third sectors) or the economic sector and sub-sector of activity (military innovation, agricultural innovation, rural innovation, silver innovation, tourism innovation, media innovation, water innovation,

*Table 5.3*    *Re-emergent and new-entrant forms of innovation (list of references in Appendix 2)*

| Innovation form/type | Innovation form/type |
| --- | --- |
| **Aesthetic innovation** | **Holistic innovation** |
| **Agricultural innovation** | **Humanitarian innovation** |
| Anchoring innovation | **ICT innovation** |
| **Artistic innovation** | **Inclusive innovation** |
| **Base-of-Pyramid/Bottom of the Pyramid innovation** | **Indigenous innovation** |
| **Below-the-radar innovation** | **Institutional innovation** |
| Biomimicry innovation | **Jugaad innovation/Gandhian innovation** |
| **Blockchain innovation** | Lean innovation |
| **Blockchain innovation commons** | **Low-cost innovation** |
| **Blowback innovation** | Luxury innovation |
| Blue innovation | Media innovation |
| **Brand-driven innovation** | **Military innovation** |
| **Business innovation/business model innovation** | **Normative innovation** |
| **Citizen innovation/civic innovation** | **Open collaborative innovation** |
| **Co-innovation** | **Open innovation** |
| **Collaborative innovation** | **Open eco-innovation** |
| **Commercial innovation** | Open inclusive innovation |
| **Common innovation/innovation commons** | **Open social innovation** |
| **Conceptual innovation** | **Paradigm-based/paradigm innovation** |
| **Constraint-based innovation** | **Peace innovation** |
| **Convergent/convergence innovation** | **Peer-to-peer innovation** |
| **Corporate Social Innovation (CSI)** | **Philanthropic innovation** |
| **Cost innovation** | **Policy innovation/political innovation** |
| **Cross-industry innovation** | **Pro-poor innovation** |
| **Cross-sector innovation** | **Public innovation/Public service innovation** |
| **Crowd innovation** | **Public social innovation** |
| Crowdfunding innovation | **Reflexive innovation** |
| **Cultural innovation** | **Regulatory innovation** |
| **Customer innovation/customer-centred** | Religious innovation |
| customer-driven innovation | **Responsible innovation/RRI** |
| **Democratic innovation** | **Reverse innovation** |
| **Design-driven innovation/design innovation** | **Rural innovation** |
| Digital-based innovation or ICT-enabled innovation | **Rural social innovation** |
| Digital Social Innovation (DSI) | **Scarcity-induced innovation** |
| **Distributed innovation** | **Service innovation** |
| **Eco-cultural innovation** | **Shanzhai innovation** |
| **Eco-innovation** | Silver innovation |
| **Economic innovation** | **Social innovation** |
| Educational innovation/education technology innovation | **Soft innovation** |
| **Employee-driven innovation** | **Stylistic innovation** |
| **Environmental innovation** | **Sustainable business model innovation** |
| **Fashion innovation** | **Sustainable innovation** |
| **Financial innovation** | Symbiotic innovation |
| Fin-tech innovation | **Technological Environmental Innovation (TEI)** |
| Food innovation | **Tourism innovation** |
| **Free innovation** | **Transformative innovation** |
| **Frugal innovation** | **Urban innovation** |
| Good-enough innovation | **Urban social innovation** |

| Innovation form/type | Innovation form/type |
| --- | --- |
| Governance innovation | **User innovation/ user-led/ user-centred/user-driven** |
| **Grassroots innovation** | **Value-chain innovation** |
| **Grassroots social innovation** | **Value innovation** |
| **Green grassroot innovation** | **Water innovation** |
| **Green innovation** | **Workplace innovation/labour innovation** |

*Note:* Academic literature highlighted in bold (SPIS, Sociolog). The list is not exhaustive as it does not include some labels due to the scarce literature or not being related to the conceptual analysis of an innovation type (e.g., empathetic innovation, ceramic/ceramic tile innovation, crop innovation, gendered innovation, GIG innovation, nano-innovation, robotic innovation, pharmaceutical innovation, poetic innovation, smart innovation, organismal innovation, etc.).

etc.). Some forms can be considered fashionable or irrelevant in giving solid arguments of their existence while others can be identified as a "branch" of broader categories. The analysis delineates the existence of three big fields[11] related to social, cultural, and technological innovation (and intertwined forms) while other types can be considered "transversal," for example, institutional innovation and the recent introduction of anchoring innovation.[12] Agreeing with previous contributions, many types can be grouped attending to different criteria, for example, the social or environmental dimensions of sustainable development (Edwards-Schachter 2016). Besides, there are ample literature proposing indicators of each innovation type (e.g., the Eco-Innovation Scoreboard,[13] and the OECD Sustainable Manufacturing Toolkit, the European Digital Social Innovation Index, etc.).

Reviews conducted by Carlborg et al. (2014) and Witell et al. (2016) reveal a huge increase in the number of service innovation articles from the 1980s onwards. They argue for the creation of a specific innovation typology for services, comprising Knowledge-Business Intensive Services (KIBS) (very similar to technology-intensive manufacturing in terms of R&D and technological intensity) and other less innovative and "low intensity" services, including services in social innovation and cultural innovation and Digital Social Innovations (DSI) (Aoyama and Izushi 2003; Hipp and Grupp 2005).

New forms of cultural innovation re-appear in the 1980s in evolutionary studies of the new cultural economy (fashion, films, videogames, publishing, advertising, performing arts, etc.) embracing Cultural and Creative Industries (CCI) and cultural movements associated to social or collective creativity closely related to urban transformation and the generation of cultural spaces in cities (Landry 2000; Florida 2002; Miles and Green 2008; O'Brien and Shennan 2010). Cultural innovation embraces aesthetic innovation, artistic innovation, stylistic and soft innovation, mixed with digital and social innovations. Stylistic innovation is defined as changes in the aesthetic and symbolic elements of a variety of industrial products and services (Cappetta et al. 2006), also linked with the fashion innovation, fashion tech industry, and luxury innovation (Behling 1992; Hoffmann and Hoffmann 2012). Another approach to cultural innovation comes from anchoring innovation that aspirates to conciliate humanist and anthropological perspectives with innovation studies of cultural change in organizations (Barnett 1953; Sluiter 2017).

Some X-innovations enlarge pre-existent types, as the case of business model innovation and further sustainable and circular business model innovations and the more general category of open business model innovation (Chesbrough 2007; Bocken et al. 2014). Also, there is a growing number of publications about business model innovation in the social economy, sharing economy, and recent trends in lean innovation (Sehested and Sonnenberg 2011).

Notwithstanding the fuzziness of "dark" innovations, publications in representative top SPIS journals can be considered as an "indicator" of legitimization. Examples are more than one hundred papers on environmental, green, and eco-innovations in *Research Policy* since decades ago[14] to date as well as some recent contributions on frugal innovation (van der Boor et al. 2014), common innovation (Foray 2015) and the introduction of social innovation as a "new research field in innovation studies" (van der Have and Rubalcaba 2016). Some recent forms are associated with the fourth industrial revolution with the expansion of the digitaliza-tion frontiers and, in particular, blockchain ("Internet of value"), which is considered a second web revolution. Blockchain constitutes an ongoing breakthrough innovation linked to other technologies such as Artificial Intelligence, Big Data, Virtual and Augmented Reality, Internet of Things, and so on. Blockchain involves a novel form of the innovation process, which can be applied as a new form of institutional innovation to improve governance and organizational structures (Allen et al. 2020).

## HOW TO EXPLAIN THE PROLIFERATION OF NEW INNOVATION INCUMBENTS?

Overall, new incumbent innovations follow what is central to any innovation: the human tendency to creatively transform reality by the introduction of novelties and problem-solving. But what explains this spectacular rise in the number of "labels" in a few decades? This heterogeneity seems to be "framed" by the confluence of different and even competing logics involving a complex dialogue between successive and fast waves of socio-cultural, economic, and technological change and policies establishing the directions of R&D, innovation and human development. The most distinctive aspects are:

- A mix of factors coming from the impact produced by globalization, the Internet and the expansion of the Knowledge economy; together financial crisis and digitalization, which in turn favored the democratization of innovation practices. Many X-innovations materialize a "new nature of innovation" fuelled by global challenges, value co-creation with users, global knowledge and collaborative networks, and the rise of the public sector demand (Prahalad et al. 2009).
- Normative and policy-framed fields, including the generation of specific programs, research calls, public procurement and other incentives. Many forms of innovation have been introduced by consultants and academics closely working with policy-makers. For instance, the generation of new policy agendas and European research of social innovation (Moulaert and Leontidou 1992; BEPA 2010), responsible innovation (Sutcliffe 2011; von Schomberg 2011), and blue innovation, linked to the new Blue Growth European strategy (Pauli 2010; Ferreira 2015; EC 2019).
- Change in innovation processes, that is, in the purposes, actors, drivers and resources, inputs, activities and outcomes, structural and institutional context, and, in particular, between knowledge producers and users (Mulgan 2006; Hutter et al. 2015; von Hippel 2005, 2017; Edwards-Schachter 2018). Changes from the so-called Schumpeter "Mark I" (innovation performed by individual heroic entrepreneurs) to Schumpeter "Mark II" (large firms) to new models of organizing collaborative innovation and "co-creation" processes by involving diverse actors, including users, communities, business, public and

civil society sectors and, more recently, networks of collaborative teams and diversity of innovation groups/communities. Other novel innovation processes are associated with new aspects and factors. For instance, Godin and Vinck (2017) describe innovation processes through withdrawal (agriculture without synthetic pesticides, withdrawal of plastic bags, food without gluten, etc.) where the withdrawal – and not the introduction of something new – is structuring the innovation process.

- Changes from the organization of individuals and isolated teams to collective innovation communities as the basis of the worldwide lab and hub revolution, more recently linked with the generalization of the start-up, innovation contests, and "hackathon" phenomena, with a variety of entrepreneurship fields (green, social, sustainable, etc.), including intra and extra-preneurship (Audretsch 2009; Edwards-Schachter et al. 2012; Johnson and Robinson 2014).
- The relevance of cultural and institutional contexts in shaping innovation processes (Rammert 2002; Hutter et al. 2015). In China, *Shanzhai* innovation represents large clusters with specific ethos and practices of manufacturing. *Shanzhai* is a term familiar to millions of Chinese that means "mountain fortress" or "mountain village," representing illegal products produced far away from the law, as an act of rebellion against the emperor in the medieval period. Nowadays *Shanzhai* is associated with China's Knockoff Economy, encompassing counterfeit, imitation, or parody products and trademark infringement of brands and companies and the subculture surrounding them (Zhu and Shi 2010).

A range of different restorative and transformative forms of innovation takes the part of "a simultaneous process of appropriation and contestation of narratives," pointing to the need for "a different kind of innovation to generate desirable social impacts – such as inclusion, sustainable development, the democratization of knowledge" (Bagattolli and Brandão 2019, p. 69). In this sense, reactive movements against hegemonic discourses on technological innovation and economic growth from the 1960s onwards[15] are a source of innovations associated with the three traditional pillars of sustainable development (ecological, social, and economic) (Fairweather 1972; Edwards-Schachter et al. 2012).

The most representative is social innovation, which has been present in several incarnations for almost two centuries (Godin 2012a), reappearing in social movements in the 1960s, with a strong revitalization after the Earth Summit in 1992 to date. Howaldt et al. (2017, p. 24) describe social innovation as "a new innovation paradigm" that becomes central to "economic, political and environmental challenges of the 21st century" due to the limitations of technological innovation "to resolving pressing social challenges" (p. 26). The controversial relationship with technologies together are the major or minor "purity" of social innovations; that is, the specific purpose, who initiates the innovation process (activists, community, government agency or corporation) and the specific context have led to the existence of numerous forms of social innovation (rural, urban, grassroots, collective, corporate, public, open, green, digital). Edwards-Schachter and Wallace (2017) identified the coexistence of social innovation as a "transformative" source of social change encompassing large societal transformations and a more "instrumental perspective" focused on the provision of services, usually involving the participation of the civil sector. Nowadays, social innovation is generally characterized by cross-sector collaborations in developing product and services innovation as a means to solve social needs. It can be associated with maker movements, networks of activists and social entrepreneurs and labs and hubs revolution. But it is also used as a label by any type of organization oriented to the production of "social impact" (which explains the appearance of open social innovation and Corporate Social Innovation).

It is worth stressing that the relevance of social innovation is recognized by SPIS mainstream literature as a "subsidiary," a "complement" or an "inductor" of technological innovation (Gershuny 1987). Social innovation counts as the "social dimension" that allows and supports the diffusion of techno-economic innovations. This perspective is present in the innovation systems approach (Lundvall 1992; Edquist 1997) and the socio-technical transitions construct (Geels 2005), where "social" constitutes a co-evolving dimension of technological innovation. Within the new policy paradigm focused on "missions" to produce "transformative change," Schot and Steinmueller (2018b) mention the term regarding the urgency of advancing a socio-technical system transformation, whose major changes "involves social innovation, since the focus is on many social elements and their relations with technological opportunities" (p. 1562).

Narratives centered in social needs also come from many "pro-poor" innovations and inclusive innovations from various decades ago. The term grass-root innovation was coined by Gupta (Gupta et al. 2003, 2019) and embodied in the Honey Bee Network (HBN) organization, created in 1989 for supporting people in Indian villages to develop technological and non-technological innovations based on indigenous and local knowledge. Inequalities and technological and development gaps across countries are addressed by inclusive innovation, being characterized by different purposes, dynamics, organization forms and actors through various theoretical approaches (Cozzens and Sutz 2012; Foster and Heeks 2013).

Inclusive innovation may adopt many forms and can be more or less linked to social or technological innovation (or both). It can be grass-root and bottom-up or take place within the context of a regional or national program to improve inclusiveness and sustainability through local action (top-down). Inclusion and inclusive innovation strongly differ from Base of Pyramid (BoP), Below-the-Radar-Innovation (BRI), frugal and "constraint-based" innovations, which refer overwhelmingly to innovation activities of MNCs in developing countries. These innovation types follow strategic changes towards a new lens of "inclusive capitalism" (Prahalad and Lieberthal 1998), where corporations compete and additionally exhibit their commitment towards sustainability through Corporate Social Responsibility and Corporate Social Innovation practices (Hart 1997; Herrera 2015). It can also be noticed that there is an increase in debates about indigenous innovation and terms such as Gandhian, Jugaad and Shanzhai innovation, with contested and differentiated narratives on "self-reliant," "self-determined" and "independent innovation" in the context of developing countries, notably in China, India and several African countries (Vinig and Bossink 2015; Gupta et al. 2019).

One of the most extended "new paradigms" in the last two decades is open innovation,[16] which facilitates collaboration and the access to external knowledge sources (from customers, providers, etc.), reducing the costs, outsourcing the risks to mitigate the consequences of possible failures and enabling a high speed of innovation, mostly technological (Chesbrough 2003). The original definition experienced some shifts across time, being adopted by other innovation forms (open and green, open and environmental, open and inclusive and cross-sector innovation). One decade later, Chesbrough and Di Minin (2014, p. 170) introduce the definition of Open Social Innovation (OSI) as "the application of either inbound or outbound open innovation strategies, along with innovations in the associated business model of the organization, to social challenges … and to the social sector," considering business' orientation of large firms and SMEs to social demands and "social" market as well as CSR practices, also adopted by public and social enterprises and NPOs.

Nevertheless, it can be said that the major shift from the late 1980s to today was the change of the user–producer relationship towards the "user-innovation" paradigm and the unceasing "democratization of innovation" (von Hippel 2005). In a few decades the role of consumers and users changed from providing feedback and/or testing of products and services to more active participation as leaders or co-creators of innovations in a variety of collaborative settings (von Hippel 2005, 2017; Edwards-Schachter et al. 2012). More recently, von Hippel (2017) talks about a new "free innovation paradigm," defining free innovation as "a functionally novel product, service, or process that (1) was developed by consumers at private cost during their unpaid discretionary time (that is, no one paid them to do it) and (2) is not protected by its developers, and so is potentially acquirable by anyone without payment – for free." Examples are crowd innovation, where thousands of people act as partner co-creators (solving scientific, technological and societal problems), peer-to-peer innovation and common innovation, carried out by the common man and woman for their benefit (Boudreau and Lakhani 2013; Swann 2014). According to Swann (2014, p. 3) common innovation "takes place quite outside the domain of business, the professions or government" and "it could also be called *vernacular*, as it is not intended for commercial use, but for the benefit of the innovators and their community." Regardless of the unfolded use of this new terminology, there are significant differences between user innovation and open innovation, peer-to-peer innovation, private–collective innovation, common and free innovation (see e.g., von Hippel 2017, p. 144). The most important is that "free innovation is defined as one that diffuses for free [...] Within the OECD, in contrast, the definition for an innovation includable in government statistics requires that it be introduced onto the market" (von Hippel 2017, p. 8).

The impressive growth of collaborative innovation[17] between a range of actors, including users, universities, firms and governments calls into question the role of firms as the central actor in the innovation process, also noticed in the modifications introduced in the latest edition of the Oslo Manual (2018). Narrowly connected with collaborative innovation is a movement around "convergent innovation" (nano, bio, cogno and emergent technologies), symbiotic innovation and the fast expansion of blockchain innovation together with associated debates on Responsible Research and Innovation (RRI), which are posing new challenges to innovation studies (Bainbridge and Roco 2006; Depledge et al. 2010; Martin 2016). Many contemporary innovation incumbents bring to the fore the need to debate different implications on aspects related to the impact of this new "technology revolution" concerning the asymmetries and fragility of knowledge sharing, intellectual property and patents, funding mechanisms and governance of the "commons" (Gächter et al. 2010; Filitz et al. 2015; Potts 2018).

Perhaps the most neglected forms of innovation are placed in the cultural sector and the so-called creative and cultural industries, taking into account the amplitude of the term "culture" (Fribourgh Declaration on Cultural Rights 2007). Cultural innovation embraces an impressive variety of technological and non-technological mixed forms related to audiovisual and multimedia, visual and performing arts, design, advertising, art crafts, publishing, fashion, architecture, and so on. In October 2018 fourteen European partners presented their European Manifesto on Supporting Innovation for Cultural and Creative Sectors addressing the role of cultural and creative sectors in the digital age, and coping with the challenges of new technologies such as artificial intelligence, virtual reality and blockchain and the starting era of quantum computing.[18]

What becomes more apparent is the hybridization of innovations due to the involvement of citizens, the public and/or business (cross-sector innovation) in social and cultural "mixed" tech-innovations or tech-cultural-social-innovations, usually associated with discourses on "technologies for social good" and the broad label "innovation with social impact." A recent report about the ongoing digital transformation (OECD 2017, p. 154) points to "mixed modes of innovation" that are reconfiguring markets and blurs boundaries between society, technology and culture (Edwards-Schachter 2018). Hybrid innovation[19] (e.g., blockchain innovation commons) represents a coexistence of many innovation types embedded in the innovation process itself, being different to "mixed" innovations.[20] Hybrid innovation describes increasingly complex, non-linear and dynamic cross-sectoral processes of knowledge creation, diffusion and use by the interplay of social, cultural and techno-economic interactions between producers and users, involving converging systems and networks.

## CONCLUSION: TOWARDS A COMPREHENSIVE VIEW OF INNOVATION?

The most remarkable result of this study is the rich semantic variety of innovation that reminds of the complexity and diversity of human creativity. Findings also shed new light on current constraints to capture many aspects of what we call the nature of innovation. In just over a decade hidden and rejected innovation types are increasingly acquiring recognition by academic and epistemic communities. At the same time, findings confirm the great disciplinary fragmentation of innovation studies.

Some X-innovations evolved from pre-existent types, others can be considered just "fads" or pertain to marginal academic literature or even can be attributed to certain opportunistic behavior of some academics and policy-makers. Some new incumbents have been progressively integrated by mainstream innovation, such as the cases of eco, environmental and open innovation as well as innovations that build upon existing activities, extending, connecting, and diversifying in new areas (e.g., sustainable business model innovation, philanthropic and crowdfunding innovation).

Many X-innovations emerge accompanying the evolving awareness of the sustainability imperative and the increased attention to the production of "impact" (environmental and social). Results suggest that technological innovation is more and more "social-oriented," "inclusive" and "environmentally friendly," and mixed in hybrid forms of innovation.

The existence of consolidated communities and academic publications indicate that the nature of innovation can no longer be limited only to the "technological nature" of innovation but to a wider and enriched vision. In this direction, modifications introduced in the latest edition of the Oslo Manual (2018) point out the need to expand boundaries and develop a more comprehensive view of innovation. Nevertheless, the persistent lack of consensus in the arena of inquiry and the rise of fragmentation and disconnection among scholar communities render the task extremely difficult.

Should we give up the possibility of building a general theory (still non-existent as such) based on a better disciplinary integration? The question is not new. Decades ago, Rowe and Boise (1974, pp. 289–90) said that the most straightforward way of accounting for "theoretical confusion" can be to reject the notion that a unitary theory of innovation exists and accept the existence of distinct types of innovations explained by a number of correspondingly distinct theories. An alternative is searching for theoretical integration or a certain conciliation degree between disciplinary scopes in describing "core" common attributes of innovation. There is an ample agreement in considering innovation as both a process and an outcome, in the words of Dodgson et al. (2015, p. 4) "a fact and an act" involving actors and their purposive motivation and actions, values, and socio-cultural practices. Inventions can be technological but also social, cultural, or mixed forms (Ogburn 1964; Coleman 1970: Whyte 1982; Linnekin 1991).

Although the SPIS field has evolved from the study of technical change to technological change and later to socio-technical change (in particular, the construct of socio-technical transitions), the notion of change should be revisited considering change produced by different forms of innovation in all spheres of life, including cultural, social, organizational, political, and technological change (Godin 2008; Hutter et al. 2015). Another aspect that can contribute to the conceptualization of innovation is the study of socio-cultural practices, and how they become institutionalized attending to different innovation forms. Formal, informal, "hard" and "soft" institutions are concerned with complex constellations of social and cultural practices by social groups and innovation communities who develop innovation, being central elements to study their relationship in the configuration of different agencies and the evolution of inter-twined clusters, networks, and systems in interrelated technological, cultural and social fields.

Some new incumbent innovations (e.g., indigenous innovation, frugal innovation, Shanzhai innovation), as well as the rise of mixed and hybrid innovations, reveal numerous changes operated with regard to knowledge sources from which innovation arises, introducing a necessary reflection on local knowledge and cultural boundaries as well as networks of "knowledge spaghetti by multi-player game" (Owen et al. 2013; Vinig and Bossink 2015). The list of new innovation types is long and research trends indicate it will become longer due to the interest in poorly studied sectors and emergent sectors that "matter": innovation linked to the economy of health, medicine, agriculture, and education and even rejected and controversial innovation linked to the religious economy, and the prosperous erotic economy and pornography industry and other illicit products and services.

In early 2020, during the completion of this chapter, the worst economic downturn (the Great LockDown) since the financial collapse during the Great Depression was announced, and there is now a new post-coronavirus world ahead.[21] After this unprecedented global disruption, a process of "creative destruction" begins to be glimpsed with changes in lifestyles, forms of production and consumption, financing mechanisms and crucial socio-political decisions on future directions of the sustainability agenda at global and national levels. COVID-19 has already sparked many innovations worldwide in a mix of top-down and bottom-up initiatives (makers movements, the proliferation of innovative gadgets, homemade masks, home transformation into co-working space, etc.).[22] Beyond promises and dilemmas posed regarding over-technification and humanization of technology, the generation of innovation incentives as well as the recognition, monitoring, and measurement of innovations that can contribute to the generation of public value as a whole become crucial.

## NOTES

1. The acronym was introduced by the U.S. Army War College to describe the world which resulted from the end of the Cold War. Since January 2020 innovation is claimed to be crucial to the "post-normal" environment introduced by the COVID-19 pandemic.
2. Gaglio et al. (2019) define X-innovation as a semantic pluralization of forms or kinds of innovation.

3.   Crossan and Apaydin (2010) conducted a systematic review of organizational innovation literature (1981–2008) identifying as the ten most relevant journals: *Research Policy, Strategic Management Journal, Journal of Product Innovation Management, Management Science, Academy of Management Journal, Organization Science, Regional Studies, Administrative Science Quarterly, Academy of Management Review* and *Rand Journal of Economics*. Thongpapanl (2012) constructed a similar list, which includes *Technovation*.

4.   Schumpeter (1934, pp. 65–6) refers to the production of change through "new combinations," which covers "(1) The introduction of a new good or a new quality of a good (2) a new method of production (3) The opening of a new market (4) The conquest of a new source of supply or raw materials or half-manufactured goods (5) The carrying out of the new organization of any industry, like the creation of a monopoly position or the breaking up of a monopoly position." Schumpeter's categorization was subsequently picked up by others, in particular Chris Freeman, who was responsible for the first OECD Frascati Manual as well as publishing the influential book *The Economics of Industrial Innovation* (Freeman 1974).

5.   Historical studies from Godin (2008) suggest that the origin of systematic studies on technological innovation owes its existence to R. Maclaurin from the Massachusetts Institute of Technology (MIT), who further developed Schumpeter's ideas and presented a seminal conceptualization of the linear model of innovation and constructed one of the first taxonomies for measuring technological innovation in the literature that led to current indicators on high technology.

6.   For a deep discussion on the recognition of new innovation types see Gault (2012, 2018).

8.   The manual dedicates a section to differentiate between collaboration, cooperation, and co-innovation.

9.   The list is very long. Examples are *Innovation Journal* (1995), *International Journal of Innovation and Sustainable Development* (2005), *International Journal of Innovation Science* (2009), *Environmental Innovation and Societal Transitions* (2011), *Innovation and Development* (2011), *Journal of Responsible Innovation* (2014), *Journal of Open Innovation: Technology, Market, and Complexity, Environmental Technology and Innovation* (2014), *Financial Innovation* (2015), *European Public & Social Innovation Review* (2016), *Journal of Frugal Innovation* (2016), *International Journal of Innovation Studies* (2017), *International Journal of Blockchains and Cryptocurrencies* (2019), *NoVAtion* (2019), etc.

10.   From this point of view "innovativeness degree," for example, constitutes a secondary attribute (e.g., radical and incremental social innovations; radical green innovation, etc.).

11.   Following the definition of Fligstein (2001, p. 108) fields are "local social orders or social arenas where actors gather and frame their actions vis-à-vis one another" that strongly influence the production of change in embedded social practices and institutionalization processes.

12.   Research agenda on anchoring innovation includes the broad scope of the "human factor" in adopting and dealing with change at different domains with a holistic view (in dialogue with Humanities) with regards to identity, group cohesion, and cultural belonging. Anchoring innovation concerns to debates on values, particularly in policy domains but also in the organizational culture, identifying "anchor" as a heuristic tool to investigate different situations and domains of adopted change: in what ways do the relevant social groups accommodate what they consider 'new' by attaching it to what they construct as already 'theirs'. What are the processes, what are the 'anchors'? (Sluiter, 2017, p. 33).

13.   https://ec.europa.eu/environment/ecoap/indicators/index_en;       http://www.oecd.org/innovation/green/toolkit/; https://www.nesta.org.uk/feature/european-digital-social-innovation-index/.

14.   Martin (2016) affirms that innovation for sustainable development emerged as a literature stream in the 1990s in parallel to the increasing awareness of environmental damages and sustainability.

15.   Environmental movements included social movements and "green" political movements in the form of activist nongovernmental organizations and environmentalist political parties around four pillars: protection of the environment, grassroots democracy, social justice, and nonviolence.

16.   By contrast, closed innovation relies on the idea that innovation comes from internal resources, usually confined to developments by R&D departments or labs, where information is kept within the confines of the firm and protected.

17.   The Sustainable Development Goal N° 17 (Revitalize the global partnership for sustainable development) explicitly address collaboration between governments, the private sector and civil society.

18. https://www.buchmesse.de/en/press/press-releases/2018-10-11-european-manifesto-supporting -innovation-cultural-creative-sectors.
19. The term "hybrid innovation" here differs from the notion of hybrid organizations (between academia, government and industry) and the hybrid "modes" or models proposed by von Hippel (2017, p. 38) involving the hybridization of single free innovators with collaborative free innovation projects involving unpaid household sector contributors and producers (a single, non-collaborating firm).
20. In the early 1950s Tupperware was a "mix" of innovations in raw material, technology process and marketing strategy, with the introduction of a business model empowering women that was a source of social innovation. Blockchain innovation commons is a technology that results from community validation to synchronize the content of ledgers replicated by multiple users, different from governance mechanisms of (private) blockchain. Blockchain innovation commons (since the creation of Bitcoin in 2009) is characterized by decentralization, point-to-point transmission, transparency, traceability, non-tampering, and data security, being "simultaneously" a technological, social and institutional innovation.
21. U.S. unemployment claims surged by over three million for the week ending March 21, 2020 to 3.283 million, a record, according the Department of Labor (DOL). On April 25, China pursued its national blockchain platform called BSN (Blockchain Service Network), establishing digitalcoin.
22. The Indian government launched the Aaroga Setu app in early April 2020 which uses GPS and Bluetooth to inform people when they are at risk of exposure to COVID-19. The app was developed by the National Informatics Center before a similar initiative from tech giants Google and Apple got off the ground.

# REFERENCES

Abernathy, William, J., & Clark, Kim B. (1985). Innovation: Mapping the winds of creative destruction. *Research Policy*, 14, 3–22.

Allen, Darcy W., Berg, Chris, Markey-Towler, Brendan, Novak, Mikayla, & Potts, Jason. (2020). Blockchain and the evolution of institutional technologies: Implications for innovation policy. *Research Policy*, 49(1), 103865.

Aoyama, Yuko, & Izushi, Hiro. (2003). Hardware gimmick or cultural innovation? Technological, cultural, and social foundations of the Japanese video game industry. *Research Policy*, 32(3), 423–444.

Audretsch, David B. (2009). The entrepreneurial society. *The Journal of Technology Transfer*, 34, 245–254.

Bagattolli, Carolina, & Brandão, Tiago. (2019). Counterhegemonic narratives of innovation. *NOvation: Critical Studies of Innovation*, (I), 39–39.

Bainbridge, William Sims, & Roco, Mihail C. (2006). *Managing Nano-Bio-Info-Cogno Innovations*. Dordrecht, the Netherlands: Springer.

Barnett, Homer Garner. (1953). *Innovation: The Basis of Cultural Change*. New York: McGraw-Hill.

Behling, Dorothy U. (1992). Three and a half decades of fashion adoption research: What have we learned? *Clothing and Textiles Research Journal*, 10(2), 34–41.

Bennett, Nathan, & Lemoine, G. James. (2014). What a difference a word makes: Understanding threats to performance in a VUCA world. *Business Horizons*, 57(3), 311–317.

BEPA (2010). *Empowering People, Driving Change. Social Innovation in the European Union*. European Commission on Social innovation.

Bocken, Nancy M., Short, Samuel W., Rana, P., & Evans, Steve. (2014). A literature and practice review to develop sustainable business model archetypes. *Journal of Cleaner Production*, 65, 42–56.

Borrás, Susana, & Edquist, Charles. (2019). *Holistic Innovation Policy: Theoretical Foundations, Policy Problems, and Instrument Choices*. Oxford: Oxford University Press.

Boudreau, Kevin J., & Lakhani, Karim R. (2013). Using the crowd as an innovation partner. *Harvard Business Review*, 91(4), 60–9.

Buckler, Sheldon A. (1997). The spiritual nature of innovation. *Research-Technology Management*, 40(2), 43–47.

Cappetta, Rossella, Cillo, Paola, & Ponti, Anna. (2006). Convergent designs in fine fashion: An evolutionary model for stylistic innovation. *Research Policy*, 35(9), 1273–1290.

Carlborg, Per, Kindström, Daniel, & Kowalkowski, Christian. (2014). The evolution of service innovation research: a critical review and synthesis. *The Service Industries Journal*, 34(5), 373–398.

Charmes, Jacques, Gault, Fred, & Wunsch-Vincent, Sacha. (2016), Formulating an agenda for the measurement of innovation in the informal economy, in Kraemer-Mbula, Erika, & Wunsch-Vincent, Sacha. (Eds.), *The Informal Economy in Developing Nations* (pp. 336–367). Cambridge: Cambridge University Press.

Chesbrough, Henry W. (2003). *Open Innovation: The New Imperative for Creating and Profiting From Technology*. Boston, MA: Harvard Business Press.

Chesbrough, Henry W. (2007). Why companies should have open business models. *MIT Sloan Management Review*, 48(2), 22–28.

Chesbrough, Henry W., & Di Minin, Alberto. (2014). Open social innovation, in Chesbrough, Henry, Vanhaverbeke, Wim, & West, Joel. (Eds.), *New Frontiers in Open Innovation* (pp. 169–188). Oxford: Oxford University Press.

Christensen Clayton M. (1997). *The Innovator's Dilemma: When New Technologies Cause Great Firms to Fail*. Boston, MA: Harvard Business School Press.

Coccia, Mario. (2006). Classifications of innovations survey and future directions. Ceris Cnr, W.P N°2, 7–19. https://core.ac.uk/download/pdf/7068477.pdf [Accessed October 12, 2019].

Coleman, James S. (1970). Social inventions. *Social Forces*, 49(2), 163–173.

Cozzens, Susan, & Sutz, Judith. (2012). *Innovation in Informal Settings: A Research Agenda*. Ottawa: IDRC.

Crossan, Mary M., & Apaydin, Marina. (2010). A multi-dimensional framework of organizational innovation: A systematic review of the literature. *Journal of Management Studies*, 47, 1154–1191.

Depledge, Michael, Bartonova, Alena, & Cherp, Aleh. (2010). *Responsible and transformative innovation for sustainable societies. Fundamental and applied research*. Report of the Environment Advisory Group, December 2010. Brussels: European Commission.

Dodgson, Mark, Gann, David M., & Phillips, Nelson. (Eds.) (2015). *The Oxford Handbook of Innovation Management*. Oxford: Oxford University Press.

Dosi, Giovanni. (1988). The nature of the innovative process, in Dosi, Giovanni, Freeman, Christopher, Nelson, Richard, Silverberg, Gerald, & Soete, Luc. (Eds.), *Technical Change and Economic Theory*. (pp. 221–238). London: Pinter Publishers.

Downs Jr, George W., & Mohr, Lawrence B. (1976). Conceptual issues in the study of innovation. *Administrative Science Quarterly*, 700–714.

Durand Thomas. (1992). Dual technology trees: assessing the intensity and strategic significance of technology change. *Research Policy*, 21, 361–380.

EC (European Commission) (2017). *New Horizons: Future Scenarios for Research & Innovation Policies in Europe. A Report from Project BOHEMIA – Beyond the Horizon: Foresight in Support of the Preparation of the European Union's Future Policies in Research and Innovation*. Brussels: European Commission, Directorate-General for Research and Innovation.

EC (European Commission) (2019). *The EU Blue Economy Report*. Luxembourg: Publications Office of the European Union.

EC, IMF, OECD & UN (2009). The World Bank System of National Accounts, 2008, United Nations, New York, USA.

Edquist, Charles. (Ed.) (1997). *Systems of Innovation: Technologies, Institutions, and Organizations*. New York: Psychology Press.

Edwards-Schachter, Mónica E., Matti, Cristian E., & Alcántara, Enrique. (2012). Fostering quality of life through social innovation: A living lab methodology study case. *Review of Policy Research*, 29(6), 672–692.

Edwards-Schachter, Mónica E. (2016). Challenges to firms' collaborative innovation facing the innovation Babel Tower, in Al-Hakim, Latif, & Jin, Xiaobo. (Eds.), *Handbook of Research on Driving Competitive Advantage through Sustainable, Lean, and Disruptive Innovation* (pp. 204–227). Pennsylvania, PA: IGI Global.

Edwards-Schachter, Mónica E., & Wallace, Matthew L. (2017). 'Shaken, but not stirred': Sixty years of defining social innovation. *Technological Forecasting and Social Change*, 119, 64–79.

Edwards-Schachter, Mónica E. (2018). The nature and variety of innovation. *International Journal of Innovation Studies*, 2(2), 65–79.

Fagerberg, Jan, Martin, Ben R., & Andersen, Esben S. (Eds.), (2013). *Innovation Studies: Evolution and Future Challenges*. Oxford: Oxford University Press.

Fairweather, George W. (1972). *Social Change: The Challenge to Survival*. Morristown, NJ: General Learning Press.

Ferreira, João. (2015). *Report on untapping the potential of research and innovation in the blue economy to create jobs and growth* (2014/2240(INI)). Committee on Industry, Research and Energy.

Filitz, Rainer, Henkel, Joachim, & Tether, Bruce S. (2015). Protecting aesthetic innovations? An exploration of the use of registered community designs. *Research Policy*, 44(6), 1192–1206.

Fligstein, Neil. (2001). Social skill and the theory of fields. *Sociological Theory*, 19, 105–125.

Florida, Richard L. (2002). *The Rise of the Creative Class: And How It's Transforming Work, Leisure, Community and Everyday Life*. New York: Basic Books.

Foray, David. (2015). Common innovation (how we create the wealth of nations) in the light of reflections on mass flourishing (how grassroots innovation created jobs, challenge, and change). *Research Policy*, 44(7), 1403–1405.

Foster, Christopher, & Heeks, Richard. (2013). Conceptualising inclusive innovation: Modifying systems of innovation frameworks to understand diffusion of new technology to low-income consumers. *European Journal of Development Research*, 25(3), 333–355.

Freeman, Chris. (1974). *The Economics of Industrial Innovation*. Harmondsworth: Penguin Books.

Freeman, Chris. (1991). *The Nature of Innovation and the Evolution of the Productive System* (pp. 303–314). Paris: OECD.

Gächter, Simon, von Krogh, Georg, & Haefliger, Stefan. (2010). Initiating private–collective innovation: The fragility of knowledge sharing. *Research Policy*, 39(7), 893–906.

Gaglio, Gérald, Godin, Benoît, & Pfotenhauer, Sebastian. (2019). X-innovation: Re-inventing innovation again and again. *Novation: Critical Studies of Innovation*, (1), 1–16.

García, Rosanna, & Calantone, Roger. (2002). A critical look at technological innovation typology and innovativeness terminology: A literature review. *Journal of Product Innovation Management*, 19(2), 110–132.

Gault, Fred. (2012). User innovation and the market. *Science Public Policy*, 39, 118–128.

Gault, Fred. (2018). Defining and measuring innovation in all sectors of the economy. *Research Policy*, 47(3), 617–622.

Geels, Frank. W. (2005). *Technological Transitions and System Innovations: A Co-Evolutionary and Socio-Technical Analysis*. Cheltenham, UK and Northampton, MA, USA: Edward Elgar Publishing.

Gershuny, Jonathan I. (1982). Social innovation: Change in the mode of provision of services. *Futures*, 14(6), 496–516.

Gershuny, Jonathan. (1987). Technology, social innovation, and the informal economy. *Annals of the American Academy of Political and Social Science*, 493, 47–63.

Godin, Benoît. (2008). In the shadow of Schumpeter: W. Rupert Maclaurin and the study of technological innovation. *Minerva*, 46(3), 343–360.

Godin, Benoît. (2012a). *Social Innovation: Utopias of Innovation from circa-1830 to the Present*. Working Paper.

Godin, Benoît. (2012b). "Innovation studies": The invention of a specialty. *Minerva*, 50(4), 397–421.

Godin, Benoît. (2015). *Innovation Contested – The Idea of Innovation Over the Centuries*. London: Routledge.

Godin, Benoît, & Vinck, Dominique. (2017). Introduction: Innovation – from the forbidden to a cliché, in Godin, Benoît, & Vinck, Dominique (Eds.), *Critical Studies of Innovation: Alternative Approaches to the Pro-Innovation Bias*. Cheltenham, UK and Northampton, MA, USA: Edward Elgar Publishing.

Godin, Benoît. (2019). *The Invention of Technological Innovation. Languages, Discourses, and Ideology in Historical Perspectives*. Cheltenham, UK and Northampton, MA, USA: Edward Elgar Publishing.

Gupta, Anil, Sinha, R., Koradia, D., Patel, Ramesh, Parmar, M., Rohit, P., ... & Chandan, A. (2003). Mobilizing grassroots' technological innovations and traditional knowledge, values and institutions: articulating social and ethical capital. *Futures*, 35(9), 975–987.

Gupta, Anil K., Shinde, Chintan, Dey, Anamika, Patel, Ramesh, Patel, Chetan, Kumar, Vipin, & Patel, Mahesh. (2019). Honey bee network in Africa co-creating a grassroots innovation ecosystem in

Africa, ZEF Working Paper Series, No. 178, University of Bonn, Center for Development Research (ZEF), Bonn. https://www.econstor.eu/bitstream/10419/206968/1/1049302001.pdf.

Hart, Stuart L. (1997). Beyond greening: Strategies for a sustainable world. *Harvard Business Review*, 75(1), 66–77.

Herrera, María Elena B. (2015). Creating competitive advantage by institutionalizing corporate social innovation. *Journal of Business Research*, 68(7), 1468–1474.

Hipp, Christiane, & Grupp, Hariolf. (2005). Innovation in the service sector: The demand for service-specific innovation measurement concepts and typologies. *Research Policy*, 34(4), 517–535.

Hoffmann, Jonas, & Hoffmann, Betina (2012). The pier framework of luxury innovation, in Hoffmann, Jonas, & Coste-Manière, Ivan. (Eds.), *Luxury Strategy in Action* (pp. 57–73). London: Palgrave Macmillan.

Hopp, Christian, Antons, David, Kaminski, Jermain, & Salge, Torsten O. (2018). The topic landscape of disruption research – a call for consolidation, reconciliation, and generalization. *Journal of Product Innovation Management*, 35(3), 458–487.

Horn, Paul M. (2005). The changing nature of innovation. *Research-Technology Management*, 48(6), 28–31.

Howaldt, Jürgen, Schröder, Antonius, Butzin, Anna, & Rehfeld, Dieter. (2017). Towards a general theory and typology of social innovation. SI-DRIVE Deliverable, 1. http://www.si-drive.eu/wp-content/uploads/2018/01/SI-DRIVE-Deliverable-D1_6-Theory-Report-2017-final-20180131.pdf.

Hutter, Michael, Knoblauch, Hubert, Rammert, Werner, & Windeler, Arnold. (2015). Innovation society today: The reflexive creation of novelty. *Historical Social Research*, 40(3), 30–47.

Johnson, Peter, & Robinson, Pamela. (2014). Civic hackathons: Innovation, procurement, or civic engagement? *Review of Policy Research*, 31(4), 349–357.

Knight, Kenneth E. (1967). A descriptive model of the intra-firm innovation process. *Journal of Business*, 40, 478–496.

Kuhlmann, Stefan, & Rip, Arie. (2018). Next-generation innovation policy and grand challenges. *Science and Public Policy*, 45(4), 448–454.

Landry, Charles. (2000). *The Creative City: A Toolkit for Urban Innovators*. London, Earthscan Publications.

Lee, Sang M., Olson, David L., & Trimi, Silvana. (2012). Co-innovation: Convergenomics, collaboration, and co-creation for organizational values. *Management Decision*, 50(5), 817–831.

Linnekin, Jocelyn. (1991). Cultural invention and the dilemma of authenticity. *American Anthropologist*, 93(2), 446–449.

Linton, Jonathan D. (2009). De-babelizing the language of innovation. *Technovation*, 29(11), 729–737.

Lundvall, Bengt-Åke. (1992). *National Systems of Innovation*. London: Pinter.

Marsili, Orietta, & Salter, Ammon. (2006). The dark matter of innovation: Design and innovative performance in Dutch manufacturing. *Technology Analysis and Strategic Management*, 18, 515–534.

Martin, Ben R. (2012). The evolution of science policy and innovation studies. *Research Policy*, 41(7), 1219–1239.

Martin, Ben R. (2016). Twenty challenges for innovation studies. *Science and Public Policy*, 43(3), 432–450.

Martin, Ben R. (2019). The future of science policy and innovation studies: Some challenges and the factors underlying them, in Simon, Dagmar, Kuhlmann, Stefan, Stamm, Julia, & Canzler, Weert (Eds.), *Handbook on Science and Public Policy* (pp. 423–542). Cheltenham, UK and Northampton, MA, USA: Edward Elgar Publishing.

Mashelkar, Raghunath A., & Prahalad, Coimbatore K. (2010). Innovation's holy grail. *Harvard Business Review*, 116–126.

Mazzucato, Mariana. (2018). Mission-oriented innovation policies: Challenges and opportunities. *Industrial and Corporate Change*, 27(5), 803–815.

Miles, Ian, & Green, Lawrence. (2008). *Hidden Innovation in the Creative Industries*. London: NESTA.

Moulaert, Frank, & Leontidou, Lila. (1992). *Local Development Strategies in Economically Disintegrated Areas: A Pro-Active Strategy against Poverty in the European Community*. Lille-Athens, intermediate report for the European Commission, DGV.

Mulgan, Geoff. (2006). The process of social innovation. *Innovations: Technology, Governance, Globalization*, 1(2), 145–162.

Myers, Summer, & Marquis, Donald G. (1969). *Successful Industrial Innovation*. Washington, DC: National Science Foundation.

Nelson, Richard, & Winter, Sydney. (1982). *An Evolutionary Theory of Economic Change*. Cambridge, MA: Belknap Press.

NESTA (2006). *The Innovation Gap: Why Policy Needs to Reflect the Reality of Innovation in the UK*. London: NESTA.

Nicholls, Alex, & Murdock, Alex. (2012). The nature of social innovation, in Nicholls Alex, Murdock, Alez. (Eds.), *Social Innovation* (pp. 1–30). London: Palgrave Macmillan.

O'Brien, Michael J., & Shennan, Stephen J. (Eds.) (2010). *Innovation in Cultural Systems: Contributions from Evolutionary Anthropology*. Cambridge, MA: MIT Press.

OECD (1991). *The Nature of Innovation and the Evolution of the Productive System. Technology and Productivity, the Challenge for Economic Policy*. Paris: OECD, pp. 303–314.

OECD (2011). *Fostering Innovation to Address Social Challenges*. https://www.oecd.org/sti/inno/47861327.pdf.

OECD (2015). *The Innovation Imperative: Contributing to Productivity, Growth and Well-Being*. Paris: OECD.

OECD (2016). *Science, Technology and Innovation Outlook 2016. Megatrends affecting science, technology and innovation*. OECD: Paris.

OECD (2017). *Science, Technology and Industry Scoreboard 2017. The digital transformation*. OECD Publishing. Paris, France.

OECD/Eurostat (2005). *Oslo Manual, Guidelines for Collecting and Interpreting Innovation Data*, 3rd edition. Paris: OECD.

OECD/Eurostat (2018). *Oslo Manual 2018: Guidelines for Collecting, Reporting and Using Data on Innovation*, 4th edition. Paris: OECD.

Ogburn, William. (1964). *On culture and Social Change: Selected Papers*. Chicago: University of Chicago Press.

Oke, Adegoke. (2007). Innovation types and innovation management practices in service companies. *International Journal of Operations & Production Management*, 27(6), 564–587.

Owen, Richard, Bessant, John R., & Heintz, Maggy. (Eds.) (2013). *Responsible Innovation: Managing the Responsible Emergence of Science and Innovation in Society*. New York: John Wiley & Sons.

Patton, Michael. (1990). *Qualitative Evaluation and Research Methods*. Thousand Oaks, CA: Sage.

Pauli, Gunter A. (2010). *The Blue Economy: 10 Years, 100 Innovations, 100 Million Jobs*. Boulder, CO: Paradigm Publishing.

Pérez, Carlota. (2010). Technological revolutions and techno-economic paradigms. *Cambridge Journal of Economics*, 34(1), 185–202.

Pérez, Carlota. (2013). Innovation systems and policy for development in a changing world, in Fagerberg, Jan, Martin, Ben. R., & Andersen, Esben S. (Eds.), *Innovation Studies. Evolution and Future Challenges* (pp. 90–110). Oxford: Oxford University Press.

Pérez, Carlota, & Murray, Tamsin M. (2018). Smart green 'European way of life': The path for growth, jobs and wellbeing. BTTR WP 2018-01. http://beyondthetechrevolution.com/wp-content/uploads/2014/10/BTTR_WP_2018-1.pdf.

Potts, Jason. (2018). Governing the innovation commons. *Journal of Institutional Economics*, 14(6), 1025–1047.

Prahalad, Coimbatore K., & Lieberthal, Kenneth (1998). The end of corporate imperialism. *Harvard Business Review*, July–August 1998.

Prahalad, Coimbatore K., McCracken, Paul, & McCracken, Ruth. (2009). *The new nature of innovation*. Report for OECD. Copenhagen: FORA.

Rammert, Werner. (2002). The cultural shaping of technologies and the politics of technodiversity, in Sorensen, Knut H., & Williams, Robin. (Eds.), *Shaping Technology, Guiding Policy: Concepts, Spaces, and Tools* (pp. 173–194). Cheltenham, UK and Northampton, MA, USA: Edward Elgar Publishing.

RICYT (2000). Bogota Manual. Standardisation of Indicators of Technological Innovation in Latin American and Caribbean Countries. Iberoamerican Network of Science and Technology Indicators (RICYT), Organisation of American States (OAS)/CYTED PROGRAM COLCIENCIAS/OCYT. http://www.ricyt.org/wp-content/uploads/2019/09/bogota_manual.pdf.

Rowe, Lloyd A., & Boise, William B. (1974). Organizational innovation: Current research and evolving concepts. *Public Administration Review*, 34(3), 284–293.

Rowley, Jennifer, Baregheh, Anahita, & Sambrook, Sally. (2011). Towards an innovation-type mapping tool. *Management Decision*, 49(1), 73–86.

Salter, Ammon, & Alexy, Oliver. (2014). The nature of innovation, in Dodgson, Mark, Gann, David M., & Phillips, Nelson. (Eds.), *The Oxford Handbook of Innovation Management* (pp. 26–52). Oxford: Oxford University Press.

Schot, Johan, & Steinmueller, W. Edward (2018a). Three frames for innovation policy: R&D, systems of innovation and transformative change. *Research Policy*, 47(9), 1554–1567.

Schot, Johan, & Steinmueller, W. Edward. (2018b). New directions for innovation studies: Missions and transformations. *Research Policy*, 47(9), 1583–1584.

Schumpeter, Joseph. (1934). *The Theory of Economic Development: An Inquiry into Profits, Capital, Credit, Interest and the Business Cycle*. Cambridge, MA: Harvard University Press.

Schwab, Klaus. (2017). *The fourth industrial revolution*. World Economic Forum.

Sehested Claus, & Sonnenberg Henrik. (2011). *Lean Innovation: A Fast Path from Knowledge to Value*. New-York: Springer.

Sluiter, Ineke. (2017). Anchoring innovation: A classical research agenda. *European Review*, 25(1), 20–38.

Smits, Ruud, & Kuhlman, Stephan. (2004). The rise of systemic instruments in innovation policy. *International Journal of Foresight and Innovation Policy*, 1(2–3), 4–32.

Subramanian, Ashok, & Nilakanta, Sree. (1996). Organizational innovativeness: Exploring the relationship between organizational determinants of innovation, types of innovations, and measures of organizational performance. *Omega*, 24(6), 631–647.

Sutcliffe, Hilary. (2011). *A report on responsible research and innovation*. UK: MATTER & the European Commission.

Swann, G.M. Peter. (2014). *Common Innovation: How We Create the Wealth of Nations*. Cheltenham, UK and Northampton, MA, USA: Edward Elgar Publishing.

Thomas Robert J., & Wind Yoram. (2013). Symbiotic innovation: Getting the most out of collaboration, in Brem Alexander, & Viardot Éric. (Eds.), *Evolution of Innovation Management* (pp. 1–31). London: Palgrave Macmillan.

Thongpapanl, Narongsak. T. (2012). The changing landscape of technology and innovation management: An updated ranking of journals in the field. *Technovation*, 32, 257–271.

van der Boor, Paul, Oliveira, Pedro, & Veloso, Francisco M. (2014). Users as innovators in developing countries: The global sources of innovation and diffusion in mobile banking services. *Research Policy*, 43(9), 1594–1607.

van der Have, Robert P., & Rubalcaba, Luis. (2016). Social innovation research: An emerging area of innovation studies? *Research Policy*, 45(9), 1923–1935.

Vinig, Tsvi, & Bossink, Bart. (2015). China's indigenous innovation approach: The emergence of Chinese innovation theory? *Technology Analysis and Strategic Management*, 27(6), 621–627.

von Hippel, Eric. (2005). *Democratizing Innovation*. Cambridge, MA: MIT Press.

von Hippel, Eric. (2017). *Free Innovation*. Cambridge, MA and London: MIT Press.

von Schomberg, Rene. (2011). *Towards Responsible Research and Innovation in the Information and Communication Technologies and Security Technologies Fields*. Brussels: European Commission.

Whyte, William F. (1982). Social inventions for solving human problems. *American Sociological Review*, 1–13.

Wijngaarden, Yosha, Hitters, Erik, & Bhansing, Pawan V. (2019). 'Innovation is a dirty word': Contesting innovation in the creative industries. *International Journal of Cultural Policy*, 25(3), 392–405.

Windrum, Paul, & Koch, Per. (Eds.) (2008). *Innovation in Public Sector Services: Entrepreneurship, Creativity and Management*. Cheltenham, UK and Northampton, MA, USA: Edward Elgar Publishing.

Witell, Lars, Snyder, Hannah, Gustafsson, Anders, Fombelle, Paul, & Kristensson, Per. (2016). Defining service innovation: A review and synthesis. *Journal of Business Research*, 69(8), 2863–2872.

Zhu, Sheng & Shi, Yongjiang (2010). Manufacturing – an alternative innovation phenomenon in China: its value chain and implications for Chinese science and technology policies. *Journal of Science and Technology Policy in China*, 1, 29–49.

# APPENDIX 1

Abernathy, William. J., & Clark, Kim B. (1985). Innovation: Mapping the winds of creative destruction. *Research Policy*, 14, 3–22.

Bessant, John. (2005). Enabling continuous and discontinuous innovation: Learning from the private sector. *Public Money and Management*, 25(1), 35–42.

Boer, Harry, & Gertsen, Frank. (2003). From continuous improvement to continuous innovation: A (retro)(per)spective. *International Journal of Technology Management*, 26(8), 805–827.

Chandy, Rajesh K., & Tellis, Gerard J. (2000). The incumbents curse: Incumbency, size, and radical product innovation. *Journal of Marketing*, 64, 1–17.

Christensen, Clayton M. (1997). *The Innovator's Dilemma: When New Technologies Cause Great Firms to Fail*. Boston, MA: Harvard Business School Press.

Damanpour, Fariborz. (1987). The adoption of technological, administrative, and ancillary innovations: Impact of organizational factors. *Journal of Management*, 13, 675–688.

Freeman, Chris. (1974). *The Economics of Industrial Innovation*. Harmondsworth: Penguin Books.

Freeman, Chris, & Pérez, Carlota. (1988). Structural crises of adjustment, business cycles and investment behavior, in Dosi, G., Freeman, C., Nelson, R., Silverberg, G., & Soete, Luc. (1988). *Technical Change and Economic Theory* (pp.39–66). Pisa, Italy: Laboratory of Economics and Management (LEM), Sant Anna School of Advanced Studies.

García, Rosanna, & Calantone, Roger. (2002). A critical look at technological innovation typology and innovativeness terminology: A literature review. *Journal of Product Innovation Management*, 19(2), 110–132.

Henderson, Rebecca M., & Clark, Kim B. (1990). Architectural innovation: The reconfiguration of existing product technologies, and the failure of established firm. *Administrative Science Quarterly*, 35, 9–30.

Kleinschmidt, Elko J., & Cooper Robert G. (1991). The impact of product innovativeness on performance. *Journal of Product Innovation Management*, 8, 240–251.

Linton, Jonathan D. (2009). De-babelizing the language of innovation. *Technovation*, 29(11), 729–737.

Moriarty, Rowland T., & Kosnik, Thomas J. (1990). High-tech concept, continuity and change. *IEEE Engineering Management Review*, 3, 25–35.

Schumpeter, Joseph. (1934). *The Theory of Economic Development: An Inquiry into Profits, Capital, Credit, Interest and the Business Cycle*. Cambridge, MA: Harvard University Press.

Teece, David J. (1984). Economic analysis and strategic management. *California Management Review*, 26(3), 87–110.

Tidd, John. (1995). Development of novel products through intraorganizational, and interorganizational networks: The case of home automation. *Journal of Product Innovation Management*, 12, 307–322.

Tuff, Geoff, & Nagji, Bansi. (2012). Managing your innovation portfolio. *Harvard Business Review*, pp. 1–9.

Wheelwright Steven C., & Clark Kim B. (1992). *Revolutionising Product Development*. New York: Free Press.

## APPENDIX 2

| Innovation form/type | Selected representative authors |
| --- | --- |
| Aesthetic innovation | (Silver, 1981; Lievrou & Pope, 1994; Eisenman, 2013; Filitz et al., 2015) |
| Agricultural innovation | (Feder et al., 1982; Ghadim & Pannell, 1999; Sunding & Zilberman, 2001; Klerkx et al., 2010; Wright, 2012) |
| Anchoring innovation | (Gupta, 2011; Pellegrini, 2014; Sluiter, 2017) |
| Artistic innovation | (Castaner & Campos, 2002; Fourmentraux, 2007; Galenson, 2008; Hutter, 2015; Montanari et al., 2016) |
| Base-of-Pyramid innovation/ Bottom of the Pyramid innovation | (Hart & Christensen, 2002; Prahalad, 2005; Anderson & Billou, 2007; Anderson & Markides, 2007; Simanis et al., 2008; Hall et al., 2014) |
| Below-the-radar innovation | (Anderson & Markides, 2009; Kaplinsky et al., 2009; Wamae, 2009; Kaplinsky, 2011; Pansera & Martínez, 2017) |
| Biomimicry innovation | (Benyus, 1997) |
| Blockchain innovation | (Liebenau & Elaluf-Calderwood, 2016; Beck & Müller-Bloch, 2017; Scott et al., 2017; Klotz, 2018; Xu et al., 2019; Allen et al., 2020) |
| Blockchain innovation commons | (Allen, 2017; Kewell et al., 2017; Rozas et al., 2018) |
| Blowback innovation | (Brown & Hagel, 2005; Von Zedtwitz et al., 2015) |
| Blue innovation | (Pauli, 2009, 2010; Varga et al., 2013; Ferreira, 2015; Mohanty, 2018) |
| Brand-driven innovation | (Abbing & van Gessel, 2008; Grant, 2011; Nguyen et al., 2015) |
| Business innovation/business model innovation | (Mitchell & Coles, 2003; Chesbrough, 2007, 2010; Osterwalder & Pigneur, 2010; Amit & Zott, 2012; Schneider & Spieth, 2013; Girotra & Netessine, 2014; Clauss, 2017; Foss & Saebi, 2017) |
| Citizen innovation/civic innovation | (Pascu & van Lieshout, 2009; Eskelinen et al., 2015; Biekart et al., 2016; Zieger, 2017; Smith & Stirling, 2018) |
| Co-innovation | (Romero & Molina, 2011; Lee et al., 2012; Castaldi et al., 2013; Bugshan, 2015; Wang et al., 2016) |
| Collaborative innovation | (Donofrio et al., 2008; Gloor, 2006; Swink, 2006; Bommert, 2010; Baldwin & von Hippel, 2010; Sørensen & Torfing, 2011; Hartley et al., 2013; Arora et al., 2016; Croby et al., 2017) |
| Commercial innovation | (Myers & Marquis, 1969; Peters, 1969; Roberts & Peters, 1981; Kline & Rosenberg, 2010; Croby et al., 2017) |
| Common innovation/ innovation commons | (Swann, 2014; Allen & Potts, 2015; Foray, 2015; Potts, 2015, 2018) |
| Conceptual innovation | (Collier & Levitsky, 1997; Lidskog & Waterton, 2016) |
| Constraint-based innovation | (Sharma & Gopalkrishnan, 2012; Pansera & Owen, 2015; Agarwal et al., 2016; Molina-Maturano et al., 2019) |
| Convergent/convergence innovation | (Dowling et al., 1998; Bainbridge & Roco, 2006; Hackin, 2007; Hacklin et al., 2009; Dubé et al., 2014; Lee, 2015; Kim, 2016) |
| Corporate Social Innovation | (Kanter, 1998; Herrera, 2015; Mirvis et al., 2016) |
| Cost innovation | (Zeng & Williamson, 2007; Williamson & Zeng, 2009; Williamson, 2010; Wan et al., 2019) |
| Cross-industry innovation (CII) | (Enkel & Heil, 2014; Levén et al., 2014; Cortimiglia et al., 2016; Hauge et al., 2017) |
| Cross-sector innovation | (Murphy et al., 2012; Gallouj et al., 2013; Jensen et al., 2013; Park, 2013; Evald et al., 2014; Butzin & Widmaier, 2016) |
| Crowd innovation | (Collm & Schedler, 2012; Boudreau & Lakhani, 2013; Mack & Landau, 2015; Chen & Zhang, 2017) |
| Crowdfunding innovation | (Nasrabadi, 2016; Eldridge et al., 2019; Wu et al., 2020) |
| Cultural innovation | (Aoyama & Izushi, 2003; Wilson & Stokes, 2005; Bettencourt & Ulwick, 2008; Chapain et al., 2010; Brandellero & Kloosterman, 2010; Eltham, 2013; Shaw, 2013; Kolodny et al., 2015; Lewandowski, 2015; Wijngaarden et al., 2019) |

| Innovation form/type | Selected representative authors |
|---|---|
| Customer innovation/customer centred/customer-oriented/ customer-driven innovation | (Thomke & von Hippel, 2002; Sawhney et al., 2005; Desouza et al., 2008; Greer & Lei, 2012; Robra-Bissantz & Lattemann, 2017) |
| Democratic innovation | (Saward, 2000; Smith, 2009; Bua, 2012; Geißel & Newton, 2012; Kössler, 2015) |
| Design-driven Innovation/ design innovation | (Walker & Roy, 1999; Verganti, 2006, 2009; Hopkins, 2010; Wu et al., 2015; Na et al., 2017; Wrigley, 2017) |
| Digital-based innovation or ICT-enabled innovation | (Markides & Anderson, 2006; Kampylis et al., 2012; Ashurst et al., 2012; Bocconi et al., 2013; Brecko et al., 2014; Briscoe & Mulligan, 2014; Fichman et al., 2014; Misuraca et al., 2015, 2016, 2018) |
| Digital Social Innovation | (Bria et al., 2014, 2015; Stokes et al., 2017) |
| Distributed innovation | (Andersen et al., 2000; Sawhney & Prandelli, 2000; Kogut & Metiu, 2001; Lakhani & Panetta, 2007; Bogers & West, 2012; Chesbrough, 2017; Joly, 2019) |
| Eco-cultural innovation | (Dieleman, 2013; Lenhart & Fitzgerald, 2017) |
| Eco-innovation | (Fussler & James, 1996; Rennings, 2000; Andersen, 2002, 2008; Faucheux & Nicolaï, 2011; Díaz-García et al., 2015; Ghisetti et al., 2015; Bossle et al., 2016) |
| Economic innovation | (Gambardella et al., 1999; Wonglimpiyarat, 2010; Lamore et al., 2013; Ramella, 2015) |
| Educational innovation/ education technology innovation | (Adams, 1981; Doodley, 1999; Fullan, 2001; Kozma & Voogt, 2003; Vincent-Lancrin et al., 2017; Nicholls, 2018; Visvizi et al., 2018) |
| Employee-driven innovation | (Kesting & Parm Ulhøi, 2010; Høyrup, 2012; Wallace et al., 2016; Lukes & Stephan, 2017; Ramus, 2018) |
| Environmental innovation | (Lanjouw & Mody, 1996; Kemp & Arundel, 1998; Rennings, 2000; Huber, 2004; Oltra & Saint Jean, 2009; Rennings & Rammer, 2011; De Marchi, 2012; Chiarvesio et al., 2015; Wijethilake et al., 2018) |
| Fashion innovation | (Behling, 1992; Crane, 1999; Zhang & Di Benedetto, 2010; Tran et al., 2011) |
| Financial innovation | (Silver, 1983; Merton, 1992; Tufano, 2003; Mishra, 2008; Engelen et al., 2010; Gennaioli et al., 2012; Khraisha & Arthur, 2018; Laperche & Burger-Helmchen, 2019) |
| Food innovation | (Leek et al., 2001; Jongen & Meulenberg, 2005; Olsen, 2015; Reade et al., 2015) |
| Free innovation | (von Hippel, 2016, 2017) |
| Frugal innovation | (Howard, 2011; Zeschky et al., 2011, 2014; Bhatti, 2012; Tiwari & Herstatt, 2012; Radjou & Prabhu, 2014; Soni & Krishnan, 2014; Simula et al., 2015; Weyrauch & Herstatt, 2017; Hossain, 2018; Lim & Fujimoto, 2019) |
| Good-enough innovation | (Hart & Christensen, 2002; Zeschky et al., 2011) |
| Governance innovation | (Hartley, 2005; Swyngedouw, 2005; Moore & Hartley, 2008; Meijer, 2015; Kanie & Biermann, 2017) |
| Grassroots innovation | (Gupta et al., 2003; Seyfang & Smith, 2007; Seyfang & Haxeltine, 2012; Phelps, 2013; Seyfang & Longhurst, 2013; Ustyuzhantseva, 2015; Hossain, 2016) |
| Grassroots social innovation | (Kirwan et al., 2013; Smith et al., 2014; Martin & Upham, 2016; Pellicer-Sifres et al., 2017) |
| Green grassroot innovation | (Gupta, 2010; Nair et al., 2017) |
| Green innovation | (Foster & Green, 2000; Parayil, 2003; Chen et al., 2006; Chen, 2008; Chang, 2011; Schiederig et al., 2012; Chang & Chen, 2013) |
| ICT innovation | (Xiao et al., 2013; Hilty & Aebischer, 2015; Kleibrink et al., 2015; Lee et al., 2016) |
| Inclusive innovation | (Utz & Dahlman, 2007; Cozzens & Sutz, 2012; Sonne, 2012; Foster & Heeks, 2013; Chataway et al., 2014; Heeks et al., 2014; Bryden et al., 2017; Levidow & Papaioannou, 2017; Schillo & Robinson, 2017; Pansera & Owen, 2018) |
| Indigenous innovation | (Lazonick, 2004; Xudong, 2009; Fu & Gong, 2011; Fu et al., 2011; Howell, 2015; Zhao et al., 2015; Huang et al., 2018) |
| Institutional innovation | (Davis & North, 1970; Ruttan & Hayami, 1984; Naughton, 1994; Hargrave & Van de Ven, 2006; Markard et al., 2016; Polzin et al., 2016) |

| Innovation form/type | Selected representative authors |
| --- | --- |
| Jugaad innovation/ Gandhian innovation | (Prahalad & Malshekar, 2010; Birtchnell, 2011; Radjou et al., 2012; Albert, 2016; Pansera & Martínez, 2017) |
| Lean innovation | (Schuh et al., 2008; Ries, 2011; Sehested & Sonnenberg, 2011; Schuh, 2013) |
| Low-cost innovation | (Singh et al., 2012; Zeschky et al., 2014; Agnihotri, 2015) |
| Luxury innovation | (Hoffmann & Hoffmann, 2012; Hoffmann & Coste-Manière, 2016) |
| Media innovation | (Habann, 2008; Storsul & Krumsvik, 2013; Bleyen et al., 2014; Dogruel, 2014; English, 2016) |
| Military innovation | (Rosen, 1988, 1994; Murray & Millett, 1998; Grissom, 2006; Adamsky, 2010; Griffin, 2017; Burke II, 2019) |
| Normative innovation | (Morris, 2005; Raymond et al., 2013; Randles & Laasch, 2016; Schlaile et al., 2017; Rehfeld, 2019; Uyarra et al., 2019) |
| Open collaborative innovation | (Baldwin & Von Hippel, 2011) |
| Open innovation | (Chesbrough, 2003, 2006; Chesbrough et al., 2006; Gassmann, 2006; Dahlander & Gann, 2010; Gassmann et al., 2010; Huizingh, 2011; Chesbrough & Bogers, 2014; West et al., 2014; Randhawa et al., 2016; Bogers et al., 2017; Ramírez-Montoya & García-Peñalvo, 2018) |
| Open eco-innovation | (Ghisetti et al., 2015) |
| Open social innovation | (Chesbrough & Di Minin, 2014) |
| Paradigm-based innovation/ paradigm innovation | (Dosi, 1982; Tidd et al., 2005; Strandburg, 2008; Sveiby et al., 2012; Howaldt et al., 2016) |
| Peace innovation | (Quihuis et al., 2015; Miklian & Hoelscher, 2017) |
| Peer-to-peer innovation | (Satzger & Neus, 2010; Bruton et al., 2015; Ryan & Gaziulusoy, 2016) |
| Philanthropic innovation | (Rana, 2012; Abrahamson, 2013; Kasper & Marcoux, 2014; Bahr, 2019) |
| Policy innovation/political innovation | (Berry, 1994, 2016; Jaffe, 2000; Carpenter, 2001; Windrum & Koch, 2008; Carstensen & Bason, 2012; Mazzucato & Pérez, 2016; Sorensen, 2016, 2017) |
| Pro-poor innovation | (Gupta, 2012; Ramani et al., 2012, 2017; Abrol, 2014; Lowe et al., 2018) |
| Public innovation/Public service innovation | (Feller, 1981; Osborne, 1998; Walker et al., 2002; Mulgan & Albury, 2003; Windrum & Koch, 2008; Fuglsang & Pedersen, 2011; Osborne & Brown, 2013; Djellal et al., 2013; De Vries et al., 2016; Crosby et al., 2017; Bason, 2018) |
| Regulatory innovation | (Kane, 1988; Parker, 2000; Black et al., 2006; Hahn & Renda, 2017) |
| Religious innovation | (Iannaccone, 1992; Finke, 1997; Dawson, 2003; Collar, 2007; Williams et al., 2011; Casanova, 2013) |
| Responsible innovation | (Sutcliffe, 2011; von Schomberg, 2011, 2013; Owen et al., 2013; Stilgoe et al., 2013; Macnaghten et al., 2014; Block & Lemens, 2015; Lubberink et al., 2017) |
| Reverse innovation | (Immelt et al., 2009; Govindarajan & Ramamurti, 2011; Govindarajan & Trimble, 2012; Ostraszewska & Tylec, 2015; Zedtwitz et al., 2015; Hadengue et al., 2017) |
| Rural innovation | (Mahroum et al., 2007; Spielman et al., 2011; Brouder, 2012; Hermans et al., 2012; Snapp & Pund, 2017) |
| Rural social innovation | (Neumeier, 2012; Bock, 2016; Bosworth et al., 2016; Lindberg, 2017) |
| Scarcity-induced innovation | (Olmstead & Rhode, 1993; Srinivas & Sutz, 2006) |
| Service innovation | (Barras, 1986; Gallouj & Weinstein, 1997; Hertog, 2000; Hipp & Grupp, 2005; Spohrer & Maglio, 2008; Lee et al., 2014; Carlborg et al., 2014; Lusch & Nambisan, 2015; Snyder et al., 2016; Witell et al., 2016; Gallouj & Djellal, 2018) |
| Shanzhai innovation | (Peng et al., 2009; Zhu & Shi, 2010; Keane & Zhao, 2012; Lee & Hung, 2014; Ming, 2014; Meriade, 2016) |
| Silver innovation | (Obi et al., 2013; Kohlbacher et al., 2015; Checcucci, 2018) |
| Social innovation[1] | (Weeks, 1932; Drucker, 1957, 1987; Fairweather, 1967; Taylor, 1970; Holt, 1971; Shipman, 1971; Banks, 1972; Mulgan, 2006; Murray et al., 2010; Howaldt & Schwarz, 2010; Nicholls & Murdock, 2011; Godin, 2012a; Moulaert et al., 2013, 2017; Cajaiba-Santana, 2014; van der Have & Rubalcaba, 2016; Edwards-Schachter & Wallace, 2017; Moulaert & MacCallum, 2019) |
| Soft innovation | (Cunningham & Higgs, 2009; Stoneman, 2010, 2015; Choi, 2011) |
| Stylistic innovation | (Schweizer, 2003; Cappetta et al., 2006; Tran, 2010; Rohrbeck et al., 2013) |

| Innovation form/type | Selected representative authors |
| --- | --- |
| Sustainable business model innovation | (Bocken et al., 2014; Carayannis et al., 2015; Antikainen & Valkokari, 2016; Evans et al., 2017; Linder & Williander, 2017; Rosca et al., 2017; Geissdoerfer et al., 2018) |
| Sustainable innovation | (Larson, 2000; Schot & Geels, 2008; Boons et al., 2013; De Medeiros et al., 2014; Nielsen et al., 2016; Sarkar & Pansera, 2017; Tukker et al., 2017) |
| Symbiotic innovation | Thomas & Wind (2013) |
| Technological Environmental Innovation | (Huber, 2008; Ziegler & Seijas Nogareda, 2009) |
| Tourism innovation | (Hjalager, 1997, 2010; Novelli et al., 2006; Moscardo, 2008; Brooker & Joppe, 2014) |
| Transformative innovation | (Phillips, 2007; Scrase et al., 2009; Haxeltine et al., 2013; Sen, 2013; O'Donohue, 2016; Avelino et al., 2017) |
| Urban innovation | (Clark, 2000; Capello, 2001; Athey et al., 2008; Dente & Coletti, 2011; Frug & Barron, 2011; Ning et al., 2016) |
| User innovation/user-led/ user-centred/user-driven | (von Hippel, 1976, 1977, 2005; Baldwin & von Hippel, 2011; Gault, 2012; Gambardella et al., 2016; Hyysalo et al., 2016a, 2016b) |
| Value-chain innovation | (Sankaran & Suchitra Mouly, 2006; Sundo, 2011; Lee et al., 2012) |
| Value innovation | (Kim & Maubourge, 1997; Dillon et al., 2005; Matthyssens et al., 2006; Reypens et al., 2016) |
| Water innovation | (Barrip et al., 2004; Partzsch, 2009; Wehn & Montalvo, 2015; Moro, 2018) |
| Workplace innovation/labour innovation | (Scott & Bruce, 1994; Black & Lynch, 2004; Pot, 2011; McMurray et al., 2013; Pot et al., 2016; Oeij et al., 2017) |

*Note:*    [1] There are more than twenty papers reviewing literature on social innovation.

# 6. Social innovation: contested understandings of social change

*Cornelius Schubert*

Social innovation is often framed as an alternative and in opposition to dominant entrepreneurial understandings of technological innovation. In these cases, the term social innovation denotes bottom-up and beneficial processes of social change. The chapter traces the emergence of this specific understanding of social innovation and how it became popular in academic and policy discourse over the last 20 years. I argue that the popularity of social innovations in the policy arena rests on an instrumental notion that strongly resembles the dominant entrepreneurial understandings of technological innovation in line with Schumpeter. This however, curtails the analytical potential of social innovation for the study of social change. Rather than following the pro-innovation bias of technological innovation, social innovations also provide opportunities to focus on processes of social change in which novel solutions are explored in order to maintain the status quo in the light of larger societal transformations.

## INTRODUCTION: SOCIAL INNOVATION AS A POPULAR CONCEPT

The term social innovation has seen a remarkable academic and policy career in the past three decades, up to the point of being decried as a fashionable, yet meaningless "buzz word" in contemporary discourse (cf. Pol and Ville 2009). Indeed, the term social innovation has been criticised as being somewhat vague, carrying diverse connotations and spanning heterogeneous fields of application (cf. Edwards-Schachter and Wallace 2017; Marques et al. 2018). This chapter seeks to unpack the recent popularity of social innovation as a concept in social science and political discourse and to analyse in how far the diverse perspectives and approaches become contested in the growing field of social innovation research. In order to outline some of the specifics within social innovation discourses, I will relate theories of social innovation to theories of technological innovation and explore to what extent the former may be considered as alternatives to the latter.

The subsequent sections are structured as follows. In section 2, I will trace the emergence of the concept social innovation by drawing on selected authors, focusing on the second half of the twentieth century when it stared to gain popularity and leading up to the present day. The section highlights how a particular normative and bottom-up understanding of social innovation emerged and became dominant within the last decades, overshadowing alternative approaches. Section 3 relates social innovations with their alleged counterpart: technological innovations. I argue that both aspects of innovation cannot be separated from each other and that their interrelations are vital to understanding the dynamics of contemporary societal change. Section 4 brings the above arguments together in order to elaborate the theories of social innovation as alternatives to dominant innovation models. Section 5 concludes the chapter.

## THEORIES OF SOCIAL INNOVATION: ORIGINS AND DEVELOPMENTS

Social innovation concepts appear to be novel additions to the study of innovation. They have gained currency since the 2000s, receiving prominent attention in both academic as well as political discourses (Grimm et al. 2013; Moulaert et al. 2013; Cajaiba-Santana 2014; Nicholls et al. 2015). This novelty of social innovations may, however, be questioned in two respects. First, as Godin (2015, pp. 122–33) has argued, social innovation was already used in the early nineteenth century as a derogatory term placed on social reformers who allegedly sought to overturn the established orders. Second, social innovations as reactions to societal needs predate the use of the term social innovations (cf. Mumford 2002). For instance, novel pre-school educational facilities emerged in the late eighteenth century and later became known as kindergartens without being labelled social innovations at the time. From this, we must conclude that neither the term nor the phenomenon are particularly new.

As Godin (2015) also shows, the use of the term social innovation in the nineteenth century is in no way canonised or consolidated. Rather, within the plurality of meanings, he identifies a gradual shift from the derogatory mode to more positive notions in the second half of the century. This already indicates that social innovation is no neutral term, but a concept that is deeply interwoven with political interests and societal transformations at the time. But it is not a concept that is systematically integrated into the simultaneously emerging discourse of innovations (or the emerging field of social science, for that matter), so the dispersed discourse on social innovation does not lead up to the formation of a theory of social innovations. Despite occasional uses of the term in the first half of the twentieth century (e.g. Ward 1903; Wolfe 1921; Noss 1944), it is only after the Second World War, that the concept was taken up more broadly again, leading to its present-day prominence.

From the nineteenth-century social innovation discourse discussed by Godin (2015), we can see that the term already harbours descriptive, analytical as well normative potential. This plurality continues until today, however, like the shift from a derogatory to a more positive connotation, the dominant understandings and uses of the term have changed during its recent propagation. In addition to an analytical and theoretical use in the social sciences from the 1960s onward, it has taken up a distinctly normative meaning in policy discourses since the 2000s, carrying with it a different set of theoretical assumptions. I will address both developments in more detail by exemplifying central issues based on selected studies. Within the limitations of this selection, the main aim is to highlight the disparate and plural nature of social innovation as a theoretical concept and as a policy instrument.

### Social Innovation as an Analytical Concept

The emerging academic interest in social innovation as an analytical concept is tightly related to questions of social change (cf. Coser 1964). Moore (1960), for instance, criticises the conservative perspectives predominant in structural-functional analysis and urges sociological inquiry to concentrate much more on the sources and triggers of social innovations as basic features of societal transformation. According to Moore, social innovations point to the reflexive re-production of social orders and the constant need for adjustment in times of rapid change. The growing occurrence of social innovations indicates that there is a mismatch between different societal spheres, yet that there also exists creative potential to cope with the resulting challenges. This resonates with social theories that emphasise change over continuity and the resulting pressures to actively engage in societal transformations that can be found in early concepts such as "cultural lag" (Ogburn 1922, pp. 200–80) or more recent ideas on "reflexive modernisation" (Beck et al. 2003).

In Moore's conception, social innovation links social theory with societal dynamics. It depicts a specific form of social change under specific societal conditions that is relevant for the sociological understanding of transformative change at the time. On the most general level, social innovation is conceived as a reflexive reaction to societal problems, a "resolution of human problems" (Moore 1960, p. 813). In line with Ogburn and Beck, Moore hints that these problems do not emerge somewhere outside society, but are unanticipated consequences of modern, industrialised societies and in many cases follow from previous technological innovations (ibid., p. 817). We can see that social innovation emerges as a sociological concept in a time of rapid social change and as a critique of the dominant assumptions about social stabilities in structural-functional theory. Thus, while it can be seen empirically as a reaction to societal dynamics, theoretically it figures as a reaction to perceived analytical shortcomings. As a sociological and analytical concept, it cannot be separated from this historical situation and the overall interest in questions of social change.

About 20 years later, social innovations caught the attention of innovation and management scholars (Lundstedt and Colglazier 1982). Their interest largely lay in discussing social innovations as a neglected phenomenon to be scrutinised more closely within the larger field of innovation studies. For instance, Brooks (1982) takes up this task by relating social and technological innovation and argues that social innovation necessarily complements technological innovation, understood as "social dimensions of invention and innovation" (ibid., p. 9). Social and technological innovations are two sides of the same coin and accordingly, they should be studied as socio-technical innovations. Even "nearly purely technical" innovations, like the transistor or the laser, become entangled with social processes as they move from the development labs into fields of application, that is from invention to innovation (ibid., p. 12). In contrast to Moore, Brooke conceives social innovation not as an alternative to established theories, but as a complement. From a managerial position, they are considered a resource that should be understood, fostered and controlled as a competitive advantage.

Again, we see that social innovation as a concept is situated within a specific context. In line with Moore, social innovation is characterised as a neglected phenomenon with significant, yet untapped, epistemic potential for innovation studies. But it is not framed as a reaction to pressing social problems, but rather as a corollary of other processes of change, especially technological innovation. The analytical value of social innovations is to trace processes of change into many different societal fields, for example markets, management, politics and institutions, primarily as novel ways of doing things. Thus, the concept of social innovation allows management and innovation studies to extend their fields of research and to generate relevant knowledge for policy-makers.

Whereas the studies above framed social innovation largely as a neglected phenomenon, current scholars increasingly react towards the popularity of the concept (Pol and Ville 2009). Recent overviews of social innovations agree that the field is marked by substantial plurality concerning theories and methods (Jessop et al. 2013; Howaldt et al. 2015; Edwards-Schachter and Wallace 2017). Hence, it is not possible to do justice to all approaches within the confines of this chapter and the following remarks necessarily remain partial. One proposition to elaborate a genuine theory of social innovation is made by Howaldt et al. (2015). They put social innovation at the centre of a more general theory of "social-technological innovation" in order to explain contemporary transformative social change. In their view, social innovation points to the social practices of innovation that have long been neglected by a narrow focus on technological innovation in innovation policy. Building upon the work of Tarde (1903), they understand social innovation as a dual process of invention and imitation that allows for the analysis of social change by focusing on the micro-level of concrete innovation practices. Similar to Moore, they see social innovation as a necessary extension to theories of social change. In line with Brooks, they maintain that social innovations cannot and should not be analysed in purely social terms, but in their relation with technological innovation in socio-material practices.

The need for conceptual clarification is also underscored by Cajaiba-Santana (2014). He contends that current social innovation research largely falls into two distinct camps: one focused on agency while the other underscores the role of structures. In order to reconcile both perspectives, Cajaiba-Santana draws on neo-institutional and structuration theory with an emphasis on social practices. From this perspective, social innovations are perceived as forms of collective action and social change, nested within existing structures of social order (cf. ibid., p. 46). When conceived as social practices, social innovations are analysed in the interplay of agency and structure with a special interest in relations of the old with the new. This way, a "practice lens" (Orlikowski 2000) on social innovations does not only illuminate the different levels on which they are performed, rather, social innovations figure as prominent exemplars of social practices themselves. Or, as Cajaiba-Santana puts it: "The study of social innovation processes has the potential to provide a comprehensive framework of how practices are created and institutionalized" (2014, p. 47).

From both Howaldt et al. and Cajaiba-Santana, we can see that one dominant perspective in theorising social innovations follows the recent "practice turn" (Schatzki et al. 2001). Within this context, social innovations figure as central societal practices and as eminent drivers of social change, just as the practice theoretical approach lends itself to the study of social innovations in several ways. I will point out two salient conceptual features of this conjuncture.

Understanding social innovations as social practices especially draws attention to two relations: the relation of the old and the new and the relation of the social and the technical. The relation of the old and the new is a guiding question of innovation research per se and the practice theoretical approach reframes this in terms of structure and agency. Social innovation is thus conceived as a reflexive collective practice, tied into the enabling and constraining features of social (and technical) structures (Giddens 1984). Because innovation as a concept includes creative inventions on the level of situations and transformative diffusion, translation or imitation on the level of structures (Rogers 2003), innovation in general can be seen as a fundamental mode in which the relation of agency and structure can be studied with respect to questions of the old and the new. This has been a main concern for placing social innovations at the core of studying processes of social change since Moore's work in the early 1960s. The relation of the social and the technical (or material) equally is a central question of social innovation studies and practice theories (Reckwitz 2002, pp. 252–53). With some exceptions (e.g. Cajaiba-Santana 2014, p. 44), the majority of social innovation scholars acknowledge the importance to not reduce social innovations to immaterial or "purely" social dynamics (Howaldt et al. 2015, pp. 37–42). Subsequently, social innovations and social practices should not be understood as being composed of a social and a technical or material side, as the terms socio-technical or socio-material tend to suggest, but the social itself should be considered to be a "material-semiotic" amalgamation (Law 2009). The study of social innovations shows that meanings and materiality take up interchangeable roles: sometimes resistance to change can be found in social norms, sometimes in material infrastructures; drivers for change may equally be considered predominantly social or predominantly material (cf. Brooks 1982). The main argument is that when conceiving social innovations within a praxeological approach, the dualism of the social and the technical or material dissolve into dynamic and relational "agencements" (Gherardi 2016).

I will conclude this brief and partial outline of social innovation as an analytical concept by highlighting two main points. First, we can see that the academic discourse is driven by the recognition of concrete societal problems and their resolution with the aim of developing social theories that are sensitive to such kinds of social change. Social innovation is then considered to be a relevant and increasingly dominant mode of transformative social change. Whereas social innovations can be traced to the earliest times of humankind, they become increasingly relevant in the fast-paced eras of differentiated modern societies that are confronted with the need for continuous re-integration of cultural and other "lags" (Ogburn 1922). Second, social innovation theories draw upon the lager theoretical debates of their times, such as the current practice turn. The affinity of social innovation and social practices is no coincidence. Social innovation research is grounded in empirical cases and especially interested in the concrete and collective activities of doing social innovations. Even though many practice theories empathise the stability of social orders, for example through embodied practices (Bourdieu 1977), a social praxis is always also a place of ambiguity and change (Dewey 1929, pp. 3–25). In sum, analytical concepts of social innovation should therefore always be considered in their empirical as well as their theoretical contexts. The analytic potential rests in the connection of empirical cases and theoretical abstractions, figuring social innovations not as a general theory of social change, but as a specific mode of social change and as a "sensitising concept" for guiding research (Blumer 1954).

### Social Innovation as a Normative Device

In contrast to an analytical understanding of social innovation as a mode of social change that does not make any value judgements towards the process or outcome, a normative understanding of social innovations holds specific assumptions about their benefits and utilities. Typically, normative concepts of social innovations today carry positive connotations often connected to issues of participation, sustainability or responsibility. The "social" in social innovation thus carries a twofold meaning, separating it from technological innovation on the one side and indicating that its purpose is orientated towards societal needs and ends. In addition to being "socially good", a second normative aspect concerns the effectiveness of social innovations as a means of social change.

### The instrumental notion

One prominent and early mention of social innovations after the Second World War can be found in Drucker's *Landmarks of Tomorrow* (1957). In the context of discussing "the power of innovation" (ibid., pp. 31–45), he advocates an instrumental notion of social innovation as a means of social change: "And there is social innovation, the diagnosis of social needs and opportunities and the development of concepts and institutions to satisfy them" (ibid., p. 32). According to Drucker, the most prominent social innovations in the twentieth century are innovations in management and business, such as mass production, mass distribution and mass consumption that have evolved into the novel institution of the business enterprise and into novel processes of management. Thus, he considers social innovation primarily as a rational and instrumental endeavour: "Above all it is a method that enables us to set objectives and to organize work for their attainment" (ibid., p. 41). And in contrast to alternative forms of social change, namely reform and revolution, social innovation "does not aim at curing a defect", like reform, nor does it "aim at subverting values, beliefs and institutions", like revolution (ibid., p. 45). In essence, social innovation is seen a means of managing social change, of creating new opportunities while building upon established values and institutions. For Drucker, social innovation is not simply one mode of social change among others, but the best way of addressing the challenges of the current "post-modern world" (ibid.).

Drucker's understanding of social innovations puts the concept in line with neighbouring terms such as social entrepreneurship or social engineering. It carries normative implications with respect to the accepted and legitimate ways of governing or managing social change that resonate with earlier ideas such as "piecemeal engineering" and "piecemeal social experiments" (Popper 1945, pp. 138–48). Popper's concern lay with identifying means of social change that account for the complexity of modern societies. In contrast to forms of "utopian engineering" (ibid., p. 138), which seeks to bring about radical change by overturning established societal arrangements, piecemeal engineering starts with small changes at the local level and seeks to scale them in case of success. Because their scope is initially limited, Popper sees them as less risky and less controversial, thus as more likely to actually produce social change than large scale and utopian efforts. In addition, because piecemeal engineering grows from the local level, it is more likely to fit into democratic processes, whereas utopian engineering requires centralised control and may subsequently lead to dictatorial patterns (ibid., p. 140). Popper's notion of piecemeal social experiments can be seen as equivalent to social innovation and he uses similar examples: "a new kind of life-insurance, of a new kind of taxation, of a new penal reform, are all social experiments which have their repercussions through the whole of society without remodelling society as a whole" (ibid., p. 143).

In sum, both Drucker and Popper advocate to govern processes of social change in terms of rational and democratic, bottom-up and creative endeavours, closely linking economic and engineering ideas for their management and control. Whereas Drucker is unequivocally convinced that rational forms of management will yield the desired solutions, Popper's notion of the experiment hints at the limits of control and the need to carefully monitor and evaluate the larger societal repercussions. But the nucleus of a normative/instrumental understanding of social innovation already shows: It is a rational and reflexive process of initiating bottom-up social change through creative and entrepreneurial agents (cf. Drucker 1987, p. 33) who tackle specific societal problems for the greater good.

This did not change as social innovation became a more popular term in the late 1960s. Studies that pushed social innovation as a concept also framed it in terms of actively introducing and maintaining social change (Fairweather 1967; Taylor 1970). However, as Taylor emphasises, introducing and maintaining social inventions holds several challenges itself. He reports on an interdisciplinary project that developed novel procedures for psychiatrists and social workers dealing with low-income clients. In contrast to many technological innovations, such as a new mouse-trap, that do not require a major change in established practices, social innovations, like penal or educational reforms, will most likely disrupt central social values and procedures, in essence posing "a particularly difficult problem in marketing" (ibid., p. 70) such social inventions. The case of an interdisciplinary research project developing social inventions again underscores their rational and reflexive aspects while at the same time pointing to the considerable resistances to change. The way to overcoming such resistances is to closely engage with those affected by the interventions and to tailor the prospective solutions to their specific needs. The effectiveness of social innovations follows from his close interaction and the bottom-up nature of the process. In addition to Drucker's emphasis on management and Popper's affinity with engineering, Taylor highlights the role of social research in creating and conducting durable social innovations. If we consider social innovations as an instrument of carefully organised social change, then these three authors make it obvious that this requires a special kind of innovative agent: an entrepreneur, an engineer or a researcher. I point this out because it shows an uncanny parallel to concepts of competitive economic innovation building upon Schumpeter (1942), where entrepreneurship is a key factor in capitalist dynamics. One main conclusion is that even though social innovation is often framed in contrast to economic innovation, they both share marked family resemblances when we consider social innovation in terms of a distinct instrument of social change.

**The beneficial notion**
In addition to social innovations being considered an appropriate, if not superior, instrument of social change, the concept more often than not is also framed as being primarily beneficial to pressing social needs. This second normative aspect of social innovations rests on the common interpretation of the adjective "social" as socially desirable, which has significantly grown since the 2000s (Edwards-Schachter and Wallace 2017, pp. 71–72). Amidst the manifold meanings of social innovations in recent academic and policy discourses, this normative shift is seen by some scholars as a defining feature of social innovations: "an innovation is termed a social innovation if the implied new idea has the potential to improve either the quality or the quantity of life" (Pol and Ville 2009, p. 881). This definition reacts on the notable difficulties of clearly discerning social innovations from technological or economic innovations, since the latter always also contain elements of former. Turning from different substrates or domains of innovation to positive societal consequences then shifts the question of what a social innovation actually *is* to what it *does*. On the one hand, the criterion of beneficial consequences narrows the analytical scope of social innovation research by disregarding those with negative consequences, while on the other hand it broadens the empirical base of social innovations by opening the concept to technological, economic, regulative, normative and cultural aspects of innovation (cf. ibid., p. 879), or, as Phills Jr. et al. (2008, p. 39) state: "a social innovation can be a product, production process, or technology [...]". However, many of the beneficial notions of social innovations – that can often be found in closer proximity to policy circles – perform a dual focus on social innovation as "innovations that are social both in their ends and in their means" (Murray et al. 2010, p. 3).

The beneficial notion of social innovations also tends to cast social change in a specific light of crisis and challenges. This is where the beneficial and the instrumental notion merge into configuring social innovations as promising solutions to pressing social needs. Again, this understanding is very prominent at the intersections of academic research and policy initiatives and has become quite popular in the recent discourse on social innovations (cf. Franz et al. 2012, p. 3). Thus, a closer look is warranted in order to unpack the broader horizons of meaning this notion is embedded in. I will especially draw upon a previous analysis of the social innovation discourse on the level of EU policy (Schubert 2018).

The close relation of social innovations with crisis and challenges in recent EU policy discourse emerges from two observations: first, that current societies are faced with multiple "grand challenges" such as ageing, climate change, unemployment or health and second, that the established political and economic institutions do not provide adequate solutions. We can trace this line of argument in many policy related publications: "The Lisbon Strategy [...] with its concomitant focus on innovation through R&D has proved inadequate to tackling the social and environmental challenges facing Europe today" (European Commission and Young Foundation 2010, p. 8) or "The classic tools of government policy on the one hand, and market solutions on the other, have proved grossly inadequate" (Murray et al. 2010, p. 3). In short, there is a gap between the pressing societal challenges and the available political and economic instruments.

Social innovations are supposed to fill this gap by tapping into the creative potential of local innovators or entrepreneurs. The main line of reasoning is that the grand challenges should be met by bottom-up initiatives rather than top-down intervention. Within the EU policy discourse, this entails an emphasis on entrepreneurial action and specifically innovative actors as agents of change: "Today, societal trends are increasingly perceived as opportunities for innovation. [...] In addition, there is a real excitement around new entrepreneurial answers and solutions to the rapidly changing challenges that these trends raise" (European Commission 2013, p. 10) and "[...] when old systems are in crisis, [...] they come to an end as the new ideas diffuse, and the innovators connect to the main sources of power and money" (European Commission and Young Foundation 2010, p. 26). The EU documents position social innovation and social innovators in close proximity with concepts such as social entrepreneurship and social enterprise (European Commission 2013, pp. 15–16).

Like classic economic entrepreneurs who engage in innovation as creative destruction (Schumpeter 1942, pp. 81–86), social innovators are seen as taking risks and investing resources – however not for economic, but for societal benefits. Such benefits may be gained from increased participation: "social innovations have empowered people and organisations to develop participative solutions to pressing societal issues" (BEPA 2011, p. 16) since "the new paradigm of social intervention embodied by social innovation offers a way to address social risks *with*, rather than *for*, stakeholders" (ibid., p. 18). The promises of bottom-up, participative social innovations lie in tapping the seemingly abundant creative potential as a resource for initiating and conducting societal change: "Fortunately, there is no shortage of social innovations in Europe" (European Commission 2010, p. 8).

However, there remain some distinct challenges to unleashing the beneficial potential of social innovations in Europe: "Many innovative projects and programmes remain small, under-funded, and are not sustainable, therefore having restricted impact" (ibid.). In addition to the gap between the pressing social needs and the inadequate political and economic solutions, this gap points to inadequate regulative structures through which social inventions on the local level can be scaled up to become social innovations with broader impact. Another gap is identified with respect to detailed knowledge about social innovation processes (Murray et al. 2010, p. 2). In contrast to large EU funding programmes on technological innovation and R&D, social innovation has not received equal attention and therefore lacks a broader knowledge base.

In sum, the normative notion of social innovations on the level of EU discourse can be seen as a distinct problem/solution package in line with an established pro-innovation bias (Godin and Vinck 2017a) and a growing innovation imperative (cf. Hutter et al. 2018), essentially turning social innovation into a normative device for addressing grand societal challenges. The normative device consists of fusing the beneficial and the instrumental notion into a promising and legitimate policy tool. As the history of social innovation as a normative concept shows, both notions have been closely interrelated at least since the works of Drucker (1957). However, critical scholars have remarked that on the level of EU policy, social innovation is predominantly framed in an entrepreneurial manner, thereby facilitating an interpretation of social innovation "in economic, indeed often in narrowly market-economic, terms" (Jessop et al. 2013, p. 110) up to a (re)legitimisation of neoliberal rationality (Fougère et al. 2017).

This does of course not entail that all normative concepts of social innovation inevitably lead towards an entrepreneurial or neoliberal bottom line, but it indicates that such a bias did emerge with the recent popularity of social innovations at the intersections of academic and policy discourses. In contrast to the differences that the concept of social innovation might suggest on the level of EU policy, it is very closely related with positive instrumental connotations of economic and technological innovation (cf. Godin 2016). In the case of EU policy, social innovation can hardly be considered a distinct alternative to techno-economic innovations, rather they appear to be extensions of a dominant entrepreneurial innovation paradigm into broader societal spheres.

## SOCIAL AND TECHNOLOGICAL INNOVATION: BEYOND THE GREAT DIVIDE

Both the analytical and the normative concepts of social innovation mentioned above have made more or less explicit references to technological innovations. And the inspection of the EU social innovation discourse showed that the differences might be less pronounced than the labels "social" versus "technological" might suggest. Indeed, while the struggle for demarcations between the two can be found in nearly all overview studies on social innovation, the most common conclusion is not to advocate a clear-cut distinction between them. This begs the question if it is at all useful to specifically speak of social innovation or if the concept is too vague to be of more interest than as a fashionable buzzword. Pol and Ville (2009) argued that the surplus of social innovation lies in focusing the attention to inventive novelty and social beneficiality: social innovations are new ideas that will improve the quality of life. I would like to suggest a different route by explicitly conjoining social and technical aspects and proposing social innovation as a "sensitising concept" (Blumer 1954) for analysing processes of social change. This concept of social innovation does not function as a "lens" for focusing on novel ideas for the public good, but as a "prism" for breaking up social and technical relations in processes of social change.

From this perspective, social innovation first of all emphasises the need to study the social processes tied in with technological innovation and social change (cf. Brooks 1982; Howaldt et al. 2015, pp. 37–42). The distinction between social and technological aspects of social change is used for analytical purposes as it allows to critically engage with biased notions of either technical or social determinisms. Processes of societal change can then be analysed in terms of the relations of social and technological innovations. For instance, early scholars of social and technical change such as Mumford (1934) have argued that in many cases social innovations precede or pave the way for technological innovations. Mumford contends that without the prior reorganising of society, mechanisation and industrialisation would not have been able to take hold and spread as quickly: "The fact is, at all events, that the machine came most slowly into agriculture, with its life-conserving, life-maintaining functions, while it prospered lustily precisely in those parts of the environment where the body was most infamously treated by custom: namely, in the monastery, in the mine, on the battlefield" (ibid., p. 36). He criticises Ogburn's (1922) thesis of a "cultural lag" for being too limited in this respect (Mumford 1934, p. 264). Rather than studying the cultural adjustments to changing material conditions, Mumford is interested in the "cultural preparation" (ibid., pp. 9–59) of mechanised societies. One of these preparations and of interest in this context, is the "duty to invent" (ibid., p. 52): "The unwillingness to accept the natural environment as a fixed and final condition of man's existence had always contributed both to his art and his technics: but from the seventeenth century, the attitude became compulsive, and it was to technics that he turned for fulfilment" (ibid.).

In addition to a more nuanced understanding of historical societal change, a sensitivity to the relations of the social and the technical within processes of social innovation resonates well with the recent turns towards practices and materialities (cf. Gherardi 2017 and see earlier discussion). Innovation scholars in science and technology studies have strongly argued for including social explanations in processes of technological innovation (Bijker et al. 1987). The blurring of boundaries between the social on the one side and the technical on the other can now serve as a fruitful input to social innovation studies (Degelsegger and Kesselring 2012). Because social innovations are not limited to immaterial ideas, values and face-to-face interactions, human bodies and material artefacts become more prominent in the analysis. Consequently, the "social" in social innovation functions as a twofold sensitising concept: First, it makes the analysis sensitive to innovations that may be considered predominantly "social" in a classic sense. Second, the "social" may then be understood as a material-semiotic assemblage (Law 2009), in which social innovations are equally sensually embodied, technically mediated and socially organised. Such a praxeological understanding of social innovations would allow for a detailed micro-analytical perspective on processes of social change and provide a conceptual base for linking local social inventions through processes of imitation, translation or appropriation with more durable social innovations (cf. Howaldt et al. 2015, pp. 61–76).

Treating social innovation as a sensitising concept for studying social change highlights the praxeological irreducibilities of continuity and change, of meaning and materiality, of the social and the technical. As a conceptual prism, it can be used to analytically unfold these relations that are irreversibly intertwined in practice. Just as science and technology studies or the works of Mumford have shown that technological innovations cannot be adequately understood without social dynamics, social innovations must be analysed in their material and technical interrelations.

## SOCIAL INNOVATIONS AND INNOVATION THEORY

Despite their recent popularity, social innovation studies are a relative late-comer to the field of innovation studies (van der Have and Rubalcaba 2016). In the context of this Handbook, this begs the question in how far they may contribute to alternative theories of innovation. After the discussion of social innovation concepts on the level of EU policy, we should at least be cautious to readily equate social innovation with new or alternative theory innovation. And because of the heterogeneous and dispersed nature of the field of social innovation studies, a uniform theoretical contribution can hardly be expected. I will address two more recent lines of thought in innovation studies, to which social innovation studies may fruitfully contribute. Both take a critical stance towards mainstream innovation studies as being embedded in paradigms of novelty and progress.

Godin and Vinck (2017a) have recently highlighted that innovation studies tend to have a "pro-innovation bias", that is, they generally perceive innovation as a good thing. Social innovation studies, especially those invoking a normative notion, are no exception. On the level of EU policies, we can even observe a twofold pro-social-innovation bias, since both "social" and "innovation" are framed in decidedly positive terms. But we can also see that social innovations are conceived as reactions to societal challenges and in many cases the aim is not to disrupt social order by means of creative destruction, but to maintain order in the face of extensive social change. Rather than being equated with sweeping novelty, social innovations are portrayed as purposive means of conservation, very much in the old sense of reform as the re-creation of a previous state of affairs. In contrast to Schumpeter-type innovations that are driven by competitive entrepreneurial action within capitalist economies, social innovations within the EU policy discourse much more resemble Ogburn-type innovations that seek to resolve the conflicts brought about by the creative destructions of capitalist economies (cf. Schubert 2019, pp. 44–46). Ogburn (1922) was concerned with understanding the repercussions of changes in the material culture on the non-material culture, which he saw as a dominant trait of modern societies. From Ogburn's point of view, transformative innovations on the one side inevitably lead to adaptive innovations on the other. In this understanding, innovation is a double-edged sword, bringing about positive as well as negative changes in society. Social innovations that function as adaptive innovations to resolve cultural lags would then conform much more with a pro-innovation bias than the disruptive (often technological) innovations of the material culture. But the close inspection of such adaptive social innovations at the same time illuminates the negative consequences of preceding innovations and hence serves to critically question the pro-innovation bias. Taken together, they underscore the increasing "duty to invent" (Mumford 1934, p. 52) up to a current innovation imperative (cf. Osborne and Brown 2011 for the UK), while at the same time, to paraphrase Merton (1936), pointing to the unanticipated consequences of purposive innovative action (see discussion above).

The critical view on the pro-innovation bias in innovation studies resonates with a recent critique of a progress bias in science and technology studies. Jackson (2014, p. 221) identifies a dominant imaginary of "progress and advance, novelty and invention, open frontiers and endless development" that has been attached to technological innovation since the nineteenth century. Yet this imaginary came under pressure in the late twentieth century, when the negative consequences of technological innovation became more obvious. Jackson argues that science and technology studies have largely followed the dominant innovation paradigm of the twentieth century, focusing their attention on scientific and technological innovations that shaped modern societies. With the advent and growing recognition of global crises however, he sees the need for a switch in perspective, one that recognises the importance of repair for the maintenance of the socio-technical infrastructures of modern societies: "From this perspective, worlds of maintenance and repair and the instances of breakdown that occasion them are not separate or alternative to innovation, but sites for some of its most interesting and consequential operations" (ibid., p. 227). Ogburn-type adaptive social innovations figure as such instances of repairing and maintaining, for instance, social cohesion within disruptive situations. The repair perspective highlights aspects that often remain implicit in innovation research, such as the problematic retention of novelty. Even though this perspective is present in evolutionary innovation studies (cf. Ziman 2000), it is neither systematically addressed in innovation research more broadly, nor is repair in innovation conceived as a relevant reflexive and creative practice.

The alternatives that social innovation research offers to mainstream innovation theory lie in questioning the biases that have emerged with the dominant, novelty driven, pro-innovation, techno-economic paradigm. By not reducing social innovations to socially beneficial, non-material types of innovation, their analysis sheds light on the intricate relations of innovations and society today. If social innovations have become a dominant mode and means of social change in current societies, this will necessarily have to be reflected in contemporary innovation theories. A number of leads for this already exist, for instance from evolutionary economics (Nelson and Winter 1977), management studies (Van de Ven 1986) or the study of failed innovations in science and technology studies (Latour 1996). Such theories of innovation emphasise the relational, precarious and open-ended character of innovation and they are sensitive to the social impacts on technological innovation. Social innovation research adds to this, because it shows how deep the pro-innovation bias has diffused into current policy concerns and how innovation as a concept has spread into diverse societal realms. Due to the popularity and plurality of social innovation as a theoretical concept, however, there is no single subsumable contribution. On the one hand, social innovation research offers novel insights into areas that are not typically covered by dominant innovation theories focusing on technological innovation. On the other hand, social innovation as a policy instrument is likely to extend a dominant technological and entrepreneurial notion of innovation into many societal areas.

## CONCLUSION: AN INNOVATION THEORY OF SOCIAL CHANGE

The sections above were written against the backdrop of social innovation as a rapidly growing field of research (Nicholls et al. 2015; van der Have and Rubalcaba 2016). The recent popularity of the term social innovation is at the same time marked by conceptual vagueness and diverse domains of application (Pol and Ville 2009; Edwards-Schachter and Wallace 2017). As Edwards-Schachter and Wallace point out, it is also characterised by a shift towards a normative understanding of social innovation as being inherently "good". In effect, many social innovation studies have added to the pro-innovation bias of innovation studies in general and "have not really changed the conventional theoretical framework and conceptual tools" (Godin and Vinck 2017b, p. 2). Despite this assessment, an innovation theory of social change may draw fruitfully upon the developments in the field of social innovations in two respects.

On the one hand, the normative notion that has become popular since the 2000s is itself an interesting case of social change. It points to a growing innovation imperative that has transgressed the domain of techno-economic innovation and has spread into broader societal fields. In this case, social innovation theories are not just descriptive and analytical concepts, but have become "performative" models for introducing change in society (cf. MacKenzie 2006). An innovation theory of social innovations treats them as reflexive means of social change that have become popular through the promotional activities of interested actors and that can themselves be analysed by studying the discourses they are embedded in. As instruments of social change, they are not neutral means, but transport specific rationalities incorporated in the problem/solution packages.

On the other hand, an analytic notion of social innovations can be used as a sensitising concept or conceptual prism to trace the different dimensions of social change in their interrelations. Within a praxeological understanding, this first of all entails to switch from dualist dichotomies between the social and the technical or the new and the old to material-semiotic irreducibility and a duality of continuity and change. Complementing the novelty-bias of innovation research with a repair perspective as suggested by Jackson (2014) helps to gain a deeper understanding of social innovation as a way of maintaining social order in face of dynamic changes. Social innovations are thus not reduced to being "good" and "new", but they are deeply interwoven with their unanticipated and unintended consequences.

Social change and innovation are no doubt closely intertwined in current societies and social innovations can be considered central drivers on an increasingly global scale. Because social innovation has become such a pervasive concept and phenomenon, innovation theories need to reflect this in order to account for the performative nature of innovation in processes of social change.

# REFERENCES

Beck, Ulrich, Wolfgang Bonss and Christoph Lau (2003), 'The theory of reflexive modernization. problematic, hypotheses and research programme', *Theory, Culture & Society*, 20 (2), 1–33.

BEPA (2011), Empowering people, driving change. Social Innovation in the European Union. Bureau of European Policy Advisers. Luxembourg.

Bijker, Wiebe E., Thomas P. Hughes and Trevor J. Pinch (eds) (1987), *The Social Construction of Technological Systems*, Cambridge: MIT Press.

Blumer, Herbert (1954), 'What's wrong with social theory?', *American Sociological Review*, 19 (1), 3–10.

Bourdieu, Pierre (1977), *Outline of a Theory of Practice*, Cambridge: Cambridge University Press.

Brooks, Harvey (1982), 'Social and technological innovation', in Sven B. Lundstedt and E. William Colglazier (eds), *Managing Innovation. The Social Dimension of Creativity, Invention, and Technology*, New York: Pergamon Press, pp. 1–30.

Cajaiba-Santana, Giovany (2014), 'Social innovation: Moving the field forward. A conceptual framework', *Technological Forecasting and Social Change*, 82, 42–51.

Coser, Lewis A. (1964), 'Durkheim's conservatism and its implications for his sociological theory', in Kurt H. Wolff (ed.), *Essays on Sociology and Philosophy*, New York: Harper & Row, pp. 211–32.

Degelsegger, Alexander and Alexander Kesselring (2012), 'Do non-humans make a difference? The actor-network-theory and the social innovation paradigm', in Hans-Werner Franz, Josef Hochgerner and Jürgen Howaldt (eds), *Challenge Social Innovation. Potentials for Business, Social Entrepreneurship, Welfare and Civil Society*, Heidelberg: Springer, pp. 57–72.

Dewey, John (1929), *The Quest for Certainty. A Study of the Relation of Knowledge and Action*, New York: Minton, Balch.

Drucker, Peter F. (1957), *The Landmarks of Tomorrow*, New York: Harper & Brothers.

Drucker, Peter F. (1987), 'Social innovation: Management's new dimension', *Long Range Planning*, 20 (6), 29–34.

Edwards-Schachter, Mónica and Matthew L. Wallace (2017), '"Shaken, but not stirred": Sixty years of defining social innovation', *Technological Forecasting and Social Change*, 119, 64–79.

European Commission (2010), This is European social innovation. European Union (Enterprise & Industry). Luxembourg.

European Commission (2013), Social innovation research in the European Union. Approaches, findings and future directions. Directorate-General for Research and Innovation (ed.). Publications Office of the European Union. Luxembourg.

European Commission and Young Foundation (2010), Study on social innovation. A paper prepared by the Social Innovation eXchange (SIX) and the Young Foundation for the Bureau of European Policy Advisors. European Comission and Young Foundation. Luxembourg.

Fairweather, George W. (1967), *Methods for Experimental Social Innovation*, New York: Wiley.

Fougère, Martin, Beata Segercrantz and Hannele Seeck (2017), 'A critical reading of the European Union's social innovation policy discourse: (Re)legitimizing neoliberalism', *Organization*, 24 (6), 819–43.

Franz, Hans-Werner, Josef Hochgerner and Jürgen Howaldt (eds) (2012), *Challenge Social Innovation. Potentials for Business, Social Entrepreneurship, Welfare and Civil Society*, Heidelberg: Springer.

Gherardi, Silvia (2016), 'To start practice theorizing anew: The contribution of the concepts of agencement and formativeness', *Organization*, 23 (5), 680–98.

Gherardi, Silvia (2017), 'Sociomateriality in posthuman practice theory', in Allison Hui, Theodore R. Schatzki and Elizabeth Shove (eds), *The Nexus of Practices: Connections, Constellations, Practitioners*, Abingdon: Routledge, pp. 38–51.

Giddens, Anthony (1984), *The Constitution of Society. Outline of the Theory of Structuration*, Berkeley: University of California Press.

Godin, Benoît (2015), *Innovation Contested: The Idea of Innovation over the Centuries*, London: Routledge.

Godin, Benoît (2016), 'Technological innovation: On the origins and development of an inclusive concept', *Technology and Culture*, 57 (3), 527–56.

Godin, Benoît and Dominique Vinck (eds) (2017a), *Critical Studies of Innovation: Alternative Approaches to the Pro-Innovation Bias*, Cheltenham, UK and Northampton, MA, USA: Edward Elgar Publishing.

Godin, Benoît and Dominique Vinck (2017b), 'Introduction: Innovation – from the forbidden to a cliché', in Benoît Godin and Dominique Vinck (eds), *Critical Studies of Innovation: Alternative Approaches to the Pro-Innovation Bias*, Cheltenham, UK and Northampton, MA, USA: Edward Elgar Publishing, pp. 1–14.

Grimm, Robert, Christopher Fox, Susan Baines and Kevin Albertson (2013), 'Social innovation, an answer to contemporary societal challenges?: Locating the concept in theory and practice', *Innovation: The European Journal of Social Science Research*, 26 (4), 436–55.

Howaldt, Jürgen, Ralf Kopp and Michael Schwarz (2015), *On the Theory of Social Innovations: Tarde's Neglected Contribution to the Development of a Sociological Innovation Theory*, Weinheim: Beltz Juventa.

Hutter, Michael, Hubert Knoblauch, Werner Rammert and Arnold Windeler (2018), 'Innovation society today: The reflexive creation of novelty', in Werner Rammert, Arnold Windeler, Michael Hutter and Hubert Knoblauch (eds), *Innovation Society Today: Perspectives, Fields, and Cases*, Wiesbaden: Springer VS, pp. 13–31.

Jackson, Steven J. (2014), 'Rethinking repair', in Tarleton Gillespie, Pablo J. Boczkowski and Kirsten A. Foot (eds), *Media Technologies. Essays on Communication, Materiality, and Society*, Cambridge: MIT Press, pp. 221–39.

Jessop, Bob, Frank Moulaert, Lars Hulgård and Abdelillah Hamdouch (2013), 'Social innovation research: A new stage in innovation analysis?', in Frank Moulaert, Diana MacCallum, Abid Mehmood and Abdelillah Hamdouch (eds), *The International Handbook on Social Innovation: Collective Action, Social Learning and Transdisciplinary Research*, Cheltenham, UK and Northampton, MA, USA: Edward Elgar Publishing, pp. 110–30.

Latour, Bruno (1996), *Aramis, or the Love of Technology*, Cambridge: Harvard University Press.

Law, John (2009), 'Actor network theory and material semiotics', in Bryan S. Turner (ed.), *The New Blackwell Companion to Social Theory*, Oxford: Wiley-Blackwell, pp. 141–58.

Lundstedt, Sven B. and E. William Colglazier (eds) (1982), *Managing Innovation. The Social Dimension of Creativity, Invention, and Technology*, New York: Pergamon Press.

MacKenzie, Donald (2006), *An Engine, not a Camera. How Financial Models Shape Markets*, Cambridge: MIT Press.

Marques, Pedro, Kevin Morgan and Ranald Richardson (2018), 'Social innovation in question: The theoretical and practical implications of a contested concept', *Environment and Planning C: Politics and Space*, 36 (3), 496–512.

Merton, Robert K. (1936), 'The unanticipated consequences of purposive social action', *American Sociological Review*, 1 (6), 894–904.

Moore, Wilbert E. (1960), 'A reconsideration of theories of social change', *American Sociological Review*, 25 (6), 810–18.

Moulaert, Frank, Diana MacCallum, Abid Mehmood and Abdelillah Hamdouch (eds) (2013), *The International Handbook on Social Innovation: Collective Action, Social Learning and Transdisciplinary Research*, Cheltenham, UK and Northampton, MA, USA: Edward Elgar Publishing.

Mumford, Lewis (1934), *Technics and Civilisation*, London: Routledge.

Mumford, Michael D. (2002), 'Social innovation. Ten cases from Benjamin Franklin', *Creativity Research Journal*, 14 (2), 253–66.

Murray, Robin, Julie Caulier-Grice and Geoff Mulgan (2010), *The Open Book on Social Innovation*, Brussels: The Young Foundation / NETSA.

Nelson, Richard R. and Sidney Winter (1977), 'In search of useful theory of innovation', *Research Policy*, 6 (1), 36–76.

Nicholls, Alex, Julie Simon and Madeleine Gabriel (eds) (2015), *New Frontiers in Social Innovation Research*, Basingstoke: Palgrave Macmillan.

Noss, Theodore K. (1944), Resistance to social innovations as found in the literature regarding innovations which have proved successful. Dissertation, University of Chicago, Chicago.

Ogburn, William F. (1922), *Social Change: With Respect to Culture and Original Nature*, New York: Viking Press.

Orlikowski, Wanda J. (2000), 'Using technology and constituing structures. A practice lens for studying technology in organizations', *Organization Science*, 11 (4), 404–28.

Osborne, Stephen P. and Louise Brown (2011), 'Innovation, public policy and public services delivery in the UK. The word that would be king?', *Public Administration*, 89 (4), 1335–50.

Phills Jr., James A., Kriss Deiglmeier and Dale T. Miller (2008), 'Rediscovering social innovation', *Stanford Social Innovation Review*, 6 (4), 34–43.

Pol, Eduardo and Simon Ville (2009), 'Social innovation: Buzz word or enduring term?', *The Journal of Socio-Economics*, 38 (6), 878–85.

Popper, Karl (1945), *The Open Society and its Enemies. Volume I: Plato*, London: Routledge.

Reckwitz, Andreas (2002), 'Toward a theory of social practices. A development in culturalist theorizing', *European Journal of Social Theory*, 5 (2), 243–263.

Rogers, Everett M. (2003), *Diffusion of Innovations*, New York: Free Press.

Schatzki, Theodore R., Karin Knorr Cetina and Eike von Savigny (eds) (2001), *The Practice Turn in Contemporary Theory*, New York: Routledge.

Schubert, Cornelius (2018), 'Social innovation: A new instrument of social change?', in Werner Rammert, Arnold Windeler, Michael Hutter and Hubert Knoblauch (eds), *Innovation Society Today: Perspectives, Fields, and Cases*, Wiesbaden: Springer VS, pp. 371–91.

Schubert, Cornelius (2019), 'Social innovations as a repair of social order', *NOvation – Critical Studies of Innovation*, (1), 40–66.

Schumpeter, Joseph A. (1942), *Capitalism, Socialism and Democracy*, New York: Harper & Row.

Tarde, Gabriel (1903), *The Laws of Imitation*, New York: Henry Holt.

Taylor, James B. (1970), 'Introducing social innovation', *The Journal of Applied Behavioral Science*, 6 (1), 69–77.

Van de Ven, Andrew H. (1986), 'Central problems in the management of innovation', *Management Science*, 32 (5), 590–607.

van der Have, Robert P. and Luis Rubalcaba (2016), 'Social innovation research: An emerging area of innovation studies?', *Research Policy*, 45 (9), 1923–35.

Ward, Lester F. (1903), *Pure Sociology: A Treatise on the Origin and Spontaneous Development of Society*, New York: Macmillan.

Wolfe, Albert B. (1921), 'The motivation of radicalism', *Psychological Review*, 28 (4), 280–300.

Ziman, John (ed.) (2000), *Technological Innovation as an Evolutionary Process*, Cambridge: Cambridge University Press.

# 7. Sustainable innovation: analysing literature lineages

*Frank Boons and Riza Batista-Navarro*

## INTRODUCTION

Sustainable innovation has been defined in a variety of ways (Boons and McMeekin 2019b). This key phrase (and similar ones such as "environmental innovation") generally denotes technological change where either the process of innovation, or its outcome, is assessed by participants or observers on normative criteria related to sustainable development. "Sustainable development", and the nominalised adjective "sustainability" are in themselves contested and multifaceted concepts (Castro 2004; Weber 2017; Connelly 2007; Mebratu 1998). The term sustainability was used in the context of natural ecosystems in the context of forestry, where a sustainable yield is defined in terms of harvesting trees without destroying the forest ecosystem. In the 1980s, the United Nations Brundtland Commission linked this term to global development. As a process of development, it refers to "development that meets the needs of the present without compromising the ability of future generations to meet their own needs" (Brundtland and Khalid 1987). In many of its conceptualisations there is a reference to three dimensions of sustainable development: economic, ecological and social. Using that framing, assessing innovations in terms of sustainability implies technological change that balances positive, or at least neutral, impact on all three dimensions.

While researchers and practitioners of sustainable innovation implicitly posit the unique nature of this form of innovation in comparison with innovation in general, there is an incomplete understanding of what constitutes this uniqueness (Boons and McMeekin 2019a). In practice, linking the notions of sustainability and innovation may invoke the promise of successfully addressing global challenges such as poverty, climate change and the accumulation of waste in natural ecologies; alternatively, it relates to contestation of technological change which is seen as a threat to achieve sustainable development.

Historically, the interest in sustainable innovation emerged in different disciplines. In part this is a consequence of the interdisciplinary nature of the field of innovation studies. But, at various points in time, sustainable innovation also became a topic of interest within environmental sociology, business and management studies and engineering. In each of these disciplines, distinct approaches to understanding the relationship between innovation and sustainability emerged. We will refer to such distinct streams of research as lineages, a concept that is further discussed below.

The multi-stranded nature of the field of sustainable innovation requires a careful mapping based in a conceptual notion of science that highlights emergence and change. Previous work has provided a qualitative longitudinal overview of the field (Boons and McMeekin 2019b). In this chapter we advance a quantitative approach which applies text mining techniques to a corpus of academic literature. This analytical approach is based in a processual understanding of academic fields. In this perspective, we seek to uncover academic work on sustainable innovation as it emerges. More specifically, we will answer two questions:

*What distinctive lineages of work make up the evolving field of sustainable innovation?*

*What language is used in these lineages to distinguish them from innovation in general?*

In this chapter we seek to empirically assess the way in which the uniqueness of sustainable innovation has been articulated over the period 1995–2017. We view sustainable innovation as a multi-stranded conceptual line of enquiry that emerged in the 1990s in academic (and policy) discourse. Using a combination of data science and interpretive analytical methods we intend to uncover how the academic part of this field emerged and sustained itself by articulations of distinctiveness.

In Section 2 we outline our perspective on science as a process, to provide a conceptual grounding for our work. Section 3 draws on earlier work and summarises the academic fields from which work on sustainable innovation emerged. Section 4 describes the way we constructed a corpus of academic research to enable a quantitative analysis, as well as the methodological approach to extract lineages and to characterise them based on language use.

Section 5 presents the results of the analysis in two parts: (a) the lineages of academic work that together constitute the field of sustainable innovation in the period 1995–2018 and (b) the distinctive language employed within those lineages to position sustainable innovation is positioned as distinct from innovation in general.

## SCIENCE AS A PROCESS

Although in this chapter we have a strong focus on empirical analysis, we want to avoid the trap of not having a grounding in an explicit conceptual perspective on how academic work on sustainable innovation comes about and evolves over time (see Cobo et al. (2011) for an example of a 'theoryless' approach to science mapping). As a distinctive strand in the study of science, evolutionary epistemology is well suited to provide insight into the emergence of scientific concepts, despite the criticism it has drawn (Bradie 1986). Our perspective (see also Mahanty et al. 2019) resonates with this, and draws on the work of Hull (2010), Toulmin (1973) and Abbott (2016).

We view as central to science the ideas and concepts, and the language used expressing them, as these are articulated in academic journals. Academics interact to produce these articulations, and the motives involved in these interactions provide the micro-mechanisms that may help explain how ideas and concepts evolve (Hull 2010; Abbott 2010). In short, academics need to strike a balance between generating original ideas and concepts, while staying close enough to the existing discourse to retain an audience that can help ensure longevity of the novelty that is proposed.

Our research questions do not require us to analyse these interactions in depth; they do however suggest that concepts and ideas evolve in distinct lineages (Abbott 2016). Theoretically, a lineage refers to an unfolding body of work produced by a community of researchers interacting with each other in a shared language. For our analysis, we operationalise a lineage as a sequence of articulations which refer to each other as indicated by using a similar profile in keywords. The emergence and prolongation of a shared language is a result of the distinct communities in which academics organise themselves (Abbott 1999; Trowler 2001). At the same time, especially in emerging areas of inquiry, the boundaries of such communities can be fluid, resulting in the cross-pollination of ideas (for instance through metaphorical use of concepts from another discipline) (Mirowski 1994).

## THE FIELDS FROM WHICH SUSTAINABLE INNOVATION EMERGES

Drawing on earlier qualitative work (Boons and McMeekin 2019b) we can describe the emergence of research on sustainable innovation as the response to increased awareness of environmental problems from academics in distinct fields. These fields include:

a.  *The economics of technological change* – economists working to endogenise technological change into economic theories and models picked up on the report from the Club of Rome, which in the early 1970s stirred a debate about the environmental limits to growth. This led scholars such as Chris Freeman to develop insights into guided innovation and the specific dynamics of green economic growth, that is, economic growth that stays within what we now call planetary boundaries.

b.  *Sociology of science and technology* – sociologists seeking to provide insight into techno-logical change and the role of technology and artefacts in shaping, and being shaped by, social action became interested in the social consequences of technology. This fed into the development of constructive technology assessment which includes sustainability criteria.

c.  *Environmental sociology* – In the 1990s theorising the environment in relation to society led to the formulation of Ecological Modernisation Theory (EMT). EMT postulates relationships between the forms of technology that are developed in highly industrialised societies in response to environmental problems, and the institutional conditions needed to develop and diffuse those technologies.

d.  *Ecological Economics and Industrial Ecology* – Environmental scientists who advocated systems approaches to understand and address environmental problems developed holistic perspectives such as ecological economics and industrial ecology. These newly emerging academic fields both build on an understanding of technological change as part of the whole system that needs to change in order to achieve more sustainable societal states.

e.  *Management and organisation studies* – The social responsibility of organisations and firms in particular, has been a long-standing topic in this field. With increasing public environmental awareness, this lineage of research underwent a series of transformations, first focusing on (operational) management of environmental issues, then moving to envi-ronmental strategies. In parallel, topics such as Corporate Social Responsibility and later sustainable business models captured the concerns of practitioners as they evolved over time.

As the diversity of these fields suggests, work on sustainable innovation draws on economics, sociology, management and organisation studies. These are all social science perspectives that, in different ways, have sought to theorise and conceptualise technological change as a social phenomenon. In addition, engineering and environmental sciences have developed the assessment of existing and new technologies along sustainability criteria. This constitutes the

*Table 7.1*     *Keywords used in retrieving articles from the Scopus database*

| Abbreviation | Subject Area | Keywords |
|---|---|---|
| ECO | Eco-Innovation | "eco innovation" OR "eco-innovation" |
| EMT | Ecological Modernisation theory | "ecological modernisation" AND technolog* |
| ENV | Environmental Innovation | "environment* innovation" |
| RES | Responsible Innovation | "responsible innovation" |
| SBM | Sustainable Business Models | "business model" AND sustaina* AND innovation |
| SNM | Strategic Niche Management | "strategic niche management" AND technolog* |
| TEN | Technological Change and Environmental Impact | "technolog* change" AND "environmental impact" |
| TGR | Technological Change & Green | "technolog* change" AND green |
| TRS | Socio-technical Transition and Sustainability | transition AND sustaina* AND "socio-technical" |
| TSU | Technological Change and Sustainability | "technolog* change" AND sustaina* |
| SI | Sustainable Innovation (in general) | "sustaina* innovation" |

breadth of the academic effort to understand what we defined in our introduction as sustainable innovation.

## METHODOLOGICAL APPROACH

The empirical analysis of evolving scientific work based on academic articles typically takes two forms. One way is to look at dynamic networks of co-citation. This approach rests on links between an article and previous work as explicitly stated by the author(s). As Callon et al. (1983, 1991) point out, this approach has some limitations and word co-occurrence analysis was proposed as an alternative. Our own approach similarly makes use of word co-occurrence information and additionally provides models representative of topics appearing in a corpus of academic articles.

### Data Collection

Our corpus consists of article titles and abstracts collected using the Scopus database.[1] First, we identified keywords for each of the following subject areas within the field of Sustainable Innovation: Eco-Innovation (ECO), Ecological Modernisation theory (EMT), Environmental Innovation (ENV), Responsible Innovation (RES), Sustainable Business Models (SBM), Strategic Niche Management (SNM), Technological Change and Environmental Impact (TEN), Technological Change and Green (TGR), Socio-technical Transition and Sustainability (TRS), Technological Change and Sustainability (TSU) and Sustainable Innovation in general (SI). The Scopus database was queried with these keywords (shown in Table 7.1), resulting in total of 4028 unique documents (i.e., titles and corresponding abstracts) published between 1970 and 2017 (inclusive).

### Topic Modelling

As a first step towards finding lineages, we sought to identify any topics that have been discussed in academic literature, while observing *when* these topics emerged, became merged with other topics, or disappeared completely. To this end, we employed *topic modelling*, a statistical approach that induces non-exclusive groupings of words (i.e., topics), based on their distribution in a corpus (Nikolenko et al. 2017). Going beyond simple co-occurrence-based analysis, it is able to uncover latent semantic structures on documents, and has the potential to track incremental change in language use over time (Jacobs and Tschötschel 2019).

The underlying assumption of topic modelling is that a document (e.g., an academic article) consists of a mixture of $k$ different topics, and that the words contained in that document were "generated" by its writer by way of discussing those topics. Algorithms for topic modelling, such as Latent Dirichlet Allocation (Blei and Jordan 2003) fit a model that produces words associated with a given topic, by observing frequency and co-occurrence patterns in a corpus of documents. Topic models can help answer analytical questions such as:

- Which words are representative of a given topic?
- Which topics (contexts) are certain keywords most strongly associated with?
- Which topics are most substantially discussed in a given document?

In this work, we are interested in identifying topics that were being discussed in academic literature as time progressed. Hence, our corpus was subdivided into smaller subsets based on article publication dates, grouping the documents according to four-year sliding time windows, for example 1970–1973, 1971–1974, 1972–1975 and 2014–2017. After elimination of subsets that contain less than 100 documents, only the 20 subsets corresponding to the sliding windows from 1995 to 2017 were retained. All of the documents in the remaining subsets were pre-processed whereby all characters were converted to lowercase (e.g., from "Climate Change" to "climate change"), punctuation symbols, numbers and stop words were removed, and words were stemmed, that is, converted to their root words (e.g., from "technology" and "technologies" to "technolog").

The pre-processed text formed the basis for building topic models for each time window (e.g., 1995–1998). The Structural Topic Model (STM) R package[2] for topic modelling was used, specifying 10 as the value of $k$ (the number of topics we wish to induce). This resulted in the identification of 10 topics across articles published within every four-year time window. Table 7.2 shows the set of 10 topics identified based on documents in each of the 2004–2007 and 2005–2008 subsets, represented in terms of the 12 words that are most associated with a topic (i.e., words with the highest probability of being produced by a topic).

**Topic Continuity Detection**

In order to bring out lineages, a method for detecting topic continuity was developed. Specifically, we automatically detected whether a topic identified in a given time window (e.g., 1995–1998) is highly similar to a topic identified in the succeeding time window (e.g., 1996–1999). This was carried out by calculating the Jaccard Similarity Index (Fletcher and Islam 2018): the ratio between the number of shared words (i.e., the intersection) between two sets of keywords, and the total number of unique words in both sets (i.e., the union). For example, Topic 5 in the 2004–2007 window and Topic 7 in the 2005–2008 window (cf. Table 2) share 9 words and have 15 unique words combined; hence the similarity between them, if calculated using the Jaccard Similarity Index, is 9/15 or 0.60. In our work, a Jaccard similarity value of at least 0.4 calculated for a topic $t_w$ in a four-year time window and a topic $t_{w+1}$ in the next time window, is taken to indicate substantial similarity between the two topics.

Topic $t_w$ is linked to any similar topics in the next time window. For example, having obtained 0.60 as the Jaccard similarity value for Topic 5 in the 2004–2007 window and Topic

*Table 7.2*     *Examples of identified topics*

| 2004–2007 | | 2005–2008 | |
| --- | --- | --- | --- |
| Topic no. | 10 most associated words | Topic no. | 10 most associated words |
| 1 | energy, technolog, emiss, polici, cost, carbon, chang, fuel, system, develop, trade, climat | 1 | polici, environment, innov, technolog, process, industri, regul, instrument, paper, pollut, new, research |
| 2 | sustain, chang, organ, new, technolog, work, paper, develop, inform, innov, studi, design | 2 | environment, industri, innov, product, technolog, firm, adopt, manufactur, paper, manag, compani, use |
| 3 | industri, technolog, environment, materi, product, sector, paper, new, sustain, ecolog, use, practice | 3 | technolog, chang, model, growth, product, econom, use, emiss, agricultur, climat, analysi, impact |
| 4 | innov, process, knowledg, sustain, system, develop, manag, network, environ, firm, technolog, compani | 4 | innov, sustain, knowledg, process, develop, network, busi, build, model, paper, need, research |
| **5** | **transit, system, sustain, manag, chang, develop, technolog, approach, institut, process, problem, econom** | 5 | technolog, chang, new, market, system, will, environ, develop, manag, growth, control, trade |
| 6 | product, environment, use, design, develop, system, paper, technolog, water, manufactur, can, recycl | 6 | product, environment, design, develop, studi, use, eco-innov, new, perform, improv, method, integr |
| 7 | environment, innov, polici, technolog, regul, product, adopt, improv, process, ecolog, develop, industry | 7 | **sustain, transit, system, develop, technolog, chang, manag, social, approach, process, innov, framework** |
| 8 | sustain, develop, social, technolog, cultur, chang, societi, scienc, discuss, can, environ, framework | 8 | technolog, develop, research, water, design, paper, system, learn, project, new, studi, evalu |
| 9 | agricultur, technolog, chang, use, water, new, product, system, land, design, increas, inform | 9 | energi, technolog, develop, emiss, chang, environment, polici, ecolog, renew, use, market, fuel |
| 10 | technolog, model, chang, econom, growth, use, develop, region, product, climat, effect, global | 10 | cultur, technolog, ecolog, scienc, develop, social, chang, area, natur, use, societi, discuss |

*Note:*     Identified topics are from within two sample time windows: 2004–2007 and 2005–2008, each represented by a ranked list of the 12 most associated (root) words. The topics highlighted in bold are considered as belonging to one lineage by way of similarity, calculated using the Jaccard Similarity Index.

7 in the 2005–2008 window, we consider them as belonging to the same lineage and connect them.

If multiple topics in the next time window are substantially similar (having a Jaccard similarity value of at least 0.4), $t_w$ is linked to all of them. If none of the topics in the next time window is similar enough (i.e., none of the topics obtained a Jaccard similarity value of at least 0.4), no connection to topic $t_w$ is made. Following the connections between topics across consecutive time windows allows for observing the extent to which a topic "persisted" over time.

## RESULTS

Our quantitative analysis produces a set of topics that link across time. These linked topics are presented in Table 7.3. We have checked the extent to which these linked topics reproduce our keyword searches; this is clearly not the case. The analysis produces something which differs from the set of articles that is defined by using the keywords we used when searching Scopus. Below we first show how our method provides a way of identifying lineages of research. We then discuss the language that is distinctive for research on innovation that deals with sustainable development.

**From Time-Linked Topics to Lineages**

Based on our interpretation of the quantitative results we identify several types of lineages. We do so by inspecting the time-linked topics that result from our method, and which are presented in Table 7.3. The 28 items in this list differ substantially in terms of the time period covered and the (root) words that characterise them.

**(a) Long-Term Lineages**

Three lineages extend over at least 9 years. The earliest is captured as item #3 in Table 7.3. It has a variegated profile of (root) words that identify it, namely environmental technology and policy in relation to industry and business. This lineage includes classical papers on environmental innovation, such as Rennings' (2000) paper on redefining innovation, work on the Environmental Kuznetz Curve (de Bruyn et al. 1998), as well as work that investigates the diffusion of environmental technology in industry (Henriques and Sadorsky 2007). Within this lineage we also find work on Ecological Modernisation Theory, of which the strand developed by Huber can be seen to fit that profile. In terms of a community of researchers that produces this work, the early period of the Greening of Industry network fits quite well.

   The second lineage is constituted by item #9 in Table 7.3, and from its (root)words is easily identifiable as work using the framing of transitions theory. Interestingly, this lineage contains a significant proportion of work that alludes to this theory, rather than focuses on it. As the first one, this lineage has a distinct community of researchers which shapes this work: ISST; it also has its dedicated journal in *Environmental Innovation and Sustainable Transitions*.

   A third long-term lineage is item #15 in Table 7.3. Here the (root)words suggest research on technological change in relation to emissions that are responsible for climate change. Unlike the previous lineage, which is defined in terms of theoretical/conceptual language, this third lineage contains modelling work relating to a specific aspect of (ecological) sustainability. Typical work in this lineage is Davis et al.'s (2010) work on modelling the future $CO_2$ emissions and climate change from existing energy infrastructure. It is much more difficult to assign this lineage to a recognisable community of research like the previous two.

**(b) Short-Term Lineages**

There is a substantial number of entries in Table 7.3 that show topic continuity over a relatively short period; this suggests that, unless they appear at the end of the period analysed, research in these items has less continuity in the field of sustainable innovation. A good example is item #14, which distinguishes itself through a unique (root)word: cultur*. This lineage contains work that investigates technological change from a social science perspective, including studies on consumption (for instance Juliet Schor's (2005) work on sustainable consumption and worktime reduction); which are distinctive as for many of the other items in Table 7.3, (root)words such as firm, industr* and busi* indicate a focus on production. This lineage also includes papers on ecological footprint analysis (i.e., Fiala 2008). Looking at all items that constitute this lineage, this work is more difficult to see as the product of a distinct community

*Table 7.3*    *Topics linked across time (N=28) in the sustainable innovation literature in the period 1995–2017, identified by top (root) words*

| item # | Top (root)words | start | end |
|---|---|---|---|
| 1 | environment; develop; econom* | 1995 | 1999 |
| 2 | technolog*; chang*; agricultur* | 1995 | 2000 |
| 3 | environment; technolog*; polici*; ecolog*; modernis* | 1996 | 2008 |
| 4 | paper; chang*; technolog* | 1996 | 2001 |
| 5 | sustain*; technolog*; develop*; polici* | 1996 | 2003 |
| 6 | environment; innov*; firm; technolog* | 1997 | 2005 |
| 7 | technolog*; chang*; agricultur* | 1998 | 2003 |
| 8 | paper; technolog*; environment; modern*; ecolog* | 1999 | 2003 |
| 9 | transit*; sustain*; system; technolog* | 2001 | 2017 |
| 10 | environment; chang*; water | 2001 | 2005 |
| 11 | sustain*; innov*; process | 2001 | 2007 |
| 12 | polici*; chang*; technolog* | 2002 | 2006 |
| 13 | environment; product; innov* | 2002 | 2006 |
| 14 | cultur*; technolog*; sustain*; social | 2002 | 2008 |
| 15 | technolog*; chang*; model; emiss*; climat* | 2003 | 2017 |
| 16 | technolog*; chang*; inform* | 2003 | 2007 |
| 17 | energi*; technolog*; polici*; emiss* | 2004 | 2009 |
| 18 | environment; innov*; industri* | 2005 | 2009 |
| 19 | product; eco-innov*; use | 2007 | 2017 |
| 20 | technolog*; research; develop*; design | 2007 | 2013 |
| 21 | develop*; econom*; sustain*; ecolog* | 2008 | 2013 |
| 22 | innov*; sustain*; busi*; model | 2008 | 2015 |
| 23 | environment; innov*; firm | 2008 | 2013 |
| 24 | energi*; electri* | 2009 | 2013 |
| 25 | respons*; social; technolog*; innov* | 2012 | 2016 |
| 26 | energi*; technolog*; electr* | 2012 | 2016 |
| 27 | environment; eco-innov*; firm; perform* | 2012 | 2017 |
| 28 | sustain*; busi*; model | 2013 | 2017 |

of researchers; instead, they capture a topic that is peripheral to the field, and only at certain moments in time has enough critical mass to show up in the analysis.

Item #25 is a lineage that is short, but constitutes a more central concern: responsible innovation. This lineage contains key articles such as Stilgoe et al. (2013) which have defined this lineage of research (see Cuppen et al. 2019); it also includes more general work looking at the ethical dimensions related to innovation. It is somewhat surprising that the lineage does not extend into the final time window; there is no apparent reason for this.

**(c) Broken Lineages**

A third category of lineages consists of work that in the analysis is broken up into several pieces; these enter as separate items in Table 7.3, but based on similarity in top (root)words, as well as similarity in the articles that constitute the basis for these items, they can be interpreted as a continuous stream of work.

An example of a broken lineage is the work on sustainable business models that deals with innovation and technological change. It appears as items #22 and #28 in Table 7.3. There is no clear interpretive reason for these items to be separate; one indicator for this is that the work of one of the key authors in this lineage, Nancy Bocken, appears in item #22 as well as in item #28. Also, in terms of Journals and Proceedings, there is no clear distinguishing feature.

**General Observations**

The three longer lineages identified constitute distinct areas of inquiry at the intersection of sustainable development and innovation/technological change. During the 1990s, the Greening of Industry network was a key platform for researchers that sought to address questions about how to improve the environmental performance of firms and industries. Technological change and processes of innovation were a key focal point in the discussions, in addition to work on policy and environmental management systems. Over time, this community dissolved as other platforms took precedence and different framings gained ground.

One such framing is to analyse change in sociotechnical systems in terms of transitions. This perspective emerged in the field of innovation studies at the end of the 1980s; much of its empirical applications are on technologies that are developed with the explicit aim to move towards more sustainable systems, that is, sustainable mobility and renewable energy systems. In the field of sustainable innovation it emerges in 2001. Its appearance as the longest separate lineage is helped by the distinct language employed, in particular the term 'transition' in relation to (socio-) technological systems (see below).

The third long-term lineage is different as it revolves around a specific sustainability issue, climate change, rather than around a community and/or theoretical approach. This is characteristic in the sense that some of the other, shorter, lineages also revolve around specific problems or even industries: item #8 for instance, which has work on environmental innovation in the paper industry, or item #2 which deals with sustainable innovation in relation to agriculture (for instance Foster and Rosenzweig 1995).

Some lineages are not constituted so much by a conceptually distinct approach; rather, they are formed by work that covers a specific empirical area, such as the agricultural sector or energy systems. While our conceptual framework was formulated to focus on theoretical concepts and ideas; there is, however, no reason why areas of application cannot form the basis for a lineage in the field of sustainable innovation, especially given its problem-oriented nature. It resonates with Toulmin's (1973) notion of problem solving as a selection context for the evolution of concepts.

**Language**

Turning our attention now to the actual ideas and concepts, we use the root words that characterise each lineage to get a sense of the specificity of separate lineages, as well as that of the field as a whole. Table 7.4 shows how often root words appear and in how many lineages, ranked from the root word occurring in the most lineages. Here we get closer to observing in what way work on sustainable innovation is distinct from innovation in general.

**Observation 1: The Study of Sustainable Innovation is Part of the Wider Field of Innovation Studies**

High-scoring root words clearly indicate how the field employs language that fits the generic topic of innovation: 'technolog', often co-occurring with 'change', 'innov' and 'system' do

*Table 7.4*      *Ranked occurrence of root words in lineages*

| (root) word | Sum of appearances | Number of lineages | (root) word | Sum of appearances | Number of lineage |
|---|---|---|---|---|---|
| technolog | 92 | 25 | effect | 14 | 9 |
| chang | 72 | 19 | adopt | 13 | 6 |
| develop | 71 | 18 | impact | 13 | 5 |
| environment | 51 | 17 | transit | 13 | 1 |
| innov | 50 | 17 | social | 12 | 6 |
| polici | 48 | 15 | design | 12 | 7 |
| sustain | 46 | 12 | agricultur | 12 | 5 |
| product | 42 | 13 | cost | 12 | 4 |
| use | 39 | 15 | eco-innov | 11 | 4 |
| paper | 37 | 16 | can | 11 | 8 |
| econom | 33 | 11 | global | 11 | 6 |
| manag | 31 | 12 | busi | 10 | 4 |
| industri | 31 | 12 | regul | 10 | 4 |
| model | 29 | 9 | compani | 9 | 6 |
| system | 28 | 10 | result | 9 | 5 |
| new | 27 | 16 | market | 9 | 5 |
| process | 26 | 9 | increas | 9 | 3 |
| energi | 23 | 8 | practic | 8 | 5 |
| studi | 21 | 13 | green | 8 | 4 |
| emiss | 20 | 7 | govern | 8 | 2 |
| ecolog | 20 | 9 | approach | 8 | 2 |
| firm | 16 | 7 | valu | 7 | 2 |
| research | 16 | 8 | perform | 7 | 3 |
| growth | 16 | 7 | water | 7 | 5 |
| climat | 15 | 6 | carbon | 7 | 5 |

not distinguish the work from innovation studies in general and signify that we are dealing with a subfield of that wider area of study. A similar argument can be made for other topics that are covered by the field: for instance, the general language of "business model innovation" is used in relation to sustainability, and this work is taken in literature reviews as part of the generic body of work on business models (Massa et al. 2017).

## Observation 2: Sustainability is Referred to Through Aspects

The root word 'sustain' relates to our key defining concept in its different forms. Interestingly, it helps defining less than half (12/28) of the lineages making up the field. This shows how the specific 'sustainability' character of lineages is often defined through aspects of sustainability. The highest scoring 'sustainability-specific' root word is 'environ', which in the form of environment refers to one of the defining distinguishing characteristics of the field, that is, the relationship between technological change and environmental impact. There is a relative absence, or at least much lower ranking, of (root) words referring to the economic and social dimension, although some lineages do refer to "growth" and "econom*". Also, in relation to firms, the root word "perform*" refers to performance, and this usually deals with questions of how environmental and economic/financial performance of firms correlate (for instance Wagner 2010). As qualitative work has shown, key lineages focus on eco-innovation and environmental innovation (Boons and McMeekin 2019b), so the lack of emphasis on the social dimension is not surprising.

**Observation 3: Root Words that Define the Specific Nature of the Field are Multi-Word Expressions Rather than Individual Root Words**

Lineages often show a specific combination of root words that distinguishes them from others. One form is by putting the word 'sustainable' in front of an otherwise generic term: this is the case for our label of the whole field, but it also happens in conjunctions such as 'sustainable business model innovation', 'sustainable transitions'. But terms such as 'green technologies' and 'responsible innovation' likewise are distinctive in their combination, rather than as individual root words. This can be taken as an indication that researchers in the field are, at least in parts, explicitly defining their work as constituting a subfield. This is corroborated by reviews of parts of the field. For instance, the work on systems innovation often uses sustainable technologies as its empirical focus; it is seen by many as an area of application of concepts of system innovation which are deemed to be generic, rather than as the specific study of innovation of sustainable systems (Bergek 2019).

**Observation 4: Few Lineages Have Unique Root Words**

Some root words appear in one lineage only, but these do not appear in the first fifty (root) words shown in Table 7.4. As noted above, this is true for the word 'culture', which occurs only in item #14. This is also the case for 'ethic', which appears only in item #25 (responsible innovation). This refers to the inclusion of ethical considerations into the conceptual framing of innovation processes. Such unique words do not occur often though; most items listed in Table 7.1 combine terms that are used in other items as well.

## DISCUSSION

Our analysis provides an insight into the field of sustainable innovation that is complementary to a qualitative interpretation of its unfolding as developed elsewhere (Boons and McMeekin 2019b). While it does justice to the multi-stranded nature of academic work that relates concerns about sustainable development to technological change, without further systematic interpretive steps it leads to a conclusion of high fragmentation. In addition to three longer term lineages of research we find a high number of shorter lineages. Closer inspection reveals that at least some of these shorter lineages constitute a single strand of research, but it can certainly not be argued for all of the shorter items in Table 7.1.

One way to interpret this is to accept that the field of sustainable innovation is quite fragmented. Such fragmentation is likely to be a product of the diversity of theoretical backgrounds (ranging from economics to environmental sociology), as well as the plurality of objectives (from theoretical to normative). This fragmentation shows itself in the range of outlet in which research appears, as well as the variegated language that is used to denote the phenomenon of interest: the way in which technological change constitutes a cause of environmental problems, as well as a possible pathway towards addressing such problems. The social dimension of sustainability is only marginally represented in this field: a distinct lineage on responsible innovation is the main evidence for this. Other social aspects, such as well-being and life styles, do not appear key (root)words in the topic modelling.

An alternative, or perhaps complementary interpretation builds on the observation of the frequent use of multi-word expressions. We saw how this can be interpreted as researchers coining terms where more generic concepts are applied to the empirical problem area of sustainability. Given this problem-orientation, a variety of disciplinary insights is brought to bear on the subject. This variety is mapped by our method.

A reflection on the method used is in order. The identification of a lineage based on a quantitative topic modelling approach demands an interpretive step. Specifically, in the presented data we used a relatively strict criterion for linking clusters across time slices. As a result, we get separate lineages that upon inspection seem to be connected, given our conceptual definition of a lineage. This suggests the need to complement the quantitative method with an interpretive step. Alternatively, the analysis of citations could be used for triangulation. This would bar further analysis of the interplay between conceptual evolution and citation patterns. As our clustering generates meaningful results that differ from the data we used as an input (i.e., sets of articles based on keyword searches in the Scopus database), the analysis adds insight by showing how some research lineages are more robust than others.

## NOTES

1. https://www.scopus.com/.
2. http://www.structuraltopicmodel.com/.

## REFERENCES

Abbott, Andrew (1999), *Department and Discipline: Sociology at One Hundred*, Chicago: University of Chicago Press.

Abbott, Andrew (2010), *Chaos of Disciplines*, Chicago: University of Chicago Press.

Abbott, Andrew (2016), *Processual Sociology*, Chicago: University of Chicago Press.

Bergek, Anna (2019), 'Technological innovation systems: a review of recent findings and suggestions for future research', in Frank Boons and Andrew McMeekin (eds), *Handbook of Sustainable Innovation*, Cheltenham, UK and Northampton, MA, USA: Edward Elgar Publishing, pp. 200–18.

Blei, David M., Andrew Y. Ng and Michael Jordan (2003), 'Latent dirichlet allocation', *Journal of Machine Learning Research*, 3 (Jan), 993–1022.

Boons, Frank A. and Andrew McMeekin (eds) (2019a), *Handbook of Sustainable Innovation*, Cheltenham, UK and Northampton, MA, USA: Edward Elgar Publishing.

Boons, Frank and Andrew McMeekin (2019b), 'An introduction: mapping the field (s) of sustainable innovation', in Frank Boons and Andrew McMeekin (eds), *Handbook of Sustainable Innovation*, Cheltenham, UK and Northampton, MA, USA: Edward Elgar Publishing, pp. 1–25.

Bradie, Michael (1986), 'Assessing evolutionary epistemology', *Biology and Philosophy*, 1 (4), 401–59.

Brundtland, Gro H. and Mansour Khalid (1987), *Our common future*, New York: United Nations.

de Bruyn, Sander, Jeroen van den Bergh and J. Hans B. Opschoor (1998), 'Economic growth and emissions: reconsidering the empirical basis of environmental Kuznets curves', *Ecological Economics*, 25 (2), 161–75.

Callon, Michel, Jean-Pierre Courtial and Françoise Laville (1991), 'Co-word analysis as a tool for describing the network of interactions between basic and technological research: the case of polymer chemistry', *Scientometrics*, 22 (1), 155–205.

Callon, Michel, Jean-Pierre Courtial, William A. Turner and Serge Bauin, S. (1983), 'From translations to problematic networks: an introduction to co-word analysis', *Information (International Social Science Council)*, 22 (2) 191–235.

Castro, Carlos J. (2004), 'Sustainable development: mainstream and critical perspectives', *Organization & Environment*, 17 (2) 195–225.

Cobo, Manuel J., Antonio G. López-Herrera, Enrique Herrera-Viedma and Francisco Herrera (2011), 'An approach for detecting, quantifying, and visualizing the evolution of a research field: a practical application to the fuzzy sets theory field', *Journal of Informetrics*, 5 (1), 146–66.

Connelly, Steve (2007), 'Mapping sustainable development as a contested concept', *Local Environment*, 12 (3), 259–78.

Cuppen, Eefje, Elisabeth van de Grift and Udo Pesch (2019), 'Reviewing responsible research and innovation: lessons for a sustainable innovation research agenda?', in Frank Boons and Andrew McMeekin (eds), *Handbook of Sustainable Innovation*, Cheltenham, UK and Northampton, MA, USA: Edward Elgar Publishing, pp. 142–64.

Davis, Steven J., Ken Caldeira and H. Damon Matthews (2010), 'Future $CO_2$ emissions and climate change from existing energy infrastructure', *Science*, 329, no. 5997, 1330–1333.

Fiala, Nathan (2008), 'Measuring sustainability: why the ecological footprint is bad economics and bad environmental science', *Ecological Economics*, 67 (4), 519–25.

Fletcher, Sam and Md Zahidul Islam (2018), 'Comparing sets of patterns with the Jaccard index', *Australasian Journal of Information Systems*, 22. https://doi.org/10.3127/ajis.v22i0.1538.

Foster, Andrew D. and Mark R. Rosenzweig (1995), 'Learning by doing and learning from others: human capital and technical change in agriculture', *Journal of Political Economy*, 103 (6), 1176–209.

Henriques, Irene and Perry Sadorsky (2007), 'Environmental technical and administrative innovations in the Canadian manufacturing industry', *Business Strategy and the Environment*, 16 (2), 119–132.

Hull, David L. (2010), *Science as a Process: An Evolutionary Account of the Social and Conceptual Development of Science*, Chicago: University of Chicago Press.

Jacobs, Thomas and Robin Tschötschel (2019), 'Topic models meet discourse analysis: a quantitative tool for a qualitative approach', *International Journal of Social Research Methodology*, 22 (5), 469–85.

Mahanty, Sampriti, Frank Boons, Julia Handl and Riza Batista-Navarro (2019, November), 'Studying the evolution of the "circular economy" concept using topic modelling', in *International Conference on Intelligent Data Engineering and Automated Learning*, Cham: Springer, pp. 259–70.

Massa, Lorenzo, Christopher L. Tucci and Allan Afuah (2017), 'A critical assessment of business model research', *Academy of Management Annals*, 11 (1), 73–104.

Mebratu, Desta (1998), 'Sustainability and sustainable development: historical and conceptual review', *Environmental Impact Assessment Review*, 18 (6), 493–520.

Mirowski, Philip (ed.), (1994), *Natural Images in Economic Thought: Markets Read in Tooth and Claw*, Cambridge: Cambridge University Press.

Nikolenko, Sergey I., Sergei Koltcov and Olessia Koltsova (2017), 'Topic modelling for qualitative studies', *Journal of Information Science*, 43 (1), 88–102.

Rennings, Klaus (2000), 'Redefining innovation – eco-innovation research and the contribution from ecological economics', *Ecological Economics*, 32 (2), 319–332.

Schor, Juliet B. (2005), 'Sustainable consumption and worktime reduction', *Journal of Industrial Ecology*, 9 (1–2), 37–50.

Stilgoe, Jack, Richard Owen and Phil Macnaghten (2013), 'Developing a framework for responsible innovation', *Research Policy*, 42 (9), 1568–80.

Toulmin, Stephen (1973), 'Human understanding, Vol. I: the collective use and evolution of concepts', *Journal for General Philosophy of Science*, 4 (2), 398–402.

Trowler, Paul. R. (2001), *Academic Tribes and Territories*, London: McGraw-Hill Education (UK).

Wagner, Marcus (2010), 'The role of corporate sustainability performance for economic performance: a firm-level analysis of moderation effects', *Ecological Economics*, 69 (7), 1553–60.

Weber, Heloise (2017), 'Politics of "leaving no one behind": contesting the 2030 Sustainable Development Goals agenda', *Globalizations*, 14 (3), 399–414.

# 8.  Challenges for responsible innovation
## *Lucien von Schomberg*

## INTRODUCTION

In launching the *International Handbook on Responsible Innovation*, René von Schomberg and Jonathan Hankins (2019) bring together renowned authors from around the globe to address the need for a paradigm shift in the innovation discourse, driving innovation away from mainstream economic interests towards societally desirable outcomes. This aspiration departs from the observation that innovations currently delivered by the market insufficiently serve the public good, which thus urges the call for a new – that is, responsible – approach to innovation in fields ranging from agriculture and medicine, to nanotechnology and robotics. Although the concept of responsible innovation is subject to a variety of perspectives and assessments, proponents generally agree that the innovation process is neither inherently good nor unmanageable. They argue that by engaging governmental bodies, industries and societal actors within the innovation process, it can be regulated in accordance with the values and expectations of society, and steered towards normative goals concerning for instance global sustainable development.

The realization of responsible innovation comes with several challenges. At an *epistemic* level it faces the complexity of anticipating the unexpected outcomes of innovation, which conflicts with the ideal of steering innovation into a predetermined direction (Grunwald 2019; Nordmann 2014). With regard to this predetermined direction, frameworks of responsible innovation have also been questioned at a *political* level for insufficiently addressing the different values and interests of stakeholders involved in the innovation process (Blok 2019; van Oudheusden 2014). Moreover, at a *conceptual* level the discourse of responsible innovation is arguably confined to an intrinsic relation between technology and the market, thereby undermining its attempt to liberate innovation from economic ends (von Schomberg and Blok 2019; Blok and Lemmens 2015).

Against this background, this chapter poses the following research question: *to what extent is the attempt of responsible innovation to develop a paradigm shift in the innovation discourse feasible?* As a first step, I elaborate on the emergence of responsible innovation, particularly in relation to its precursors. I then report on the three aforementioned challenges – the epistemic, political and conceptual challenge, respectively – and discuss to what extent they bring the feasibility of responsible innovation into question. Finally, in light of the conceptual challenge, I suggest that the epistemic and political implications of responsible innovation relate to the widely presupposed concept of technological innovation and commercialized innovation. In this vein, I argue that the ideal of responsible innovation is inhibited by the overarching incentive of technological and economic progress and, as such, has difficulties to achieve the paradigm shift in question. This in turn leads me to conclude with a call for future research to explore an alternative concept of innovation which addresses the public good beyond mainstream economic thought.

## THE EMERGENCE OF RESPONSIBLE INNOVATION

Over the past decade the concept of responsible innovation has taken a central place in the discourse on science and emerging technologies. Several research funding bodies, such as the Netherlands Council for Research (NWO), have dedicated entire programs on the subject matter. The UK Engineering and Physical Sciences Research Council (EPSRC) continuously show interest in the field, along with the US National Science Foundation (NSF) which supported the construction of a range of projects, including for instance the Virtual Institute of Responsible Innovation (VIRI) and the Program on Responsible Innovation and Corporate Social Responsibility (SAMANSVAR). Likewise, China has included responsible innovation as a formal policy in their latest five-year plan on science, technology and innovation, which resulted in ongoing initiatives such as the introduction of ecological sea ports (Wang and Yan 2019). The reality of today's global issues has also urged the European Commission to introduce the concept of Responsible Research & Innovation (RRI), which was presented as a cross-cutting issue under the European Union's Framework Program for research and innovation 'Horizon 2020' (European Commission 2015).

The emergence and increased usage of the term responsible innovation implies that innovation is not always that responsible. Particularly the imperative of economic growth inherent in innovation is said to be fundamentally at odds with the imperative of solving today's societal and environmental issues. To be sure, innovation understood as "the development of new ideas into marketable products and processes" (Stoneman 1995, p. 2) – primarily focusing on delivering value to customers (Carlson and Wilmot 2006) – is arguably one of the main sources of today's increasingly unequal distribution of wealth (cf. Rolston III 2012; Naudé and Nagler 2016), and as "the root cause of many environmental problems" it stands "in direct conflict with sustainability" (Huesemann and Huesemann 2011, p. 256). For this reason, the discourse on responsible innovation calls for innovation processes to exceed the mere purpose of generating commercial value. Instead, they should primarily focus on generating the right impact, particularly with regard to today's global issues. In response to the complexity of these issues and to the indeterminacy of the right impact, a frequently cited and particularly influential definition of responsible innovation calls for "a transparent, interactive process by which societal actors and innovators become mutually responsive to each other with a view on the (ethical) acceptability, sustainability and societally desirability of the innovation process and its marketable products" (von Schomberg 2013, p. 63). A commonly used framework of responsible innovation builds on this definition by featuring four specific dimensions: anticipation, reflexivity, inclusion and responsiveness. In this view, innovators and organizations are to anticipate the future outcomes of innovation processes, reflect on what responsibilities they have as moral agents, engage with a broad variety of stakeholders, and respond to the values and changing circumstances of society. These dimensions present several governance mechanisms and management practices that claimed to enable more responsible innovation (Owen et al. 2012; Stilgoe et al. 2013).

The call to institutionalize ethical and social dimensions of science and technology is not entirely new. Risk identification and analysis, as a separate activity that is executed by unbiased professionals, dates back to late 1960s (Evers and Nowotny 1987). The call for such an activity became particularly urgent in the early 1970s when it became clear that there was no immediate solution for the storage of nuclear waste, which would eventually lead to many efforts on nuclear disarmament (Kevles 1995). Likewise, already at the initial stages of recombinant DNA research many ethical concerns were raised both within and beyond biology circles (Krimsky 1984). In addition to this, many public debates on food crises related to genetically modified organisms and other food products involving nanotechnology resulted in the early adoption of the precautionary principle, which highlights "the importance of informing people and policy makers about what is known and where uncertainty persists" (Commission of the European Communities 2001, p. 19). Later efforts of governing science and technology include technology assessment (Rip et al. 1995), science and technology studies (Hackett et al. 2007), anticipatory governance (Guston 2014), and research on ethical, legal and social implications (ELSI) – or aspects (ELSA) – of emerging technologies.

Responsible innovation does not emerge as a response to its pre-history, but much rather as an incremental reform that further builds on it (Guston and Valdivia 2015). Specific to this reform is distributing responsibility "*throughout* the innovation enterprise, locating it even at the level of scientific research practices" (Fisher and Rip 2013, p. 165, original emphasis), which differs from previous approaches that focus much more readily on the interference of democratic institutions, for example technology assessment. Moreover, responsible innovation moves "beyond an ethics of constraints (for example, focusing on what should be prohibited or limited) to an ethics of construction" in which professional bodies "look into the type of outcomes we want to achieve from research and innovation processes" (von Schomberg 2019, p. 29). For this reason, the rise of responsible innovation can be understood in relation to the precautionary principle, which does not merely point to the negative consequences of innovation, but also accounts for its potential benefits, thereby providing an incentive that paves the way for new research and development trajectories (European Environmental Agency 2002). In this respect, Miles Brundage and David Guston (2019) characterize the rise of responsible innovation as part of a scientific/intellectual movement that features "significant contestation of the knowledge core that they are oriented towards, and that much of their activity involves (re)articulating that core" (p. 106). In other words, responsible innovation is not an isolated process, but is instead inevitably intertwined with prior debates on similar topics.

Although the intimate relation between responsible innovation and its precursors evokes the impression that responsible innovation is merely an umbrella term used to denote any activity pertaining to discourse on science and emerging technologies, in some respects it promises to be much more revolutionary and fundamental. For example, responsible innovation takes a critical position against the dominance of mainstream economics, and calls for a radical transformation of the entire research and innovation system. In practice, however, responsible innovation still widely adheres to current techno-economic practices and ideologies (von Schomberg and Blok 2019). The duality at stake points to the ambiguous conceptual stance of responsible innovation, to which I will return later in this chapter.

## THE EPISTEMIC CHALLENGE OF RESPONSIBLE INNOVATION

The notion of responsible innovation does not only cover an ethical dimension, as implied by the term 'responsible', but it also includes an *epistemic* dimension. Decisions made throughout innovation processes require values and criteria as well as valid and reliable knowledge of the outcomes and impacts of these decisions. Grunwald (2019) points out that without this knowledge "any ethics of responsibility may well fail, lead to arbitrary conclusions (Hansson 2006) or end up in political rhetoric and appeals without practical impacts" (p. 326). A crucial characteristic of innovation is that its outcomes cannot always be known (Rammert 1997), which thus conflicts with the ideal of responsible innovation to *steer* innovation into a responsible and desirable direction. This concern is illustrated in cases such as the development of biofuel, in which the involved stakeholders concluded that since biofuel is inherently renewable, locally produced and less polluting, its introduction to the market promises responsible and desirable outcomes. However, as a result of the higher demand for biofuels, farmers needed to grow more crops for biofuel production, which in turn led to an increase in the food prices. An increase in the price of food was not initially anticipated and now raises the question if introducing biofuels was in fact responsible and desirable, especially considering that people in developing countries were harmed by this unforeseen outcome (Blok and Lemmens 2015).

The epistemic challenge at stake – also identified by the literature as the challenge of 'epistemic insufficiency' (Blok and Lemmens 2015) – is particularly prominent in the field of new and emerging science and technology (NEST). While the effects of technologies such as synthetic biology are unpredictable, they are said to be deep-ranging and revolutionary (Ilulissat Statement 2007). The result of this radical uncertainty is that the 'responsibility' of responsible innovation no longer has reasonable purpose (Bechmann 1993) or at most becomes principally limited:

> If the output of responsible innovation processes is characterized by a fundamental uncertainty, which means that our *knowledge* of the impact of our innovations is not only limited but principally *insufficient*, the presupposed 'foresight' of responsible innovation becomes questionable. In other words, our knowledge is principally insufficient to assess the impact of innovation processes and there will always be unintended consequences of our innovations which can be harmful. (Blok and Lemmens 2015, p. 28)

The unexpected outcomes of innovation thus bring into question the extent to which innovation can be steered, let alone into a responsible and desirable direction. This in turn hinders the attempt of responsible innovation to develop a paradigm shift in the innovation discourse, which for a great part consists precisely in the call for steering innovation.

The discourse of responsible innovation acknowledges the above problematic, but specifies that while "an ethics focused on the intentions and/or consequences of actions of individuals is not appropriate for innovation," an ethics of "collective co-responsibility" is (von Schomberg 2013, p. 59). Here R. von Schomberg stresses that since modern innovations are not intentionally created by a single actor, the unforeseen effects are more likely the result of collective action. In a recent article, R. von Schomberg (2020) provides further substance to this view by articulating the ethics of responsibility that underlies the notion of responsible innovation. Departing from Karl-Otto Apels' diagnosis of the shortcomings of philosophical ethics, particularly those concerning individual accountability, he demonstrates how these shortcomings currently prevail in the context of an ecological crisis and socio-technical change. To this end, he suggests that under the sway of responsible innovation a further social evolution of the systems of science, economy and law will enable the institutionalization of collective co-responsibility. Accordingly, responsibility should not be assigned to the individual, but instead shared by all stakeholders involved in the innovation process. This goes in line with the philosophical argument that plurality rather than singularity provides "the remedy for unpredictability, for the chaotic uncertainty of the future" (Arendt 1998, p. 213). This idea is implicit throughout the literature on responsible innovation as reflected in arguments such as the following:

Embedding iterative risk (and benefit) analysis with technology assessment and public/stakeholder engagement approaches within innovation research proposals was seen as offering a mechanism that considers technical risk issues and associated uncertainties, but that could also provide opportunities for identifying as yet unforeseen effects (economic, societal and ethical) as these emerge. It may also facilitate upstream engagement with stakeholders and the public as to how these emerging impacts are received. (Owen and Goldberg 2010, p. 1705)

In other words, in the face of uncertainty, an inclusive approach to innovation can still ensure the uptake of societal values and concerns.

In a similar vein, Nordmann (2010) warns us that responsibility debates concerning new and emerging technologies should focus on wishful futures rather than on speculative antic-ipations. This enables more visionary and critical ideas for improving the future. Likewise, Grunwald (2019) argues that responsible innovation must accept the thesis that anticipation is impossible on any sound epistemic ground and should therefore move beyond consequen-tialist modes of orientation. In this respect, responsible innovation can gain further insight from earlier proposals, such as vision assessment (Ferrari et al. 2012), explorative philosophy (Grunwald 2010) and the various hermeneutic responses given to the unpredictable nature of emerging technologies (e.g. van der Burg 2014).

## THE POLITICAL CHALLENGE OF RESPONSIBLE INNOVATION

The uncertainty of the future results in many disagreements among stakeholders as to what the problem is and how to solve it (Kreuter et al. 2004; Batie 2008; Rittel and Webber 1973). These conflicts are often the result of opposing agendas and motives of, for example, for-profit and non-profit corporations (Yaziji and Doh 2009). In the procedure of responsible innovation, initiatives such as RRI Tools attempt to account for the different viewpoints by organizing debates with all types of stakeholders, ranging from civil society organizations to business and industry. However, power imbalances among these stakeholders – the funders of an innovation process tend to have the upper hand – often contribute to more disparities. Hence, a collective solution is difficult to reach (Bryson et al. 2006). In this respect, the epistemic challenge of responsible innovation brings us to a range of *political* questions: Who defines the grand chal-lenges? On the basis of which values and criteria should these challenges be confronted? What are the outcomes responsible innovation aims for?

There are at least three reasons for why stakeholder engagement plays a crucial role in the implementation of responsible innovation. First, in relation to the epistemic challenge, the conflicting interests and value frames of the involved stakeholders allow for a better assess-ment of the future impact of innovation processes like biotechnology and nanotechnology (Chilvers 2008). Second, it enables stakeholders to learn from each other according to which shared objectives and decisions are easier reached (Gould 2012). Third, it helps them to better understand each other's roles and interests, thereby stepping one step closer to collectively determine the direction of innovation processes (Jackson et al. 2005). Consequentially, stake-holders share knowledge and values (von Schomberg 2013), attempt to reach common objec-tives (Flipse 2012) and thus take co-responsibility for the outcomes of innovation processes (Owen et al. 2012).

However, stakeholder engagement conceptualized in light of this ideal of unity fails to embrace systemic and political issues; it lacks a second-order reflexivity that broadens normative standpoints and policies (Owen and Pansera 2019). Precisely for this reason Michiel van Oudheusden (2014) criticizes the discourse of responsible innovation because it largely ignores questions about the constitution and contestation of power:

> How do actors "co-create" outcomes? How do they deliberate? On whose terms is participation (i.e. deliberation) established and why? What, in fact, is "public" about the "public interest," "public expectations," and "the public," and whose definition of the public counts? (van Oudheusden 2014, p. 73)

Noticeably, in a recent series of workshops on the challenges of responsible innovation, scholars showed similar concern with regard to the presupposed harmony and transparency among stakeholders.[1] They pointed to a certain naivety, raising questions as to how to deal with the different values and interests of stakeholders. They pointed to problems of inclusion as well as to power imbalances that undermine shared viewpoints and mutual responsiveness and understanding. Likewise, Richard Owen and Mario Pansera (2019) demonstrate that stakeholder engagement is "usually narrowly configured to include a limited range of (internal and sometimes external) stakeholders, and that second-order reflexivity and the political are almost entirely beyond scope, or at least deeply tacit" (p. 41). Also from a business perspective the presupposed harmony and transparency can be seen as naive because it is undermined by the competitive advantage a new innovation needs in order to succeed on the market (Blok and Lemmens 2015). To achieve this competitive advantage, companies rely on information asymmetries, that is, additional knowledge they have about business opportunities that other companies are oblivious of. In the context of responsible innovation, companies pursue such information with regard to discovering new solutions for existing and anticipated grand challenges. However, transparency among the involved stakeholders naturally implies a reduction of these information asymmetries, thereby taking away the main source of competitive advantage. For reasons as these, the ideal of achieving harmonious and transparent collaboration among all stakeholders may be perceived as rather unrealistic.

Contrary to traditional conceptualizations, Vincent Blok (2019) proposes a non-reductive and ethical approach to stakeholder engagement which "does not presuppose a direct or indirect ideal of harmony and alignment [...] but acknowledges and appreciates the role of difference and constructive conflict without allowing only bridgeable or complementary difference among multiple stakeholders" (Blok 2019, p. 255). Instead of a priori conceptualizing stakeholder engagement in light of a unity among stakeholders, Blok suggests to depart from their radical differences. Contrary to the argument that today's global issues demand a dialogue in which multiple stakeholders should depart from a *common goal* (Gorman et al. 2009), he explains that for stakeholders to engage in a dialogue they first of all need to have their *own* definition of and solution to the problem, according to their own interest and value frames (van Huijstee et al. 2007). Inspired by the philosophy of Emmanuel Levinas, he further argues that stakeholder engagement serves as a platform for stakeholders to combat each other's viewpoints and values in the ultimate attempt to deconstruct them. This point can be illustrated by the engagement between Shell and two human rights organizations, Amnesty International (AI) and Pax Christi International (PCI) (Lawrence 2002). The deliberation did not depart from a common ground, where the company's concern for their reputation met the request of the organizations for cooperative support for human rights. Instead, AI and PCI essentially deconstructed the business operations and core values of Shell.

The political challenge of responsible innovation, and the significance of the above debate, is best captured in the following question:

> Are innovation, and responsible innovation, always destined to be bedfellows of a market-based Schumpeterian model of competitive, creative destruction, or can they – and should they – allow space for other alternatives of innovation and responsibility based on other political beliefs, ways of organizing, ways of distributing power, ways of relating to each other and ways of being; a quality deliberation that favors the confrontation of various arguments and conceptions of the good? (Owen and Pansera 2019, p. 41)

This question reflects a political challenge, but it also demands the discourse of responsible innovation to take a clear conceptual stance. For this we turn to the next section.

## THE CONCEPTUAL CHALLENGE OF RESPONSIBLE INNOVATION

In 'Pathways to Transformation', a recent conference of responsible innovation organized by two EU-funded projects NUCLEUS and RRI-Practice, the concept of responsible innovation was continuously addressed by the general public – and more remarkedly by many of the invited speakers – as "vague and unclear." Numerous times the *conceptual* question was raised: what *is* responsible innovation? The discussions that followed generally revolved around how to employ 'the responsibility dimension' of responsible innovation. In this respect, Phil Macnaghten provided a thorough overview of different understandings and implementations of responsible innovation, hinting at tensions between responsible innovation working within the current political landscape and responsible innovation working towards transforming this political landscape, between applying responsible innovation locally and applying responsible innovation globally, and between responsible innovation as incremental change and responsible innovation as disruptive change. These tensions reflect what has also been called the mainstream challenge of responsible innovation, in which the discourse has to decide whether to continue business as usual or to take a radical stance against it. While the concept of responsible innovation has revolutionary potential, it also contains conservative force. Which way is it heading towards?

Noticeably, throughout visions, frameworks and policies of responsible innovation much focus is dedicated to the governance of innovation processes, while little thought goes to what innovation itself means conceptually (von Schomberg and Blok 2018). This remains the case even after the release of the latest handbook, as noted in a review by Robert Frodeman (2019). Therefore, responsible innovation still requires to critically consider the concept of innovation as an object of reflection. What is meant by innovation? When innovation is said to be the chief mission of universities and of the European Union, what presupposed understanding of innovation underpins this mission? What implications does this presupposed understanding of innovation have for the realization of responsible innovation?

To this end, my hypothesis is that while the revolutionary potential of responsible innovation is illustrated in the call for innovation to generate a good impact rather than mere commercial value, the conservative power lies in the way the discourse uncritically presupposes the concept of innovation as technological innovation and commercial innovation. This is particularly reflected in the exclusive focus on economically beneficial technologies, such as synthetic biology, nanotechnology, and information and communications technology (ICT). Conversely, other forms of innovation like social innovation (e.g. fair trade) and attitudinal or behavioral innovation (e.g. lifestyle interventions) receive minimal attention. It is also worth considering that the broader policy context within which EU projects on responsible innovation operate is characterized by the overarching goal to become a genuine innovation union that turns "great ideas into products and services that will bring growth to our economy and create jobs" (European Commission 2014, p. 3). Analysis thus shows that while the dimensions of responsible innovation are broad and varied, innovation processes coupled with these dimensions are subject to a technological and commercial context. The question is whether this context is at all compatible with the dimensions that the discourse of responsible innovation so eagerly endorses.

> To what extent is it possible to operationalize the dimensions of responsible innovation within a context where innovation is understood in light of an intrinsic relation between technology and the market? For instance, reflecting upon the ethical significance of technologies could be jeopardized by the self-interested pursuit of economic welfare. Similarly, inclusion and deliberation may proceed strategically in function of maximizing one's own profit, while responsiveness may easily amount to window dressing. (von Schomberg and Blok 2019, p. 315)

In this respect, the epistemic implications and political implications of responsible innovation could be understood in relation to the presupposed concept of technological innovation and commercialized innovation. That is to say, the lack of foresight and transparency in innovation processes is specifically the case when these processes are limited to a mainstream economic understanding of what it means to innovate.[2] So long as this understanding is left uncriticized, responsible innovation will struggle to realize its ideal.

In other words, it seems problematic for responsible innovation to achieve a paradigm shift when it is merely applied to the existing concept of innovation. What is needed is a radical transformation of the concept of innovation itself. To this end, I urge future research on responsible innovation to explore an alternative concept of innovation which addresses the public good beyond the current privatization wave. The political origins of the concept of innovation (cf. Godin 2015), along with the political ends that the responsible innovation literature explicitly prioritizes, suggests that we should inquire into a political orientation of innovation. It is in this direction that Blok (2021) develops a political dimension of innovation in which the direction of the innovation process is essentially determined by a political agenda. In this view, the innovation process is no longer set by commercial ends, but rather by, for example, the Paris Agreement on mitigating global warming and the UN Sustainable Development Goals. This enables a more encompassing understanding of innovation that could also, for instance, draw attention to social innovations that are currently overshadowed by their commercial alternative. Instead of, for example, limiting the discussion of the over-consumption of meat to the possible benefits and implications of in vitro meat, this broader concept of innovation may also include considering innovative ways of simply empowering non-meat protein sources and may further enlarge the scope to apply, for instance, user-based innovations, open source and peer-to-peer (p2p) innovation strategies. Hence, by employing a political understanding of innovation, societal and environmental issues would no longer have to solely depend on technological and commercial solutions, thereby enabling RRI to primarily respond to its political ideals.

## CONCLUSION

In this chapter I departed from the observation that the emergence of responsible innovation explicitly calls for a paradigm shift in the innovation discourse, away from private interests and towards the prioritization of public interests. We learned that, to a certain extent, this objective is the result of a longer history of efforts and movements whereby the development of emerging technologies was initially left without the interference of democratic institutions, but then increasingly became a topic of wider concern. However, unlike its more nuanced precursors, the paradigm shift at stake in responsible innovation promises to overthrow mainstream economic thought, along with a radical transformation of the current research and innovation system.

I outlined three main challenges which bring the feasibility of responsible innovation into question. First, the epistemic challenge suggests that innovation as defined by its unpredictable nature simply cannot guarantee desirable and responsible outcomes, even if dimensions of responsibility are incorporated into the preceding process. Second, the political challenge reflects the difficulty of responding to the conflicting values and interests of different stakeholders, facing particular issues with recurring power imbalances. Third, the conceptual challenge demonstrates that the societal purpose of responsible innovation fundamentally conflicts with the imperative of maximizing economic growth inherent in the widely presupposed concept of technological innovation and commercialized innovation. The latter challenge ties all three together and essentially encourages us to rethink what it means to innovate. To what extent does innovation necessarily relate to the market? Is it possible to develop an alternative concept of innovation that is separated from economic ends? How can we intellectualize and implement, for example, a political understanding of innovation? In light of these closing questions, I conclude with the call for responsible innovation to explore an alternative, perhaps more political, route of innovation.

## NOTES

1. For a summary report of the workshops see: https://renevonschomberg.wordpress.com/challenges -for-responsible-innovation/.
2. For a more detailed account of the relation between the limitations of responsible innovation and the presupposed techno-economic concept of innovation see von Schomberg and Blok (2019).The central point in this chapter is to illustrate the ambiguous position of responsible innovation, where its ideal to exceed the market in order to serve society conflicts with today's general adherence to a techno-economic view on innovation.

## REFERENCES

Arendt, Hannah (1998), *The Human Condition*, Chicago: The University of Chicago Press.
Batie, Sandra (2008), 'Wicked problems and applied economics', *American Journal of Agricultural Economics*, 90 (5), 1176–91.
Bechmann, Gouhard (1993), 'Ethische Grenzen der Technik oder technische Grenzen der Ethik?', *Geschichte und Gegenwart*, 12 (3), 213–25.
Blok, Vincent (2019), 'From participation to interruption: Toward an ethics of stakeholder engagement, participation and partnership in corporate social responsibility and responsible innovation', in René

von Schomberg and Jonathan Hankins (eds), *International Handbook on Responsible Innovation: A Global Resource,* Cheltenham, UK and Northampton, MA, USA: Edward Elgar Publishing, pp. 243–58.

Blok, Vincent (2021), 'What is innovation? Laying the ground for a philosophy of innovation', *Techne: Research in Philosophy and Technology,* 25 (1), 72–96.

Blok, Vincent and Pieter Lemmens (2015), 'The emerging concept of responsible innovation. Three reasons why it is questionable and calls for a radical transformation of the concept of innovation', in Bert-Jaap Koops, Jeroen van den Hoven, Henny Romijn, Tsjalling Swierstra and Ilse Oosterlaken (eds), *Responsible Innovation 2: Concepts, Approaches and Applications,* Dordrecht: Springer, pp. 19–35.

Brundage, Miles and David Guston (2019), 'Understanding the movement(s) for responsible innovation', in René von Schomberg and Jonathan Hankins (eds), *International Handbook on Responsible Innovation: A Global Resource,* Cheltenham, UK and Northampton, MA, USA: Edward Elgar Publishing, pp. 102–21.

Bryson, John, Barbara Crosby and Melissa Stone (2006), 'The design and implementation of cross-sector collaborations: Propositions from the literature', *Public Administration Review,* 66 (s1), 44–55.

Carlson, Curts and William Wilmot (2006), *Innovation: The Five Disciplines for Creating What Customers Want,* New York: Crown Business.

Chilvers, Jason (2008), 'Environmental Risk, Uncertainty, and Participation: Mapping an Emergent Epistemic Community', *Environment and Planning,* 40 (2), 2990–3008.

Commission of the European Communities (2001), *European Governance: A White Paper.* Brussels.

European Commission (2014), *The European Union Explained: Research and Innovation.* Resource document. https://europa.eu/european-union/sites/europaeu/files/research_en.pdf.

European Commission (2015), *Horizon 2020: Work Programme 2016–2017: Science with and for Society.* Resource document. http://ec.europa.eu/research/participants/data/ref/h2020/wp/2016_2017/main/h2020-wp1617-swfs_en.pdf.

European Environmental Agency (2002). *The Precautionary Principle in the Twentieth Century: Late Lessons From Early Warnings.* http://www.rachel.org/lib/late_lessons_from_early_warnings.030201.pdf.

Evers, Adalbert and Helga Nowotny (1987), *Uber den Umgang mit Unicherheit,* Frankfurt am Main: Suhrkamp Verlag.

Ferrari, Arianna, Christopher Coenen and Armin Grunwald (2012), 'Visions and ethics in current discourse on human enhancement', *NanoEthics,* 6 (3), 215–29.

Fisher, Erik and Arie Rip (2013), 'Responsible innovation: Multi-level dynamics and soft intervention practices', in Richard Owen, Maggy Heintz and John Bessant (eds), *Responsible Innovation,* London: Wiley, pp. 165–83.

Flipse, Steven (2012), *Considerations of Social and Ethical Aspects in Industrial Life Sciences & Technology.* PhD thesis, Delft University.

Frodeman, Robert (2019), 'International handbook on responsible innovation. A global resource', *Journal of Responsible Innovation,* 6 (2), 255–7.

Godin, Benoît (2015), *Innovation Contested: The Idea of Innovation over the Centuries,* New York: Routledge.

Gorman, Michael, Patricia Werhane and Nathan Swami (2009), 'Moral imagination, trading zones, and the role of the ethicist in nanotechnology', *NanoEthics,* 3 (3), 185–95.

Gould, Robert (2012), 'Open innovation and stakeholder engagement', *Journal of Technology Management & Innovation,* 7 (3), 1–11.

Grunwald, Armin (2010), 'From speculative nanoethics to explorative philosophy of nanotechnology', *NanoEthics,* 4 (2), 91–101.

Grunwald, Armin (2019), 'Responsible innovation in emerging technological practices', in René von Schomberg and Jonathan Hankins (eds), *International Handbook on Responsible Innovation: A Global Resource,* Cheltenham, UK and Northampton, MA, USA: Edward Elgar Publishing, pp. 326–38.

Guston, David (2014), 'Understanding anticipatory governance', *Social Studies of Science,* 44 (2), 218–42.

Guston, David and Walter Valdivia (2015), Responsible innovation: a primer for policy makers, Executive summary, Washington DC.

Hackett, Edward, Olga Amsterdamska, Michael Lynch and Judith Wajcman (2007), *The Handbook of Science and Technology Studies*, Cambridge: MIT Press.

Hansson, Sven (2006), 'Great uncertainty about small things', in Joachim Schummer and Davis Baird (eds), *Nanotechnology Challenges: Implications for Philosophy, Ethics and Society*, Singapore: Springer, pp. 315–25.

Huesemann, Michael and Joyce Huesemann (2011), *TechNoFix. Why Technology Won't Save Us or the Environment*, Gabriola Island: New Society Publishers.

Ilulissat Statement (2007), Synthezising the future. A vision for the convergence of synthetic biology and nanotechnology, Ilulissat.

Jackson, Roland, Fiona Barbagallo and Helen Haste (2005), 'Strengths of public dialogue on science-related issues', *Critical Review of International Social & Political Philosophy*, 8 (3), 349–58.

Kevles, Daniel (1995), *The Physicists: The History of a Scientific Community in Modern America*, Cambridge, MA: Harvard University Press.

Kreuter, Marshall, Christopher De Rosa, Elizabeth Howze and Grant Baldwin (2004), 'Understanding wicked problems: A key to advancing environmental health promotion', *Health Education & Behavior*, 31 (4), 441–54.

Krimsky, Sheldon (1984), 'Epistemic considerations on the value of folk-wisdom in science and technology', *Review of Policy Research*, 3, 246–63.

Lawrence, Anne (2002), 'The drivers of stakeholder engagement: reflections on the case of Royal Dutch/ Shell', *Journal of Corporate Citizenship*, 6 (15), 71–85.

Naudé, Wim and Paul Nagler (2016), *Is Technological Innovation Making Society More Unequal?* Resource document. https://unu.edu/publications/articles/is-technological-innovation-making-society-more-unequal.html.

Nordmann, Alfred (2010), 'A forensics of wishing: Technology assessment in the age of technoscience', *Poiesis & Praxis*, 7 (1–2), 5–15.

Nordmann, Alfred (2014), 'Responsible innovation, the art and craft of anticipation', *Journal of Responsible Innovation*, 1 (1), 87–98.

Owen, Richard and Nicola Goldberg (2010), 'Responsible innovation: A pilot study with the UK Engineering and Physical Sciences Research Council', *Risk Analysis*, 30, 1699–707.

Owen, Richard and Mario Pansera (2019), 'Responsible innovation. Process and politics', in René von Schomberg and Jonathan Hankins (eds), *International Handbook on Responsible Innovation: A Global Resource*, Cheltenham, UK and Northampton, MA, USA: Edward Elgar Publishing, pp. 35–48.

Owen, Richard, Phil Macnaghten and Jack Stilgoe (2012), 'Responsible research and innovation: From science in society to science for society, with society', *Science and Public Policy*, 39 (6), 751–60.

Rammert, Werner (1997), 'Innovation im Netz. Neue Zeiten für Innovation: heterogen verteilt und inter-aktiv vernetzt', *Soziale Welt*, 48 (4), 394–416.

Rip, Arie, Thomas Misa and Johan Schot (1995), *Managing Technology in Society: The Approach of Constructive Technology Assessment*, London: Pinter.

Rittel, Horst and Melvin Webber (1973), 'Dilemmas in a general theory of planning', *Policy Sciences*, 4 (2), 155–69.

Rolston, Holmes III (2012), *A New Environmental Ethics: The Next Millennium for Life on Earth*, New York: Routledge.

Stilgoe, Jack, Richard Owen and Phil Macnaghten (2013), 'Developing a framework for responsible innovation', *Research Policy*, 42(9), 1568–80.

Stoneman, Paul (1995), *Handbook of the Economics of Innovation and Technological Change*, Oxford: Wiley-Blackwell.

van der Burg, Simone (2014), 'On the hermeneutic need for future anticipation', *Journal of Responsible Innovation*, 1 (1), 99–102.

van Huijstee, Mariëtte, Mara Francken and Peter Leroy (2007), 'Partnerships for sustainable development: A review of current literature', *Environmental Sciences*, 4 (2), 75–89.

van Oudheusden, Michiel (2014), 'Where are the politics in responsible innovation? European governance, technology assessments, and beyond', *Journal of Responsible Innovation*, 1 (1), 67–86.

von Schomberg, Lucien and Vincent Blok (2018), 'The turbulent age of innovation', *Synthese*. https://doi.org/10.1007/s11229-018-01950-8.
von Schomberg, Lucien and Vincent Blok (2019), 'Technology in the age of innovation: Responsible innovation as a new subdomain within the philosophy of technology', *Philos. Technol.* https://doi.org/10.1007/s13347-019-00386-3.
von Schomberg, René (2013), 'A vision of responsible research and innovation', in Richard Owen, Maggy Heintz and John Bessant (eds), *Responsible Innovation*, London: Wiley, pp. 51–74.
von Schomberg, René (2019), 'Why responsible innovation?', in René von Schomberg and Jonathan Hankins (eds), *International Handbook on Responsible Innovation: A Global Resource*, Cheltenham, UK and Northampton, MA, USA: Edward Elgar Publishing, pp. 12–32.
von Schomberg, René (2020), 'In memory of Karl-Otto Apel: The challenge of a universalistic ethics of co-responsibility', *SSRN Electronic Journal*.
von Schomberg, René and Jonathan Hankins (2019), *International Handbook on Responsible Innovation: A Global Resource*, Cheltenham, UK and Northampton, MA, USA: Edward Elgar Publishing.
Wang, Qian and Ping Yan (2019), 'Responsible innovation: Constructing a seaport in China', in René von Schomberg and Jonathan Hankins (eds), *International Handbook on Responsible Innovation: A Global Resource*, Cheltenham, UK and Northampton, MA, USA: Edward Elgar Publishing, pp. 441–55.
Yaziji, Michael and Jonathan Doh (2009), *NGOs and Corporations: Conflict and Collaboration*, Cambridge: Cambridge University Press.

# PART IV

# ALTERNATIVE TYPES OF INNOVATION

# 9. User-centred innovation: from innovative users to user centred programmes

*Bastien Tavner*

## INTRODUCTION

Focusing on user-centred innovation (UCI), this chapter does not simply present a formal theory but rather a set of approaches whose aim is to organize and guide the innovation process. These approaches have grown in popularity since the end of the 2000s and can be defined as the voluntary involvement of "real" users in the innovation process and the integration of specific skills dedicated to the analysis of uses within innovation groups. Extending beyond the borders of the R&D sectors of large corporations, UCI is progressively permeating the prescriptive frameworks of public programmes for innovation guidance and support in a whole host of sectors ranging from education to health, energy, mobility, and so on. This chapter will first provide a genealogical overview of the theoretical bases having allowed users to progressively attain the status of key actors of innovation, in representations and at times in practices (section 1). It will then propose a definition of UCI as an approach that finds itself at the crossroads of theoretical and operational considerations stemming from the breaking down of silos and the participation of citizens in innovation processes (section 2). Finally, the last part of the chapter discusses programmes and organizational modes developed in response to this injunction to participate.

## THE CHANGING FIGURE OF THE USER AS AN "ACTOR" OF INNOVATION IN SHS LITERATURE

Before finding themselves at the centre of the latest innovation management trends as potentially disruptive entities, users were for many years, and indeed continue to be at times, associated with homogeneous and passive groups. This is still notably the case in the media during crises or social movements that result in journalists and political leaders describing the "anger" of "public service users" (underground users, health service users, energy users, etc.). Although still emerging in the media world, this recognition of the "active" and multifaceted aspect of users is something that the social sciences have been advocating for over fifty years. Among the many research movements concerned, several should be cited here: the sociology of cultural practices and the attention given to the "poaching of meaning" (de Certeau 1980), communication theories and the *uses and gratifications* school of thought (Katz and Lazarsfeld 1955), media sociology and reception theory through the focus on the appropriation of ICT's and the domestication of technology (Lull 1988; Livingstone and Bovill 2001), or the French sociological theory of uses developed in the 1980s. Many entrepreneurial, political and associative initiatives have been inspired by this research just as many targeted definitions of users have emerged (citizens, pupils, inhabitants, patients, beta-testers, etc.), together with the processes set up to involve these users in efforts to guide and develop specific products, services or measures. However, before being specifically identified as a stakeholder of these processes, the figure of the user was encapsulated in rhetoric where "needs" and "demand" were recognized as the real drivers of innovation. Traces of this can be found as early as the end of the 1960s, notably in the positions of the *US Department of Commerce* and the speeches of Herbert Hollomon (Godin 2020). Since then, representations of the user and representations of the innovation process have undergone a series of reciprocal adjustments and theoretical proposals and counter-proposals.

## Diffusion of Innovation Theory and the Role of "Adopters"

By placing the decision to adopt innovations at the centre of the analysis, the room for manœuvre granted to the recipients of innovation in *diffusion studies* programmes is limited at best. Moreover, these recipients are not yet designated as "users" but as "*adopters*", more often than not grouped into "units of adoption" – echoing the contagion clusters described by epidemiologists – where the social career of the innovation in question is outlined according to its perceived newness:

> An innovation is an idea, practice, or object perceived as new by an individual or other unit of adoption. (Rogers 1962, p. 35)

Hence, the focus in these studies is on the key propagation factors linked to the "novelty perceived" within specific social groups who often belong to a specific geographic area. This is the case of the pioneering study carried out by Ryan and Gross on the spread of the use of hybrid corn in the 1940s in two Iowa farming communities (Ryan and Gross 1950). Here, the dissemination of this innovation is largely explained by the interindividual level of information exchanges between farmers about their use experience. Beyond this restricted role as regulators of the speed of innovation diffusion, *adopters* gradually saw themselves being given a certain corrective power in the 1970s with the introduction of the concept of *reinvention*:

> Reinvention is the degree to which an innovation is changed by the adopter in the process of adoption and implementation after its original development. (Rice and Rogers 1980, p. 500)

Far from celebrating the creative ingenuity of Michel de Certeau's "poachers of meaning" (de Certeau 1980), or Eric von Hippel's lead users (von Hippel 1986), here the reinvention by adopters is above all a necessary response to an array of problems (technical complexity, lack of anticipation, adaptation to local conditions, etc.), hindering the spread of innovations in their initial state. Furthermore, the concept of reinvention is not extended to all cases, as conceded by Rogers who, in subsequent editions of *Diffusion of Innovations*, limits its use to innovations whose flexibility results in different implementation modes.[1]

In the context of the 1980s and the increasing importance of innovation and design in the industrial sector, an incredible proliferation of standard patterns for the innovation process along with its key phases, from innovation to market, can be observed. Among these, Rothwell and Gardiner's proposal, based on the study of the development of the *Hovercraft* in the United Kingdom, stands out since it confers an active role on users before the innovation's dissemination (Rothwell and Gardiner 1985). This is often limited to the expression of their needs during preliminary market research ("invention" phase), although users can also be involved sporadically in development work during the functional tests performed during the so-called "innovation" phase, hence participating in a number of technical adjustments. But it is in the third phase, that is, *reinnovation*, when users really offer a more significant contribution via a series of developments leading to incremental innovations. This direct user involvement is presented as a means of reducing mismatches between the original innovation design and its implementation context. Presented as "iterative" by the authors, this representation of the innovation process is nevertheless still fairly linear in terms of the sequencing of steps to be overcome for an innovation to be disseminated, among which commercial and technical viability are portrayed as pre-requisites to user participation:

> Finally, the hovercraft story illustrates that while potential users can make a significant contribution to the development of a radical innovation, once its technical and commercial viability has been established, initially they may have a negative influence through resisting it, or simply through ignoring it. (Rothwell and Gardiner 1985, p. 185)

Sharing the same overall vision of the innovation process as a "collision" between technical opportunities and user needs, Bengt-Åke Lundvall concentrates on exploring the modalities of interaction between *innovation units* and *user units* (Lundvall 1985). More specifically, Lundvall differentiates between *consumers* and *professional users* and focuses on the latter whose goals are seen to be more circumscribed in relation to the developed innovation and who are more likely to invest more actively in its development. This involvement is reflected in three user–producer interaction modalities: exchange of products, exchange of information and cooperation, defined as a "learning-by-interacting" process. Although the spotlight here is on a form of reciprocal dependence between users (who need information in order to adapt and adopt innovations), and producers (who need to know users' needs in order to direct the innovation process), this amendment to the diffusion theory perpetuates a fairly mechanical vision of the encounter between technical innovations and needs. Dimensions such as the temporal, social and cultural conditions of the development of "real" uses, as well as the methods through which they can be investigated with a view to producing innovation, are underexploited if not ignored entirely.

### Lead Users as Targets

With the progressive recognition of the active role that some users can play in the social career of new products/services, it became essential to identify the individuals or social groups eager to participate in such novel encounters. This identification contributed to the success of the ideal type of *lead user* within innovation networks. Forged by Eric von Hippel, the notion of *lead user* designates an "advanced" user, characterized by experience and specific expertise in a field and hence benefiting from an ideal position in order to anticipate relevant solutions to cover needs not yet satisfied on the market (von Hippel 1986). Unlike the ordinary users highlighted through SCOT theory, sociology of innovation or the *Common Innovation* concept (see Chapter 12 by Swann in this book), *lead users* have peculiar characteristics that allow them to play both sides of the fence, that is, the side of invention and that of the market. This is why they are sometimes called upon by those having taken over von Hippel's concept as an "updated" version of the Schumpeterian entrepreneur in a personified and bottom-up view of the diffusion of innovation, that is, from use to market. But *user innovation*, which is not exploited by von Hippel as a general "theory" but rather as a set of activities varyingly linked to commercial operations, does not only imply identifying separate individuals. Far from restricting itself to the intrinsic qualities of these innovative users, research on "bottom-up" or "horizontal" innovation processes (depending on who is writing), often underlines the anchoring of users in a community and the successive innovation circles through which user innovation meets the market (Cardon 2006). Here can be found numerous studies about communities of practice structured around a common discipline, just like the 1970s community of Californian surfers (von Hippel 1986), or climbing enthusiasts (Akrich 1998), but also the emergence of a new technical infrastructure such as WiFi (van Oost et al. 2008). The horizontality of the resulting innovation-through-use networks hence allows for a militant reading of *user innovation* as a collective innovation organization mode ("by the user" and "for the user") free from the dictates of technology-centred markets.

**Users' Say in Non-linear Approaches to Innovation**

In the 1980s, Trevor J. Pinch and Wiebe E. Bijker criticized the arbitrary nature of linear innovation models (and their random number of "key phases"), as well as their focus on an *ex ante* explanation of their success – as opposed to an in-depth analysis of their failures – and went on to outline a constructivist approach called the *Social Construction Of Technology* (SCOT). SCOT theory strives to define the development of technological artefacts as a "multidirectional" process (Pinch and Bijker 1984) through which social groups give specific meaning to technology and hence define an array of specific problems giving rise to a multitude of possible answers (or *solutions*). Faced with the *interpretative flexibility* of new technical artefacts, *relevant groups*, including various emerging categories of users, began to play a crucial role as they fuelled controversies and injected into them various considerations (material, sociocultural, moral, etc.) until the artefacts were *stabilized*. In the Bijker and Pinch approach, "society is an environment or a context in which technologies develop" (Winner 1993, p. 366). This separation between society and technology, and the resulting domination of the latter by the former, is not as clear-cut for all researchers critical of the diffusionist theory. Among these, scholars representing what is called the sociology of "translation", who popularized the *actor-network* theory, call this division into question. The innovation actor decompartmentalization programme, launched by the SCOT movement, was then substantially extended with the introduction of the principle of symmetry between "human" and "non-human" actors within "sociotechnical networks" (Callon 1986; Latour 1996; Akrich et al. 2002). From this viewpoint, innovation follows a *whirlwind model* fed by iterations (or "chains of translation") that reach beyond the tight framework of adjustments allowed by the interactions between producers and users described in *diffusion studies*. In the "*diversified crowd of actors*" making up sociotechnical networks (Akrich et al. 2002), users do not always benefit from the same consideration in case studies that are based upon the proponents of translation sociology. In *Aramis, or the Love of Technology* , Bruno Latour uses the emblematic failure of a new mode of automated transport at the end of a long and costly project to build the principles of the sociology of translation (Latour 1996). The project's path was chaotic, which eventually led to it being abandoned, while the users were relegated to the role of hastily selected spokespersons taken from a "representative sample" forming a "mockup public" with a limited say (Latour 1996, p. 188). In contrast, Madeleine Akrich takes the "active user" as her research subject, and not the "represented user", as she strives to describe how users act on existing technical devices (Akrich 1998). Focusing on the material characteristics of objects and their prescriptions for use, she distinguishes four ways in which the initial function and use of objects can be changed by users: displacement, adaptation, extension and creep. According to Akrich, far from confining themselves to incremental and marginal adjustments, in some cases users can propose radical modifications and hence become fully blown *innovator users*, similar to *lead users*.

## UCI AT THE CROSSROADS OF DECOMPARTMENTALIZED AND INCLUSIVE APPROACHES TO INNOVATION

"User-Driven Innovation", "User-Centred Design", "innovation-through-use" and so on: among the many names describing user innovation, it is still difficult to know where to place the cursor between "by", "for" and "with" the user. The term *User-Centred Innovation* (UCI) is, to my mind, the one that best expresses the diversity of logics and end purposes where the focus is collectively set on users. This chapter argues that this ambivalence creates a shared operational direction inspired by multiple theories.

### Definition of UCI

While the user is a multifaceted being subject to multiple categorizations, UCI itself acts as an *intermediary object* (Trompette and Vinck 2009) allowing different groups to work together and share a common vision. Although the contours of UCI are flexible and the contexts in which it is carried out vary, this does not mean that its very specific nature cannot be defined. As an operational approach, UCI can be defined by operational principles. Although, as we shall see later, methods and tests vary depending on the case, three basic and interdependent pre-requisites can be carved out to circumscribe UCI-related initiatives:

(1)    the involvement of a population of "real" users during the design and/or experimentation phases of an innovative product/service;
(2)    the integration of skills dedicated to the observation and analysis of uses so that the product/service can be developed and adapted;
(3)    the sharing and discussion of knowledge produced about uses within a multidisciplinary group.

By channelling these requirements into the spheres of innovation, UCI marks its difference with respect to the logic of "projected uses", which cuts corners by not testing in real-life situations, as well as with respect to rhetoric extolling the "taking into account of users", which is actually unsupported by any specific skills. It also extends beyond the confines of usability tests performed in the R&D sectors of major firms since it allows users to have a genuine say in the (re)creation of the products/services. The UCI model is thus gaining ground as a purposefully inclusive, multidisciplinary and methodologically "equipped" approach that can be adapted to a variety of business sectors, development phases and types of innovation governance.

**Innovation Embedded in the "Participatory Turn"**

Although we can consider that *user innovation* (or *bottom-up innovation*) is the archetype of innovation "use approaches", *user-centred innovation* (UCI), such as I propose to define it, represents a hybrid form to a certain degree since it is "filtered" by organizations. The UCI approach is nevertheless more or less linked to a set of theories with which it shares a certain "family likeness" (Kuhn 1996), without necessarily faithfully representing any specific traits. Furthermore, its popularization can be interpreted as a translation within the "participatory turn" innovation networks observed by the STS in technoscientific spheres (Jasanoff 2003; Wynne 2007). Since the end of the 1990s, this expression designates the progressive institutionalization of citizens' participation in arenas for the debate of technoscientific programmes. This can be notably observed in the United States with the multiplication of public reviews financed by public funds, then gradually in the European Commission's guidelines for governance and, more specifically, with respect to the involvement of citizens within institutional risk assessment bodies (Jasanoff 2003). However, far from bringing actors in line with a shared vision of why public participation is justified, this trend is leading to diffuse representations and aspirations:

> Further, the interpretations of this rising interest in as well as the proliferation of participatory activities remain rather uneven. For some, it is simply a process of inevitable participatory democratization expanding into closed technoscientific realms. For others it is a move to restore shaken trust in science, thus also scientific public authority, so as to foster the credibility of policy commitments to technoscientific innovations. For still others it is a unique way to change the very way of producing innovation and making policies, through new forms of mutual understanding and intersection of science and society. And for a very few, the explicit acknowledged aim is to conduct a reflexive and accountable review of encultured institutional understandings and related practices concerning science, innovation and governance. The common theme across these different approaches is that it no longer seems legitimate to think and work within the classical policy framework without including some kind of stakeholder involvement or public participation. (Felt and Wynne 2007, p. 56)

Like the measures developed at national and supranational level to strengthen citizen participation in the governance of technosciences, a set of programmes for supporting innovation focused on the notion of user *involvement* was created (or perhaps renewed) at the end of the 2000s. The ambivalence of users' attachment to the "participatory turn" injected unequally shared aspirations into UCI. These were strategic (linked to the opening of innovation processes), sociopolitical (linked to the *empowerment* of citizens/users), and even social and community-related (linked to the support of collective horizontal innovation processes). From an economic and strategic point of view, the UCI approach maintained ties with the theoretical model of *open innovation*[2] (Chesbrough 2003; Chesbrough et al. 2008) by directing innovation groups towards the conquest of new sources and/or new openings linked to existing or projected ecosystems of use. UCI emerged as an alternative means of standing out in a market as it projected the image of an increasingly peopled and checkered sociotechnical matrix. In a number of its translations, the UCI approach also echoed the notion of *social innovation* (Moulaert 2013) by making it possible to investigate the needs of targeted communities of use with a view to improving their living conditions. Yet, in spite of the diversity of its theoretical roots and its similarities with the participatory turn observed within technoscientific governance, UCI has remained an operational approach in which the central figure of the *user* never completely merges with that of the *citizen*. In this way, UCI acts as an instrument and hence stands out from other approaches such as *participatory design*, which can be associated with "a value-centred design approach because of its commitment to the democratic and collective shaping of a better future" (Van der Velden and Mörtberg 2014).

## OPERATIONAL TRANSLATIONS OF UCI AND OVERLAPPING CHALLENGES

### UCI in Public Programmes

At the crossroads of these theoretical influences, which may or may not be explicitly applied, the user moves beyond the confined realm of R&D and is propelled into innovation policy framing statements. At times, the uses may simply represent one direction among others. This is for example the case of research financing programmes. For instance, in France, calls for projects launched by the French National Research Agency[3] (ANR) include "thematic directions" specifically dedicated to the analysis of uses, especially in the digital and information technologies sector. Although still largely centred on technology, clusters of companies in the digital sector are also increasingly covering the "theme" of use in their strategic focus. In France, this is notably the case of the "competitiveness clusters"[4] that began to emerge in 2005. In local or regional calls for projects, the involvement of users is sometimes a major concern. This is reflected in specifically dedicated selection criteria. More broadly speaking, programmes to support and promote user-centred innovation in all sectors are being developed. This is the case of the European Network of Living Labs (ENoLL) whose members today come from well beyond the European borders. Launched in 2006 under the Finnish presidency of the European Union, the "Living Lab" label is based on the "4P" model (standing for "Public-Private-People-Partnerships"). The Living Labs are here defined as environments for testing and experimentation in real situations (real-life tests) where users and co-designers co-build products/services according to user-driven open innovation involving a wide range of ICT's, fields of activity and skills. The scope of practices promoted in Living Labs bearing the European Network of Living Labs (ENoLL) label covers four main activities: the co-creation of innovations by users and companies, the exploration of emerging markets and/or uses, experimentation within user communities and the assessment of the concepts and products/services developed.

Stakeholders of the UCI model more or less explicitly agree to play the game of innovation as an "iterative" or "whirlwind" process as described by the sociology of translation (Akrich et al. 2002). According to this perspective, UCI exposes innovation groups to a set of tests during which the technological assemblies, targeted publics and use scenarios are collectively (re)defined and adjusted based on the effects of product knowledge via the observation and analysis of uses. In fact, the application of this use-centred iterative logic only rarely covers the entire innovation process. The user may be involved in the ideation, co-design, experimentation or validation phases of a product/service before it is marketed. The frequency of targeted user involvement is also likely to vary from one structure to the next. Users might be systematically asked to participate or only rarely, and interactions with them might be constant or more episodic, regular or random. The duration of measures and actions set up to observe and study uses can also vary, ranging from one month to "validate the acceptability" of innovations[5] to several years to develop research programmes. Numerous and, at times, connected disciplines are invited to help carry out use studies and fuel these interest-creating ("*interessment*") and retroactive loops. As of the 1980s, these might include marketing specialists and ergonomists, the "official" user spokespersons of large firms (Boullier 2002), who see new opportunities for collaboration with new actors (geographic areas, associations, start-ups, etc.) in the expansion of user-centred approaches. Representatives of the social sciences with diverse disciplinary backgrounds and thematic interests can also be found: sociologists of use, anthropologists of health, psycho-sociologists specialized in communication interfaces, and so on. Many agencies specializing in service design or UX design and other independent "use consultants" have also appeared. Besides coming from a wide range of disciplinary backgrounds, these professional user representatives are invited along to take part in a sort of competition between programmes supporting innovation. The aim here is to promote postures that are decidedly geared towards users whereas the rhetoric applied is very often disconnected from concrete experimental and investigation practices. In practice, however, this distinct taste for user-centred approaches, in both public programmes and companies focused on innovation, is dampened by the level of complexity encountered, which can subject innovators to a series of disillusions:

> Amid all this buzz, commercial design firms are increasingly winning international development contracts [...] and development practitioners are increasingly disappointed with the results. (Lee 2015)

Considering the diversity of collective processes driven by this shared desire to innovate with users, the ensuing sections of this chapter describe three overlapping challenges peppered with specific tensions: the determination of users, their recruitment and the analysis of the effects of product knowledge during use experiments.

## Determining Users: Between Public Typologies and Use Ecosystems

In a project, deciding on the populations of users to be placed at the centre of the UCI process is rarely limited to the start. On the contrary, one of the major challenges of UCI is being able to progressively determine target user populations as potential "markets" to be addressed, stakeholders of specific use scenarios and "allies" of their making. Yet, the first operations performed to determine users are often crucial as they drive – and/or are often driven by – the directions followed by project teams, the use contexts investigated and the relevant points of contact for the recruitment of *co-designer* or *experimenter* populations. Similarly, the actors in charge of determining who the users will be are not selected once and for all when a project is launched. Although "use specialists" or specific service providers (sociologists, designers, ergonomists, etc.) are obviously called upon, they are often invited to participate after a long list of other actors have already inscribed their ideas, concerns and requirements in the framing of the common project: "technologists", partner structures, regulators (notably in the case of infrastructure-related or controversial innovations), regions, and so on. These "use specialists" are not only expected to come up with new ideas, their role in this context is to translate the different wishes expressed into shared criteria for defining user populations. In order to do this, they generally apply, either together or separately, two recurring principles for circumscribing users (see Figure 9.1). The first principle involves drawing up a *public typology* for the innovation. Based on the use scenarios developed by the project team, which may be relatively sophisticated, this typology differentiates between sub-groups of users characterized by traditional sociodemographic criteria (sex, age, socioprofessional category), and/or more targeted considerations. The latter may be the result of preliminary surveys on use contexts, which shed light on key variables: level of technical skill or equipment, sensitivity to theme, belonging to a specific community of practice, etc. They may also be inspired – from a reflexive or critical point of view – by pre-existing typologies. In the digital technology sector, there are many typologies covering stereotypical categories such as "digital natives", "connected seniors", "geeks", "technophobes", and so on. The second user categorization principle stems from a more interactive logic. It requires identifying *use ecosystems*, defined as a circuit of interdependence between different categories of users and actors essential to the proper operation of the innovative product/service being developed. The categories concerned refer to a definition of the user in the broad sense of a set of stakeholders in a defined use scenario: the end users in direct contact with the product/service, the various professional users who deploy and/or administer it, the prescriptive users who specify the use framework, the intermediary users who allow others to appropriate it, and so on.

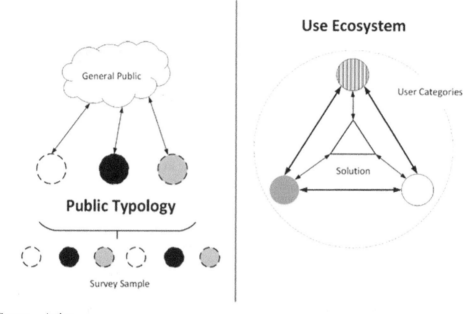

*Source:*    Author.

*Figure 9.1*    *User categorization principles*

By gearing efforts to adapt products/services to the diversity of interactions developed between different user categories around and/or through the artefact developed, the use ecosystem model calls for *situation-centred design practices* to provide "an understanding of the 'user' as more than just one homogeneous demographic" (Lee 2015). In practice, these two user categorization principles are not exclusive options. On the contrary, project teams' propensity to link them together during the innovation process makes for a progressive and cross-cutting investigation of the usability conditions of the innovation, its appropriation contexts and its marketing openings.

**Recruiting Users**

The user recruitment phase in the UCI process constitutes a determining step. Indeed, the challenge is not only to adhere as faithfully as possible to the chosen definition criteria but also to offer target users a framework for interaction that is at once acceptable, attractive and relevant with respect to the design and/or adaptation objectives of the sociotechnical device. The user recruitment mode varies according to a set of parameters. Besides the specific methods for finding recruits (random or from person to person, face-to-face or remote, etc.), three determining, albeit non-exhaustive factors apply: the use experts' level of autonomy when in contact with the target populations, the degree of flexibility of the investigation methods followed and the availability of the populations concerned. In some cases, field access conditions can lead to the "direct" recruitment of users by researcher experts. In other cases, mediation by intermediary actors is required to obtain a list of eligible candidates and contact them. This is especially the case in the health field where creating a user (focus) group often requires a complex and protocol-based validation phase (local health department, partner organization, service concerned by the experiment, etc.), which limits use experts' room for manoeuvre. Concerning the investigation methods, these contribute to recruitment conditions in that they determine the project teams' propensity to change (or not change) user populations "as they go". For example, methods stemming from multi-criteria sampling, when there is a need for statistic representativeness, require users with known profiles to be recruited. Acquiring such specific profile knowledge means carrying out preliminary surveys (based on questionnaires or access to sociodemographic databases), or calling on professional and/or institutional intermediaries who may be private, public, associative, and so on. Besides the question of access, the availability of the user populations concerned also has an impact on their recruitment mode. Here, I am not referring to contingent individual availability, which is obviously important, but to users being disposed to volunteer for the type of experiment proposed by the UCI project. By building up different UCI projects and anchoring these in a specific geographic area, some Living Labs are able to turn to previously created user groups that are then simply "activated" for specific projects. For instance, this is the case of the *Citilab*[6] (Cornellà de Llobregat, Spain), which, through its various innovation focuses (work, education, media, etc.), runs communities of "regulars" in a specific place (active professionals/unemployed, young people/seniors, etc.), who can be called upon for public–private partnership projects. These "to-hand" user communities provide project teams with a certain advantage since they require less effort to create interest (*interessment*) compared with a situation where the team has to "start from scratch" in order to set up a group of experimenters unfamiliar with the challenges of UCI projects. This role as a "supplier" of users and facilitator of communities can also be carried out by associations. In France since the 2000s this is notably one of the functions of "Cyber-base" spaces, today referred to as *NetPublic* spaces. These are public places dedicated to digital use inductions and often deployed in local UCI projects. The prior (and more or less objectifiable) creation of user communities does not necessarily guarantee successful recruitment phases. Some user ecosystems (such as health professionals) are over-solicited for testing while innovators sometimes imagine that a community will be more enthusiastic about experimenting with their products/services than it actually is (notably in the digital platform sector), all of which creates bias when it comes to the effective recruitment of volunteers.

**Analyzing Uses**

While the variety of methods based on the UCI model makes it impossible to distinguish a standard format for the analysis of uses, the appropriation of results within project teams is subject to a set of transverse tensions relating to collaboration dynamics (both from an interdisciplinary and interindividual point of view). Besides the effect of project leaders' backgrounds on their receptiveness to non-technical approaches to innovation, the way results are used raises the question of interaction spaces (steering meetings, workshops, work side by side, intermediary reports and presentations, etc.), organized as part of projects so that viewpoints and epistemic references can be compared and contrasted and uses thereby appraised. Furthermore, the scope for use studies to be appropriated by project leaders as well as by possible future users (according to an *Open Innovation* logic) is closely tied to the technical, organizational and even marketing-based directions followed by projects. While UCI projects aim to base such (re)-directions on the analysis of uses, many considerations, requirements and opportunities may exclude these uses from the decision-making process. There are many UCI practitioners who, as a project leader becomes drawn to the latest, fashionable technology, see months of data collection focusing on a specific use context spoiled by the sudden re-assessment of sociotechnical arrangements. This is where one of the major difficulties of putting UCI principles into practice arises since these principles are not simply based on the idea of "unravelling the mysteries" of use. The reach of the UCI process lies in the interaction between two heterogeneous groups driven by multiple interests. One is set up beforehand but its cohesion is ever fragile (project team), while the other has to be built and consolidated (use ecosystem).

## CONCLUSION

"Reducing time-to-market", "anticipating obstacles to use", "guaranteeing the social acceptability of technologies" and so on: considering the challenges and tensions evoked in this chapter, it is surprising that many adopt a highly optimistic attitude to the strategic influence of UCI. This difference in vision is partly due to the frenzied application of the latest terminology ("open innovation", "user innovation", "social innovation" and so on), which, as we have seen, actually refers to specific representations of the figure of the user as an "actor" of innovation. Not only are innovation actors exposed to much theoretical confusion when they engage with user-centred approaches to uses, they may also be haunted by the spectre of the *approver user*, instrumentalized to validate already technically and financially "packaged" solutions. The theoretical bases underpinning the UCI model must not therefore be reduced to a list of aggregate "ingredients". As already highlighted in this chapter, UCI is circumscribed by operational principles – involving a population of real users, the application of skills dedicated to the analysis of uses, and a multidisciplinary organization – and can engender projects that do not necessarily stem from a "bottom-up" innovation process or be guided by aspirations for *social change*. These projects may also provide the opportunity to link different perspectives in an original and, of itself, innovative manner. While UCI does not allow the aspirations and interests of all the actors involved to mechanically converge together, it can contribute to a deep and relatively unexpected redefining of significations, objectives, properties, recipients and openings for innovations. In the connectionist world of the *city by projects* (Boltanski and

Chiapello 2007), UCI nevertheless takes on some apparently ambiguous end goals. On the one hand, investigating use-related issues can provide an opportunity to widen the network of innovation process contributors via the integration of users and their spokespersons (researchers, usability experts, associations and so on). On the other hand, however, the constant revision of projects and the reconfiguration of lead teams can undermine the in-depth study of the conditions for these innovations to be *embedded* (Granovetter 1985) in determined contexts.

## NOTES

1.  With reference to the study by Ryan and Gross, Rogers stipulates that: "Other innovations are more difficult or impossible to reinvent; for example, hybrid seed corn does not allow a farmer much freedom to re-invent, as the hybrid vigor is genetically locked into seed (for the first generation) in ways that are too complicated for a farmer […]" (Rogers 1983, p. 17).
2.  See Chapter 10 by Tiago Brandão in this book.
3.  Agency created in 2005 to support public and partnership-based research in France.
4.  In France, the closer cooperation between industrial and academic worlds in different regions has been promoted by the competitiveness cluster policy since 2005. Developed around strategic research and development (R&D) projects, these clusters are structured around fields as varied as transport, agrifood or CIT's.
5.  Developed at the start of the 2000s by the French sociologist Philippe Mallein, the CAUTIC® (use-assisted design for technologies, innovation and change) offers companies a qualitative method for assessing innovations by focusing on uses.
6.  Citilab is a Spanish *Living Lab* opened in 2008 and bearing the ENoLL label. In the city of Cornella de Llobregat (Catalonia), the lab's activities take place on the emblematic site of a former brick factory where, today, digital equipment is now housed. The site provides free access to computers and modular co-working spaces and offers the possibility of producing web TV programmes (https://www.citilab.eu/).

## BIBLIOGRAPHY

Akrich, Madeleine (1995), 'User representations: Practices, methods and sociology', in Arie Rip (ed.), *Managing Technology in Society: The Approach of Constructive Technology Assessment*, London: Pinter, 167–84.

Akrich, Madeleine (1998), 'Les utilisateurs, acteurs de l'innovation', *Education Permanente*, (134), 79–90.

Akrich, Madeleine, Michek Callon, Bruno Latour and Adrian Monaghan (2002), 'The key to success in innovation part II: The art of choosing good spokespersons', *International Journal of Innovation Management*, 6 (2), 207–25.

Bijker, Wiebe E., Thomas P. Hughes and Trevor J. Pinch (eds) (1987), *The Social Construction of Technological Systems: New Directions in the Sociology and History of Technology*, Cambridge, MA: MIT Press.

Boltanski, Luc and Ève Chiapello (2007), *The New Spirit of Capitalism*, London: Verso.

Boullier, Dominique (2002), 'Les études d'usages: entre normalisation et rhétorique', *Annales des télécommunications*, 57 (3–4), 190–209.

Callon, Michel (1986), 'The sociology of an actor-network: The case of the electric vehicle', in Michel Callon, John Law and Arie Rip (eds), *Mapping the Dynamics of Science and Technology*, London: Palgrave Macmillan, pp. 19–34.

Cardon, Dominique (2006), *La Trajectoire Des Innovations Ascendantes : Inventivité, Coproduction et Collectifs Sur Internet*, Innovations, usages, réseaux, November 2006, Montpellier.

Chesbrough, Henry W. (2003), *Open Innovation: The New Imperative for Creating and Profiting from Technology*, Boston, MA: Harvard Business School Press.

Chesbrough, Henry, Wim Vanhaverbeke and Joel West (2008), *Open Innovation: Researching a New Paradigm*, Oxford: Oxford University Press.

de Certeau, Michel (1980), *L'invention du quotidien. 1, Arts de faire*, Paris: Gallimard.

Felt, Ulrike and Bryan Wynne (eds) (2007), *Taking European Knowledge Society Seriously: Report of the Expert Group on Science and Governance to the Science, Economy and Society Directorate*, Directorate-General for Research, European Commission, Luxembourg: Office for Official Publications of the European Communities.

Flichy, Patrice (1995), 'Socio-technological action and frame of reference', *Réseaux*, 3 (1), 9–30.

Flichy, Patrice (2003), *L'innovation technique: récents développements en sciences sociales, vers une nouvelle théorie de l'innovation*, Paris: Éd. La Découverte.

Godin, Benoît (2020), *The Idea of Technological Innovation: A Brief Alternative History*, Cheltenham, UK and Northampton, MA, USA: Edward Elgar Publishing.

Granovetter, Mark (1985), 'Economic action and social structure: The problem of embeddedness', *American Journal of Sociology*, 91 (3), 481–510.

Hyysalo, Sampsa, Torben E. Jensen and Nelly Oudshoorn (2016), *New Production of Users: Changing Innovation Collectives and Involvement Strategies*, London: Routledge.

Jasanoff, Sheila (2003), 'Technologies of humility: citizen participation in governing science', *Minerva*, 41 (3), 223–44.

Katz, Elihu and Paul Lazarsfeld (1955), *Personal Influence: The Part Played by People in the Flow of Mass Communications*, Glencoe: Free Press.

Kuhn, Thomas S. (1996), *The Structure of Scientific Revolutions*, 3rd ed, Chicago, IL: University of Chicago Press.

Latour, Bruno (1996), *Aramis, or the Love of Technology*, Cambridge, MA: Harvard University Press.

Lee, Panthea (2015), 'Before the backlash, let's redefine user-centered design', *Stanford Social Innovation Review*, (26 August), https://ssir.org/articles/entry/before_the_backlash_lets_redefine _user_centered_design.

Livingstone, Sonia M. and Moira Bovill (eds) (2001), *Children and Their Changing Media Environment: A European Comparative Study*, Mahwah, NJ: Erlbaum.

Lull, James (ed.) (1988), *World Families Watch Television*, Newbury Park: Sage.

Lundvall, Bengt-Åke (1985), *Product Innovation and User–Producer Interaction*, Aalborg: Aalborg University Press.

Moulaert, Frank (ed.) (2013), *The International Handbook on Social Innovation: Collective Action, Social Learning and Transdisciplinary Research*, Cheltenham, UK and Northampton, MA, USA: Edward Elgar Publishing.

Pinch, Trevor J. and Wiebe E. Bijker (1984), 'The social construction of facts and artefacts: Or how the sociology of science and the sociology of technology might benefit each other', *Social Studies of Science*, 14 (3), 399–441.

Rice, Ronald E. and Everett M. Rogers (1980), 'Reinvention in the innovation process', *Knowledge*, 1 (4), 499–514.

Rogers, Everett M. (1962), *Diffusion of Innovations*, New York: Free Press.

Rogers, Everett M. (1983), *Diffusion of Innovations*, 3rd edn, New York: Free Press.

Rothwell, Roy and Paul Gardiner (1985), 'Invention, innovation, re-innovation and the role of the user: A case study of British hovercraft development', *Technovation*, 3 (3), 167–86.

Ryan, Brice and N. Gross (1950), 'Acceptance and diffusion of hybrid corn seed in two Iowa communities', *Research Bulletin (Iowa Agriculture and Home Economics Experiment Station)*, 29, (372).

Tavner, Bastien (2015), *L'innovation centrée usagers dans la cité par projets: ethnographie de l'appropriation d'une consigne plurivoque dans le secteur numérique: le cas du programme PACA Labs*, thèse de doctorat en sociologie, Télécom ParisTech.

Tremblay, Hélène P. (2007), 'Innovation sociale et société innovante. Deux versants d'une nouvelle réalité', in Juan-Luis Klein and Denis Harrisson (eds), *L'innovation sociale, Emergence et effets sur la transformation des sociétés*, Presses de l'Université du Québec, pp. 329–41.

Trompette, Pascale and Dominique Vinck (2009), 'Revisiting the notion of boundary object', *Revue d'anthropologie des connaissances*, 3 (1), 3–25.

van der Velden, Maja and Christina Mörtberg (2014), 'Participatory design and design for values', in Jeroen van den Hoven, Pieter E. Vermaas and Ibo van de Poel (eds), *Handbook of Ethics, Values, and Technological Design*, Dordrecht: Springer Netherlands, pp. 1–22.

van Oost, Ellen, Stefan Verhaegh and Nelly Oudshoorn (2008), 'From innovation community to community innovation: User-initiated innovation in wireless Leiden', *Science, Technology & Human Values*, 34 (2), 182–205.

von Hippel, Eric (1976), 'The dominant role of users in the scientific instrument innovation process', *Research Policy*, 5 (3), 212–39.

von Hippel, Eric (1986), 'Lead users: A source of novel product concepts', *Management Science*, 32 (7), 791–805.

von Hippel, Eric (1988), *The Sources of Innovation*, New York and Oxford: Oxford University Press.

von Hippel, Eric (2006), *Democratizing Innovation*, Cambridge, MA: MIT Press.

Winner, L. (1993), 'Upon opening the black box and finding it empty: Social constructivism and the philosophy of technology', *Science, Technology, & Human Values*, 18 (3), 362–78.

Wynne, Bryan (2007), 'Public participation in science and technology: Performing and obscuring a political–conceptual category mistake', *East Asian Science, Technology and Society*, 1 (1), 99–110.

# 10. Open innovation: the open society and its entrepreneurial bias

*Tiago Brandão*

## INTRODUCTION

Open innovation represents a catchword that captured the imaginaries of our societies through striving to point out desirable and transformative changes in our political economy. 'Open innovation' is now viewed as 'the next word' in innovation studies and management practices. Here, we shall nevertheless see how there are historical and genealogical facets behind the apparent novelty of this framework. Based on propositions drawn from management studies, open innovation discourses evoke old and longstanding liberal ideas, revamped by a new ultraliberal ethos based on technological and organizational firm-centred trajectories long since heralded by a previous generation of scholars.

An authoritative definition is invariably attributed to Chesbrough who broadly defined Open Innovation as "…the use of purposive inflows and outflows of knowledge to accelerate internal innovation, and expand the markets for external use of innovation, respectively" (Chesbrough et al. 2006, p. 1). By exploiting both the inbound and outbound channels to the market of companies (Dahlander and Gann 2010), open innovation was from its very beginnings presented as "the new imperative for creating and profiting from technology" (Chesbrough 2003a). In a post-industrial era marked by globalized and corporate economic hegemony, open innovation has striven to become the "the new paradigm for understanding industrial innovation" (Chesbrough et al. 2006, p. 1).

Ever since its inception in 2003, open innovation has stood out as a significant trend in the innovation studies field. As expressed by van de Vrande et al. (2010), "[r]esearch on open innovation has been *mushrooming* ever since, and the scope has been broadened in different directions" (p. 221). According to Kovács et al. (2015, p. 1): "The concept of open innovation has attracted considerable attention since Henry Chesbrough first coined it to capture the increasing reliance of firms on external sources of innovation". Several bibliometrics studies reveal the magnitude of this growth (Kovács et al. 2015; van de Vrande et al. 2010; Dahlander and Gann 2010). (Figure 10.1)

These bibliometric reviews (Kovács et al. 2015; Dahlander and Gann 2010; Huizingh 2011) have conveyed how open innovation has spawned special issues, numerous books and hundreds of papers. As pointed out by West et al. (2014, p. 808), an epistemic community has taken root around the concept, with a series of workshops, conferences, doctoral training programs and other events to catalyze and connect researchers working in this field. Lobbying activities also became increasingly institutionalized, gathering academics and practitioners, not just in North America but spread out globally through corporate and political forums. Thus, a small community of management researchers all of a sudden developed into an established research field (Gassmann et al. 2010, p. 213).

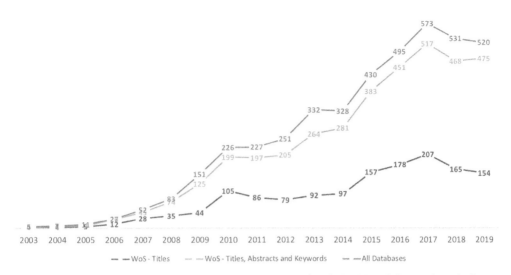

*Note:*   This search took place looking for the words 'Open Innovation' in the title and abstract, the author's keywords and the KeyWords Plus® index. All databases mean not just the Web of Science Core Collection (WoS), but also Current Contents Connect, Derwen Innovation Index, KCI-Korean Journal Database, MEDLINE®, Russian Science Citation Index and SciELO Citation Index.
*Source:*   Web of Science, www.webofknowledge.com.

*Figure 10.1*    *Open innovation bibliometric trend*

The open innovation concept has gained increasing popularity across leading industries, penetrating pioneering high-tech industries such as software, electronics, telecom, pharma and biotech. Indeed, the software and electronics industries have been built around the open innovation management framework. Today, many electronic suppliers drive open innovation at the strategic level. Further contextualization, however, might nevertheless still help in understanding just why open innovation became such a trend and why this trend so particularly applies to computer industries.

As notoriously exploited by management authors proposing open innovation as a new business model and a management framework, at the heart of this open innovation phenomenon are the revolutions in telecommunications and data processing. As identified by the historian David Reynolds (2000), these technological trajectories are inherently entwined with the Cold War, which enabled the electronics revolution through federal research funding for the solid-state physics that generated technological breakthroughs in electronic components and materials. They formed the foundations for the technological breakthroughs that gave birth to the digital era in the 1990s and 2000s (Chandler 2005 [2001]; Castells 2010 [1996]) This historical background merges with our contemporary 'digital transformation' that we are living through and which is bringing the 'third wave'[1] (Bogers et al. 2018, p. 8).

That lies at the core of Manuel Castells' (2010 [1996]) informational economy, a central pillar of the so very commonly mentioned new economy. That knowledge economy is based on the ICTs – Information and Communication Technologies. The "development and application of ICTs have become intertwined in a new era of innovative capitalism" (Cantwell 2001, p. 25). For example, new ICTs, especially the Internet, "accelerated the knowledge diffusion process and increased the personal mobility of knowledge workers", with many specialized knowledge workers (e.g., freelancers, consultants or part-time engineers) making a living as portfolio workers (Gassmann 2006, p. 224). This reflects one major outcome of the current techno-economic paradigm, which relies on flexibility through computerization (Cantwell 2001, p. 21).

As formulated by David Reynolds: "the information became a commodity, to be packaged and sold like toothpaste or automobiles – whether to big corporations in the early days of mainframe computing, or to the ordinary consumer when the PC came of age" (Reynolds 2000, p. 515). The information society context shapes the 'open innovation' framework and the technological trajectories of the computer and Internet culture in addition to open innovation discourses being a product of the 'network enterprise,' viewed as the 'new organizational logic' of contemporary societies (Castells 2010, pp. 163–64).

For the optimistic, open innovation might involve the ultimate conceptual *framework* of reference, an integrative view of managing knowledge within the context of innovative firms (Lichtenthaler 2011, p. 80). While Teece declares that "[a] framework is less rigorous than a model[,] as it is sometimes agnostic about the particular form of theoretical relationships which may exist" (Teece 2006, p. 1138), our purpose here is to reveal how open innovation is neither quite so 'agnostic' nor absent of its own political and ideological presumptions.

This chapter is divided into three sections, beyond the introduction and conclusion, where some contextualization and critical points are resumed and summed up. In the first section, the aim is to understand the liberal ideas from where open innovation discourse thrives, its primary authors, as well as the values (e.g., open, free-market) that shape our market-oriented societies. This provides, in fact, the common ground of our present-day liberal mainstream, whose values and worldviews are not always reflexively dealt with by the management literature, the primary field where open innovation emerges. In the second section, we cast our attention over the precursors, the classical authors, including some otherwise already forgotten, as well as several accounts from the 1970s to the 1990s, including fields other than management that have also been dealing with contextual changes in the firm nature that became fundamental to building the presumptions behind the open innovation framework. Finally, we deal with the explicit emergence of this business model and management framework, its major features, describing the essential entrepreneurial bias on which the open innovation framework is based.

## LIBERAL IDEAS AND THE OPEN SOCIETY

Today, open innovation is perceived as reaching beyond some managerial fad or fashionable concept. Nevertheless, there have been controversial debates over the actual value of open innovation frameworks on occasion over recent decades. These similarly encapsulate a matter of assessing "whether open innovation is a sustainable trend rather than a management fashion" (Lichtenthaler 2011, p. 75). Trott and Hartmann (2009) provide possibly the most critical account of open innovation in warning against "sloppy thinking and the uncritical adoption of concepts" (Trott and Hartmann 2009, p. 729).

This is not merely a critical stance as its sympathizers display clear concerns: "Once a field grows rapidly, there is a danger that it may become a short-term fashion and hype" (Gassmann et al. 2010, p. 213). That is per se a reason for us to understand its ideological embeddedness. Furthermore, its societal and political texture is indeed fully embedded within the modern and contemporary liberal creed.

We can trace back many of the features that characterize the 'open innovation' ethos from within the liberal political field. From the classics of political liberalism, there emerges John Locke (1632–1704) – property rights – or Adam Smith (1723–1790) – self-interest and the 'invisible hand' of the market – to the *homo economicus* of John Stuart Mill (1806–1873), with the latter two classical authors more in the scope of economic liberalism. Heading through the twentieth century brings forth liberal thinkers such as Ludwig von Mises (1881–1973) – the market society – Michael Polanyi (1891–1976) – and his thesis on the qualities of the spontaneous organization of the Republic of Science – passing Joseph Alois Schumpeter (1883–1950) – the innovative entrepreneur – and Friedrich von Hayek (1899–1992) – the usage and distribution of knowledge – before then arriving at Karl Poppers' (1902–1994) *Open Society*. One may indeed recognize the values and ideas that shape our contemporary market-oriented societies within this tradition of thought.

Many aspects require clarification when speaking of open innovation. One facet is "to clarify the definition of 'openness' as currently used in the literature on open innovation" (Dahlander and Gann 2010, p. 699). Hence, one of the central concepts to situating 'open innovation' amid the liberal political ideology derives from its understanding of 'open.' Despite the prevailing general lack of awareness over its underlying assumptions and implications, "Institutional openness is becoming increasingly popular in practice and academia: open innovation, open R&D and open business models" (Gassmann et al. 2010, p. 213). On the one hand, scholarly utilization of openness has been so loose that we are dealing with a construct that means different things (Dahlander and Gann 2010, p. 706). On the other hand, market-oriented usages of the openness concept are indeed well documented in the "Corporate Capitalism's Use of Openness" (Lund and Zukerfeld 2020).

This ambiguity around the 'openness' concept has also been subject to recent discussion by the literature. As noted by Pomerantz and Peek (2016) in their suggestively entitled article "Fifty shades of open", "ambiguity leads to misinterpretation", and its increased usage leads to dubious practices. The critical literature on 'open' and 'openness' duly reports that "the intuitive appeal of open" most probably interrelates with the fact that "… everyone likes free things — free as in *gratis*". 'Open' is similarly frequently deployed to indicate that a resource is accessible "for no monetary cost".

The free software and open-source movements brought about definitions that respond to a particular understanding of "free", thus conveying how free in software is about liberty, not price – "you should think of 'free' as in 'free speech,' not as in 'free beer,'" highlight Pomerantz and Peek (2016) This then gave rise to certain other definitions, articulating an openness to everything, from hardware to knowledge, free access, or coding access to improve software, democratization for hardware that entail a wide range of economic and societal implications. For instance, recently, in 1993, billionaire George Soros, inspired by the Popperian political philosophy of openness, launched the Open Society Institute, which was, for example, responsible for promoting the 2002 Budapest meeting. This initiative launched the open-access movement in academic circles,[2] with extraordinary progress in recent years earning support from international organizations and undergoing integration into national policies and funding agencies. Henceforth, 'open access' gave rise to a wide range of other 'opens' whether approaching scholarship, publication or cultural heritage in general (Pomerantz and Peek 2016). Open science and citizen science movements,[3] for example, came to merge assumptions inherited from open-source communities and the more recent open-access public policies for scientific publishing.

Indeed, policy usages explain another portion of this ongoing ambiguity. For instance, when the European Commission RISE Group put together an idiosyncratic report entitled *Europe's Future: Open Innovation, Open Science, Open to the World – Reflections of the RISE Group* (European Commission 2018), this came to be known as the 'three opens' report, demonstrating how liberal thinking and its 'openness' discourse is fully embedded in EU policy-making circles. At the European policy level in particular, 'open innovation' seems to incorporate a wide array of expectations. According to the ideas of Martin Curley, an Intel Labs Europe Director and one of the authors responsible for coining the term 'open innovation 2.0' – OI2 (Curley and Salmelin 2018), OI2 is perceived as a new paradigm or 'new mode' (of innovation) that "blurs the lines between universities, industry, governments and communities", the core mission of the OISPG[4] – a 'think tank' group set up in 2010 and working closely with the European Commission, which Curley chairs. This provides a case in which an entrepreneurial business model approach seems to be in the process of merging with public domain values, public policy and democratic service delivery to citizens, social responsibility, non-profit, open access – to knowledge, from education to science publishing.

Nevertheless, the 'open society' term is believed to have entered usage prior to World War I "to indicate a society tolerant of religious diversity" (Pomerantz and Peek 2016). Furthermore, the philosophical stance was put forward by Karl Popper and his 1945 book *The Open Society and its Enemies.* On the eve of post-World War II reconstruction, authors such as F. A. Hayek and his *The Road to Serfdom* (1944) feared the rise of the Soviet regime and, in particular, the influence of Marxism and its then growing presence in European and American universities. Furthermore, Schumpeter, with his book *Capitalism, Socialism, and Democracy* (1942), also sought to claim the superiority of liberalism and their market-oriented societies.

Popper, in turn, provided the philosophical grounds for challenging Communist regimes (and Marxism in particular). Since then, however, the Cold War ideological narrative became hegemonic, shaping and dividing the world history into totalitarian regimes, on the one side, and liberal democracies, on the other. The ideological aim was to merge all the leftist revolutionary historical experiences with regimes such as Italian fascism, German Nazism and Soviet Bolshevism, regardless of their differences – which indeed does represent a liberal way of looking at the world! This all arises from the literature on the so-called question of totalitarianism, an intellectual debate that profoundly shaped the culture of the twentieth century (Traverso 2001). This also represents the worldview embodied in managerialism and in different economic currents, including Schumpeterian and Neo-Schumpeterian. Nevertheless, it still remains remarkable how Popper himself perceived 'managerialism' as a surrogate of totalitarianism (Popper 1966 [1945], p. 14).

The Popperian 'open society' embodies historical aspirations, including the rejection of the absolute authority "of the merely established and the merely traditional" (Popper 1966 [1945], p. 6). Such an open society is based on human reason and freedom and indeed duly praised for the "competition for status among its members". The closed society, on the other hand, deems its institutions as sacrosanct, taboo. Thus, the transition, from a closed to an open society, was driven by commerce and particularly by a new class engaged in trade and seafaring. Most meaningfully for Popper, the closed society holds very deep fears of 'commercialism' with the development of communications and commerce identified as the forces that enable the rise in the open society. The open society in itself entails a doctrine of commercialism and entrepreneurship with trade and commercial initiative emerging as "the forms in which individual initiative and independence can assert itself" (Popper 1966 [1945], p. 180).

In a similar vein, but further back in time, for a liberal philosopher such as John Stuart Mill, an 'open society' was already a commercial society. Openness was achieved by means of commerce, the ways of developing new branches of trade with openness constituting the ideal framework for profits and fulfilling economic interests (Mill 2000 [1844]). In turn, for Ludwig von Mises 'open' means the free market, open to commerce with the gate open to innovations (von Mises 2002 [1927], p. 95), which increase the productivity of labour and thereby bring about a higher standard of living. The open society is therefore "a system of unhampered competition *open* to all" (von Mises 2002 [1927], p. 186), a system in which entrepreneurs prosper.

Another liberal author interested in the principles of a free and open society was Friedrich von Hayek, especially in his work *The Constitution of Liberty* (1960). The point of departure in Hayek is again, as with any liberal thinker, the primacy of the individual (and the private) over the state: "It is one of the characteristics of a free society that men's goals are open, that new ends of conscious effort can spring up, first with a few individuals, to become in time the ends of most" (Hayek 1960, p. 87).

Clearly, the concept of openness is inherently not agnostic and even when we try to perceive openness "as social praxis": Smith and Seward (2017) wish to believe that open is "more than just a set of social practices predicated on technological affordances". However, they also acknowledge a lack of clarity over whether openness is "in contrast to the dominant market paradigm". Common sense lacks any understanding of the ideological implications behind those usages, especially as regards the liberal correlation between openness and freedom and entrepreneurial and market biases.

Schumpeter's own work, for example, does not develop a philosophical stance on 'openness' as others undoubtedly did, most especially the so-called Schumpeterian authors (e.g., Diamond 2019). However, Schumpeter does use 'open' as 'open market,' 'open trade,' in the sense of 'opening up of new markets' for trade opportunity. According to Schumpeter, there is no margin for doubt that 'openness' is to mean the "opening up of new markets [foreign or domestic] or of new sources of supply" (Schumpeter 1939, pp. 80–83). Like this, since its first inception in western politics (and economics), the use of 'open' (and openness based narratives) has carried the weight of an ideology, with its norms and values of an entrepreneurial ethos running rampant in more recent decades.

Understandably, certain expectations have arisen and often evoking past utopias. For instance, the origins of the open-source movement in the 1960s and 1970s gave an aura that still remains present in some of the more intuitive interpretations of open innovation accounts. The 'open' agenda might exude a "subversive, even emancipatory" and "activist air" (Söderberg 2017, p. 117). Most certainly, movements based on open source experiments were the drivers behind the advent of the open access, open science and citizen science agendas. However, the genesis of the 'openness' movement, the 'openness' of open innovation primarily traces its roots to the Liberal tradition of thought, including the Schumpeterian and entrepreneurial understanding of 'openness.' Openness therefore provides a way of reaching (and sustaining) profit, triggered by innovation dynamism and not necessarily aiming to disseminate free knowledge or gratis products or services.

From an internal perspective of the field of open innovation, Dahlander and Gann (2010, p. 706), in particular, define different types of openness, dividing 'inbound' and 'outbound' innovation and 'pecuniary' and 'non-pecuniary' interactions. This was nevertheless still drawn from a firm-centric and commercial exchange perspective. Hence, openness is thereby sketched according to the dichotomic logic of commercial 'advantages driving openness' and 'disadvantages driving closeness.' The role of openness in this view evolves around "the way firms go about organizing the search for new ideas that have commercial potential" (Laursen and Salter 2006, p. 131). Thus, when we set aside the more intuitive understanding, we end up sticking with the firm-centric and entrepreneurial bias.

Moreover, there has certainly been no lack of effort over theorizing and gathering evidence on the correlation between liberalism and 'openness' and particularly by redeeming the entrepreneurship spirit. For example, Audretsch et al. (2009, pp. 1–2) strive to prove that "[e]ntrepreneurial activity can thrive (…) only in 'open' settings in which consumers and firms are free to transact with one another"; without freedom, they say, entrepreneurs do not take risks and would not even launch new enterprises. Therefore, 'openness' and 'entrepreneurship' correspondingly mutually reinforce each other.

From a liberal economic perspective, openness means governments have lowered or erased legal barriers to trade and the movement of capital and people – a much-praised trend after World War II, enhancing trade and commerce across national borders as was once also praised by Adam Smith in the very earliest days of economic liberalism. However, openness also means firms are able to respond "to shifts in customer tastes or to adopt new modes of production", as well as workers who adapt to new job requirements… In that regard, there is acknowledgement that "openness is likely to be viewed more as a curse than a blessing" (Audretsch et al. 2009, p. 2). Openness is connected to entrepreneurship, innovation and economic growth and, above all, is believed to favour the innovation cycle. More than mere declarations of faith, openness is here being theorized. However, this also mainly takes place in a purely entrepreneurial fashion, which is very much coherent with its liberal traditions.

On the usefulness of being aware of these ideological entanglements, and against those who assume that "openness is 'a non-controversial argument'" (Dahlander and Gann, (2010, p. 703) John Maynard Keynes (1936) once issued a piece of provocative advice: "practical men, who believe themselves to be free from any intellectual influences, are generally slaves to some extinct economist" (Keynes 1936, p. 340). Unequivocally, today that became the case not only with neoclassical economists but also with the evolutionists and the Neo-Schumpeterian accounts, supposedly less orthodox when compared to the so-called mainstream economists.

## THE PRECURSORS OF OPEN INNOVATION

When liberal democratic establishments now compare themselves with the patterns of Soviet Union technoscience, especially after the debacle of the Communist regime perestroika initiative, this provides North Americans with a sense of superiority, envisaging the U.S. as a far more complex and much more competitive economy than their giant rival which, above all, according to the liberal creed, derived from being a more open society in which freedom and the entrepreneurial ethos were intrinsic to the 'American way of life.' As expressed by historians, the Soviet Union simply could not match the "entrepreneurial consumer culture" (Reynolds 2000, p. 538).

That *entrepreneurial | consumer* logic was indeed based on the material culture provided by the computer and electronics revolution. With microprocessors and the 'computer on a chip' concept, electronic machines became personal (e.g., IBM PC), followed by the dissemination of user software (nowadays 'applications' for increasingly user-friendly platforms, such as smartphones, tablets). The commercialization of PCs (i.e., personal computers) first emerged in the California youth culture of the Vietnam War period when ideals such as 'openness' and 'democratization' went hand in hand in the counterculture movement.

For example, the Apple company slogan was 'one person – one computer,' very much attuned to a generation of counterculture and anti-establishment ideals. One participant of that generation, Alan Kay, a well-known code programmer and key technological lead-user since the 1960s, did indeed claim that "there were people like me, and many people in this research community, who wanted to improve the world".[5] Others, including the charismatic salesman Steve Jobs (1955–2011), nevertheless sensed the business opportunity posed by this philosophy of democratizing computer use.

Open-source software communities gave strong inspirational insights to the generation that built up the open innovation narrative: "the technological success of open-source software, such as Linux and Apache, has played an important role in spreading open innovation thinking" (Gassmann 2006, p. 223). Indeed, it was from looking at those milieus that one of the precursors of the open innovation narratives, MIT Sloan School of Management professor Eric von Hippel, based his pioneering work.

Today's classical account entitled *The Sources of Innovation* (von Hippel 1988) was responsible for introducing the concept of user-centred (and used-driven) innovation, that is, the role of users, further developed over the years (von Hippel 2005). The narrative runs as follows: *aided by improvements in computer and communications technology, users can now develop new products and services for themselves*. This scope was enabled by the software and information products responsible for the emergence of a *system* of *user-centred* innovation.

Even von Hippel acknowledges that "many of the advantages associated with user innovation communities also apply to open information networks and communities" (von Hippel 2005, p. 17). Chesbrough (2003a, p. 56), also, reflects on the role of *customers*, who he maintains "have important information that can be vital to open innovation". Additionally, "[t]he most advanced, most demanding customers often push your products and services to the extreme", thereby positioning them as 'innovators themselves,' "what Eric von Hippel calls lead users" (Idem, p. 56). From its very beginning, open innovation was proposed as a paradigm shift heading into a digital world in which "innovation is no longer a linear process but one where the user is feeding back to the producer what innovation is needed" (Bogers et al. 2018, p. 10).

Appropriating user ideas and experiences emerged as a recurrent theme within the innovation literature. West et al. (2014) concur: "Open innovation builds upon the research (…) in user innovation, and there is considerable overlap between the two bodies of research" (p. 808). The difference might stem from how the open innovation literature appears as a more firm-centric theory of innovation. Also, open innovation results to a greater extent from an explicit top-down strategy. Indeed, there are authors from the open innovation field who duly acknowledge that users are not well known as sources for radical innovation and entrepreneurship and, above all, that the open source software story, despite telling of an exciting version of open innovation, "cannot be transferred one to one to an average industrial environment" (Gassmann 2006, p. 227).

Whether or not accepting the reach of 'user-centric innovation' over current open innovation literature, the Kovács et al. (2015, p. 966) analysis reinforces how open innovation research "is primarily (…) focus[ed] on the user-centric perspective", "which distinguishes the open innovation literature from the broader literature on R&D collaboration" even while both streams are based on a previous scholarship from the 1980s and 1990s. In turn, the user innovation literature has constituted an autonomous stream of studies, influencing not just directly open innovation scholars but also other disciplinary fields beyond management studies.[6]

Another unmistakeable precursor stream arises from the literature on the nature of the firm. Hayek first raised the question of managing knowledge as well as identifying how firms could access and extract the benefits of knowledge. Hayek dealt with the problem of knowledge and its uses within a rational economic order. He in fact also pointed to "the unavoidable imperfection of man's knowledge and the consequent need for a process by which knowledge is constantly communicated and acquired" (Hayek 1945, p. 530). Hayek held the understanding that "scientific knowledge is not the sum of all knowledge" (Hayek 1945, p. 521) in line with another renowned liberal thinker, the more philosophically grounded distinction that Michael Polanyi made between personal and tacit knowledge (Polanyi 1958, 1966).

One indeed does encounter here a theme concealed by the open innovation literature, the evolving nature and growth of the firm. For instance, Edith Penrose (1914–1996), in her classic work *The Theory of the Growth of the Firm* (2009 [1959]), provides a hallmark in this regard. In particular, authors throughout the 1950 and 1960s made outstanding contributions in following on from the steps taken by Schumpeter. For instance, Penrose had as her thesis supervisor another critical author, Fritz Machlup (1902–1983), who, in turn, studied economics under Ludwig von Mises and Friedrich Hayek at the University of Vienna in the 1920s. Machlup's comprehensive study on *The Production and Distribution of Knowledge* (1962) does indeed rank among the precursors. Still remarkable in many aspects, but especially for his comprehensive concept of knowledge, based on a rationale that approaches knowledge from an economic standpoint, but "including all kinds of knowledge, not only scientific knowledge" (Godin 2008, p. 7) – including information, the flow of ideas, R&D, innovation.

In turn, Penrose's work framed the role of the firm in terms of managing the dynamics of knowledge and innovation. According to Penrose, the firm, or the multinational enterprise, was not just a set of supply and demand functions (as perceived by classical economics) and rather represented a place where the real economy happened, a 'flesh and blood' organization. Thus, a firm should not be defined in terms of its products but rather of its resources, including its technological and entrepreneurial capabilities. Those resources, persisting within the firm's shell, are susceptible to recombination to enable other expansion opportunities. Furthermore, the external environment should be grasped by entrepreneurs and company *cadres* as they continue searching for possible dynamic interactions between the internal and external environments and potential opportunities for leverage. Above all, in the view of Penrose, a firm constitutes a place in which it is possible to pursue the expansion of innovation by devising and deploying strategies (business models), holding precisely that aim.

These ideas and accounts generated significant implications for managerial practices and are influential over several of the streams that preceded open innovation theory. Ever since, the literature has expanded "by leaps and bounds" (Pitelis 2009, p. xxxi), through fruitful currents of literature within the strategic management field (Kovács et al. 2015; Dahlander and Gann 2010). Based on the resource-knowledge and dynamic capabilities perspectives, these accounts directly shaped the open innovation framework, especially during the 1990s in the period prior to the appearance of the 'open innovation' narrative.

Even in the 1980s, the organizational economics literature was already pointing to major structural challenges as regards in-house R&D management (Teece 1986). One of the landmarks pointed out by this literature is project SAPPHO (Rothwell et al. 1974), an industrial innovation study conducted by scholars at the University of Sussex (Curnow and Moring

1968). This involved a systematic comparison of successful and unsuccessful innovations in companies from different economic sectors (computers, scientific instruments). Based on those surveys, Rothwell (1994, p. 19) states that "accessing external know-how has long been acknowledged as a significant factor in successful innovation" with the network model of innovation (Rothwell and Zegveld 1985) identified as core to what got termed the 'fifth-generation innovation process.'

Since then, an ongoing debate has taken place, encompassing more technical management studies and with many scholars making contributions to better understanding this process. On the one hand, the debate has pointed to the precedents of the network enterprise that embody the open innovation paradigm, with several practices and policies already in effect, including a variety of contractual forms and different degrees of organizational integration. Nuances already existed in the literature, such as the notion that in-house development was not an exclusive strategy of the large corporate firm: alliances, partnerships, technological transfers and licensing were indeed strategies adopted by all kinds of companies. For instance, earlier contributions, such as Hagedoorn (1993), Rothwell and Dodgson (1991), among others, had already demonstrated how the flow of information and networking matters. As noticed by Powell et al. (1996), "the locus of innovation is found within the networks of interorganizational relationships that sustain a fluid and evolving community" (p. 142). Nevertheless, synergies were also found within firms, between their departments, by 'technological brokering' (Hargadon and Sutton 1997), by recombining ideas, presenting knowledge and skills, including tacit knowledge of previously established hierarchies. The Teece (1986; Teece et al. 1997) and March (1991) 'organizational learning' approaches also convey how important it is for firms to develop their assets, which are then expressed as the 'absorptive capabilities' (Cohen and Levinthal 1990) necessary to dealing with an open innovation business model. Organizational routines and (core) competencies may then extend outside of the firm to embrace alliance partners (Teece et al. 1997, p. 516). March (1991), in turn, did make a distinction between 'exploration' and 'exploitation' in organizational learning: "Exploration includes things captured by terms such as search, variation, risk-taking, experimentation, play, flexibility, discovery, *innovation*. Exploitation includes such things as refinement, choice, production, efficiency". The point made immediately afterwards was that adaptative processes, "by refining exploitation more rapidly than exploration, are likely to become effective in the short run but self-destructive in the long run" (March 1991, p. 71).

In sum, several management rationales regarding private R&D and the relation of corporate multinationals with the dynamics of innovation were already in place well before the emergence of the so-called open innovation paradigm. Teece et al. (1997), inclusively, expressed early concern about "the fascination with strategic moves and Machiavellian tricks [that] will distract managers from seeking to build more enduring sources of *competitive advantage*" (p. 513). This might well be the case with open innovation and some of its current practices. Meanwhile, open innovation emerged and was presented as the only way to keep pace with the innovative dynamics ongoing in our enterprise-biased society, speeding up businesses and accelerating innovation for its own sake.

## THE EMERGENCE OF 'OPEN INNOVATION' – ITS ENTREPRENEURIAL AND MARKET BIAS

The explicit emergence of 'open innovation,' both the concept and its discourse, has received praise as a clear shift in the management literature in which the 'stakeholder citizen' and 'employee participation' (also 'non-R&D experts') become valuable actors in the innovative dynamics ongoing in our society (Segercrantz et al. 2017, p. 283; Bogers et al. 2017 [2016]). The innovation process would thus be 'opened up' and with open innovation becoming a facet of managerial praxis, seemingly 'universal' for every organization not just on rhetorical grounds but also endowing practical recommendations for ambitious CEOs.

Even before the seminal contribution from Henry Chesbrough, two partners at Bain & Company, a notoriously secretive American management consultancy in Boston, had published an article entitled 'Open-Market Innovation' (2002) in the *Harvard Business Review*. Chris Zook and Darrell K. Rigby were then managers and consultants and were arguing for a 'free trade agenda' and, above all, posed the question of "How can we reach outside our own four walls for the ideas we need?".

However, it was Chesbrough's writings that became famous and launched a massive wave of open innovation studies and happenings. A 'small club' of devoted managerial practitioners soon became a global phenomenon. (Gassmann et al. 2010, p. 213). Presently at the Haas School of Business (Berkeley), Henry Chesbrough was then teaching (1997–2003) at Harvard Business School (Boston) and proposed his 'open innovation' concept in 2003, with a book entitled *Open Innovation: The New Imperative for Creating and Profiting from Technology*, which was, as he explicitly states, written "to managers of companies struggling with the innovation process in their own firms" (p. xiv). In this seminal work, presently, with more than twenty thousand citations, he provided a detailed description of what he calls the 'open innovation model.' By open innovation, Chesbrough means...

> ...that valuable ideas can come from inside or outside the company and can go to market from inside or outside the company as well. This approach places external ideas and external paths to market *on the same level of importance as that reserved for internal ideas* and paths to market during the Closed Innovation era. (Chesbrough 2003a, p. 43)

Even though this inherently believes in knowledge, Chesbrough is, however, striving for a more specific 'knowledge landscape' in which firms, despite the competitive environment prevailing, exchange ideas and expertise. Furthermore, universities stand out as central actors in this landscape whenever fully embracing an open innovation (i.e., entrepreneurial) mindset. In essence, they are full of professors and graduate students "clearly eager to apply [their] science to business problems" (Chesbrough 2003a, p. 44). The "rise of excellence in university scientific research" is thus deemed to involve the entrepreneurial and business-oriented perspectives that were already a trend in American top universities.

One motive for this situation arising stems from the decline in government funding for basic scientific research, compelling research milieus to be "more astute about the needs and problems of industry" (Chesbrough 2003a, p. 45) while simultaneously opening up their knowledge niches to the market and 'ending of the knowledge monopolies.' This "new logic of innovation" thus incorporates the idea that "the research game is over" (Chesbrough 2003a, p. 51). Additionally, open innovation enhances the entrepreneurial bias regarding the exclusion of public administration, which played a key role in each breakthrough innovation (e.g., computing, nuclear). Only firms, users and universities, in that order, still remain relevant actors while we know from the history of technology how public authorities played major roles and also interrelated to the pursuit of a global vision. Today, in fact, most innovation takes place with minimal reference to public authorities.

This new logic should turn "old assumptions on their head", which particularly implies companies restructuring their R&D capabilities and "structure themselves to leverage" the aforementioned "knowledge distributed landscape" instead of pursuing "their own internal research agendas". Instead of 'hoarding' research projects and the internal development of technologies, companies now should contemplate a new business model in which grasping the immediate opportunities for making money should be the priority. This involved not simply the integrating of external knowledge and partnerships and downsizing internal R&D teams and portfolios but also "leveraging multiple paths to market for your [in-house developed] technology". For instance, "[i]nstead of managing intellectual property (IP) as a way to exclude anyone else from using your technology, you manage IP to advance your own business model and to profit from your rivals' use" (Chesbrough 2003a, pp. 51–52). Indeed, Zook and Rigby (2002) had previously explained the benefits of deploying IP not as a way to shield innovators and block wider commercialization (preserving technological monopolies) but rather as a means of obtaining immediate profits.

This indeed changes, for example, the researcher career paths inside R&D firms. A new rationale for internal R&D then emerged, to be followed and improvised by eager, visionary managers, and "[w]ith this Open Innovation approach to knowledge, research managers must evaluate researchers' performance in different ways" (Chesbrough 2003a, p. 52). In this view, within walls R&D departments should organize themselves: (i) 'to identify, understand and select' external knowledge to the company as well as find ways of connecting to these innovative groups; (ii) to integrate internal and external knowledge in new and more complex combinations, building new *systems and architectures*; and, last but not least, (iii) "to generate additional revenues and profits from selling research outputs to other firms for use in their own systems" (Idem, p. 52). In particular, research (and knowledge in itself) no longer constitutes the priority: "Research takes a long time to deliver useful outcomes, and company strategies change at a far faster rate than the rhythm of basic research" (Idem, p. 53). In this regard, there is also a mismatch between the two dimensions of any R&D group, the laboratory, on the one hand, "operating as a cost centre", and the development group, on the other hand, "operating as a profit centre" (Idem, p. 58). R&D retains a critical role but the bias is mostly entrepreneurial. With this new approach, research (and researchers) become 'more manageable,' and the company architecture became more suitable for business.

According to Chesbrough though, open innovation therefore means a purposeful 'paradigm shift,' from an era of 'closed innovation' to a new era (or paradigm) of open innovation. The idea of a closed innovation era, where firms had their R&D departments, within their walls, where they conducted internal experiments without any external linkage, represents one of the facets most criticized by the literature. This furthermore claimed that internal R&D no longer constitutes the strategic asset it once was (Chesbrough 2003b, p. 36). In all likelihood, this is not especially accurate as highlighted by several authors not just providing historical accounts but also including some open innovation proponents, who swiftly began advocating in favour of avoiding dichotomic and antithetical reasoning (Huizingh 2011, p. 3; Trott and Hartmann 2009, p. 715), including Enkel et al. (2009) who warn that "[t]oo much openness can negatively impact companies' long-term innovation success", leading to loss of control and core competencies. Nevertheless, from within the field, there is an understanding that a closed innovation approach does not serve "the increasing demands of shorter innovation cycles and reduced time to market" (Enkel et al. 2009, p. 312).

Initially, however, Chesbrough was less explicit than some authors think he should have been (Groen and Linton 2010) in terms of the extent to which his analysis drew upon previous scholarship. Beyond his earlier mention of Eric von Hippel's work and his indebtedness to Christensen's work on the concept of 'disruptive innovation,' it is only very recently that Chesbrough (2018, p. 930) came to acknowledge "the long stream of economics literature". Indeed, several such streams clearly precede open innovation theory as we have already set out above. Nevertheless, the lack of due acknowledgement of Chesbrough in history represents a common feature of management studies in general (Boltanski and Chiapello 2007 [1999], p. 62), whose intellectual gurus are biased by design to deal more with the "*business model* of the future" (Braudel 1983 [1979], p. 453).

In this regard, a deeper historical perspective needs adopting to reveal how, by the late 1960s, a new discourse was already emerging and challenging what was previously a sort of 'conventional wisdom' in practitioner milieus as regards the relationships between research and technology. A more experienced, pragmatic and entrepreneurial mind ethos was already coming to challenge the intuitive understanding of one particular way of looking at innovation processes that came to be known as the 'linear view' of innovation. Over the 1970s and 1980s, this then spread to research-oriented firms (and to corporate businesses as well) with a "growing sophistication about technological innovation", which the following crystal-clear words of organizational theorist Donald Schon succinctly summarize: "Technological innovation is only one route to growth" – and it may not be the most effective, that is, besides knowledge (and technology) other production factors might be strategically allocated (labour, capital). And also, according to Schon, "Scientific research is only one route to technological innovation", which thus already meant that "invention or engineering, without a base in science or, at any rate, without a corporate base in science, can be as effective" (Schon 1967, pp. 113–14). This is what Chesbrough later came to term as the new "increasing metabolism of knowledge" (Chesbrough 2003a, p. 58), which was, in fact, not new.

There is indeed plenty of literature conveying the notion that the R&D departments of large corporations were already opening up to external collaborations and networking long before the open innovation discourse of our time. As acknowledged by Lichtenthaler, "there are many influential studies in the general field of interorganizational innovation that enhance our understanding of open innovation" (Lichtenthaler 2011, p. 77). According to scholars such as David C. Mowery, those same corporate strategies were already under implementation from the late nineteenth century onwards (Mowery 2009); and, according to Rothwell and Zegveld (1985) as well as Trott and Hartmann (2009, p. 723), throughout the twentieth century, 'late entrants' surpassed pioneering companies. Furthermore, Mowery (1983) also explains how internal R&D also emerged in response to the lower costs of in-house organizations when compared with acquiring ideas and resources from the marketplace.

There are nevertheless differences in those streams of literature. When approaching user-driven innovation and open innovation, which we considered in the previous section, significant differences arise from the fact that the latter is firm-centric. In terms of the management and economics authors mentioned above, the open innovation literature argues they have a balanced view between internal and external knowledge, while the latter streams consider external knowledge as an additional input to internal innovation. Open innovation (OI) gurus argue that those new 'OI' strategies (for example, design, market intelligence, crowdsourcing, co-innovation, globalized R&D networks, etc.) enable firms to create a 'competitive advantage' (Porter 1990) in unprecedented ways and that an open innovation approach "requires a different organisational structure and *mindset* than the more traditional R&D approaches do" (van de Vrande et al. 2010, p. 231). Furthermore, while both the management economics and open innovation literature streams are firm-centric, it is our perception that economics and economic history, for instance, adopt a more socially responsible view towards human resources and R&D departments, less performative in terms of the profit-oriented and business model proposition than that held by open innovation.

## CONCLUDING REMARKS

Open innovation is undergoing widespread application across the academia, business and policy-making milieus (Table 10.1). The open innovation literature and its 'best practices' are not just 'mushrooming' but scholars and practitioners have also been 'flirting' with its increasing ambiguities. Why is this so? As acknowledged by Porter and Kramer (2011), the "Capitalist system is under siege" thereby reflecting how this kind of transformative framework holds an increasingly strong appeal not just within the system but also through resonating old reforming and utopian ideals.

Open innovation proponents claim the concept "…has a clear managerial impact because it has further strengthened awareness of the importance of innovation in many firms" (Lichtenthaler 2011, p. 79). However, open innovation has also been subject to criticism on the grounds of being, for instance, 'old wine in new bottles' (Trott and Hartmann 2009). We have here set out how the prior literature had already made most of the points put forward by open innovation advocates. Furthermore, as the Kovács et al. (2015, pp. 2–4) literature survey points out, the "lack of conceptual clarity" and "a certain degree of subjectivity and bias, since they rely on (…) idiosyncratic views and perspectives".

Several assumptions also underlie this open innovation literature, in particular: (i) society, every society is market-driven; (ii) capitalism has been undebatable ever since the collapse of 'real socialism' historical experiments in the 1980s; (iii) competition (i.e., 'competitive advantage,' Porter 1990) is the organizing principle of the business model and, therefore, presumably of all society; (iv) 'openness' is a non-controversial value and the new utopia for market-driven societies long since evoked by a much-venerated liberal tradition; and, last but not least, (v) innovation is always good and irrespectively of the unintended consequences potentially arising from Schumpeterian 'creative response' and its acceleration (i.e., the unsurmountable tension between 'creative destruction' versus 'destructive creation').

There is indeed "a limited understanding about the costs of openness" (Dahlander and Gann 2010, p. 707) and its unattended consequences. 'Open innovation' might be perceived as a strategy to 'crowding out' R&D human resources as well as deepening the difficulties that private finance poses to the needs of firms for long-term investment for innovation. As "the short-termism of shareholder capitalism brings about negative incentives for companies seeking to invest in (uncertain) innovation" (Mazzucato 2013, p. 855), management gurus have unveiled the 'open innovation' framework as the new 'silver bullet' for innovationists all over the globe. Apart from conjecture over its soundness and practices, open innovation might in itself constitute a get-out clause to the unwillingness of corporate governance structures *to take risks* and thereby shunning investments *in the type of risks that innovation entails* (and in the proper infrastructures required by genuinely democratic societies). Avoiding the cost burdens imposed by R&D investments might appear a means of 'maximizing shareholder value,' one pinnacle in the prevailing managerial ideology of late capitalism and what does matter according to proponents of open innovation is this very same appropriation of profits from innovation (Huizingh 2011, p. 7).

As recognized by von Hippel, "Open, distributed innovation is 'attacking' a major structure of the social division of labour" (von Hippel 2005, p. 2). On his behalf, Chesbrough also acknowledges that "the pattern of high labour mobility is unlikely to return to the earlier pattern of long-term, 'lifetime' employment" (Chesbrough 2003a, p. 49). This means then that instead of hiring the best engineers, companies broke from outside talent their need for human resources, allocated in temporary projects. Even enthusiasts of the field acknowledge that one part of its success, and "the greater interest" in open innovation, relates to "its greater popularity among managers, either as a cost-reduction measure or because more firms are in a position to use technology than to create it" (West et al. 2014, p. 808), which first, means cuts to human resource and precarious labour bonds and, second, a lack of interest in research among medium and large firms.

Thus, open innovation becomes a 'common currency' mainly because "professionals seek portfolio careers rather than a job-for-life with a single employer" (Dahlander and Gann 2010, p. 699), reflecting social and economic changes in working patterns. This indeed interlinks with broader changes in society and industry that have led to "increased mobility of knowledge workers and the development of new financial structures such as venture capital, causing the boundaries of innovation processes" (Elmquist et al. 2009, p. 327).

The 'future of Open Innovation' is then encapsulated by this simple but crucial question; "is this good for the economy?" Or even, does open innovation really work? This kind of critical inquiring especially applies because "[b]alancing risk-taking and promoting cumulative innovation are challenging social questions" (Gassmann et al. 2010, p. 219). Therefore, while there might be managers that are fascinated by the value of cooperative R&D for higher innovation rates, there is no doubt that R&D outsourcing has been reduced to cost savings in most companies (Gassmann 2006, p. 225). In this regard, one may argue that open innovation provides a kind of outwards transfer of risk. Thus, it is not at all clear that "looking externally, opening up the companies, is always positive" (Lang et al. 2017, p. 2); in this regard, as Lichtenthaler recommends, "academics and managers should not oversimplify" (2011, p. 90).

Another aspect has to do with the ongoing extrapolations and general looseness associated with the concept. As Clayton M. Christensen points out – most influential business theorist, whose inspirational contribution were praised by Henry Chesbrough when launching his 'open innovation' model in 2003:

> ...an imprecise definition not only makes *open innovation* more difficult to understand (…) but it also makes it more difficult to implement, because there are a lot of people who claim to talk about 'open innovation' but are actually talking about something else. (Christensen 2012)

More generally, however, one might argue whether it is not the ambiguity (and even confusion) around 'open innovation' that better explains its diffusion and circulation. Despite this historical survey, which does reveal how open innovation should not appear as totally novel, there is nowadays a huge interest in presenting open innovation as a 'framework' for different agendas and exploiting the ambiguity around 'openness.'

Conceptual management frameworks are particularly prone to this outcome and, in the case of open innovation, entangled by management studies, become a "self-reinforcing circle" (Segercrantz et al. 2017, p. 291). Such frameworks have indeed represented "a strategy of reducing complexity by deflecting attention from precursors" in which there are some signs of "entrepreneurial self-promotion" interspersed with the "tactical devaluation of alternative management frames" and "the creation of a peer group around a buzzword". Moreover, and above all, in a critique from which one may not easily deviate, there is a considerable temptation – for anyone who uses the 'open innovation' conceptualization – that "the strength of [that] rhetoric (…) permits us (if we buy this rhetoric) to do two incongruous things at once: understand ourselves as change-agents contributing to nothing less than the glorious reinvention of [a fair] capitalism, and stay firmly within today's capitalism's comfort zone without essentially changing anything" (Kettner 2017, p. 153). This seems to be in line with the concerns of Groen and Linton (2010) when they equated whether open innovation was in itself a field or "are we creating false barriers that inhibit communication between different groups of academics" (p. 554), deflecting, for instance, all the previous and ongoing literature on development.

## NOTES

1. The first wave established the Internet infrastructure while the second wave surfed social networking and smartphone apps.
2. Budapest Open Access Initiative. https://www.budapestopenaccessinitiative.org/read. Access on 27 April 2020.
3. Naturally, it is not the aim to deal with 'open science' in this chapter in terms of its ideological and philosophical features. However, one should clarify that open innovation appeared slightly earlier (2002/2003) and has today an evident tradition of thought in Liberal and Schumpeterian ideas. In turn, Open Science, as it is explicitly and frequently deployed today in policy discourses, is comparatively more recent (e.g., Suber 2012). Although this recent emergence being related to benefits from the open-source movement and broader recognition of open-access practices, regarding scientific peer-reviewed publications, while looking at open science, one could also evoke some previous traditions of thought, such as the Republic of Science values or the Mertonian sociology. Even so, depending on the definition attributed to the concept, previous S&T policy history from the second post-war, at least, could provide several precedents of collaborative and synergetic efforts. Open science, however, might be not only open to society but also open to market opportunities (Chesbrough 2015), which could include lessons from the history of Big Science contracts, the university–private corporation military complex, among different experiments that were pursued by several different national scientific 'systems.'
4. OISPG – Open Innovation Strategy and Policy Group. https://ec.europa.eu/digital-single-market/en/open-innovation-strategy-and-policy-group. Accessed on 25 March 2020.
5. *Explained*, an American documentary television series that premiered on Netflix on 23 May 2018, produced by Ezra Klein and Joe Posner. Episode "Coding", a history of the evolution of computer coding, and predictions for how coding can shape the future (October 2019).
6. For a more detailed presentation of the *User Innovation* theory, see Chapter 9 by Bastien Tavner in this book.

*Table 10.1*   Main points of criticism to open innovation

| Critical stance | Area Studies | Some key references |
| --- | --- | --- |
| Liberal traditions of thought and their ideological weighting \| the entrepreneurial and market-oriented bias. | Intellectual History; Critical Studies of Innovation. | |
| The ambiguity of 'openness'. | Interdisciplinary Studies; Conceptual History; Internet and Computer Studies. | Enkel et al. 2009; Dahlander and Gann 2010; Pomerantz and Peek 2016; Smith and Seward 2017; Lund and Zukerfeld 2020. |
| Management fad or fashionable concepts. | Management Studies; Critical Studies of Innovation. | Abrahamson 1991; Trott and Hartmann 2009; Gaglio et al. 2019. |
| Insufficient acknowledgement of history in general and twentieth-century corporate enterprise history in particular. | Economics Literature; Economic History. | Penrose 1959; Mowery 1983, 2009; Teece 1986; Teece et al. 1997; March 1991; Cohen and Levinthal 1990; Huizingh 2011. Among others. |
| Acknowledgement on the precursors of open innovation. | Management Studies. | von Hippel 1988; Christensen 2012. |
| Lack of conceptual clarity in general. | Management Studies; Open innovation Studies; Critical Studies of Innovation. | Christensen 2012; Kovács et al. 2015. |
| Dichotomic and antithetical reasoning. | Open innovation Studies. | Trott and Hartmann 2009; Enkel et al. 2009; Huizingh 2011. |
| Corporate short-term and downsizing strategies for maximizing shareholder value. | Open innovation Studies; Innovation Studies. | Huizingh 2011; Mazzucato 2013. |
| Labour precariousness. | Sociology, User-Driven and Open innovation Studies. | von Hippel 2005; Dahlander and Gann 2010; Elmquist et al. 2009; West et al. 2014. |
| Outward transfer of risk. | Open innovation Studies. | Gassmann et al. 2010; Lang et al. 2017; Lichtenthaler 2011. |
| Corporate capitalism's comfort zone. | Economics; Ethical Economy. | Porter and Kramer 2011; Kettner 2017; Lund and Zukerfeld 2020. |
| Inhibiting communication between different groups of academics. | Management Studies; Development Economy. | Groen and Linton 2010; Kettner 2017. |
| On the the performative nature of management studies. | Sociological Studies. | Boltanski and Chiapello 2007 [1999]; Segercrantz et al. 2017. |

*Source:*   Compiled by the author. This table does not intend to summarize all stances and authors, just the ones mentioned in this chapter.

# REFERENCES

Abrahamson, Eric (1991), 'Managerial fads and fashions: The diffusion and rejection of innovations', *The Academy of Management Review*, 16 (3), Jul., 586–612.

Audretsch, David B., Robert Litan and Robert Strom (2009), *Entrepreneurship and Openness. Theory and Evidence*, Cheltenham, UK and Northampton, MA, USA: Edward Elgar Publishing.

Bogers, Marcel et al. (2017 [2016]), 'The open innovation research landscape: Established perspectives and emerging themes across different levels of analysis', *Industry and Innovation*, 24 (1), 8–40.

Bogers, Marcel, Henry Chesbrough and Carlos Moedas (2018), 'Open innovation: Research, practices, and policies', *California Management Review*, 60 (2), 5–16.

Boltanski, Luc and Eve Chiapello (2007 [1999]), *The New Spirit of Capitalism*, trans. Gregory Elliot, London, New York: Verso.

Braudel, Fernand (1983 [1979]), *Civilization and Capitalism, 15th–18th Century*, Vol. 2 – *The Wheels of Commerce*, trans. Sian Reynolds, London: William Collins Sons & Co Ltd., Book Club Association.

Cantwell, John (2001), *Innovation, Profits and Growth: Schumpeter and Penrose*, Reading: Henley Business School, University of Reading.

Castells, Manuel (2010 [1996]), *The Rise of the Network Society*, volume 1, Oxford: Wiley-Blackwell.

Chandler, Alfred D. (2005 [2001]), *Inventing the Electronic Century. The Epic Story of the Consumer Electronics and Computer Industries*, Cambridge, MA and London: Harvard University Press.

Chesbrough, Henry (2003a), *Open Innovation: The New Imperative for Creating and Profiting from Technology*, Boston, MA: Harvard Business School Press.

Chesbrough, Henry (2003b), 'The era of open innovation', *MIT Sloan Management Review*, 44 (3), 35–41.

Chesbrough, Henry, Wim Vanhaverbeke and Joel West (eds) (2006), *Open Innovation: Researching a New Paradigm*, Oxford: Oxford University Press.

Chesbrough, Henry (2015), *From Open Science to Open Innovation*, Berkeley, CA: ESADE – Institute for Innovation and Knowledge Management.

Chesbrough, Henry (2018), 'Value creation and value capture in open innovation', *Journal of Product Development & Management Association*, 35 (6), 930–38.

Christensen, Clayton (2012), 'Open innovation and getting things right'. Personal Webpage. http://claytonchristensen.com/open-innovation/ Accessed on 22 March 2020.

Cohen, Wesley M. and Daniel A. Levinthal (1990), 'Absorptive capacity: A new perspective on learning and innovation', *Administrative Science Quarterly*, 35 (1), 128–52.

Curley, Martin and Bror Salmelin (2018), *Open Innovation 2.0. The New Mode of Digital Innovation for Prosperity and Sustainability*, Dordrecht: Springer.

Curnow, R. C. and G. G. Moring (1968), '"Project SAPPHO": A study in industrial innovation', *Futures*, 1 (2), 82–90.

Dahlander, Linus and David M. Gann (2010), 'How open is innovation?', *Research Policy*, 39, 699–709.

Diamond Jr., Arthur (2019), *Openness to Creative Destruction. Sustaining Innovative Dynamism*, New York: Oxford University Press.

Elmquist, Maria, Tobias Fredberg and Susanne Olila (2009), 'Exploring the field of open innovation', *European Journal of Innovation Management*, 12 (3), 326–45.

Enkel, Ellen, Oliver Gassmann and Henry Chesbrough (2009), 'Open R&D and open innovation: Exploring the phenomenon', *R&D Management*, 39 (4), 311–16.

European Commission (2018), *Europe's Future: Open Innovation, Open Science, Open to the World – Reflections of the RISE Group*, Brussels: Directorate-General for Research and Innovation.

Gaglio, Gérald, Benoît Godin and Sebastian Pfotenhauer (2019), 'X-Innovation: Re-inventing innovation again and again', *NOvation: Critical Studies of Innovation*, 1 (June), 1–16.

Gassmann, Oliver (2006), 'Opening up the innovation process: towards an agenda', *R&D Management*, 36 (3), 223–28.

Gassmann, Oliver, Ellen Enkel and Henry Chesbrough (2010), 'The future of open innovation', *R&D Management*, 40 (3), 213–21.

Godin, Benoît (2008), *The Knowledge Economy: Fritz Machlup's Construction of a Synthetic Concept*, Project on the History and Sociology of S&T Statistics, Working Paper No. 37.

Groen, Aard J. and Jonathan D. Linton (2010), 'Is open innovation a field of study or a communication barrier to theory development', *Technovation*, 30, 554.

Hagedoorn, John (1993), 'Understanding the rationale of strategic technology partnering: Interorganizational modes of cooperation and sectoral differences', *Strategic Management Journal*, 14, 371–85.

Hargadon, Andrew and Robert I. Sutton (1997), 'Technology brokering and innovation in a product development firm', *Administrative Science Quarterly*, 42 (4), 716–49.

Hayek, Friedrich A. (1945), 'The use of knowledge in society', *The American Economic Review*, 35 (4), 519–530.

Hayek, Friedrich A. (2006 [1944]), *The Road to Serfdom*, London, New York: Routledge.

Hayek, Friedrich A. (2011 [1960]), *The Constitution of Liberty, The Definitive Edition*, Chicago, London: Chicago University Press.

Huizingh, Eelko K. R. E. (2011), 'Open innovation: State of the art and future perspectives', *Technovation*, 31, 2–9.

Kettner, Matthias (2017), 'Between enthusiasm and overkill. Assessing Michael Porter's conceptual management frame of creating shared value', in J. Wieland (ed.), *Creating Shared Value – Concepts, Experience, Criticism*, Springer, pp. 153–68.

Keynes, John M. (1936), *The General Theory of Employment, Interest and Money*, New York: Harcourt, Brace & World.

Kovács, Adrian, Bart van Looy and Bruno Cassiman (2015), 'Exploring the scope of open innovation: A bibliometric review of a decade of research', *Scientometrics*, 104, 951–83. https://link.springer.com/article/10.1007/s11192-015-1628-0.

Lang, Alexander, Anna-Teresa Tesch and Udo Lindemann (2017), 'Opening up the R&D process is risky – how far do you have to go in order to beat your competitors?', *International Journal of Innovation Management*, 21 (5), 1–21.

Laursen, Kield and Ammon Salter (2006), 'Open for innovation: The role of openness in explaining innovation performance among U.K. manufacturing firms', *Strategic Management Journal*, 27, 131–50.

Lichtenthaler, Ulrich (2011), 'Open innovation: Past research, current debates, and future directions', *Academy of Management Perspectives*, 25 (1), 75–93.

Lund, Arwid and Mariano Zukerfeld (2020), *Corporate Capitalism's Use of Openness. Profit or Free?*, London: Palgrave Macmillan.

Machlup, Fritz (1962), *The Production and Distribution of Knowledge in the United States*, Princeton, NJ: Princeton University Press.

March, James (1991), 'Exploration and exploitation in organizational learning', *Organization Science*, 2 (1), 71–87.

Mazzucato, Mariana (2013), 'Financing innovation: Creative destruction vs. destructive creation', *Industrial and Corporate Change*, 22 (4), 851–67.

Mill, John Stuart (2000 [1844]), *Essays on Some Unsettled Questions of Political Economy*, Kitchener: Batoche Books.

Mowery, David C. (1983), 'The relationship between intrafirm and contractual forms of industrial research in American manufacturing 1900–1949', *Explorations in Economic History*, 20, 351–74.

Mowery, David C. (2009), '*Plus ca change*: Industrial R&D in the "third industrial revolution"', *Industrial and Corporate Change*, 18 (1), 1–50.

Penrose, Edith. (2009 [1959]), *The Theory of the Growth of the Firm*, Oxford, New York: Oxford University Press.

Pitelis, Christos N. (2009), 'Introduction', in Edith Penrose (ed.), *The Theory of the Growth of the Firm*, Oxford, New York: Oxford University Press, pp. ix–xlvii (first published in 1959).

Polanyi, Michael (1958), *Personal Knowledge: Towards a Post-critical Philosophy*, London: Routledge.

Polanyi, Michael (1966), *The Tacit Dimension*, New York: Doubleday.

Pomerantz, Jeffrey and Robin Peek (2016), 'Fifty shades of open', *First Monday. Peer-Review Journal on the Internet*, 21 (5).

Popper, Karl (1966 [1945]), *The Open Society and its Enemies*, 2 volumes, London: Routledge.

Porter, Michael E. (1990), 'The competitive advantage of nations', *Harvard Business Review*, March/April, 73–91.

Porter, Michael E. and Mark R. Kramer (2011), 'Creating shared value', *Harvard Business Review*, January–February issue. https://hbr.org/2011/01/the-big-idea-creating-shared-value. Accessed on 25 March 2020.

Powell, Walter W., Kenneth W. Koput and Laurel Smith-Doerr (1996), 'Interorganizational collaboration and the locus of innovation: Networks of learning in biotechnology', *Administrative Science Quarterly*, 41 (1), 116–45.

Reynolds, David (2000), *One World Divisible: A Global History since 1945*, London: Allen Lane.

Rothwell, Roy, Christopher Freeman, A. Horseley, V.T.P Jervis and J. Townsend (1974), 'Sappho updated – Project Sappho Phase II', *Research Policy*, 3, 204–25.

Rothwell, Roy and Walter Zegveld (1985), *Reindustrialization and Technology*, Harlow: Longman.

Rothwell, Roy and Mark Dodgson (1991), 'External linkages and innovation in small and medium-sized enterprises', *R&D Management*, 21 (2), 125–37.

Rothwell, Roy (1994), 'Towards the fifth-generation innovation process', *International Marketing Review*, 11 (1), 7–31.

Schon, Donald (1967), *Technology and Change: the New Heraclitus*, New York: Dell Publishing.

Schumpeter, Joseph. A. (1939), *Business Cycles. A Theoretical, Historical, and Statistical Analysis of the Capitalist Process*, Volume I, New York: McGraw-Hill.

Schumpeter, Joseph A. (1942), *Capitalism, Socialism and Democracy*, New York: Harper & Row.

Segercrantz, Beate, Karl.-Erik Sveiby and Karin Berglund (2017), 'A discourse analysis of innovation in academic management literature', in Benoît Godin and Dominique Vinck (eds), *Critical Studies of Innovation: Alternative Approaches to the Pro-Innovation Bias*, Cheltenham, UK and Northampton, MA, USA: Edward Elgar Publishing, pp. 276–95.

Smith, Matthew Longshore and Ruhiya Seward (2017), 'Openness as social praxis', *First Monday. Peer-Review Journal on the Internet*, 22 (4).

Söderberg, Johan (2017), 'Comparing two cases of outlaw innovation: File sharing and legal highs', in Benoît Godin and Dominique Vinck (eds), *Critical Studies of Innovation: Alternative Approaches to the Pro-Innovation Bias*, Cheltenham, UK and Northampton, MA, USA: Edward Elgar Publishing, pp. 115–36.

Suber, Peter (2012), *Open Access*, Cambridge, MA: MIT Press.

Teece, David J. (1986), 'Profiting from technological innovation: Implications for integration, collaboration, licensing and public policy', *Research Policy*, 15, 285–305.

Teece, David J., Gary Pisano and Amy Shuen (1997), 'Dynamic capabilities and strategic management', *Strategic Management Journal*, 18 (7), 509–33.

Teece, David J. (2006), 'Reflections on "profiting from innovation"', *Research Policy*, 35, 1131–46.

Traverso, Enzo (2001), *Le Totalitarisme. Le XXe siècle en débat*, Paris: Seuil.

Trott, Paul and Dap Hartmann (2009), 'Why " open innovation" is old wine in new bottles', *International Journal of Innovation Management*, 13 (4), 715–36.

van de Vrande, Vareska, Wim Vanhaverbeke and Oliver Gassmann (2010), 'Broadening the scope of open innovation: Past research, current state and future directions', *International Journal of Technology Management*, 52 (3/4), 221–35.

von Hippel, Eric (1988), *The Sources of Innovation*, New York, Oxford: Oxford University Press.

von Hippel, Eric (2005), *Democratizing Innovation*, Cambridge, MA: MIT Press.

von Mises, Ludwig (2002 [1927]), *Liberalism: In the Classical Tradition*, 2 volumes, San Francisco and New York: Cobden Press.

West, Joel, Ammon Salter, Wim Vanhaverbeke and Henry Chesbrough (2014), 'Open innovation: The next decade', *Research Policy*, 43 (5), 805–11.

Zook, Chris and Darrell K. Rigby (2002), 'Open-market innovation', *Harvard Business Review*, October. https://hbr.org/2002/10/open-market-innovation. Accessed on 23 March 2020.

# 11. Disruptive innovation: an organizational strategy and a technological concept

*Darryl Cressman*

> It is important to remember that disruption is a positive force.
> - "Disruptive Innovation," The Clayton Christensen Institute

> Disruption used to be a luxury. Today it is essential to survive.
> - André Loesekrug-Pietre (2018), in a speech on behalf of the
> Joint European Disruptive Initiative (JEDI)

> The reality of disruption is the loss of reason.
> - Bernard Stiegler (2019), p. 38

## INTRODUCTION

Disruptive innovation is inching towards what the conceptual historian Reinhart Koselleck (2004 [1969]) called "sloganizing ubiquity." Koselleck's ubiquitous slogan was not disruptive innovation, but "revolution," which he declared was one of those "widely used forceful expressions whose lack of conceptual clarity is so marked that they can be defined as slogans" (p. 43). Conceptual clarity was lost, Koselleck argues, after the French Revolution. Prior to 1789, revolution did not imply a new beginning or an abrupt break with the past. Originating in astronomy and the circular movements of celestial bodies, it referred to a natural cycle of recurring types of political organization that follow one another with the predictability and inevitability of the seasons. After 1789, revolution no longer implied a recurrence of pre-given conditions, but the opening up of an unknown future. It is this meaning of revolution that has endured since that time and has led to its unproblematic application to "morals, laws, religion, economy, countries, states, and portions of the earth – indeed, the entire globe" (p. 48).

Of course, disruptive innovation does not come near the litany of that which has been declared or done in the name of revolution. Yet, there are a few passing familiarities in the trajectories of these two concepts. As the epigraphs of this chapter demonstrate, disruptive innovation is, like revolution, a flexible and general concept "that means at least something anywhere in the world, but which in a more precise sense fluctuates enormously from country to country and from one political camp to another" (Koselleck 2004 [1969], p. 44). It is a management theory and strategy for economic growth (Christensen 1997; Christensen and Raynor 2003), it is a metaphor for sociotechnical change (Hesmondhalgh and Meier 2018), the basis for a critical social theory (Lepore 2014; Stiegler 2019), a case study for exploring ways to regulate the sharing economy (Gautrais 2018; Dudley, Banister and Schwanen 2017), it is used to promote a circular economy,[1] to debate transitions to sustainable energy,[2] and to describe and strategize the transformation of institutions like health care and education (Christensen, Grossman and Hwang 2009; Christensen, Horn and Johnson 2008; Levina 2017; Sharon 2016; Sims 2017). In short, disruptive innovation is used across the contemporary political and socioeconomic spectrum for any number of purposes. It strongly resonates with the experience of early twenty-first century sociotechnical culture while still maintaining an ambiguity, or ambivalence, towards any one particular variation of this culture.

This raises the question of what to make of disruptive innovation. Is it an empty buzzword or a formal theory of innovation that has been diluted by overuse? Should it be studied as a political and socioeconomic strategy or pop culture ephemera? Following Koselleck, who sought to draw out some of the features that characterized the conceptual field of revolution, in what follows I attempt to do the same for disruptive innovation by looking for shared commonalities that can be found across its numerous iterations. I begin by tracing the etymological and theoretical history of disruptive innovation. The aim of this exercise is to provide some clarity to a concept that is simultaneously a cohesive management theory, evocative metaphor and empty buzzword. This history will begin by tracing the path by which the pejorative term "disrupt" was redeemed as "disruptive innovation" by the management theorist Clayton Christensen. After years of being debated and refined by management theorists, disruptive innovation moved beyond its business school origins and began being used to describe processes by which networked digital technologies and platforms are endowed with the capability to transform what are seen as anachronistic and inefficient industries and institutions. By the second decade of the twenty-first century, disruptive innovation was becoming more of a metaphor than a theory as it was widely used to describe, promote and critique a variety of contested, and at times incommensurable, initiatives. Inspired by Koselleck's success at finding coherence in the concept of "revolution," I suggest that disruptive innovation can be characterized by two features: first, it refers to an organizational strategy that prioritizes small start-ups as opposed to large corporations; and second, as disruptive innovation moved from theory to metaphor, it became increasingly associated with a particular narrative about the history, pace and trajectory of technological change that culminates in a concept of technology that is fixed to that which is new and emerging and made meaningful through what I term "entrepreneurial flexibility."

## FROM *DISRUMPĔRE* TO DISRUPTIVE INNOVATION

When subjected to analytical tools like Google n-grams or Web of Science keyword searches, disruptive innovation returns graphs which, after a prolonged period of hovering around 0, begin a steady and rapid jump around the late 1990s (see Figure 11.1).

In what follows, I work backwards from our contemporary moment to better understand the etymology of disruptive innovation and its usage prior to and during the 1990s to better explain the rapid growth that occurred in the twenty-first century.

Disrupt emerged as an adjective in the eighteenth century and a verb in the nineteenth century. It comes from the Latin words *disruptus* and *disrumpĕre*, meaning to break (dis) apart (rumpere), shatter, or separate forcibly. More specialized usages followed but maintained the original meaning of breaking apart. Electrical engineers began referring to "disruptive discharges" in the nineteenth century to describe an excessive increase in current that would rupture its insulating medium[3] and throughout the twentieth century disrupt and disruptive were used to denote occurrences that unexpectedly broke apart order and control, such as students who disrupted a classroom or the disruption of a fixed and predictable train schedule.

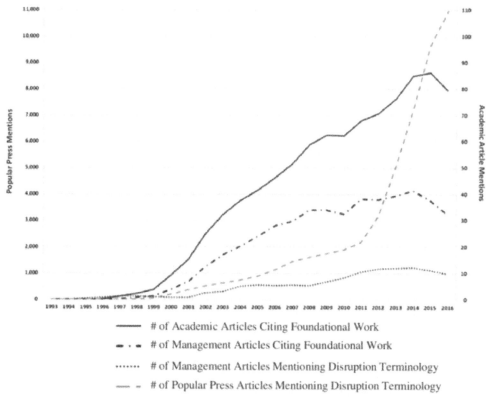

*Source:*    Taken from Christensen, McDonald, Altman and Palmer (2018, p. 1046).

*Figure 11.1    Scholarly and popular press citations of disruptive innovation 1993–2016*

The expression "disruptive innovation" appeared occasionally in English language publications throughout the twentieth century where it often described threats to stable conservativism, be it aristocratic, aesthetic, or organizational. In 1917, it was used in the context of an anticipated conservative reaction to Home Rule for traditional vassal states in a future European League of Nations (Brailsford 1917, p. 550); in a 1968 essay reporting on innovation in the aerospace technology, disruptive innovations are mentioned as essential for industries to thrive, but exist in conflict with the stability that industries also need to thrive (Emme 1968, p. 437); a 1973 obituary for the cubist sculptor Jacques Lipchitz celebrated his assimilation of the disruptive innovations of Constructivists into his work (Kramer 1973); in 1975 it appeared in a United States Senate Subcommittee report to describe an innovation that threatens economic harm to specific groups and so, when threatened with a disruptive innovation, the usual response is regulation, delay and compromise.[4] In the 1980s, the term entered the world of business schools and management theorists when, in a book about American Industrial policy, William J. Abernathy, Kim B. Clark and Alan M. Kantrow (1983) contrast it with conservative innovations (p. 97) while also expanding its scope and magnitude, writing that by its very nature, "epochal or disruptive innovation – whatever its degree of technical novelty – makes

obsolete existing capital equipment, labor skills, materials, components, management expertise, and organizational capabilities" (p. 28).

The sporadic and varied uses of disruptive innovation abruptly stopped in the mid-1990s
when management theorist Clayton M. Christensen re-defined disruptive innovation through
a series of articles and books that developed out of his doctoral research (Bower and Christensen
1995; Christensen 1997). The question that inspired this work was, why do successful, competitive and well-managed firms fail? Christensen addressed this question through the case
study of the hard disk drive industry because the rate of change in this industry was so fast and
so unrelenting that one could study business cycles over months that in other industries would
take years, what he called a kind-of drosophila for management theorists (Christensen 1997,
p. 3). Christensen's original hypothesis was that the disk drive industry consisted of firms that,
although successful, inevitably failed because they could not keep up with the pace of technological change. He called this the technology mudslide hypothesis, "coping with the relentless
onslaught of technology was akin to trying to climb a mudslide raging down a hill. You have
to scramble with everything you've got to stay on top of it, and if you ever once stop to catch
your breath, you get buried" (1997, p. 8). Research revealed that this hypothesis was incorrect.
Neither the pace nor the complexity of technological change led firms to fail. Indeed, in some
cases successful firms not only managed to stay on top of technological change, but also
prospered and grew when confronted with technological change. However, in other instances
of technological change, these same firms failed. The problem was not technological change
per se, but distinguishing between two different types of technological change: sustaining and
disruptive.[5]

Successful incumbent firms, Christensen argues, focus on their most profitable customers
and so tend towards developing sustaining technologies that improve those features or attributes that are prioritized by their existing consumers. Examples include more comfortably
spaced seats in airplanes and trains for regular travelers, professional-level camera features
on phones, and in the disk drive industry, memory and processing speed. The trajectory of
sustaining technologies can be plotted along a rate of improvement that uses performative
attributes like these as benchmarks, meeting both the expectations of customers and enabling
a predictable trajectory of improvement towards which technological innovations should
aim. Importantly, Christensen also recognized another characteristic of sustaining technologies – they overshoot the actual needs of their customers (Christensen 1997, pp. 10–14). The
functional capacity of a professional level camera on a phone, for example, regularly exceeds
the needs of most consumers.

In theory, incumbent successful firms should listen to their existing customers and improve
the products preferred by these customers. But, because they focus on sustaining innovations
that tend to overshoot the needs of these customers, they fail to meet the needs of other customers.[6] This is where disruptive technologies, or disruptive innovations, enter. New firms
that prove to be disruptive begin by successfully targeting overlooked consumers, delivering
similar functionality that incumbents do, but with technologies that tend to be cheaper, smaller,
less durable and more convenient: an inexpensive phone that takes good enough photos or airlines that forego comfort for lower prices and more frequent flights. Incumbent firms, chasing
higher profits, tend not to respond to these disruptive entrants and focus on the attributes
demanded by their existing customers, ignoring the attributes that make the entrants' products
appealing. Over time, entrants begin to move upmarket, delivering the functional capabilities
that the incumbent firm's mainstream customers require while preserving the advantages that

drove their early success, like lower prices or greater convenience. When mainstream custom-ers start adopting the new entrants' offerings in volume, disruption has occurred. Successful examples celebrated as disruptive innovations include Netflix and Blockbuster,[7] Uber and the taxi industry and low-cost airlines.

Christensen developed the theory of disruptive innovation through a number of case studies,[8] including mechanical excavators, the steel industry, motorcycles, retail department stores, photocopiers, and his original case, the disk drive industry. As noted, in the disk drive industry, customers valued memory and processing speed and so incumbent firms directed technical innovation to improve these functions along a predictable trajectory of steady increases. In the late 1970s, the market was dominated by firms that produced 8-inch drives with storage capac-ities between 10 and 40 MB (which was expected to increase along a trajectory of 40% MB per year). The consumers of these disk drives were manufacturers of minicomputers, objects that because of their price and size were largely restricted to the state, industry and universities. In 1980, a 5.25-inch drive was introduced that had a storage capacity of 5 to 10 MB, which was of no use for minicomputer manufacturers who, following a trajectory of sustaining technologies, required 40 to 60 MB drives at this time. The 5.25-inch drive, though, had attributes (size, price) that appealed to a new market, personal computer manufacturers. The 5.25-inch drives "offered a different package of attributes valued only in emerging markets remote from, and unimportant to, the mainstream" (Christensen 1997, p. 16). In time, the firms that manufac-tured 8-inch drives were supplanted by firms that manufactured the 5.25-inch drives because the memory capacity of these latter drives improved such that customers of the 8-inch drive found the 5.25-inch drive more appealing. This process was then repeated, from 5.25-inch drives to 3.5-inch drives to 2.5-inch drives to 1.8-inch drives (Christensen 1997, pp. 10–25).

In its original form, the theory of disruptive innovation developed out of case studies that emphasized the distinction between sustaining and disruptive innovations and used this dis-tinction to explain why successful firms failed when confronted with technological change. From this early theoretical and empirical work, a few hallmarks of disruptive innovation can be identified. First, disruptive innovations exist only in relation to sustaining innovations. There is nothing inherent to a technology or process that makes it disruptive. The same tech-nology or process can be both disruptive and sustaining, as in the case of 3.5-inch disk drives, which were disruptive relative to 5.25-inch drives and sustaining relative to 2.5-inch drives, or the mp3, which was disruptive relative to retail outlets that sold compact discs but sustaining relative to streaming music services.

The second hallmark of Christensen's theory is that large organizations are particularly susceptible to disruptive innovations and so it is imperative that these organizations recognize and defend themselves against disruption. This defensive perspective corresponds with the anticipated audience for Christensen's early work – managers of successful firms. The ques-tion of why successful firms fail, which inspired the theory of disruptive innovation, quickly led to asking "why disruptive technologies are such vexatious phenomena for good managers to confront successfully" (Christensen 1997, p. xxi). Thus, in its early iterations, disruptive innovation was largely understood as a threat that could be warded off through the constant vigilance of astute managers. For example, Christensen concludes *The Innovator's Dilemma* by outlining a scenario in which electric vehicles are presented as a potential disruptive inno-vation for the traditional automobile industry. This scenario was told from the perspective of an automotive company executive who must confront this disruptive innovation to ensure the continued success of an incumbent industry, and not, as perhaps one would expect, from the perspective of a disruptive entrepreneur who, working outside of traditional automotive companies, sought to disrupt an incumbent industry (Christensen 1997, pp. 235–52).

Both the aim and the scope of disruptive innovation expanded in the early 2000s when Christensen refined the theory. First, what were originally called disruptive innovations were analytically separated into two different phenomena: low-end disruptive innovations and new-market disruptive innovations (Christensen and Raynor 2003; Christensen 2006; Christensen, Raynor and McDonald 2015).[9] New-market disruptions appeal to non-consumers and create a market where none existed before, while low-end disruptions take hold in those segments of the market that are overlooked by incumbents who, while overshooting the needs of their most profitable customers, ignore those customers who require a "good-enough" product. Examples of the first kind include personal computers and transistor radios, both of which opened up new markets for consumers who were overlooked by manufacturers of minicomputers and tube radios. Low-end disruptive innovations include discount retailers like Wal-Mart, which siphoned off consumers from more expensive retailers, and Netflix, which in its original iteration appealed to cinephiles who were not interested in Blockbuster's focus on new releases.

More important than this analytical refinement was the shift away from technology as the locus of disruptive innovation. Turning away from the technical artifact itself, management theorists began to argue that the reason why successful firms succeed or fail is not technological change, but the business models through which technological change is managed: "The disruptive business model in which the technology is deployed paralyzes the incumbent leader" (Christensen 2006, p. 43). Because organizational size and culture are intertwined and the economic reality that "successful organizations can only naturally prioritize innovations that promise improved profit margins relative to current cost structure" (Christensen and Raynor 2003, p. 198) it is not enough to try to develop a disruptive innovation; rather, what is essential is a business model based on the model of a small start-up organization, through which a technology or process is developed, designed and marketed. Considered from this perspective, disruptive innovations can only occur through what Christensen calls "autonomous business units," which should be given "unfettered freedom to forge a very different business model appropriate to the situation" (Christensen 2006, p. 43).[10] Following this logic, the starting point for a successful disruptive innovation is not a technology, but rather the start-up business model: small, dynamic and flexible teams that can be easily organized and dissolved.[11]

With this organizational strategy at the forefront of disruptive innovation, Christensen and his colleagues began to think of disruptive innovation as a strategy for social change and promoted it as the solution for transforming health care (Hwang and Christensen 2008) and education (Eyring and Christensen 2011; Pistone and Horn 2016) by "providing good-enough solutions to inadequately addressed social problems" (Christensen et al. 2006, p. 1).[12] At the close of the first decade of twenty-first century, disruptive innovation was no longer defensive and inward-looking but aggressive and outward-looking, or more accurately, by means of an organizational strategy, it became a ready-made solution that translated any situation into a problem that disruptive innovation could solve.

This period – the mid-1990s to the early 2000s – marks the first stage of disruptive innovation. Over the course of its theoretical and methodological refinement, it went from describing a threat that large firms needed to defend themselves against to a business model through which small flexible teams were equated with disruption by virtue of being free from the conservativism of larger firms. Despite the numerous transformations that disruptive innovation would undergo, this organizational strategy would remain one of its key characteristics.

## FROM STRATEGY TO METAPHOR

By the time of the 2008 economic crisis disruptive innovation was becoming a flexible and generalizable concept that was used with varying degrees of adherence to Christensen's theory. Unsurprisingly, this semantic expansion irritated management theorists who claimed disciplinary rights over the term's origins and use (Christensen et al. 2018; Christensen, Raynor and McDonald 2015; Gans 2016). Theoretical fidelity, though, is of little concern for a term that is equally adept as a strategy for economic growth, a metaphor for technological change and a category of social critique.

The shift from management theory to metaphor erased some of the methodological rules that had come to characterize the theory. Notably, the idea that an innovation is disruptive only relative to sustaining innovations largely disappeared. A European Research Council (ERC) call for funding from 2018 titled "Transformative Impact of Disruptive Technologies in Public Services" is telling in this regard. Independent of any sustaining innovations, the ERC points to objects and processes such as block-chain, Internet of Things, AI and big data analytics that, by virtue of their functional potential, are defined as disruptive.[13] Using disruptive innovation independently of sustaining innovations has allowed the concept to flourish, as has the tendency to (again) associate it with technologies alongside business models. This focus on technologies characterizes the Joint European Disruptive Initiative (JEDI), a French–German public–private initiative.[14] JEDI took shape during French President Emmanuel Macron's Sorbonne Speech (September 27, 2017) in which he charted a path for a post-Brexit Europe that included a proposal that "over the next two years, we create a European agency for disruptive innovation in the same vein as the Defense Advanced Research Projects Agency (DARPA) in the United States."[15] For Macron and his allies, JEDI was to pursue innovative "moonshots," on par with DARPA's development of the Internet and GPS.[16]

As both fodder for parody (Lyons 2016)[17] and an "ethos that shapes our cultural understanding of the relationship between innovation, technology, and capitalism" (Levina 2017, p. 549), the metaphorical use of disruptive innovation has become aligned in the popular imagination with the technologies and business practices associated with Silicon Valley.[18] Used to promote an ideology that combines techno-utopianism with a culture of libertarian techno-Darwinism, the Silicon Valley iteration of disruptive innovation has been celebrated for its entrepreneurial freedom (Ester and Maas 2016) and critiqued as legitimating digital oligopolies that encourage immoral labor practices and threaten democracy (Taplin 2017; Tiku 2018). Despite ambiguity and backlash, this iteration of disruptive innovation continues to collect aspirants. Coventry University (UK), for example, is home to the Disruptive Media Learning Lab, which is described as, "a semi-autonomous cross-University experimental unit whose remit is specifically to drive innovation of teaching, learning and practice forward (in the 'Google model': to break and remake existing ways of doing higher education) so that the University can re-model its own practices."[19]

Pushed further, the Silicon Valley model has endowed disruptive innovation with a sense of inevitability that legitimates an entrepreneurialism in which market-based initiatives are promoted as ideologically neutral and universally beneficial by virtue of consumer empowerment. This is the argument of Eric Schmidt, the former executive chairman for Google, who in a short essay titled "Embracing a New Digital Era in Europe" writes:

> Europe needs to accept and embrace disruption. The old ways of doing things need to face competition that forces them to innovate. Uber, for example is shaking up the taxi market – for the good. It offers riders convenience and cheaper fares, Understandably, the incumbent taxi industry is unhappy.[20]

This attitude was also expressed by then Mayor of London Boris Johnson, who, in 2015, referred to taxi drivers that were protesting Uber as Luddites. Johnson was not aiming for historical accuracy with his intended slur; rather, he sought to paint cab drivers as irrational romantics whose fate was already sealed by the appearance of new and emerging disruptive technologies.

Although there are many examples of more careful uses of disruptive innovation that avoid the boosterism of Silicon Valley – The "New York Times Report on Innovation" (2014)[21] and the special issue of *Energy Research & Social Science*[22] dedicated to sustainable energy transformation are two such examples – a large number of the technologies, businesses and organizations described as disruptive innovations since 2008 are done so metaphorically and not theoretically.

A consequence of this ubiquity has been increased empirical and historical scrutiny. Historian Jill Lepore (2014), writing in *The New Yorker*, critiqued Christensen's method of "handpicked case studies" as a "notoriously weak foundation on which to build a theory," and after reviewing these case studies found that his sources "are often dubious and his logic questionable" (see also King and Baatartogtokh 2015; Sampere, Bienenstock and Zuckerman 2016). Political economists are quick to point out that processes described as disruptive innovation bear an uncanny resemblance to previous descriptions of capitalism. In *The Communist Manifesto* Marx and Engels (1994 [1848]) describe nineteenth-century capitalist labor processes in terms that could be applied to Christensen's disk drive industry research: "Constant revolutionizing of production, uninterrupted disturbance of all social conditions, everlasting uncertainty and agitation…" (p. 161; see also Berman 1982). Other readers have recognized Joseph Schumpeter's (2010 [1942]) idea of creative destruction in disruptive innovation (Gans 2016; Levina 2017; Schneider 2017) while Schram (2015) succinctly notes that ordinary capitalism always already is disruptive, especially to those on the lower end of the socioeconomic ladder (p. 11).

Disruptive innovation has also proven to be a useful heuristic for critiquing contemporary neoliberal ambitions and policies. In her analysis of UberHEALTH in the United States, Marina Levina (2017) argues that discourses of disruptive innovation serve to depoliticize health care by encouraging an idea of health care in which concerns of systemic inequalities are pushed aside in the drive for individual and personal solutions. A similar critique is developed by Christo Sims (2017) whose ethnographic research revealed that disruptive innovation in an educational setting reified class and power relations through reliance on deterministic notions of technology's social autonomy. This attempt to disrupt education revealed that embedded class and race relations were not accounted for by those who argued for technical fixes for educational problems. Studied in messy and complex real-world settings (especially public institutions like schools and hospitals), it becomes obvious that disruptive innovations are co-constituted with, not distinct from, embedded power relationships regarding class, politics and socioeconomic status (see also Hall 2016). This attention to de-politicization through a rhetoric of consumer empowerment and the reification of historically contingent social relations is a forceful rebuttal to the pleas of Eric Schmidt, noted previously, for Europe to accept the inevitability of disruption:

> What tech enthusiasts call "disruption" is in fact almost always directed at forms of organization that preserve a modicum of workers' control over knowledge and the products of labor. Because London taxicabs are controlled by people who have built up impressive maps of one of the world's most complex cities in their brains, they ought to be replaced by self-driving cars operating on Google maps ... automation isn't a neutral, inevitable part of capitalism. It comes about through the desire to break formal and informal systems of workers' control – including unions – and replace them with managerially controlled and minutely surveilled systems of piecework. (After Capitalism 2016, p. 10)

From Clayton Christensen's emphasis on disruptive innovation as an organizational strategy, it was transformed through various iterations such that disrupt, disruptive and disruptive innovation have come to endow whichever initiative or technological object it is attached to with an expectation of radical, consumer-friendly and inevitable sociotechnical change.

## FROM METAPHOR TO A CONCEPT OF TECHNOLOGY

With an organizational strategy in place as a unifying commonality, disruptive innovation expanded beyond business schools to become a metaphor used to describe sociotechnical change. As its use expanded, another commonality began to emerge: a shared concept of technology that informed ideas about the artifactual and functional dimensions of technology, the history and pace of technological change and the trajectory of technological change.[23]

By the second decade of the twenty-first century, disruptive innovation became more aligned with a narrative in which the pace of technological change is consistently frantic and accelerating. Christensen's technological mudslide thesis, although refuted as an explanation for the success or failure of firms, was never questioned as a theory of technological change amongst the theorists of disruptive innovation who take as their starting point that the pace of technological change is not only fast, but "pervasive, rapid, and unrelenting" (1997, p. 3; see also Cressman 2019). This is evident in the rhetoric used to promote the Joint European Disruptive Initiative. While Andrei Loesekrug-Pietri, the head of JEDI, announces that "What matters is speed...be the one that sets the speed and you will set the norms. If Europe doesn't change its rhythm it will become irrelevant," French President Emmanuel Macron pushes for an imperative to move fast so as to not be left behind, "we are not in the middle ages, we are in the global race."[24] Similarly, the narrative of the "New York Times Report on Innovation" (2014) is legitimated through the acceptance of a frantic media landscape: "the pace of change is so fast that solutions can quickly seem out of date" (p. 58).

Complementing this is a history of technology characterized by long periods of stability that are abruptly interrupted by disruptive innovations. From academic conferences that encourage participants to reconsider the history of technology as the history of socially disruptive technologies, "...such technologies include the printing press, the steam engine, electric lighting, the computer, and the Internet,"[25] to the historically momentous and socially autonomous technological moonshots idealized by the Joint European Disruptive Initiative, the history of technology is presented as a series of tremendous socio-historical ruptures. For historians and philosophers, though, a social history of technology characterized by a series of abrupt breaks with the past is detrimental to historical consciousness. Lepore (2014) suggests that disruptive innovation's trajectory from management theory to concept corresponds with a shift in the theory's implicit theory of history, from a millenarian theory of history "founded on a profound anxiety about financial collapse, an apocalyptic fear of global devastation," to the disavowal of history altogether. Lepore's skill at drawing out a theory of history from disruptive innovation parallels the insights of philosopher Bernard Stiegler (2019) who defines disruption as "reticulated disintegration" (p. 7) and argues that the reality of disruptive innovation is a permanent and intense present in which individual desire and will are algorithmically determined while the transmission of knowledge between generations is unravelling. The result is a generation incapable of imagining a future that is different than the present that they currently inhabit.

These ideas about the history and pace of technological change parallel ideas about how technology as a material artifact is considered through the lens of disruptive innovation. As conceptual historians of technology have demonstrated, the world consists of innumerable artifacts and objects, only some of which are defined as technology. From this, it is evident that there is no inherent distinction between that which counts as technology and that which does not (Kline 1995; Oldenziel 1999; Schatzberg 2019). Interpreted through disruptive innovation, technology is becoming increasingly circumscribed along artifactual and functional lines such that what counts as technology are digital networked technologies that use increasing processing speeds, big data, personalization and predictive analytics to transform existing ways of producing, distributing and consuming goods and services. Indeed, from the perspective of the second decade of the twenty-first century, describing excavators and photocopiers as disruptive innovations seems quaint.

From this, disruptive innovation orients the technological imagination towards the new and emerging at the expense of the old and unchanging. This a priori towards the new and emerging opens up a space for "entrepreneurial flexibility" in which technology is no longer considered socially autonomous, but imagined as flexible towards different ends and the agent who shapes and directs technology towards these varied ends is the entrepreneur. "Disrupt" tinged with an entrepreneurial aggressiveness became "disruptive innovation," meaning to break apart with willful and exuberant defiance within the confines of a market economy. These dimensions of technology – entrepreneurial flexibility and a conceptual bounding to the new and emerging – are intertwined. For technology to be considered contingent and flexible, albeit in a limited scope, it is necessary for it to be considered new and emerging. Or, to put it differently, those artifacts that are old and unchanging do not count as technology because they are neither flexible nor contingent and thus cannot be directed by entrepreneurial ambitions.

## CONCLUSION

The relatively short history of disruptive innovation began when Clayton Christensen asked why some successful and well-managed firms fail when confronted with technological change. From this question, he developed a theory by which the pejorative meanings of disrupt and disruptive, which were used to describe threats to the stable conservativism of steady and predictable growth, became rehabilitated as disruptive innovation, which was now aligned with the promise of radical growth and rapid scalability. From this, disruptive innovation became characterized as an organizational strategy that could be applied to any organization and directed towards any number of ends. This led to a shift from disruptive innovation as a theory to disruptive innovation as a metaphor, which allowed it to be endowed with a concept of technology that informed ideas about the history and pace of technology, the trajectory of technological change, and what counts, artifactually, as technology.

This chapter began by suggesting that the work of Reinhart Koselleck could provide a way to answer the question of what to make of disruptive innovation. Koselleck recommends that one should be wary when confronted with a general and flexible concept but this should not encourage researchers to dismiss these concepts on the basis of their ubiquity. The ubiquity of disruptive innovation has certainly warranted a dismissive attitude, yet, as I have attempted to draw out in the preceding chapter, there is more to disruptive innovation than its presumed superficiality. For Koselleck, the significance of revolution was its intertwinement with the experience of historical time and an idea of the future. The preceding study is an attempt to situate disruptive innovation somewhere within the long shadow *begriffsgeschichte* by considering that across its many different iterations and variations there are shared commonalities that provide the outlines of what is an increasingly rich semantic field: an organizational strategy and a horizon of expectations that informs ideas about the pace and history of technological change and the scope and meaning of technical agency.

## NOTES

1.  The annual Disruptive Innovation Festival sponsored by the Ellen MacArthur Foundation "aims to shift mindsets and inspire action towards a circular economy." https://www.ellenmacarthur foundation.org/our-work/activities/dif.
2.  On disruptive innovation and energy transitions, see the special issue of *Energy Research & Social Science* 37 (2018).
3.  James Clerk Maxwell (1831–1879), for example, writes of disruptive discharges in his *An Elementary Treatise on Electricity* (2011 [1881], pp. 109–11).
4.  The Competition Improvements Act of 1975, S.2028.
5.  "Sustaining" technologies, or "sustaining" innovations, as used by Christensen, should not be mistaken with "sustainable" innovation. The latter is typically characterized as foregoing economic growth in favor of environmental concerns while the former is firmly rooted in the pursuit of economic growth.
6.  Hence the dilemma in the title of Christensen's book, *The Innovator's Dilemma*: "Simply put, when the best firms succeeded, they did so because they listened responsively to their customers and invested aggressively in the technology, products, and manufacturing capabilities that satisfied their customer's next-generation needs. But, paradoxically, when the best firms subsequently failed, it was for the same reasons – they listened responsively to their customers and invested aggressively in the technology, products, and manufacturing capabilities that satisfied their customers' next-generation needs" (Christensen 1997, p. 3).
7.  Netflix began in 1997 on the wave of a new technical format, DVDs, which were smaller and lighter than VHS tapes. This enabled Netflix to use a combination of online tools and postal delivery instead of a bricks and mortar retail outlet. At this time, Netflix was a niche service that appealed to non-users of Blockbuster, largely those who did not have access to retail outlets or cinephiles who were not satisfied with Blockbuster's emphasis on new releases of mainstream popular films. In the early 2000s Netflix changed their business model to a subscription-based service that allowed consumers to pay a flat monthly rate allowing them access to all of the films they wanted without late fees. Blockbuster did not consider the desired attributes of the customers who were drawn to Netflix and instead focused on sustaining innovations for their existing, and most profitable, customers who wanted new releases and other impulse purchases. Sustaining innovations, in this case, were an increase in the quantity of new releases, even guaranteeing their availability. Disruption occurred when Netflix shifted to an online streaming service built on its subscription model. Very

quickly, Netflix captured a market that was once dominated by Blockbuster (Christensen, Raynor and McDonald 2015, pp. 48–49; Gans 2016, pp. 13–22).

8. On the transformation from case studies to concept and theory, see Vinck and Gaglio, Chapter 22 in this volume.

9. There may be a tendency to equate the distinction between low-end and new-market disruptions with the distinction between incremental and radial innovation, but both low-end and new market disruptions are quite radical in the sense that both seek to disrupt a market. Incremental innovations seek to make small improvement to existing products (Christensen would call this a sustaining innovation) and in this sense neither low-end nor new-market disruptive innovations are intended to describe incremental innovations.

10. The same argument, albeit with less strategic precision, is made in Bower and Christensen (1995), "How then can an established company probe a market for a disruptive technology? Let start-ups – either ones the company funds or others with no connection to the company – conduct the experiments. Small, hungry organizations are good at placing economical bets, rolling with the punches, and agilely changing product and market strategies in response to the feedback from initial forays into the market" (p. 51).

11. Interestingly, this trajectory from technologies to organizational strategy parallels the history of the concept of innovation, which, as Godin (2015) points out, characterized the trajectory of the concept of innovation in the post-war period.

12. See also the Clayton Christensen Institute, whose self-stated mission is to improve the world through disruptive innovation: https://www.christenseninstitute.org/.

13. http://ec.europa.eu/research/participants/portal/desktop/en/opportunities/h2020/topics/dt-transformations-02-2018-2019-2020.html.

14. https://jedi.group/.

15. The full text of Macron's speech (in English) can be found here: http://international.blogs.ouest-france.fr/archive/2017/09/29/macron-sorbonne-verbatim-europe-18583.html. For commentary on the proposal for European agency for disruptive innovation, see: https://www.delorsinstitut.de/en/all-publications/how-an-eu-agency-for-disruptive-innovation-could-look-like/.

16. As the director of JEDI André Loesekrug-Pietri puts it, moonshots are "projects that are massively risky but that could potentially completely disrupt an industry and/or lay the technological foundations for a completely new sector." https://www.bundestag.de/blob/556394/ff7f0a1f37e430410961b15ceb58e2b4/3--jedi-en-fr-data.pdf.

17. See also the television program *Silicon Valley* (HBO 2014–2020) and popular press articles with titles like, "Why it's time to Retire 'Disruption,' Silicon Valley's Emptiest Buzzword," *The Guardian*, January 11, 2016.

18. The Silicon Valley iteration of disruptive innovation has inspired Christensen and his colleagues to clarify and re-claim disciplinary rights over the term's use by using the case of Uber to explain why this quintessential Silicon Valley company is not a disruptive innovation (Christensen, Raynor and McDonald 2015). This attempt at elucidation was critiqued for not being able to properly account for Platform businesses like Uber, which are different than product-based businesses and thus require a new theory of disruptive innovation. Platforms, it was argued, are able to create new connections (and potential exchanges) between users through predictive algorithms while also collecting user data that can be commodified and used for the creation of both low-end and new markets (Moazed 2016).

19. https://dmll.org.uk/about/.

20. This essay is part of a series that was sponsored by the European Commission called *Digital Minds for a New Europe*. https://lisboncouncil.net/publication/publication/118-digital-minds-for-a-new-europe-.html.

21. See also Massing (2015) for more on disruptive innovation and print media.

22. See endnote 1.

23. This follows from work in the philosophy of technology (Herf 1984; Zimmerman 1990) and the conceptual history of technology and innovation (Oldenziel 1999; Godin 2015; Long 1991; Marx 1997; Schatzberg 2006).

24 These quotes are taken from: https://www.bloomberg.com/news/articles/2018-03-27/european-technology-irrelevance-feared-as-u-s-china-dominate.

25.  https://ethicsandtechnology.eu/news/4tu-ethics-bi-annual-conference-thursday-7th-friday-8th
     -november-2019-tu-eindhoven/. This is not the only example that describes the printing press as
     a disruptive innovation. The title of John Naughton's (2014) book *From Gutenberg to Zuckerberg:
     Disruptive Innovation in the Age of the Internet* points to a historical trajectory that confirms the
     historiography of disruptive innovators.

## REFERENCES

Abernathy, William J., Kim B. Clark and Alan Kantrow (1983), *Industrial Renaissance: Producing
    a Competitive Future for America*, New York: Basic Books.
After Capitalism, *n+1*, 24 (Winter 2016), p.10.
Berman, Marshall (1982), *All that is Solid Melts into Air: The Experience of Modernity*, New York:
    Penguin.
Bower, Joseph and Clayton Christensen (1995), 'Disruptive Technology: Catching the Wave', *Harvard
    Business Review*, 43–53.
Brailsford, Henry Noel (H.N). (1917), 'Nationality as Culture', *The World Court*, 548–51.
Christensen, Clayton (1997 [2011]), *The Innovator's Dilemma*, New York: Harper Business.
Christensen, Clayton (2006), 'The Ongoing Process of Building a Theory of Disruption', *Journal of
    Product Innovation Management*, 23, 39–55.
Christensen, Clayton, Baumann, Heiner, Ruggles Rudy and Sadtler, Thomas (2006), 'Disruptive
    Innovation for Social Change', *Harvard Business Review*, 1–7.
Christensen, Clayton, Grossman, Jerome H. and Hwang, Jason (2009), *The Innovator's Prescription:
    A Disruptive Solution for Health Care*, New York: McGraw-Hill.
Christensen, Clayton, Horn, Michael B. and Johnson, Curtis W. (2008), *Disrupting Class: How
    Disruptive Innovation will Change the way the World Learns*, New York: McGraw-Hill.
Christensen, Clayton, McDonald, Rory, Altman, Elizabeth and Palmer, Jonathan (2018), 'Disruptive
    Innovation: An Intellectual History and Directions for Future Research', *Journal of Management
    Studies*, 55 (7), 1044–78.
Christensen, Clayton and Raynor, Michael (2003), *The Innovator's Solution: Creating and Sustaining
    Successful Growth*, Boston: Harvard Business School Press.
Christensen, Clayton, Raynor, Michael and McDonald, Rory (2015), 'What is Disruptive Innovation?',
    *Harvard Business Review*, December, 44–53.
Cressman, Darryl (2019), 'Disruptive Innovation and the Idea of Technology', *Novation*, 1, 17–39.
Dudley, Geoffrey, Banister, David and Schwanen, Tim (2017), 'The Rise of Uber and Regulating the
    Disruptive Innovator', *The Political Quarterly*, 88 (3), 492–99.
Emme, Eugene M. (1968), 'Aeronautics, Rocketry, and Astronautics', *Technology & Culture*, 9 (3),
    436–55.
Ester, Peter and Arne, Maas (2016), *Silicon Valley: Plant Startup, Disruptive Innovation, Passionate
    Entrepreneurship and Hightech Startups*, Amsterdam: Amsterdam University Press.
Eyring, Henry and Christensen, Clayton (2011), *The Innovative University: Changing the DNA of Higher
    Education from the Inside Out*, San Jose: Josey Bass.
Gans, Joshua (2016), *The Disruption Dilemma*, Cambridge: MIT Press.
Gautrais, Vincent (2018), 'The Normative Ecology of Disruptive Innovation', in D. McKee, F. Makela
    and T. Scassa (eds), *Law & the Sharing Economy: Regulating Online Market Platforms*, Ottawa:
    University of Ottawa Press, pp. 115–47.
Godin, Benoît (2015), *Innovation Contested: The Idea of Innovation over the Centuries*, New York:
    Routledge.
Hall, Gary (2016), *The Uberfication of the University*, Minneapolis: University of Minnesota Press.
Herf, Jeffrey (1984), *Reactionary Modernism: Technology, Culture, and Politics in Weimar and the
    Third Reich*, Cambridge: Cambridge University Press.
Hesmondhalgh, David and Meier, Leslie (2018), 'What the Digitalization of Music Tells us
    About Capitalism, Culture, and the Power of the Information Technology Sector', *Information,
    Communication & Society*, 21 (11), 1555–70.

Hwang, Jason and Clayton Christensen (2008), 'Disruptive Innovation in Health Care Delivery: A Framework for Business Model Innovation', *Health Affairs*, 27 (5), 1329–35.

King, Andrew and Baatartogtokh, Baljir (2015), 'How Useful is the Theory of Disruptive Innovation', *MIT Sloan Management Review*, 57 (1), 77–90.

Kline, Ronald (1995), 'Construing "Technology" as "Applied Science": Public Rhetoric of Scientists and Engineers in the United States, 1880–1945', *Isis,* 86, 194–221.

Koselleck, Reinhart (2004 [1969]), 'Historical Criteria of the Modern Concept of Revolution', in R. Koselleck (ed.), *Futures Past: On the Semantics of Historical Time*, New York: Columbia University Press, pp. 43–58.

Kramer, Hilton (1973), 'The Achievement of Jacques Lipchitz', *The New York Times*, June 10, 155.

Lepore, Jill (2014), 'The Disruption Machine: What the Gospel of Innovation gets Wrong', *The New Yorker*, June 23.

Levina, Marina (2017), 'Disrupt or Die: Mobile Health and Disruptive Innovation as Body Politics', *Television & New Media*, 18 (6), 548–64.

Long, Pamela (1991), 'Invention, Authorship, "Intellectual Property," and the Origins of Patents: Notes toward a Conceptual History', *Technology & Culture*, 32 (4), 846–84.

Lyons, Dan (2016), *Disrupted: My Misadventure in the Start-Up Bubble*, New York: Hachette Books.

Marx, Karl and Engels, Friedrich (1994 [1848]), 'The Communist Manifesto', in Lawrence Simon (ed.), *Selected Writings*, Indianapolis: Hackett Publishing Co., pp. 157–87.

Marx, Leo (1997), 'Technology: The Emergence of a Hazardous Concept', *Social Research*, 64 (3), 965–88.

Massing, Michael (2015), 'Digital Journalism: How Good is it?', *The New York Review of Books*, LXII (10), 43–45.

Maxwell, James Clerk (2011 [1881]), *An Elementary Treatise on Electricity*, Cambridge: Cambridge University Press.

Moazed, Alex (2016), 'Why Clayton Christensen is Wrong about Uber & Disruption', https://techcrunch .com/2016/02/27/why-clayton-christensen-is-wrong-about-uber-and disruptiveinnovation/?guc-counter=1&guce_referrer_us=aHR0cHM6Ly93d3cuZ29vZ2xlLmNvbS8&guce_referrer_cs=O-jgHTkxzIzTMqd6Qg2tC5w.

Naughton, John (2014), *From Gutenberg to Zuckerberg: Disruptive Innovation in the Age of the Internet*, London: Quercus

'New York Times Report on Innovation' (March 24, 2014), https://www.presscouncil.org.au/uploads/52321/ufiles/The_New_York_Times_Innovation_Report_-_March_2014.pdf.

Oldenziel, Ruth (1999), *Making Technology Masculine: Men, Women and Modern Machines in America, 1870–1945*, Amsterdam: Amsterdam University Press.

Pistone, Michele R. and Horn, Michael B. (2016), 'Disrupting Law School: How Disruptive Innovation Will Revolutionize the Legal World', *Clayton Christensen Institute for Disruptive Innovation*.

Sampere, Juan Pablo, Bienenstock, Martin, and Zuckerman, Ezra (2016), 'Debating Disruptive Innovation', *MIT Sloan Management Review*, 57 (3), 26–30.

Schatzberg, Eric (2006) 'Technik Comes to America: Changing Meanings of Technology before 1930', *Technology & Culture*, 47(3), 486–512.

Schatzberg, Eric (2019), *Technology: Critical History of a Concept*, Chicago: The University of Chicago Press.

Schmidt, Eric (2014), 'Embracing a New Digital Era in Europe', https://www.giplatform.org/sites/default/files/Embracing%20a%20new%20digital%20era%20in%20Europe%20E%20Schmidt.pdf.

Schneider, Henrique (2017), *Creative Destruction & the Sharing Economy: Uber as Disruptive Innovation*, Cheltenham, UK and Northampton, MA, USA: Edward Elgar Publishing.

Schram, Sanford (2015), *The Return of Ordinary Capitalism: Neoliberalism, Precarity, Occupy*, Oxford: Oxford University Press.

Schumpeter, Joseph A. (2010 [1942]), *Capitalism, Socialism and Democracy*, London: Routledge.

Sharon, Tamar (2016), 'The Googlization of Health Research: From Disruptive Innovation to Disruptive Ethics', *Personalized Medicine*, 13 (6), 563–74.

Sims, Christo (2017), *Disruptive Fixation: School Reform and the Pitfalls of Techno-Idealism*, Princeton: Princeton University Press.

Stiegler, Bernard (2019), *The Age of Disruption: Technology and Madness in Computational Capitalism*, Cambridge: Polity.
Taplin, Jonathan (2017), *Move Fast & Break Things: How Facebook, Google, and Amazon Cornered Culture and Undermined Democracy*, New York: Little Brown.
Tiku, Nitasha (2018), 'An Alternative History of Silicon Valley Disruption', *Wired*, October 22, https://www.wired.com/story/alternative-history-of-silicon-valley-disruption/.
Zimmerman, Michael (1990), *Heidegger's Confrontation with Modernity: Technology, Politics, Art*, Bloomington: Indiana University Press.

# 12. Common innovation: the oldest species of innovation?[1]

*G.M. Peter Swann*

## INTRODUCTION

This chapter considers *common innovation*. The following definition captures the essence of this (Swann 2014, p. 3):

> Common innovation is carried out by 'the common man and woman' for their own benefit. It takes place quite outside the domain of business, the professions or government. It could, indeed, be described as non-business innovation, to emphasise its essential difference from business innovation. It could also be called vernacular, as it is not intended for commercial use, but for the benefit of the innovators and their community.

To broaden the definition slightly, we can say that common innovation is innovation by the common man and woman, by families and by small communities, *for themselves*. I shall argue that common innovation is the oldest of all forms of human innovation, because stone-age innovations, from stone tools onwards, are all examples of common innovation.

The sceptical reader may immediately have a series of questions. Is common innovation really innovation? Is this just another example of an unnecessary proliferation of different categories of innovation? Is *common innovation* really any different from *user innovation*? Is *common innovation* really any different from *social innovation*?

I shall consider these questions and others in this chapter. First, for the sake of clarity, I start with a brief statement of the definition of innovation that underpins this chapter. Second, I provide a justification for my assertion that common innovation is the oldest form of innovation. Third, I consider some modern examples of common innovation. Fourth, I consider the scientific function of *X-Innovation*: this is the term coined by Gaglio et al. (2017) to describe the practice of adding an adjective before the word innovation. Fifth, I reflect on the relationship between common innovation and other categories of innovation – notably, user innovation and social innovation. And finally, I reflect on the argument that some types of innovation are becoming dysfunctional, in the sense that they may benefit the innovator, but also produce many undesirable side-effects that impinge on others. This is becoming quite common with business innovation, but not, I think, with common innovation.

One point should be stressed at the start. I am an economist who studies innovation. I have some knowledge of several related disciplines that also study innovation – other social sciences, other areas of business studies, and some areas of technology – but cannot claim a scholar's knowledge of these. At times, readers from other disciplines may find that I linger over points that are commonplace in their own discipline. I apologise if that is frustrating, but would point out that it is only comparatively recently that the ideas around common innovation have entered the economics of innovation.

## INNOVATION OR INVENTION?

In everyday speech, the words invention and innovation are sometimes used interchangeably, but to those who study innovation, there is an important distinction: innovation implies the *application* of new ideas (or the *new* application of *old* ideas), while invention does not imply application. So, for example, Charles Babbage's invention of the *Analytical Engine* was first described in 1837, but he was not able to turn this into a working innovation during his lifetime. Indeed, it was not until the late 1940s that it became possible to make general purpose computers based on Babbage's invention.[2]

Some would place a further condition on the use of the term, innovation. To them, innovation is the *commercial* application of inventions, and strictly speaking, therefore, the application of new ideas *outside* business and commerce doesn't count. From that perspective, they would exclude some sorts of user innovation (those made by non-commercial users) and all sorts of common innovation. This is true of most economists, especially those schooled in the Schumpeterian tradition, but also, as Godin (2019) has documented, many others – notably, practitioners, engineers, managers and policy-makers.

The application of this condition may reflect different disciplinary thinking and priorities but, in my view, this additional condition is inappropriate in economics. In mainstream economics (Becker 1965) and in anthropology (Sahlins 1972) it is common to talk of household production, or the domestic mode of production.[3] If households can produce, then they can also innovate, and therefore economists should certainly include household and other non-business innovation within the general definition of innovation. And, moreover, scholars from other disciplines, especially anthropologists such as Barnett (1953), have done this for many decades: see Godin (2015, 2017) for a survey. In what follows, innovation is the application of ideas to enhance wealth and wellbeing – wherever those ideas are applied. And, from that viewpoint, common innovation is certainly innovation.

## THE OLDEST SPECIES OF INNOVATION

Many histories of invention and innovation (for example, Fagan 2004; Lilley 1948; Singer et al. 1954) start with the earliest stone tools made around 2 500 000 years ago. Fagan (2004, p. 21) describes the transition from using natural objects as tools and the creation of new tools. He starts with this observation:

> It is likely that the first stone tools were discovered, rather than invented. An unmodified nodule of stone can be a very effective tool, providing either a hammer, a sharp edge for cutting or a missile for throwing. Not surprisingly, therefore, in the earliest known archaeological sites of East Africa stones are often found that are unmodified but have been transported from their source.

But he continues:

> When a stone nodule is used as a hammer it often breaks in half, or flakes become detached.
> These flakes can provide sharp edges and would have been useful for cutting apart animal carcasses … It may therefore not have been long (in an evolutionary sense) before hominids started knocking rocks together with the deliberate intention of producing sharp flakes to use in butchery. That knocking, or 'knapping' as archaeologists call it, made the difference between a simple discovery and the invention of a stone tool.

The reader will note that Fagan talks only of invention, not innovation. But our Stone Age ancestors did not stop at invention. They did not experiment in this way just for the pleasure of making inventions, with no regard to what they might do with them. On the contrary, they depended on application of these innovative tools for their very survival.

Moreover, I shall argue that Stone Age innovations, from stone tools onwards, are all examples of *common innovation*. In short, I believe that they are all consistent with my definition above: "innovation by the common man and woman, by families and by communities, *for themselves*". To support this assertion, we need to consider the three principal eras within the Stone Age.

Let us start with the most recent, the *Neolithic* era, which in Western Europe corresponds roughly to the period, 7 000 BP to 4 000 BP.[4] In this era, people lived in permanent settlements, often small villages, and practised agriculture which allowed them to abandon the mobile lifestyle of their ancestors. Sahlins (1972) argues that economic activity in this era followed the domestic mode of production. Families, kinship groups and small communities produced for their own consumption, and not for trade – which did not start until the Bronze Age era (after 4 000 BP). If production was done by the family for the family, then it is probable that innovation was also done by the family for the family. We might therefore speak of a domestic mode of innovation, and that is clearly consistent with the definition of *common innovation* above.

Sahlins (1972, p. 80) describes the essential character of the relationship between man and tool in this context:[5]

> The primitive relation between man and tool is a condition of the domestic mode of production. Typically, the instrument is an artificial extension of the person, not simply designed for individual use, but as an attachment that increases the body's mechanical advantage (for example, a bow-drill or a spear thrower), or performs final operations (for example, cutting, digging) for which the body is not naturally well equipped. The tool thus delivers human energy and skill more than energy and skill of its own.

Next, consider the two previous eras, the *Mesolithic* (roughly 15 000 BP to 7 000 BP in Western Europe) and *Palaeolithic* (roughly 3 000 000 BP to 15 000 BP).

In these eras, people lived as tribes and bands of peripatetic hunter-gatherers. For our purposes, the main distinctions between these two eras are: the sophistication of their tools; the intensity of hunting and gathering; the degree of mobility; and the nature of their habitation. In the *Palaeolithic* era, tools were not sophisticated, and this meant that it was not practical to carry out intensive hunting and gathering in a limited area. As a result, the tribes were highly mobile, and lived (for the most part) in caves or other suitable naturally occurring sites. In the *Mesolithic* era, tools were more sophisticated and more varied, and this made it possible to hunt and gather more productively and intensively. As a result, while the tribes were still mobile, they sometimes occupied temporary villages at suitable locations.

This is not exactly the same sort of economy as described in Sahlin's domestic mode of production, but nevertheless, similar economic principles applied. Production (hunting and gathering) was done by the tribe for the benefit of the tribe, and in the same way, innovations were made by members of the tribe for the benefit of the tribe. This is clearly consistent with the definition of *common innovation* above.

Table 12.1 gives eighteen examples of Stone Age innovations. Roughly speaking, the first six are *Palaeolithic*, the next six *Mesolithic*, and the last six *Neolithic*. Innovation in the *Palaeolithic* era was infrequent and related to basic tools. Innovation in the *Mesolithic* era was

*Table 12.1    Approximate dates of some Stone Age innovations*

| Innovation | Origin | Date BP |
|---|---|---|
| Earliest stone tools | Tanzania | 2 500 000 |
| Earliest camp fires | Kenya | 1 600 000 |
| Wooden spears | Germany | 400 000 |
| Composite stone tools | Africa | 250 000 |
| Cave paintings | Europe | 30 000 |
| Tools made from bone | Europe | 25 000 |
| Pottery | Japan | 15 000 |
| Twined textiles | Turkey | 12 000 |
| Bow and arrows | Germany, Denmark | 11 000 |
| Baskets | North America | 9 500 |
| Boats made from logs | Netherlands | 9 200 |
| Board games | Middle East, Africa | 9 000 |
| Potters kiln | Mesopotamia | 7 000 |
| Potter's wheel | Mesopotamia | 5 500 |
| Wheels for transport | Switzerland, Slovenia | 5 500 |
| Stone furniture | Orkney (Scotland) | 5 100 |
| Sewers | Orkney (Scotland) | 5 100 |
| Ovens | Mesopotamia | 5 000 |

*Source:*    Author's table; dates from Fagan (2004).

somewhat more frequent and related to higher quality tools and artefacts. Innovation in the *Neolithic* era was more frequent still, and brought forward a much wider range of artefacts.

Benjamin Franklin memorably said that mankind is a "tool-making animal",[6] and it is true that many histories of invention and innovation are preoccupied with tools for production. But Mumford (1952, p. 35) has conjectured that, "Man was perhaps an image maker and a language maker, a dreamer and an artist, even before he was a toolmaker."[7] I think Mumford's point is exceptionally important one, and in the discussion of innovation, and especially common innovation, it is essential that we cast our net much wider than simply looking at innovative tools. I shall pay particular attention to that point in the next section.

After the Stone Age came a more sophisticated model of economic development in the Bronze Age, with the growth of urban centres, the division of labour, craft production and trade. In that era, production was not only for domestic consumption by the tribe, but for trading with different communities (Earle 2002). At that stage, innovation is no longer common innovation alone, but the earliest forms of business innovation start to emerge. Therefore, we might say that if common innovation emerged in the Stone Age, then business innovation emerged in the Bronze Age. However, this does not mean that common innovation disappeared from the Bronze Age onwards – nor does it mean that common innovation is absent today.

## MODERN EXAMPLES OF COMMON INNOVATION

In this section, we provide a brief introduction to some modern forms of common innovation. It is not surprising to find that these are very different in scope from those described in the last section. For we should remember that, in the modern era, the role of common innovation is to fill some of the gaps that are not covered by business innovation.

A good place to start is with what development economists[8] have observed about innovations taking place in the informal sector.[9] For example, Elkan (2000, pp. 23–4), drawing on his fieldwork in Africa, describes some of the useful things that are made from waste materials for people who cannot afford machine made articles:

> What is produced is usually rather crude, for example beds are wooden and are given a springy base made from straps laboriously cut from old lorry tyres that are past re-treading. Charcoal stoves are made from scrap iron. Aluminium pots use smashed up cars for raw material and paraffin lamps are made from empty oil cans retrieved from petrol stations. None can compete in glossiness with machine made articles. But unlike the latter, they are made from recycled materials that would otherwise go to waste.

Much of this is common innovation, in the sense that it is done by members of a small community for the good of that community, and there is no formal trade. Some of it is done by microenterprises in the informal sector, which does perhaps count as business innovation – though it is very different from what most students of innovation have in mind when they discuss business innovation.

In my book on *Common Innovation* (Swann 2014), I have given an introduction to some of the many types of common innovation that are found in the most developed countries. Here I summarise a few of the examples discussed in that book. A good place to start is to consider the many examples of common innovation in the home. The creative but impecunious cook turns ordinary and inexpensive ingredients into a delicious meal. The creative amateur can use his or her imagination and effort to arrange and decorate the home to make it a better and happier place to live. The creative gardener can do something similar in the garden. The creative parent can encourage their children to choose hobbies, and thus enhance the wellbeing and development of the family – whether it involves painting, writing, music, collecting, or something else. In all of these cases, members of the family use common innovation to enhance the wellbeing of the family.

Moving outside the home to the local community, an important form of common innovation is found when members of the community work to enhance their local natural environment. Swann (2014, Chapter 17) describes a remarkable case study of how a derelict industrial site in the UK was transformed into a nature reserve. A decisive step in this case was when local people fought a vigorous campaign to prevent the industrial site being used to store waste, and set out instead their vision of how nature (with some human help) could transform the site from a wasteland into a nature reserve. That vision was endorsed by naturalists from further afield, and today the reserve is recognised as a Site of Special Scientific Interest (SSSI). It would be wrong to attribute all of this to common innovation: to complete the project required support from business, local government, charities and environmental agencies. But the formative step involved common innovation by local people.

A related form of common innovation is directed at enhancing the local social environment. This obviously overlaps with the idea of social innovation which is the subject of another chapter. Examples of such common innovation include:

(a)  Microcredit: making very small loans to borrowers who would be rejected by traditional lenders and can only borrow at very high rates of interest.

(b)    Community currencies: these are currencies that can only be used in a small local area, and aim to serve the needs of people who will always be short of conventional (national) currency.

(c)    Online help forum: the internet contains many sites that offer free advice on how to repair cars, computers and household goods, or advice on health problems, financial problems and many other topics. In this context, the term 'local' does not mean users are geographically close together, but refers to the fact that their online community shares a common interest or concern.

(d)    Pro bono: this refers to work done by professionals, at no cost, for the benefit of those in the community who could not afford professional fees.

Common innovation can play an important part in education. I shall start with an example that was already mentioned in the last section. *Oware* (its name in Ghana) is one of the 'pit and pebble' (or *Mancala*) games, considered to be the oldest board game that is still widely played. It is more than just a game: it is an important tool to teach children skills in calculation and strategy. In the same way, the 'school trip' is an enlightened way of encouraging children to learn things outside the classroom. Destinations for the school trip include farms, zoos, historic buildings, museums, art galleries, factory visits and Christmas markets. And in universities, we find students supplement their formal classes with mutually supportive self-help groups, who meet together to learn from each other, but without the involvement of a university tutor.

While those who aspire to work in the arts and sciences must achieve a high standard if they are to earn a living as artists and scientists, there is still an important role for common innovation in the arts and sciences. For several centuries, many classical composers have drawn on themes from folk music, but expressed within the conventional forms of classical music. But in the twentieth century, in particular, composers started to invent new forms of music which have their very roots in the style of folk music. And in the sciences, many citizen scientists have made valuable contributions to knowledge. An exceptional example was Gregor Mendel, the Augustinian monk, and Abbot of Brünn (now Brno, Czech Republic), in the 1860s. Mendel carried out experiments breeding peas to be grown in the monastery gardens, and his empirical results led him to a new theory of genetics. It was ignored for forty years, but by the 1940s, it provided the foundations for the new science of genetics.

While I only cite a few examples here, my conclusion in Swann (2014) was that common innovation is still widespread in the most developed countries, even if its scope is very different from business innovation. However, I have noticed a striking difference in how the idea of common innovation is received by economists in different parts of the UK. In London, and the south-east of England more widely, I find that many economists consider the concept of common innovation is peculiar, trivial and barely worthy of consideration.[10] By contrast, in the north midlands, the north of England, and many parts of Scotland, many consider the concept of common innovation to be commonplace and important.

Why the distinction? IPPR (2018, p. 19) noted that the UK is, by European standards, a very unbalanced economy – in geographical terms, at least. London is the richest region in Northern Europe, and one of the richest in the world, but the UK also has six of the poorest regions in North-Western Europe. In these poorest regions, it is not surprising to hear that quite a few people consider that the economy is simply not working for them. In such an environment, it is hardly surprising if ordinary people resort to the same approach to common innovation described in Elkan (2000).[11]

# FUNCTION OF X-INNOVATION

The practice of adding an adjective before the word innovation has become quite common in recent years. Gaglio et al. (2017) have coined the term, *X-Innovation*, to describe this phenomenon:

> ... it is only during the second half of the twentieth century that innovation became a fashionable concept and turned into a buzzword. It gave rise to a plethora of terms like technological innovation, organizational innovation, industrial innovation and, more recently, social innovation, open innovation, sustainable innovation, responsible innovation and the like. We may call these terms X-innovation. How can we make sense of this semantic extension? Why do these terms come into being? What drives people to coin new terms? What effects do the terms have on thought, on culture and scholarship and on policy and politics?

What is the scientific function of this X-Innovation phenomenon? I shall argue that the proliferation of different types of X-innovation is the equivalent of the classification of species developed by Carl Linnaeus. In that classification, it is considered essential that the scientist should identify the similarities between different species *and* the differences between them. Linnaeus achieved this by defining a hierarchical taxonomy, which identified different species within a particular *genus*, different *geni* within a particular family, different families within a particular order – and so on up the hierarchy. The fact that different species exist within a particular *genus* implies that they have some important similarities, but also some important differences.

To illustrate the point, consider the simple example of *genus Quercus* – the oak tree. There are currently almost 600 different species within *genus Quercus*. They are all oak trees, and have important similarities. But there are also important differences between the different species, and therefore, good reasons for recognising that multitude of species. The landscape gardener needs to know which species will provide the most suitable tree for a particular space, landscape and climate. The carpenter needs to know the particular species from which a consignment of oak timber comes, to know whether it is fit for a particular purpose. For example, Douglas (1914) said of *Quercus virens* that it is the best of all oak timbers from which to build boats (p. 40), while he described *Quercus rubra* as (p. 47), "one of the most magnificent of trees ... as far as regards appearance; its timber is, however, of very inferior quality".

What happens if the scientist ignores the differences between different species? In this context, we would be at risk of finding that a particular oak tree does not have the characteristics we are looking for. That is, *we underestimate differences.* And what happens if the scientist ignores the fact that the different species belong to a common genus? In this context, we would lose the information that the different species do share some important common characteristics. That is, *we overestimate differences.* Both of these outcomes are bad, because they violate one of the most basic principles of the science of classification and measurement, as stated by Rabinovich (1993): *we should neither overstate nor understate the accuracy of a measurement.*

From this point of view, X-Innovation plays an important scientific role. It recognises that students of innovation continue to identify 'new' species of innovation,[12] and that it is helpful to recognise these as having enough in common with other species of innovation to share membership of the genus innovation. Nevertheless, it also recognises that there are important differences between different species of innovation, and to avoid confusion we need to identify each species using a distinct adjective.

Consider what is liable to happen if we do not use the X-Innovation approach. There is a salutary lesson here from research on technological clusters. Until the 1990s, research on agglomeration and geographical clusters was mainly located within the geography discipline (especially, economic geography). In the 1990s, that started to change following influential studies by Porter (1990) and Krugman (1991). Following on from these, a variety of economists and other business school academics started researching in this field, and came up with different examples of, and ideas about, clustering phenomena.

Two geographers, Martin and Sunley (2003) wrote a much-cited article which expressed regret at this influx, and the way it was turning the term 'cluster' into a chaotic concept. Some economists responded that, even so, this influx of researchers from a different tradition was discovering important things that were related to the idea of clusters – albeit in different ways.

So, who was right, and who was wrong? I would say that both sides were half right (and half wrong). Martin and Sunley were right to express concern that this influx meant that one word was being used to describe some quite different phenomena. But the economists were right to say that their findings brought something new and valuable to the clusters debate. The influx was a good thing, but the chaos caused was not. The *X-Innovation* approach offers a natural solution to this problem: it maintains conceptual clarity while allowing scholars from one discipline to learn from the influx of ideas from another discipline. In the language of the business school, that is a 'win–win'. Instead, in the cluster case, the absence of such a solution led to an episode of academic turbulence, where different traditions did not learn to coexist with each other.

In conclusion, I would say the phenomenon of X-innovation is a 'good thing'[13] – so long as the appearance of new species of innovation does not simply reflect the vanity of researchers who wish to be associated with a particular species. Ultimately, as the experience of genus *Quercus* (oak trees) demonstrates, the Linnaeus method of classification is self-correcting,[14] because duplicate species are identified and deleted, redundant species are identified and deleted, and misclassifications are corrected. If it is considered useful to differentiate between some 600 species of oak tree, then we should not be uncomfortable at recognising many different types and purposes of innovation. There may be some duplication along the way, and some taxonomic errors that need to be resolved later on, but this is normal for taxonomy.

## IS COMMON INNOVATION DISTINCT FROM OTHER SPECIES?

Some people have suggested to me that common innovation is no different from the much more familiar concept of user innovation, which has its origins in von Hippel's pioneering work (1988, 2005) – see also Chapter 9 by Tavner in this book. I have two comments to make.

First, in their edited volume on user innovation, Harhoff and Lakhani (2016, p. xi) describe user innovation as, "a new paradigm in innovation research". But I have argued above that common innovation is certainly not a new phenomenon – and we must be careful not to confuse a new concept with the phenomenon, which is much older. If these two propositions are accurate, then user innovation and common innovation *cannot* be 'the same thing'.

Second, using a simple figure inspired by the Venn diagram, we can show that common innovation and user innovation occupy different places in 'innovation space'. Consider Figure

12.1. The term, 'user innovation', implies that the innovator is using something, so we should ask: what is that something? It is a useful item, certainly, but one where the user makes it even more useful by applying innovations of his/her own. The rows of Figure 12.1 denote different sorts of item: rows 1 and 2 are inexpensive or free items, while rows 3 and 4 can be moderately or very expensive. The columns of Figure 12.1 denote the different identities of the 'user'.

|   |                            | A                            | B                                  |
|---|----------------------------|------------------------------|------------------------------------|
|   |                            | Innovation by business users | Innovation by common man and woman |
| 1 | Waste products             |                              | **CI**                             |
| 2 | Inexpensive raw materials  | UI                           | **CI** + UI                        |
| 3 | Traditional products       | **UI**                       | UI + CI                            |
| 4 | Innovative products        | **UI**                       | **UI** + CI                        |

*Source:*   Author.

*Figure 12.1*     *User innovation and common innovation compared*

Where are common innovation and user innovation located in this space? Figure 12.1 uses a simple notation to answer that question. The presence of CI in a cell implies that common innovation is found in that cell, and the presence of UI in a cell implies that user innovation is found in that cell. The use of **bold** font implies that type of innovation is **frequently** found in that cell, while normal font implies that type of innovation is sometimes found in that cell.

Let us start with common innovation. By definition, common innovation is strictly confined to column B (innovation by the common man and woman). That is a tautology – but a useful one, because it helps us to see the difference between common innovation and user innovation. Moreover, much common innovation is located in rows 1 and 2: it takes unremarkable raw materials and waste products, and turns these into useful (if modest) products. Most common innovation, therefore, is located in cells B1 and B2. Some, however, may also be found in cells B3 and B4. For example, when we talk about common innovation creating online helplines, that really belongs in cell B4, as it represents the application of common innovation on an innovative platform (the internet).

Now let us turn to user innovation. Users in user innovation can be commercial or business users as well as the common man and woman, and therefore user innovation is found in columns A *and* B. Much of the early work on user innovation focused on case studies where users were using highly innovative products (containing a great deal of business innovation), and yet they still saw fit to make further innovations of their own. The narrative of some early case studies seems to be this: however, much business innovation is embodied in an innovative product, users want to add more. These case studies are located in cells A4 and B4.

As the discussion of user innovation has developed, however, I detect that the concept is becoming broader in scope. User innovators are not just creative people who can improve upon even the most innovative products. They are also people who improve on traditional products with user innovation. This means that user innovation is also found in cells A3 and B3.

I think it fair to say that the bulk of user innovation is located in cells A3, A4, B3 and B4. Some may also be found in cells A2 and B2 – users making innovations to enhance the value of the most basic raw materials – though I am not clear to what extent scholars from the user innovation school would call this user innovation.

However, I think it is problematic to include cells A1 and B1 in user innovation: at the very least, it presents a linguistic puzzle. The term, user innovation, implies that the user uses something, and that in turn implies that this something must be useful – or at least, usable. Can that include waste products? How can a waste product be useful? For if it is useful, why is it a waste product? In short, I am suggesting that however far the concept of user innovation may expand, I think it is problematic to include row 1.

What can we conclude from this? We can safely say that user innovation and common inno-vation are *not the same*: the former is found in columns A and B, while the latter is only found in column B. I believe we can also conclude that common innovation is *not* a proper subset of user innovation, because common innovation is frequently found in row 1, while user innova-tion is not found there. However, I think it is safe to say that there is a clear overlap between the two species of innovation: both types are found in cells B2, B3 and B4.

Some may disagree with me, and argue that user innovation *can* be found in row 1. In that case, common innovation would be a proper subset of user innovation. In that case, we may ask: is there any sense in reserving a distinct name (common innovation) for something that is a proper subset of a better-known category (user innovation)? In my view, the answer is yes: common innovation tends to work with raw materials and waste products, while much user innovation works with finished products which may or may not embody a lot of business innovation, but can be moderately or very expensive. This is a very important difference and suggests that much common innovation is quite different in character to much user innovation. Moreover, the underlying philosophy of the Linnaeus approach to taxonomy is this: if there is sufficient diversity amongst the species within a genus, then it is always worthwhile to recog-nise that diversity, for example, by X-Innovation.

Some have suggested that common innovation is barely any different from *social inno-vation* – as discussed by: Schubert, Chapter 6 in this book; Douthwaite 1996; Mulgan 2007; Moulaert et al. 2013.

Is there a real distinction? Yes, I think so, but to see this, we need to recognise that there are two very different concepts of social innovation. The first concept refers to the identity of the innovators. The innovators are a broad social group – not an individual, not a family, not a company, nor any other narrowly defined organisation. The second concept refers to where the effects of the innovations are felt, and does not identify the innovators: to echo Barnett (1953), innovation is the impetus for cultural and social change.

If we are concerned with the first idea of social innovation, then the distinction between common innovation and social innovation is one of scale: how large is the group of innova-tors? In this case, the boundary between common innovation and social innovation will never be a *hard* boundary: if the group is small, it looks like common innovation, but if the group is large, then it looks like social innovation.

On the other hand, if you are concerned, as I am, with the second idea of social innovation, then the distinction is clear. Common innovation describes innovations that are carried out by individuals, families and small communities, and where the benefits of those innovations accrue, in the main, to those same individuals, families and small communities. In contrast, while the benefits of social innovation accrue to large communities, or societies, the identity of the innovators is ambiguous: the innovators could be a large social group, or a much smaller group, or professional innovators.

These two types of X-innovation are perhaps the closest to common innovation. However, some would say that common innovation overlaps with *many* of the other species of X-innovation, and this overlap reduces the usefulness of common innovation as a concept. I agree with the first part of this proposition and disagree with the second. Yes indeed, it does overlap, but I see no problem with that. Above all, I certainly don't think it implies any redundancy, because many of the other species of innovation are quite different in scope. The term common innovation refers to who does the innovation (common man and woman) and who benefits (common man and woman). Many of the other labels discussed in this book (and elsewhere) describe other characteristics of the innovation process.

The reader of this book will easily find many other overlaps for themselves. So, for example, much common innovation is indeed open, but that does not mean common innovation is the same as the species of open innovation described by Chesborough (2003) and Brandão (Chapter 12 in this book). In the same way, much common innovation is frugal, but that does not mean common innovation is the same as the species of frugal innovation described by Radjou et al. (2012), and Cholez and Trompette (Chapter 14 in this book). We can consider there is also some overlap between common innovation and (*inter alia*): responsible innovation (von Schomberg, Chapter 8 this book); soft innovation (Stoneman 2010), localised technological change (Atkinson and Stiglitz 1969; Antonelli 1995, 2008); and sustainable innovation (Boons and Batista, Chapter 7 in this book). We could say that these different species 'cut the cake' in different ways, and I see no problem in doing that.

## COMMON INNOVATION IS A BENIGN BREEZE, NOT A PERENNIAL GALE

Several of the papers in Godin and Vinck (2017) argue that there is a pervasive pro-innovation bias in many debates about innovation, which suggests that the community of innovation scholars is in denial about the dysfunctional side of business innovation. I think they are right. Business innovation serves at least two quite distinct purposes: the purposes of the innovators, who wish to enhance their businesses by innovation, and the purposes of the users. Healthy business innovation manages to satisfy both at the same time. But we often find, especially in the later stages of a product life cycle, that while business innovations may satisfy the innovator's purpose, they do little for the user.

As Schumpeter (1954) put it, business innovation is about *creative destruction*. The innovators do damage to other businesses that are less competitive, but with luck, the beneficial effects to the innovators and consumers offset that damage. From this point of view, the destructive effects of business innovation are often accepted as the unavoidable collateral damage of technological progress. But many now consider that this view is too complacent: the destructive side of business innovation deserves much more serious attention than that.

In Swann (2014, Chapters 7–13), I considered seven case studies of the destructive side of business innovation. I showed that it is over-optimistic to think that the negative effects of these innovations only impact on other businesses. We also find negative effects on education, art, the quality of the marketplace, the socio-economic environment, the natural environment, consumption and health. Surely, if business innovation has destructive side-effects on this scale, we have a duty to ask whether it has, in reality, become dysfunctional, in the sense discussed earlier, or *pathological* (Swann 2019, Chapter 15). A striking example of this is provided in The *New York Times* (2018), where it was observed that, in Silicon Valley, "Technologists know how phones really work, and many have decided they don't want their own children anywhere near them."

I would argue that common innovation is not nearly so susceptible to these problems. We do not have to compute the balance of interest across various interested parties, because in common innovation, the innovator and the user are the same. If the common innovator considers the innovation is worthwhile, then it is worthwhile for the common innovator in his/her capacity as innovator and user.

There is another reason why common innovation is not susceptible to the same criticism as business innovation. As all students of the economics of innovation know, Joseph Schumpeter (1954) likened innovation to a 'perennial gale of creative destruction'. Business innovation achieves its objectives by turbulence, and it is right to ask whether our society and environment can continue to cope with that. In contrast, as I put it in Swann (2014, p. 21), common innovation is not a perennial gale; a better metaphor would be a, 'gentle and benign breeze of creativity'. Common innovation involves a very large number of small and local initiatives to enhance wellbeing. It works by creating where there is little or nothing. It is innovation in the spirit intended by Schumacher (1974).

## CONCLUSION

This chapter has been concerned with one particular type of X-Innovation. I have used the term common innovation to describe innovation by the common man and woman, by families and by small communities, for themselves. It is the oldest species of innovation, and is still with us[15] – even if modern examples of common innovation are very different from stone-age innovations. For that reason, we could say it has an essential place in the taxonomy of innovation types.

I used the adjective 'common' to describe this particular example of X-innovation because it is done by the common man and woman. But while I believe this expression is a new one, academic discussion of this phenomenon is definitely not new: archaeologists and anthropologists, in particular, have studied this sort of innovation for many, many decades. In view of that, we might expect that there is a much older term to describe this specific phenomenon, but, as far as I know, there is not.

I have discussed the scientific function of X-Innovation, and suggest that it is the equivalent of the classification of biological species developed by Carl Linnaeus. In that classification, it is considered essential that the scientist should identify the similarities between different species *and* the differences between them. This means that a comprehensive classification may require us to identify many, many different species. The fact that different species exist within a particular *genus* implies that they have some important similarities, but also some important differences. In the same way, it is hardly surprising that many types of X-innovation are being identified, where there are some similarities between different types, but also some important differences. Common innovation may overlap with some of the other types of X-innovation discussed in this book – notably user innovation, social innovation and informal innovation – but I believe it is not the same as any of them.

## NOTES

1. I am grateful to the editors for helpful comments and suggestions on an earlier version of this chapter.
2. Wikipedia (2019).
3. See also Marx (1867/1974, pp. 173–76).
4. BP denotes years *before present*, where the standard date for 'the present' is 1950. Therefore, 5000 BP is approximately 3000 BC (or BCE).
5. This relationship changes fundamentally in the machine age. Marx observed (1867/1974, p. 398): "In handicrafts and manufacture, the workman makes use of a tool; in the factory, the machine makes use of him." Veblen (1918, p. 307) said much the same: "… the machine process makes use of the workman".
6. Bronowski (1973, p. 116). Bronowski also observed that, "we find tools today made by man before he came man", and Fagan (2004, p. 21) asserts that our hominid ancestors were making tools from stone around 2,500,000 years ago.
7. Mumford (1952, p. 39) goes on to make this remarkable observation (dates added): "One has only to compare the cave paintings of the Aurignacian hunters (40,000 BP) with the tools that they used to see that their technical instruments, even if eked out with wood and bone instruments that have disappeared, were extremely primitive, while their symbolic arts were so advanced that many of them stand on a par, in economy of line and esthetic (sic) vitality, with the work of the Chinese painters of the Sung dynasty (960–1279 AD)."
8. It is not really surprising that discussion of these sorts of innovation entered the discipline of economics through the field of development economics. There are two likely reasons for this. First, common innovation is typically more frequent in less-developed regions – see below. Second, dialogue between anthropologists and development economics is (or, at least, *was*) much more frequent than dialogue between anthropologists and mainstream economists.
9. In his editorial comments on my paper, Benoît Godin stressed that discussion of innovation in the informal sector has produced one more form of X-innovation in the last few years: *informal innovation*. This also has much in common with my definition of common innovation, but I don't think they are quite the same.
10. Indeed, when I discussed the idea of consumers as innovators (e.g. Swann 1999), I found the reaction was similar.
11. Benoît Godin considers this a good example of the evaluative, normative and programmatic use of a concept. I consider Godin and Gaglio (2019) a good example of the same, in the context of sustainable innovation.
12. In comments on this chapter, one of the editors, Dominique Vinck, argued that the proliferation of terms corresponds not only to a proliferation of innovation species, but also to a dynamic where new scholarship takes different perspectives, and highlights different aspects of innovation. I agree with that point. Nevertheless, my present view is that all of the examples of X-innovation in this Handbook are distinct innovation species.
13. Some readers may consider this normative statement claims too much, and it would be better if I made the more modest claim that X-innovation can stimulate intellectual debate. But I stand by my claim. I believe it is timely for researchers to start using these different examples of X-innovation to understand how innovations evolve in response to socio-economic challenges – for example, the challenges posed by COVID-19.
14. To be precise, I mean that the *algorithm* used in the Linnaeus method of classification is self-correcting, in the sense that it deals with duplication, redundancy and misclassification. That is not to deny that the advent of new scientific approaches (e.g. genetics) may lead researchers to contest the Linnaeus classification. I am grateful to Dominique Vinck for this point.

15. A recent survey by the Office for National Statistics (2019) found that, "one in three graduates is over-educated for their current role". Rather than say that people are "over-educated", I would prefer to say that these jobs, "under-use the talents of their employees". In view of that, it is hardly surprising if people deploy their unused talent in the form of common innovation.

# REFERENCES

Antonelli, Cristiano (1995), *The Economics of Localized Technological Change and Industrial Dynamics*, Dordrecht: Kluwer Academic Publishers.

Antonelli, Cristiano (2008), *Localised Technological Change: Towards the Economics of Complexity*, London: Routledge.

Atkinson, Anthony B. and Joseph E. Stiglitz (1969), 'A New View of Technological Change', *Economic Journal*, 79, 573–78.

Barnett, Homer G. (1953), *Innovation: The Basis of Cultural Change*, New York: McGraw-Hill.

Becker, Gary S. (1965), 'A Theory of the Allocation of Time', *Economic Journal*, **75**, 493–517.

Bronowski, Jacob (1973), *The Ascent of Man*, London: British Broadcasting Corporation.

Chesborough, Henry (2003), *Open Innovation: The New Imperative for Creating and Profiting from Technology*, Boston, MA: Harvard Business School Press.

Douglas, David (1914), *Journal Kept by David Douglas during his travels in North America, 1823–1827*, London: William Wesley and Son.

Douthwaite, Richard (1996), *Short Circuit: Practical New Approach to Building More Self-Reliant Communities*, Totnes: Green Books.

Earle, Timothy (2002), *Bronze Age Economics: The Beginning of Political Economies*, Boulder, CO: Westview Press.

Elkan, Walter (2000), 'Manufacturing Microenterprises as Import Substituting Industries', in H. Jalilian, M. Tribe and J. Weiss (eds), *Industrial Development and Policy in Africa: Issues of De-Industrialisation and Development Strategy*, Cheltenham, UK and Northampton, MA, USA: Edward Elgar Publishing, pp. 22–29.

Fagan, Brian M. (ed.) (2004), *The Seventy Great Inventions of the Ancient World*, London: Thames and Hudson (with contributions from 41 other authors).

Gaglio, Gérald, Benoît Godin and Sebastian Pfotenhauern (2017), *X-Innovation: Re-inventing Innovation Again and Again*, Montreal: Project on the Intellectual History of Innovation.

Godin, Benoît (2015), *Innovation Contested: The Idea of Innovation Over the Centuries*, New York and London: Routledge.

Godin, Benoît (2017), *Models of Innovation: The History of an Idea*, Cambridge, MA: MIT Press.

Godin, Benoît (2019), *The Invention of Technological Innovation: Languages, Discourses and Ideology*, Cheltenham, UK and Northampton, MA, USA: Edward Elgar Publishing.

Godin, Benoît and Gérald Gaglio (2019), 'How does Innovation Sustain "Sustainable Innovation"', in Frank Boons and A. McMeekin (eds), *Handbook of Sustainable Innovation*, Cheltenham, UK and Northampton, MA, USA: Edward Elgar Publishing, pp. 27–37.

Godin, Benoît and Dominique Vinck (eds) (2017), *Critical Studies of Innovation: Alternative Approaches to the Pro-Innovation Bias*, Cheltenham, UK and Northampton, MA, USA: Edward Elgar Publishing.

Harhoff, Dietmar and Karim R. Lakhani (eds) (2016), *Revolutionizing Innovation: Users, Communities and Open Innovation*, Cambridge, MA: MIT Press.

Institute for Public Policy Research (IPPR) (2018), *Prosperity and Justice: A Plan for the New Economy, Report of the IPPR Commission on Economic Justice*, Cambridge: Polity Press.

Krugman, Paul (1991), *Geography and Trade*, Cambridge, MA: MIT Press.

Lilley, Samuel (1948), *Men, Machines and History: A Short History of Tools and Machines in Relation to Social Progress*, London: Cobbett Press.

Martin, Ron and Peter Sunley (2003), 'Deconstructing Clusters: Chaotic Concept or Policy Panacea?', *Journal of Economic Geography*, 3 (1), 5–35.

Marx, Karl (1867/1974), *Capital: A Critical Analysis of Capitalist Production*, Volume 1, London: Lawrence and Wishart.

Moulaert, Frank, Diana MacCallum, Abid Mehmood and Abdelillah Hamdouch (2013), *The International Handbook on Social Innovation: Collective Action, Social Learning and Transdisciplinary Research*, Cheltenham, UK and Northampton, MA, USA: Edward Elgar Publishing.

Mulgan, Geoff (2007), *Social innovation: what it is, why it matters & how it can be accelerated*, https:// youngfoundation.org/wp-content/uploads/2012/10/Social-Innovation-what-it-is-why-it-matters-how -it-can-be-accelerated-March-2007.pdf (accessed 31 May 2021).

Mumford, Lewis (1952), *Art and Technics*, London: Oxford University Press.

*New York Times* (2018), 'A Dark Consensus About Screens and Kids Begins to Emerge in Silicon Valley', 26 October, www.nytimes.com/2018/10/26/style/phones-children-silicon-valley.html (accessed 6 June 2019).

Office for National Statistics (2019), *One in Three Graduates Overeducated for their Current Role*, www .ons.gov.uk/news/news/oneinthreegraduatesovereducatedfortheircurrentrole (accessed 6 June 2019).

Porter, Michael (1990), *The Competitive Advantage of Nations*, London: Macmillan.

Rabinovich, Semyon (1993), *Measurement Errors: Theory and Practice*, New York: American Institute of Physics.

Radjou, Navi, Jaideep Prabhu and Simone Ahuja (2012), *Jugaad Innovation: Think Frugal, Be Flexible, Generate Breakthrough Growth*, San Francisco: Jossey Bass.

Sahlins, Marshall (1972), *Stone Age Economics*, London: Tavistock Publications.

Schumacher, Ernst (1974), *Small is Beautiful*, London: Abacus/Sphere Books.

Schumpeter, Joseph A. (1954), *Capitalism, Socialism and Democracy*, 4th edition, London: Unwin University Books.

Singer, Charles, Eric J. Holmyard and Alfred R. Hall (eds) (1954), *A History of Technology, Volume I: From Early Times to Fall of Ancient Empires*, London: Oxford University Press.

Stoneman, Paul (2010), *Soft Innovation: Economics, Product Aesthetics and the Creative Industries*, Oxford: Oxford University Press.

Swann, G.M. Peter (1999), 'Marshall's Consumer as an Innovator' in S. Dow and P. Earl (eds), *Economic Organisation and Economic Knowledge: Essays in Honour of Brian Loasby*, Cheltenham, UK and Northampton, MA, USA: Edward Elgar Publishing.

Swann, G.M. Peter (2014), *Common Innovation: How We Create the Wealth of Nations*, Cheltenham, UK and Northampton, MA, USA: Edward Elgar Publishing.

Swann, G.M. Peter (2019), *Economics as Anatomy: Radical Innovation in Empirical Economics*, Cheltenham, UK and Northampton, MA, USA: Edward Elgar Publishing.

Veblen, Thorstein (1918), *The Instinct of Workmanship: And the State of Industrial Arts*, New York: B. W. Huebsch.

von Hippel, Eric (1988), *The Sources of Innovation*, Oxford: Oxford University Press.

von Hippel, Eric (2005), *Democratizing Innovation*, Cambridge, MA: MIT Press.

Wikipedia (2019), *Analytical Engine*, https://en.wikipedia.org/wiki/Analytical_Engine, (accessed 6 June 2019).

# 13. Grassroots innovation: mainstreaming the discourse of informal sector

*Fayaz Ahmad Sheikh and Hemant Kumar*

## INTRODUCTION

Despite decades of extensive research, innovation discourse is devoid of careful discussion on informal sector grassroots innovations (henceforth "GIs"). Producer innovation has remained at the heart of innovation research (Godin and Vinck 2017; von Hippel 2017) and is increasingly regarded as a holy grail for economic growth and a potent policy tool to overcome the pressing human challenges (Habiyaremye et al. 2019). However, the producer innovation paradigm has been criticized for perpetuating exclusion and widening inequality (Heeks et al. 2014; Santiago 2014). To serve the underprivileged and facilitate the process of inclusive policymaking, non-dominant innovation models are beginning to be encouraged (Bhaduri 2016; Bhatti 2012). Lately, hybrid models of innovation have been proposed to tackle global concerns such as poverty, inequality, and climate change (UNCTAD 2017). As the demand for social justice gains momentum worldwide, inclusive innovation models such as "democratization of innovations" (von Hippel 2005), "rethinking business" (Ventresca and Nicholls 2011), and giving up "affluence" and "abundance" (Prahalad and Mashelkar 2010) are gaining ground.

Against this backdrop, inclusive models of innovation have captured development scholars' attention. GIs, which are mainly undertaken by unlettered people, artisans, farmers and users, represent one such category of inclusive innovations. These innovations are developed outside of formal organizational structures, without the engagement of recognized firms or planned R&D.

The long-standing contributions of informal sector innovations notwithstanding, indiscriminate romanticism of the concept in modern times has spawned the belief that GIs were only recently discovered and have no historical context. Yet the history of GIs boasts compelling examples of the substantial contributions of unlettered and anonymous communities to various technological revolutions (McCloskey 2016; Mokyr 2002; Phelps 2013; Rosenberg 1976). Moreover, the significance of innovations emanating from the informal economy has been documented by some of the most prominent economic historians.

However, the mainstream innovation literature reflects a bias towards formal sector innovations. The discourse has been dominated by debates around R&D, profitability, formal market structures, exchange value, scalability and large-scale commercialization of formal sector innovations. Excessive focus on innovations in the formal sector has seriously undermined GIs' potential. Moreover, interest in GIs has been further dampened by their inability to generate economic value. As Bhaduri (2016) argued in his Prince Claus Chair address on Development and Equity, the current discourse on knowledge has essentially focused on "formal sectors" of economies in the Global South. He maintained that "informal economies" have remained largely excluded "due to the belief that innovative activities by agents of the

informal economy are [the] exception rather than the rule." He further argued that such exclusion "becomes unsustainable in light of the expansion of informal economies in the Global
South." Today, approximately 1.8 billion people – half of the world's working population
– are engaged in the informal economy, which could be valued at roughly US $10 trillion
(Neuwirth 2011).

Innovation-related discussions have primarily been led by scalable innovations in the formal
sector; accordingly, this chapter attempts to scrutinize factors leading to marginalization
of informal sector GIs. The primary objective of this chapter is to deconstruct the "elitist
portrayal" of innovations and examine the "pro-market" innovation narrative, which has
perpetually undermined the contribution of "subaltern innovators." By critically synthesizing
the extant literature using a multidisciplinary approach, this chapter highlights the political
and economic reasons underlying GI marginalization. Proactive measures to develop hybrid
innovation models and formulate sustainable innovation policies are under way; this chapter
seeks to contribute to the existing narrative. For simplicity's sake, we consider GIs inclusive of
individual and community-led innovations developed and diffused without explicit input from
the state, state-run research institutions, or firms.

This chapter is divided into five sections. Section 1 elaborates upon the concept and genesis
of GIs. In addition to discussing their characteristics, Section 1 summarizes GIs' publication timeline. Section 2 presents a brief review of various theories used to study GIs, such
as paradox theory, sociotechnical theory and systems theory. Section 3 includes a detailed
account of the historical contributions of GIs to scientific knowledge. Section 4 highlights
the political, social and economic concerns undermining GIs' potential. Section 5 presents
a roadmap for further study of GIs.

## GRASSROOTS INNOVATIONS: CONCEPT AND GENESIS

Throughout the past two decades, scholars have sought to understand the varied facets of GIs
(Hossain 2016; Kumar 2020). Compared to Schumpeterian models of innovation, this literature has assumed a different perspective on GIs. Seyfang and Smith (2007) defined GIs as
"a network of activists and organizations generating novel bottom-up solutions for sustainable
development and sustainable consumption; solutions that respond to the local situation and
the interests and values of the communities involved" (p. 585). Dey et al. (2019) identified
six dimensions of GIs: affordable cost, indigenous knowledge, informal innovations, local
fit, sustainability and adaptability. Based on these attributes, the authors conceptualize GIs
as innovations originating from economically disadvantaged people, in an informal market
ecosystem, creatively deploying their native skills and local knowledge to resolve their
community problems in an affordable and sustainable way (Dey et al. 2019). However, such
a framing limits GIs to individual and community-led innovations (Jones et al. 2021); the roles
of governmental and non-governmental agencies are omitted in this case. By contrast, Fressoli
et al. (2014) and Cozzens and Sutz (2014) posited that GIs may include interaction and
activities with and by governments, R&D institutions and aid agencies working to empower
marginalized groups. This perspective counters that of Jones et al. (2021), who viewed GIs as
"networks of activists, practitioners and academics among others who aim to develop alternative forms of knowledge creation and processes for innovation. These networks harness local
ingenuity aimed at local development" (p. 3).

In the existing GI literature, multiple terminologies have been used to describe these non-market-driven innovations; terms such as "GIs," "frugal innovations," "*jugaad*," "inclusive innovations," "user-driven innovations," "autonomous innovations," and "informal sector innovations" (Kumar 2020) have been used interchangeably. Local vocabulary has occasionally been adopted to describe these non-for-profit innovations as well: "*jua kali*" (meaning "under the sun") is used in Kenya, "*Gambiara*" in Brazil, "folklore innovations" in China, "DIY" (do-it-yourself) in the USA, "*Système D*" in France and "GIs" in the UK and India (Kumar and Bhaduri 2014). Despite differences in nomenclature, research on GIs shares a common characteristic: GIs represent an outcome of efforts made by individuals and local communities to overcome problems encountered in daily life.

In view of the surrounding context, resource constraints and local needs of such individuals and communities, Hoffecker (2018) defined an informal sector innovation as the "process and the product of developing and introducing into use new and improved ways of doing things compared to existing practice within a specific local context, which involve local people and resources in addressing challenges and opportunities present within that context" (p. 4). Hoffecker (2018) maintained that local innovation can also be perceived as "vernacular innovation," namely because most of these innovations thrive in settings that are archetypal and reflective of local idiosyncrasies.

As mentioned above, GIs are radically different from the Schumpeterian paradigm of innovation, primarily because the former are not profit-driven. Affordability and sustainability are chief characteristics of GIs. As no monetary transactions are involved in the development or diffusion of GIs, they can be correctly classified as "free innovations." According to a recent theory by Eric von Hippel (2017, p.1), free innovations "involves innovations developed and given away by consumers as a 'free good,' with resulting improvements in social welfare." He asserted that free innovations are simple, transaction-free, grassroots-level innovations that involve tens of millions of people in informal organizational structures, far removed from the authority of large corporations and formal R&D. According to von Hippel (2017), GIs cannot be associated with money; he instead defined free innovation as a "functionally novel product, service, or process" (p. 1) that has been developed by consumers incurring private costs during unpaid discretionary time. In other words, these individuals are seldom compensated for their ingenuity, nor are their inventions protected by developers; they are available to anyone free of charge. Von Hippel (2017) further contended that GI innovators seek gratification from altruistic acts. He argued that unlike Schumpeterian innovations, GIs do not yield monetary gains. Such ideas resonate with the empirical findings of a study by Bhaduri and Kumar (2011), which addressed GI innovators' intrinsic and extrinsic motivations.

As noted earlier, an emerging body of literature in the Global South has aimed to understand informal GIs in greater detail. After comprehensively reviewing trends in the literature, we identified four distinct focus areas: (1) grassroots/informal sector innovations; (2) farmers' innovations/knowledge appropriation and IP laws; (3) social technologies; and (4) GIs for sustainability.

Many Asian scholars, particularly in China and India, have explored the nuances of GIs. Their work has predominantly concerned individual grassroots innovators, their motivations, appropriation and commercialization prospects (if any). Notable works in this realm include those by Bhaduri and Kumar (2011); Kumar and Bhaduri (2014); Muchie et al. (2016); Gupta (2016); Jain and Verloop (2012); Abrol and Gupta (2014); Basole (2014); Zhang and Mahadevia (2014); and Sheikh (2014a, 2014b).

Similarly, research on farmers' and metalworkers' innovations in the informal economy has gradually emerged from Africa. The broad foci of this literature are knowledge appropriation within the informal economy and evaluating the possibility of integrating informal and formal economies. Important works include those by King (1996); Daniels (2010); de Beer et al. (2013); Kraemer-Mbula and Wunsch-Vincent (2016); and Muchie et al. (2016).

In Latin America, studies have pertained to social technologies, which are innately related to informal innovations (see, for instance, Smith et al. 2014). The fourth and last strand of literature on the subject has emerged from Europe, particularly from England. This literature revolves around GIs and their roles in ecological sustainability (see, for example, Smith et al. 2016).

In the Indian context, the origins of GI discourse can be traced to the early twentieth century when MK Gandhi and Rabindranath Tagore spearheaded the movement on GIs. However, at the turn of the 1990s, a few GI enthusiasts along with Professor Anil K. Gupta at IIM-Ahmedabad launched the Honey Bee Network, an informal GI movement intended to explore GI ingenuity in India (Smith et al. 2016). Later, a host of bodies were established (e.g., the Society for Research and Initiatives for Sustainable Technologies and Institutions [SRISTI] and the National Innovation Foundation [NIF]) that led to the creation of a formal institutional ecosystem on GIs in the country. Governmental and non-governmental organizations both proactively contributed to support such entities (Kumar 2012; Ustyuzhantseva 2015). So far, SRISTI and NIF have undertaken several activities related to GI diffusion and dissemination; more than 3 lakh GIs have been recorded to date.

**Publication Timeline**

As reflected in the preceding sections, interest has recently surfaced in understanding GIs in the informal economy. The keyword "grassroots innovations" was entered into the Scopus database and yielded 213 publications between 1997 and 2019. Similar to the methodology used by Hossain (2016) and Kumar (2020), data were analyzed based on the country of publication, funding source, study topic, authors and indexed keywords. Our search results appear in Figure 13.1.

We chose 1997 as the starting year because the first use of the term "grassroots innovation" was found in a publication by Professor Anil K. Gupta (1997) from India. However, the GI literature only began to boom in 2010. Given that GIs typically thrive in developing countries, we were surprised to see that the top five countries publishing on GIs were the United Kingdom (61), United States (35), India (31), the Netherlands (17) and Germany (11) followed by other developed countries such as Italy, Sweden, Australia and China. These results indicate that a substantial proportion of the GI literature can be attributed to developed nations where the share of the informal sector is nearly negligible.

Figure 13.2 depicts certain factors responsible for the burgeoning literature on GIs in advanced economies. Most agencies that fund research on GIs are in developed countries. For example, the Economic and Social Research Council and the University of Sussex – both in the UK – have funded several research projects on GIs. Project outcomes have been documented in seven publications. This pattern is also evident in Figure 13.3, which shows publications by affiliation. Although India is endowed with a rich GI heritage, only four institutions (Jawaharlal Nehru University-New Delhi, Indian Institute of Management-Ahmedabad,

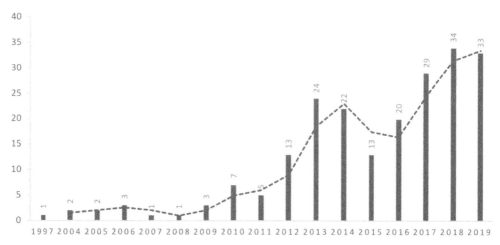

*Source:*   Scopus (through October 2019).

*Figure 13.1     Publications trends with keyword "grassroots innovations"*

Central University of Gujarat and Indian Institute of Technology, Guwahati) have contributed to the shortlisted publications.

With respect to subject areas, research on GIs has primarily involved the social sciences (23.9%), followed by environmental science (19.9%); business, management and account-ing (12.4%); energy (8.4%); engineering (7.9%); and economics, econometrics and finance

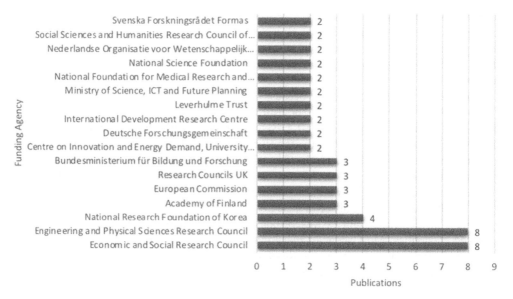

*Source:*   Scopus (through October 2019).

*Figure 13.2     Publications by funding source*

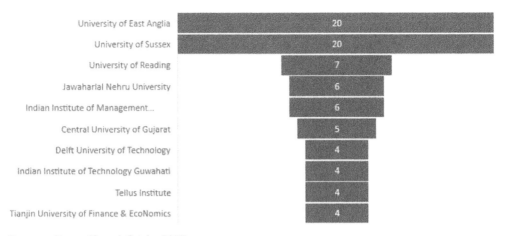

*Source:* Scopus (through October 2019).

*Figure 13.3    Publications by affiliation*

(7.0%). The prevalence of research in these domains implies that interest in GIs is growing rapidly, especially around the context of GIs and how such innovations can be used to address problems facing local communities.

## THEORIES AND GRASSROOTS INNOVATION

Our comprehensive analysis of publications on GIs revealed that current discourse involves multiple theories. Numerous perspectives and concepts have been used to explore the various facets of GIs. As no single theory is currently available to understand GIs, scholars have adopted eclectic approaches by borrowing from existing theories. We applied a search filter using publication keywords and the terms authors used to index their articles. We then plotted the results in VOSviewer software (Figure 13.4).

The data showed that scholars have levied multiple concepts and theories to understand GIs. Two major strands of literature were identified. Strand I consists of individual innovators from the informal sector. Theoretical frameworks have primarily revolved around individual ingenuity and the local contexts in which such innovations emerge. Diverse theories have been used to study GIs in this strand (Table 13.1).

Intrinsic and extrinsic factors and decision theories have been adopted to identify motivations behind GIs (see, for example, Bhaduri and Kumar 2011). Researchers such as Abrol and Gupta (2014) tested theories on the diffusion of innovation. Using social network analysis, scholars such as Kumar (2014) have aimed to understand network dynamics and their interplay in the GI process. Social exchange theory, psychological contract theory and diffusion theory have been leveraged to capture grassroots innovators' actual experiences (e.g., Joshi et al. 2015). Sheikh and Bhaduri (2020) used value theories to understand the value that such innovations create. Using value theory from economics, philosophy and sociology, the authors argued that GIs generate a diverse set of values, ranging from use value to socially embedded reciprocal exchange value to different forms of relational and non-relational intrinsic value.

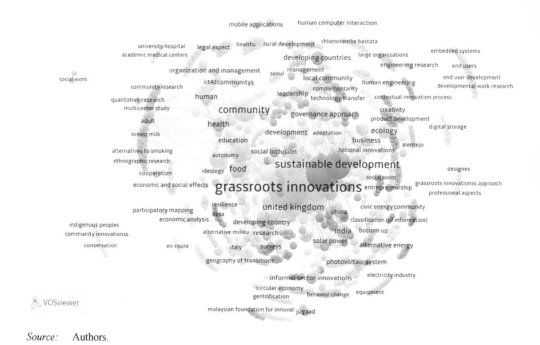

*Source:*    Authors.

*Figure 13.4    Indexed keywords on grassroots innovations*

From a methodological perspective, most authors in Strand I conducted qualitative research; others used quantitative methods and case studies.

Strand II has analyzed GIs from a community practice standpoint (Table 13.2). The focus of the strand is on sustainable development practices (Seyfang and Smith 2007; Smith and Stirling 2018) and environmental concerns (Hielscher et al. 2013). Most studies have attempted to understand community innovations in European settings. In a non-European context, Sheikh (2014b) evaluated community-level GIs and knowledge appropriation mechanisms by taking Kashmiri pashmina shawl communities in India as a case in point. In a related vein, Daniels (2010) and King (1996) studied metal-working communities in Africa. Data from 2019 showed that community-led GIs have continued to take center stage in the GI literature.

Multiple theories have been used to interpret GIs at a community level. For instance, Roysen and Mertens (2019) adopted social practice theory to analyze ecovillages in Brazil. Lehtonen and de Carlo (2019) applied the fundamentals of ecological economics to explore trust, mistrust and distrust relative to energy in the UK. Some scholars (e.g., Geels 2011) suggested using multi-level theoretical approaches, which involve sociotechnical landscapes, sociotechnical regimes and niche innovations. To understand barriers inhibiting GI growth, researchers such as Jones et al. (2021) have considered paradox theory. He applied this theory to explore how paradoxes from differences in social, commercial and cultural logic have led to tension and exacerbated the social–commercial–cultural trilemma, particularly in terms of GI evolution in remote Australian indigenous art centers.

*Table 13.1*   Grassroots innovations as an individual approach (Strand 1)

| Sr. No. | Scholar(s) | Theory/Focus Area | Methodology | Case | Country |
| --- | --- | --- | --- | --- | --- |
| 1 | Gupta (2016) | Grounded theory | Qualitative | Grassroots innovation | India |
| 2 | Bhaduri and Kumar (2011) | Motivation theory | Mixed method | Grassroots innovation | India |
| 3 | Abrol and Gupta (2014) | Diffusion theory | Mixed method | Grassroots innovation | India |
| 4 | Kumar (2014) | Network analysis | Case study | Grassroots innovation | India |
| 5 | Sahay et al. (2014) | Green biopesticide | Case study | Grassroots innovation | India |
| 6 | Joshi et al. (2015, 2016) | Social exchange theory, psychological contract theory, diffusion theory | Case study based on lived experiences | Grassroots innovation | India |
| 7 | Ustyuzhantseva (2015) | Institutionalization process | Qualitative | Grassroots innovation | India |
| 8 | Pattnaik and Dhal (2015) | Sustainable technologies | Qualitative | Grassroots innovation | India, South |
| 9 | Smith et al. (2016) | Grassroots innovation movements | Qualitative | Grassroots innovation, *fablab*, hackerspace, makerspace, The People's Science Movement, the appropriate Technology network, the appropriate technology movement, the movement for socially useful production | America, Europe |
| 10 | Devadula et al. (2017) | Social innovation and sustainable service design | Case study | Community workshop | India |
| 11 | Boruah and Das (2017) | Design support | Case study | Grassroots innovation | India and UK |
| 12 | Sharma and Kumar (2018a, 2018b) | Intellectual property rights, utility model | Case study | Grassroots innovation | India |
| 13 | Singh et al. (2018) | Structural equation modelling | Quantitative | Grassroots entrepreneurs | India |
| 14 | Sharma and Kumar (2019) | Commercialization process | Mixed method | Grassroots innovation | India |
| 15 | Kumar (2020) | Scientometric analysis | Quantitative | Grassroots innovation | India |
| 16 | Dey et al. (2019) | Climate resilient entrepreneurial pathways | Qualitative | Grassroots innovation | India |
| 17 | Sheikh and Bhaduri (2020) | Value theory | Qualitative | Grassroots innovation | India |

*Table 13.2*    *Grassroots innovations as a community approach (Strand II)*

| Sr. No. | Scholar(s) | Theory/Focus Area | Methodology | Case | Country |
|---|---|---|---|---|---|
| 1 | Seyfang (2005) | Sustainable development and community action | Case study | Community currency | UK |
| 2 | Seyfang and Smith (2007) | Sustainable development and community action | Qualitative | Community action | UK |
| 3 | Seyfang (2008, 2010); Seyfang and Longhurst (2013a, 2013b, 2016) | Sustainable development and community action, diffusion of innovation | Case study | Low carbon housing, community currency | UK |
| 4 | Smith (2016) | Strategic niche management, niche policy advocacy, critical niches | Case study | | UK |
| 5 | Seyfang and Haxeltine ( 2012); Martiskainen (2017); Pellicer-Sifres et al. (2017) | Sustainable development and community action | Case study | Food and health | UK |
| 6 | Seyfang et al. (2013) | Sustainable development and community action | Case study | Energy | UK |
| 7 | Smith et al. (2016) | Grassroots movements | Qualitative | Grassroots innovations, *fablab*, hackerspace, makerspace, The People's Science Movement, the Social Technology network, the appropriate technology movement, the movement for socially useful production | India, South America, Europe |
| 8 | Feola (2014); Feola and Nunes (2014); Feola and Him (2016); Feola and Butt (2017); Nicolosi and Feola (2016) | Sustainability, transition network movement and diffusion, success and failure | Case study, spatial analysis | Social organization and housing | Italy, France, Germany, UK, USA |
| 9 | Zhang and Mahadevia (2014) | Grounded theory | Case study | Grassroots organization of farmers | China |
| 10 | Hossain (2016) | Scientometric analysis | Quantitative | Grassroots innovation | Europe |
| 11 | Sheikh (2014b) | Community innovation and knowledge appropriation | Case study | Pashmina shawl in Kashmir | India |
| 12 | Daniels (2010) | Community action | Case study | Jua Kali | Africa |
| 13 | King (1996) | Community action | Case study | Metal work | Africa |
| 14 | Roysen and Mertens (2019) | Social practice theory | Case study | Eco village | Brazil |
| 15 | Geels (2019) | Multi-level perspective, sustainability transformation | Qualitative | Energy, transport, housing, agro-food | Europe |
| 16 | Jones et al (2021) | Paradox theory | Qualitative | Art centers | Australia |

# HISTORICAL CONTRIBUTIONS OF GRASSROOTS INFORMAL SECTOR INNOVATIONS

The preceding sections summarize burgeoning academic and non-academic interest in the nuances of GIs. This sudden attention to non-firm-level innovations implies that GIs represent a novel concept with no historical continuity. Between 1912 and 1945, the producer innovation model proposed by Joseph Schumpeter, which focused on profit maximization and private property, dominated innovation scholarship. The producer innovation paradigm has been widely accepted by economists, businesspeople, and policymakers worldwide. Even 60 years after Schumpeter's work, economists such as Teece (1996), Romer (1990), and Baumol (2002) continued to subscribe to the notion that Schumpeterian innovation is crucial for economic growth and human development. Rightly argued by von Hippel (2017), compensated transactions determine the conception, launch, profit, and success of inventive pursuits in producer innovations. Other innovation forms, like GIs, have remained virtually outside the purview of the Schumpeterian paradigm.

Non-formal innovations such as grassroots innovations do not involve transaction costs and are not safeguarded from appropriation. Inclusivity, autonomy, affordability, and sustainability are key GI features. These innovations originate from collaborative thinking in communities at the bottom of the social and economic pyramid. The core motivation for these communities is to mitigate market-based economic and social divides (Bhaduri and Kumar 2011). However, as seen above, the ingenuity of grassroots innovators and the potential of these inventions to address economic and social challenges have begun to attract attention in policy and academia. Many international organizations, including the United Nations (UN), have even begun mentioning GIs in their policy documents (UNCTAD 2017).

This section seeks to upend the narrative that non-firm-level innovations including GIs constitute a novel concept without substantial historical economic and non-economic contributions. We adopted a multidisciplinary approach by synthesizing the literature on the history of inventions. Notably, although diverse terms have been used to refer to knowledge produced by people at the grassroots level, the essence of GIs has remained intact. Bhatti (2012) argued that since the invention of Neanderthal hand tools, the fundamental characteristics of GIs (e.g., inclusiveness, frugality, and collectiveness) have remained constant.

In this chapter, we label all individual, community, and independent innovative pursuits as non-formal GIs, which emerge organically without support from formal structures (e.g., state or state-run entities, including universities and research institutions). To the best of our knowledge, GIs also receive little to no help from fortified firms, government-affiliated or otherwise. Vernacular innovations and household innovations are included in our definition as well.

The history of inventions holds ample evidence of independent innovators, bereft of formal support, devising breakthrough technologies and applying local solutions to solve community problems. Polydore Vergil (1470–1555), the Italian humanist scholar and historian who was the first to show interest in documenting the origins of inventions in his book *De Inventoribus Rerumin* (1499), listed many inventions of his time that could be classified under today's free innovation paradigm or as non-formal GIs. In his magnum opus *The Wealth of Nations*, Adam Smith (1776), a leading proponent of classical economics, explicitly referred to the inventions of common men. He was also the first to discuss the economic significance of independent innovations born outside formal structures. Not only did Smith highlight the value of the common man's ingenuity, but he also distinguished it from inventions by "men of speculation":

> All the improvements in machinery, however, have by no means been the inventions of those who had occasion to use the machines. Many improvements have been made by the ingenuity of the makers of the machines…and some by that of those who are called philosophers or men of speculation, whose trade it is not to do anything, but to observe everything (18) … and inventions of common workmen who being each of them employed in some very simple operation. (Smith 1776, pp. 17–18)

More than two centuries later, Joel Mokyr (2010) argued that during the industrial and pre-industrial revolutions, unlettered innovators and unidentified communities played major roles in knowledge advancement. He pointed out that prior to 1800, communities in Europe and other parts of the world could design useful techniques with no knowledge of why or how these techniques functioned. According to Mokyr (2010), societies could manufacture steel, brew beer, develop a vaccine for smallpox, and practice crop rotations despite lacking an understanding of metallurgy, yeasts, genetics, immunology, or soil chemistry. Ignorance therefore did not deter these societies from making important scientific advances (2002). Additionally, Mokyr (2010) stated that a "culture of making improvements" prevailed in traditional societies and, according to Wrigley (2004), enabled these societies to realize major scientific improvements.

During the first and second industrial revolutions, guilds, artisans, and unlettered communities made vast contributions. Mechanics, highly skilled clock and instrument makers, metalworkers, woodworkers, toymakers, and glasscutters pioneered the development of science and planned innovations during these times. As argued by Hilaire-Pérez (2007) and Berg (2007), skilled craftsmen could inspire improvements without assistance. According to Mokyr (2002), although James Watt had no knowledge of thermodynamics or the laws of physics, he managed to improve his dexterity in a mechanic shop. Richard Roberts, widely acclaimed as the most versatile mechanic of the industrial revolution, had never studied science – but he invented the self-acting mule in 1825, which was later used to operate spinning machines in British-owned cotton factories (Mokyr 2002). John Mercer (1791–1866), another key innovator and one of Lancashire's most successful colorists and dye specialists, was self-taught and nominated as a Fellow of the Royal Society in England in 1852. After the first industrial revolution, innovators with limited scientific background laid a foundation for the second. In *Gifts of Athena*, Mokyr (2002) argued that the Bessemer steelmaking process of 1856 materialized out of initiatives by a man with limited understanding of iron metallurgy.

Likewise, Edmund S. Phelps, who received the Nobel Prize for Economics in 2006, noted in *Mass Flourishing: How Grassroots Innovation Created Jobs, Challenge, and Change* (2013) that some nations' current economic prosperity is a product of past indigenous innovations, made possible through economic dynamism and a desire to innovate. Repudiating the theory that all material advances in a country are driven by forces of science and entrepreneurs with sophisticated technical backgrounds, Phelps (2013) affirmed that scientism and exogenous innovations could not have been driving forces behind the nineteenth-century explosion of economic knowledge. He stated that all inventors, including those who rose to prominence during the industrial revolution, were neither trained scientists nor well-educated. He also argued that Britain flourished in large part due to the people's will, capacity, and inspiration to innovate.

To some scholars, GIs represent technological disobedience and non-compliance. For example, Korea's grassroots "self-made inventors" tied incremental local innovations to debates on self-reliance and Korean nationalism during the 1920s and 1930s. According to Lee (2013), this grassroots movement altered discourse on innovations; social elites opined that invention is an indigenous concept rather than a foreign one, which is entrenched in culture and upheld by perseverant individuals fascinated by trial and error. After the Cuban revolution, local inventors demonstrated defiance and non-compliance by considering technology radically different from inventors of the West: local innovators began breaking and re-creating items from scrap. In *Technological Disobedience*, Ernesto Oroza (2015) stated that in stark contrast to the ideas of innovation introduced by the Western concept of mass production, "reparation, refunctionalization, and reinvention" have manifested in several novel forms.

Innovations by common men have made economic contributions, helped society surmount everyday problems, and played substantial roles in science. Joseph Needham (1954) asserted in *Science and Civilization in China* that during the pre-industrial revolution period, people from disadvantaged backgrounds contributed to path-breaking technologies. As a science historian, he disapproved of the theory that modern science and all technologies emerged in European cultures; instead, he contended that different cultures and unsung individuals have contributed uniquely to the development of science in Europe and elsewhere.

Conner (2005) offered additional evidence of underprivileged people's contributions to science and innovations in *Peoples History of Science: Miners, Midwives and Low Mechanics*. Referring to Isaac Newton's proposition "Standing on the shoulders of giants," Conner remarked that Newton should have credited his ability to "see further" to thousands of grassroot artisans; more specifically, "standing on the backs of untold thousands of illiterate artisans" (p. 2) would have been more appropriate to acknowledge such innovators. Conner (2005) emphasized the contributions of hunter-gatherers, peasant farmers, sailors, miners, blacksmiths, and folk healers to the expansion of modern-day science and technology. He asserted that the Amerindians discovered the use of cinchona tree bark to treat malaria, and an American slave named Onesimus uncovered the first steps toward smallpox inoculation. Conner (2005) also argued that while the discovery of vaccinations has been widely attributed to Dr. Edward Jenner, the original inventor of vaccinations was a farmer named Benjamin Jesty. Through such cases, Conner (2005) indicated that unlettered innovators generated a large body of knowledge preceding Europe's scientific and industrial revolution.

## REASONS FOR THE UNDERVALUATION OF INFORMAL SECTOR INNOVATIONS

Extensive literature has considered the roles of grassroots and unlettered innovators in advancing knowledge in different societies. However, an exhaustive review of the mainstream innovation literature suggests that original contributions of the common man are being increasingly substituted by firms and producer innovations; the knowledge generated by unlettered innovators and anonymous communities is rarely acknowledged. Most disenchanting is the fact that economic theories that earned post-Schumpeterian recognition excluded GIs. Several factors, ranging from an inadequate policy focus to skewed power relations, have led to a gradual degeneration of GIs. We explore these factors in greater detail in this section.

Nathan Rosenberg (1976) argued that while economic scholars investigate technology and its impacts, an implicitly "hierarchical conceptualization" of different types of knowledge persists. He pointed out that scientific knowledge, which presents itself as the most generic or inclusive form, is well reputed. Rosenberg (1976) also asserted that technological knowledge seldom receives the respect it deserves. According to him, this bias against technological knowledge has led to a nearly complete dismissal of marginal improvements, culminating in a flawed understanding of the relationship between technical change and economic growth. Rosenberg (1976) further contended that neglecting marginal advances in the innovation process, along with economists' fixation on radical and purely scientific innovations, has limited overall appreciation for the correlation between technological change and economic growth.

Scholars such as Rosenberg (1976), von Hippel (2017) and Godin (2017) stated that this dominant interpretation of innovations is grounded in Schumpeter's work. Minor innovations and non-firm inventions have remained largely outside the scope of Schumpeterian analyses (Rosenberg 1976). Von Hippel (2017), and Godin and Vinck (2017) argued that the influence of Schumpeterian theories has been so pervasive that until today, every model and policy document on innovation has been based on Schumpeter's definition. According to Godin and Vinck (2017), this belief has two major drawbacks: first, it excludes several important forms of knowledge that contribute substantially to economic growth and development; and second, it undermines the productive debate following any new innovation.

Even now, a pro-innovation bias towards firm-level innovations is readily apparent. In Godin's (2016) opinion, firm-level innovations are in the limelight due to technological innovations, R&D and systematic innovation policies. International organizations like the UN, World Bank and Organisation for Economic Co-operation and Development (OECD) have almost exclusively focused on market-driven innovations. These organizations often assess the significance of these innovations based on market implementation statistics. Comparatively, innovations that are non-marketable and generate non-economic value have suffered political and social neglect. According to Godin and Vinck (2017), these factors have sapped the enthusiasm of craftsmen engaged in the development of such innovations, whose economic and societal contributions are seldom recognized.

Other factors contributing to the marginalization of GIs generally emanate from unequal power relations. Conner (2005) noted that in the "romantic narratives" of history, which have been influenced by power relations, contributions from subalterns, peasants, farmers and the voiceless were deliberately disregarded and destroyed. He further opined that the important scientific contributions made by less empowered and literate populations have not been systematically documented. In the absence of proper records, notable scientific contributions from GIs remain dark. Conner (2005) particularly asserted that these "Great Men of science" are the proverbial "elephant in the parlor": they cannot be ignored, but their stories have traditionally been narrated "from the perspective of the ruling elites" (p. 18). He argued that those who perform skilled manual tasks have long been looked upon with contempt by those who refrain from such activities. Conner sarcastically noted that such innovators' contributions went unnoticed for a long period of history, as knowledge of Latin was the most sought-after "skill" and a crucial means of distinguishing the "learned from the vulgar" and the "elite from the popular." Therefore, political and social factors have both been responsible for the gradual depletion of knowledge generated by underprivileged communities.

Steven Shapin (1994), in *A Social History of Truth*, cited elitism as the pre-condition to define truth. According to Shapin (1994), in seventeenth-century England, only sophisticated gentlemen were considered truthful. In fact, these gentlemen with elitist backgrounds were widely relied on for their perceptions and wisdom about humankind and the natural order. Arcane code systems and complex methodologies were used to make sophisticated forms of knowledge almost inaccessible to people of humble backgrounds. In *Science in History, Volume 1: The Emergence of Science*, J. D. Bernal (1954) shared similar views on the elite's conceited tendency to generate knowledge:

> …where science has been kept a mystery in the hands of a selected few, it is inevitably linked with the interests of the ruling classes and is cut off from the understanding and inspiration that arise from the needs and capacities of the people. (p. 3)

By virtue of their elitist backgrounds and strong political and social networks, ruling classes enjoyed the luxury of creating knowledge and reinforcing their status, irrespective of the needs and competencies of the masses. The methodologies used by the elite and resourceful intellectuals to gather, test and evaluate knowledge were inherently biased against the knowledge of less privileged people. Such views echo the description of Baroque culture in William Eamon's *Science and the Secrets of Nature* (1994). Eamon explained how expensive artifacts and appearances guided the collection and display of "secrets" in Baroque culture. He mentioned that the gathering and display of objects in the courtly culture of early modern Europe was primarily driven by the conceitedness of the ruling classes. Members of such classes claimed to be the discoverers of "secrets of nature" and believed they could understand these secrets better than those who were less knowledgeable. Such claims enabled these classes to formulate effective strategies for collecting secrets. Thus, while the interpretation of nature was strictly the responsibility of the elite, underprivileged innovators' discoveries and secrets were granted less value.

Notwithstanding the discrimination faced by subaltern innovators, it is important to reiterate that most breakthrough technologies have been invented by individuals from modest backgrounds. As articulated in 1925 by Charles Jenkins, a prominent American inventor himself:

> Had anyone noticed, the curious fact that a great laboratory, despite its inestimable contribution to science and engineering, has never yet brought forth a great, revolutionary invention which has subsequently started a new industry, like the telegraph, the telephone and telescope; motion picture, typecasting and talking machines; typewriter, bicycle and locomotive; automobile, flying machine and radio vision [an early form of television]. It has always been a poor man to first see these things, and as a rule the bigger the vision the poorer the man. (Weightman 2015, p. 85)

Although exclusion and relegation have assumed novel forms, GI marginalization still runs rampant in the contemporary world. During our 10 years of ethnographic research in different parts of India, we have gathered data confirming that social hierarchies embedded in caste and religion continue to undermine the knowledge produced by less privileged people at the grassroots level (Ilaiah 2009). Further, although policymakers have adopted top-down strategies to upscale GIs, these strategies are often futile as they tend to be conceived and implemented without the involvement of grassroots innovators (Sheikh and Bhaduri 2020). Our research has reaffirmed the view that patriarchy, capitalism and male self-interests have contributed substantially to the relegation of GIs.

## CONCLUSION AND THE WAY FORWARD

As mainstream growth models continue to attract criticism for widening the economic gap between the rich and poor, perpetuating exclusion and creating hierarchies, the potential of new innovation and growth models is being explored at length (Lowrey 2016). Research on alternative innovation models within the last few years has provided evidence of the credibility of GIs in bridging the economic divide created by dominant innovation models. Although studies in support of GIs have often been opposed and undermined by proponents of dominant innovation models, there is a growing consensus that with sustained research, GIs could offer viable alternatives to modern innovation models. For example, Edmund Phelps (2013) contended that only sustained efforts to promote GIs can ensure a country's prosperity. Phelps appeared especially skeptical of the resource-intensive focus of classical and neoclassical growth models (Lowrey 2016). According to Phelps, economic dynamism (i.e., "people's desire, ability and intent" to innovate) is the sole determinant of sustained economic growth. Participation by people at the grassroots level is the *sine qua non* for fostering economic growth and individual well-being (Lowrey 2016).

It is important to note that the contributions of GIs to innovation discourse have been acknowledged by international organizations, which has helped GI innovators gain a foothold. For instance, in a document published in 2017 by the UN, member states require innovative approaches for achieving the organization's ambitious Sustainable Development Goals by 2030. Moreover, the OECD recently assumed an important role in the context of GIs by expanding the scope of the erstwhile definition. The essence of the former definition, adopted in 1992, was market implementation of the product; it overlooked human-centric aspects of these innovations. The new OECD definition has narrowed this definitional lacuna on GIs, making the new definition inclusive and broader in scope.[1] Even though the definition neither explicitly references GIs nor provides a roadmap to mainstreaming these innovations, it does provide a timely and nuanced understanding of GIs for academics and policymakers.

Furthermore, although countries such as India, Malaysia, England, Brazil, China and Kenya have made several attempts to adopt and upscale GIs due to challenges related to social justice, unemployment, inequality and climate change, old hierarchical structures and bloated bureaucracy have failed to yield desired outcomes. In most cases, knowledge created by grassroots innovators is evaluated by elite scientists and policymakers without the actual innovators' participation. In addition to undermining innovators' efforts, excluding innovators from evaluation exercises dampens the inherent purpose of such pursuits.

In this context, the Cistern Programme for Rainwater Harvesting in Brazil and the Community Energy Projects in England merit attention (UNCTAD 2017). In Brazil's semi-arid northeastern region, the national government's attempts to upscale local GIs for rainwater harvesting went in vain. It is important to trace the origins of this GI. Manoel Apolonio, an illiterate grassroots innovator from the same region, was engaged in a bricklaying job in the 1950s. While building a swimming pool, he wondered how a huge water tank could be useful for his community during drought seasons. Using local raw material, he designed cisterns that could collect rainwater from the roofs of houses, enough to sustain a family's needs through a drought (Fressoli et al. 2014). In 2003, the Brazilian Semi-Arid Articulation[2] chose to scale this GI and sought federal government support to do so. This action marked a turning point in several aspects: strengthening community relationships; enabling people to build, modify and use technology; and promoting the empowerment of local governments and water suppliers.

Nearly 600 000 rainwater harvesting units were constructed, including some innovative variations. However, once the federal government took over the project to upscale to 1 million cisterns, the local network was left behind and community participation ceased (Fressoli et al. 2014). Ignoring the complex set of factors driving the ecological rationality of this innovation, the government bought 300 000 plastic cisterns at almost twice the price of the original cement version. In addition to benefitting for-profit firms, the federal government's decision led to a disappointing technological solution as the plastic cisterns crumbled under intense heat during the summer. More importantly, the upscale process led to absolute exclusion and disempowerment of local communities and institutions. Similarly, in the UK, the government's intervention and support schemes to scale community innovation initiatives ignored several social and political objectives of the original innovators (Bhaduri et al., forthcoming).

Therefore, if GIs are to be considered potential alternatives to dominant innovation models, it is important to understand the context of innovation as well as innovators' perspectives. At this juncture, we present several other observations gathered through our field research. First, the value of an innovation is not merely commensurate with its economic value. As informal sector innovations are primarily conceived for wider societal gains, using a market-centric approach to gauge their value only undermines their scope and potential. In terms of eliminating ambiguities, utmost caution should be exercised while selecting an approach to innovation narratives.

Second, in replicating formal innovation models, several organizations have endorsed modern forms of knowledge appropriation (e.g., patents and copyrights) to protect informal knowledge. Given that the rationale behind GIs is to overcome social and economic challenges faced by communities, the idea of safeguarding GIs through patents and copyrights opposes the premise of such innovations. A sound approach, which aligns with GI principles, is thus required to incentivize grassroots innovators. Third, rather than depriving these individuals of economic dividends, respecting the autonomy, agency and dignity of informal innovators is important. Fourth, the purpose of many GIs is to subvert the status quo, delay planned obsolescence, and question the legitimacy of existing innovation narratives. Thus, maintaining tunnel vision while studying GIs could lead to misleading interpretations.

Therefore, despite profound contributions to numerous innovations throughout history, GIs continue to be decried for their inability to generate profit, wealth and power. Unlike formal or producer innovations, GIs have not been afforded the status they deserve. As discussed, several factors – ranging from an inadequate policy focus to unequal power relations – have led to GIs' decline. To explore the history of GIs, truth and power must be separated. As argued by Harari (2018), "Truth and power cannot meet. To spread truth, we have to renounce power and for power we have to spread fictions." Fifth, because GIs are one category of free innovations and emerge in a paradigm distinct from Schumpeterian and market-driven innovations, it would be useful to explore alternative theoretical foundations. At present, scholarship on GIs has adopted an eclectic theoretical lens, drawing upon theories from other disciplines; a more vivid picture of GIs should emerge if new theories are applied to investigate the topic in detail.

## NOTES

1.  The new definition of innovation (OECD/Eurostat 2018) states, "Innovation is more than a new idea or an invention. An innovation requires implementation, either by being put into active use or by being made available for use by other parties, firms, individuals or organisations." Implementation is defined thusly: "In order for a new idea, model, method or prototype to be considered an innovation, it needs to be implemented. Implementation requires organizations to make systematic efforts to ensure that the innovation is accessible to potential users, either for the organization's own processes and procedures, or to external users for its products." Similarly, the value of innovation is explained as follows: "Value is therefore an implicit goal of innovation, but cannot be guaranteed on an ex ante basis because innovation outcomes are uncertain and heterogeneous. The realization of the value of an innovation is uncertain and can only be fully assessed sometime after its implementation. The value of an innovation can also evolve over time and provide different types of benefits to different stakeholders."
2.  ASA is a network of more than 700 institutions, social movements, non-governmental organizations, and farmers' groups. It originated from popular mobilization against the "industry of drought" (Fressoli et al. 2014).

## REFERENCES

Abrol, Dinesh and Ankush Gupta (2014), 'Understanding the diffusion modes of grassroots innovations in India: A study of honey bee network supported innovators', *African Journal of Science, Technology, Innovation and Development*, 6 (6), 541–52.

Basole, Amit (2014), 'The informal sector from a knowledge perspective', *Yojana*, 58, 8–13.

Baumol, William J. (2002), *The Free-Market Innovation Machine: Analyzing the Growth Miracle of Capitalism*, Princeton: Princeton University Press.

Berg, Maxine (2007), 'The genesis of useful knowledge', *History of Science,* 45 (Pt. 2, 148), 123–34.

Bernal, John D. (1954), *Science in History, Volume 1: The Emergence of Science*, Cambridge, MA: MIT Press.

Bhaduri, Saradindu (2016), 'Frugal innovation by "the Small and the Marginal": an alternative discourse on innovation and development', Inaugural Lecture Prince Clause Chair in Development and Equity, Erasmus University Rotterdam, The Hague, May 23.

Bhaduri, Saradindu and Hemant Kumar (2011), 'Extrinsic and intrinsic motivations to innovate: Tracing the motivation of "grassroot" innovators in India', *Mind & Society*, 10 (1), 27–55.

Bhaduri, S., A. Corradi, H. Kumar and F.A. Sheikh (forthcoming), 'Frugality in innovation processes: A heuristics based perspective from the "informal" economy', in Cees van Beers, Saradindu Bhaduri, Peter Knorringa and André Leliveld (eds), *Handbook of Frugal Innovations*, Cheltenham, UK and Northampton, MA, USA: Edward Elgar Publishing.

Bhatti, Yasser Ahmad (2012), 'What is frugal, what is innovation? Towards a theory of frugal innovation', 1–45. SSRN: https://ssrn.com/abstract=2005910 or http://dx.doi.org/10.2139/ssrn.2005910.

Boruah, Dipanka and Amarendra Kumar Das (2017), 'Exploring grassroots innovation practices and relationships to drivers of start-up success in a multicultural context', 569–583. doi:10.1007/978-981 -10-3518-0_50.

Conner, Clifford D. (2005), *Peoples History of Science: Miners, Midwives and Low Mechanics*, New York: Nation Books.

Cozzens, Suzan and Judith Sutz (2014), 'Innovation in informal settings: Reflections and proposals for a research agenda', *Innovation and Development*, 4 (1), 5–31.

Daniels, Steve (2010), *Making Do: Innovation in Kenya's Informal Economy*, San Francisco: Creative Commons, Analogue Digital.

de Beer, Jeremy, Kun Fu and Sacha Wunsch-Vincent (2013), 'The informal economy, innovation and intellectual property: concepts, metrics and policy considerations' (WIPO Economic Research Working Paper No. 10), Geneva, WIPO, 1–76.

Devadula, Suman, Kiran Ghadge, Saritha Vishwanathan, Shuk Han Chan, Quinn Langfitt, David Dornfeld, Anil Gupta, Sudarsan Rachuri, Gaurav Ameta and Amresh Chakrabarti (2017), 'Supporting social innovation: Application of indeate tool for sustainable service design – case study of community workshops', in A. Chakrabarti and D Chakrabarti (eds), *Research into Design for Communities, Volume 2*. ICoRD 2017. Smart Innovation, Systems and Technologies, vol 66. Springer, Singapore. https://doi.org/10.1007/978-981-10-3521-0_12.

Dey, Anamika, Anil Gupta and Gurdeep Singh (2019), 'Innovation, investment and enterprise: Climate resilient entrepreneurial pathways for overcoming poverty', *Agricultural Systems*, 172, 83–90.

Eamon, William (1994), *Science and the Secrets of Nature: Books of Science in Medieval and Early Modern Culture*, Princeton: Princeton University Press.

Feola, Giuseppe (2014), 'Narratives of grassroots innovations: A comparison of voluntary simplicity and the transition movement in Italy', *International Journal of Innovation and Sustainable Development*, 8 (3), 250–69.

Feola, Giuseppe and Anisa Butt (2017), 'The diffusion of grassroots innovations for sustainability in Italy and Great Britain: An exploratory spatial data analysis', *Geographical Journal*, 183 (1), 16–33.

Feola, Giuseppe and Mina Rose Him (2016), 'The diffusion of the transition network in four European countries', *Environment and Planning A*, 48 (11), 2112–15.

Feola, Giuseppe and Richard Nunes (2014), 'Success and failure of grassroots innovations for addressing climate change: The case of the transition movement', *Global Environmental Change*, 24 (1), 232–50.

Fressoli, Mariano, Elisa Arond, Dinesh Abrol, Adrian Smith, Adrian Ely and Rafael Dias (2014), 'When grassroots innovation movements encounter mainstream institutions: Implications for models of inclusive innovation', *Innovation and Development*, 4 (2), 277–92.

Geels, Frank W. (2011), 'The multi-level perspective on sustainability transitions: Responses to seven criticisms', *Environmental Innovation and Societal Transitions*, 1 (1), 24–40.

Geels, Frank W. (2019), 'Socio-technical transitions to sustainability: A review of criticisms and elaborations of the multi-level perspective', *Current Opinion in Environmental Sustainability*, 39, 187–201. doi:10.1016/j.cosust.2019.06.009.

Godin, Benoît (2016), 'Technological innovation: On the emergence and development of an inclusive concept', *Technology & Culture*, 57 (3), 527–6.

Godin, Benoît (2017), *Models of Innovation. The History of an Idea*. Cambridge: MIT Press.

Godin, Benoît and Dominique Vinck (2017), *Critical Studies of Innovation: Alternative Approaches to the Pro-Innovation Bias*, Cheltenham, UK and Northampton, MA, USA: Edward Elgar Publishing.

Gupta, Anil K. (2016), *Grassroots Innovation: Minds on the Margin are not Marginal Minds*, Gurgaon: Random House India.

Habiyaremye, Alexis, Glenda Kruss and Irma Booyens (2019), 'Innovation for inclusive rural transformation: The role of the state', *Innovation and Development*, 10 (2), 155–68. doi: 10.1080/2157930X.2019.1596368.

Harari, Y.N. (2018), *Sapiens: A Brief History of Humankind* (Reprint ed.), New York: Harper Perennial.

Heeks, Richard, Christopher Foster and Yanuar Nugroho (2014), 'New models of inclusive innovation for development', *Innovation and Development*, 4 (2), 175–85.

Hielscher, Sabine, Gill Seyfang and Adrian Smith (2013), 'Grassroots innovations for sustainable energy: Exploring niche-development processes among community-energy initiatives', in Cohen, Maurie, Halina Brown and Philip Vergragt (eds), *Innovations in Sustainable Consumption: New Economics, Socio-technical Transitions, and Social Practices*, Cheltenham, UK and Northampton, MA, USA: Edward Elgar Publishing, pp. 133–56.

Hilaire-Pérez, Liliane (2007), 'Technology as public culture', *History of Science*, 45 (Pt. 2, 148), 135–53.

Hoffecker, Elizabeth. (2018), Local innovation: What it is and why it matters for developing economies (D-Lab Working Papers, NDIR Working Paper 1), Cambridge, MIT D-Lab.

Hossain, Mokter (2016), 'Grassroots innovation: A systematic review of two decades of research', *Journal of Cleaner Production*, 137, 973–81.

Ilaiah, Kancha (2009), *Post-Hindu India: A Discourse in Dalit-Bahujan, Socio-Spiritual and Scientific Revolution*, New Delhi: Sage.

Jain, Ashok and Jan Verloop (2012), 'Repositioning grassroots innovation in India's S&T policy: From divider to provider', *Current Science*, 103 (3), 282–5.

Jones, Janice, Seet Pi-Shen, Acker Tim and Whittle Michelle (2021), 'Barriers to grassroots innovation: The phenomenon of social-commercial-cultural trilemmas in remote indigenous art centres', *Technological Forecasting & Social Change*, 164, 119583. https://doi.org/10.1016/j.techfore.2019.02.003.

Joshi, Rajul G., John Chelliah and Veeraraghavan Ramanathan (2015), 'Exploring grassroots innovation phenomenon through the lived experience of an Indian grassroots innovator', *South Asian Journal of Global Business Research*, 4 (1), 27–44.

Joshi, Rajul G., John Chelliah, Suresh Sood and Stephen Burdon (2016), 'Nature and spirit of exchange and interpersonal relationships fostering grassroots innovations', *The Journal of Developing Areas*, 50 (6), 399–409.

King, Kenneth (1996), *Jua Kali Kenya Change and Development in an Informal Economy 1970–1995*, Athens: Ohio University Press.

Kraemer-Mbula, E. and Sacha Wunsch-Vincent (eds) (2016), *The Informal Economy in Developing Nations: Hidden Engine of Innovation?*, Cambridge: Cambridge University Press.

Kumar, Hemant (2012), *Exploring Motivation, Collaboration and Linkages for Grassroot Innovation Systems in India*, PhD Thesis Submitted to Jawaharlal Nehru University, New Delhi.

Kumar, Hemant (2014), 'Dynamic networks of grassroot innovators in India', *African Journal of Science, Technology, Innovation and Development*, 6 (3), 193–201.

Kumar, Hemant (2020), 'Publication trends in the informal sector innovation research', *Journal of Scientometric Research*, 9 (2s), s5–s13. doi: 10.5530/jscires.9.2s.32.

Kumar, Hemant and Saradindu Bhaduri (2014), 'Jugaad to grassroot innovations: Understanding the landscape of the informal sector innovations in India', *African Journal of Science, Technology, Innovation and Development*, 6 (1), 13–22.

Lee, Jung (2013), 'Invention without science: "Korean Edisons" and the changing understanding of technology in colonial Korea', *Technology and Culture*, 54 (4), 782–814.

Lehtonen, Markku and Laurence de Carlo (2019), 'Community energy and the virtues of mistrust and distrust: Lessons from Brighton and Hove energy cooperatives', *Ecological Economics*, 164. doi:10.1016/j.ecolecon.2019.106367.

Lowrey, Ying (2016), *The Alibaba Way: Unleasing Grassroots Entrepreneurship to Build the Worlds Most Innovative Internet Company*, McGraw-Hill Education.

Martiskainen, Mari (2017), 'The role of community leadership in the development of grassroots innovations', *Environmental Innovation and Societal Transitions*, 22, 78–89.

McCloskey, Deirdre (2016), *Bourgeois Equality: How Ideas, Not Capital or Institutions, Enriched the World*, Chicago: University of Chicago Press.

Mokyr, Joel (2002), *The Gifts of Athena: Historical Origins of the Knowledge Economy*, Princeton, NJ: Princeton University Press.

Mokyr, Joel (2010), 'The contribution of economic history to the study of innovation and technical change: 1750–1914', in B.H. Hall and Nathan Rosenberg (eds), *Handbook of the Economics of Innovation*, Burlington: Elsevier, pp. 11–50.

Muchie, Mammo, Amare Desta and Mentensnot Mengesha (2016), *Science, Technology and Innovation: for Sustainable Future in the Global South*, Trenton, NJ: Red Sea Press.

Needham, Joseph (1954), *Science and Civilisation in China*, Cambridge, UK: Cambridge University Press.

Neuwirth, Robert (2011), *Stealth of Nations: The Global Rise of the Informal Economy*, New York: First Anchor Books.

Nicolosi, Emily and Giuseppe Feola (2016), 'Transition in place: Dynamics, possibilities, and constraints', *Geoforum*, 76, 153–63.

OECD/Eurostat (2018), Oslo Manual 2018: Guidelines for Collecting, Reporting and Using Data on Innovation, 4th Edition, The Measurement of Scientific, Technological and Innovation Activities, OECD Publishing, Paris/Eurostat, Luxembourg. https://doi.org/10.1787/9789264304604-en.

Oroza, Ernesto (2015), *Technological disobedience*. https://assemblepapers.com.au/2017/04/28/technological-disobedience-ernesto-oroza/.

Pattnaik, Binay Kumar and Debajani Dhal (2015), 'Mobilizing from appropriate technologies to sustainable technologies based on grassroots innovations', *Technology in Society*, 40, 93–110.

Pellicer-Sifres, Victoria, Sergio Belda-Miquel, Aurora López-Fogués and Alejandra Boni Aristizábal (2017), 'Grassroots social innovation for human development: An analysis of alternative food networks in the city of Valencia (Spain)', *Journal of Human Development and Capabilities*, 18 (2), 258–74, DOI: 10.1080/19452829.2016.1270916.

Phelps, Edmund S. (2013), *Mass Flourishing: How Grassroots Innovation Created Jobs, Challenges, and Change*, Princeton, NJ: Princeton University Press.

Prahalad, Coimbatore Krishnarao and Raghunath A. Mashelkar (2010), 'Innovations Holy Grail', *Harvard Business Review*. https://hbr.org/2010/07/innovations-holy-grail.

Romer, Paul M. (1990), 'Endogenous technological change', *Journal of Political Economy*, 98 (5), S71–S102.

Rosenberg, Nathan (1976), 'On technological expectations', *Economic Journal*, 86 (343), 523–35.

Roysen, Rebeca and Frédéric Mertens (2019), 'New normalities in grassroots innovations: The reconfiguration and normalization of social practices in an ecovillage', *Journal of Cleaner Production*, 236. doi:10.1016/j.jclepro.2019.117647.

Sahay, Nirmal Shankar, Chetankumar J. Prajapati, Keyur A. Panara, Jayshree D. Patel and Pawan K. Singh (2014), 'Anti-termite potential of plants selected from the SRISTI database of grassroots innovations', *Journal of Biopesticides*, 7, 164–9.

Santiago, Fernando (2014), 'Innovation for inclusive development', *Innovation and Development*, 4 (1), 1–4.

Seyfang, Gill (2005), 'Shopping for sustainability: Can sustainable consumption promote ecological citizenship?', *Environmental Politics*, 14 (2), 290–306.

Seyfang, Gill (2008), 'Avoiding Asda? Exploring consumer motivations in local organic food networks', *Local Environment: The International Journal of Justice and Sustainability*, 13 (3), 187–201.

Seyfang, Gill (2010), 'Community action for sustainable housing: Building a low-carbon future', *Energy Policy*, 38 (12), 7624–33. doi:10.1016/j.enpol.2009.10.027.

Seyfang, Gill and Alex Haxeltine (2012), 'Growing grassroots innovations: Exploring the role of community-based initiatives in governing sustainable energy transitions', *Environment and Planning C: Government and Policy*, 30 (3), 381–400. doi:10.1068/c10222.

Seyfang, Gill and Noel Longhurst (2013a), 'Desperately seeking niches: Grassroots innovations and niche development in the community currency field', *Global Environmental Change*, 23 (5), 881–91. doi:10.1016/j.gloenvcha.2013.02.007.

Seyfang, Gill and Noel Longhurst (2013b), 'Growing green money? Mapping community currencies for sustainable development', *Ecological Economics*, 86, 65–77. doi:10.1016/j.ecolecon.2012.11.003.

Seyfang, Gill and Noel Longhurst (2016), 'What influences the diffusion of grassroots innovations for sustainability? Investigating community currency niches', *Technology Analysis and Strategic Management*, 28 (1), 1–23. doi:10.1080/09537325.2015.1063603.

Seyfang, Gill and Adrian Smith (2007), 'Grassroots innovations for sustainable development: Towards a new research and policy agenda', *Environmental Politics*, 16 (4), 584–603. doi:10.1080/09644010701419121.

Seyfang, Gill, Jung Jin Park and Adrian Smith (2013), 'A thousand flowers blooming? An examination of community energy in the UK', *Energy Policy*, 61, 977–89. doi:10.1016/j.enpol.2013.06.030.

Shapin, Steven (1994), *A Social History of Truth – Civility & Science in Seventeenth-Century England*, London: University of Chicago Press.

Sharma, Gautam and Hemant Kumar (2018a), 'Exploring the possibilities of utility models patent regime for grassroots innovations in India', *Journal of Intellectual Property Rights*, 23 (2–3), 119–30.

Sharma, Gautam and Hemant Kumar (2018b), 'Intellectual property rights and informal sector innovations: Exploring grassroots innovations in India', *The Journal of World Intellectual Property*, 21, 123–39. doi: 10.1111/jwip.12097.

Sharma, Gautam and Hemant Kumar (2019), 'Commercialising innovations from the informal economy: The grassroots innovation ecosystem in India', *South Asian Journal of Business Studies*, 8 (1), 40–61. https://doi.org/10.1108/SAJBS-12-2017-0142.

Sheikh, Fayaz A. (2014a), 'Science, technology and innovation policy 2013 of India and informal sector innovations', *Current Science*, 106, 21–3.

Sheikh, Fayaz A. (2014b) 'Exploring informal sector community innovations and knowledge appropriation: A study of Kashmiri pashmina shawls', *African Journal of Science Technology Innovation and Development*, 6 (3), 203–12.

Sheikh, Fayaz A. and Saradindu Bhaduri (2020), 'Perspectives on grassroots innovations in the informal economy: Gleaning insights from value theory', *Oxford Development Studies*, 48(1), 85–99.

Singh, Sonal H., Birud Sindhav, Dale Eesley and Bhaskar Bhowmick (2018), 'Investigating the role of ICT intervention in grassroots innovation using structural equation modelling approach', *Sadhana Academy Proceedings in Engineering Sciences*, 43 (7). doi:10.1007/s12046-018-0909-8.

Smith, Adam (1776), *An Inquiry into the Nature and Causes of the Wealth of Nations*, London: W. Strahan and T. Cadell.

Smith, Adrian (2016), 'Alternative technology niches and sustainable development: 12 years on', *Innovation: Organisation & Management*, 18 (4), 485–8. doi:10.1080/14479338.2016.1241153.

Smith, Adrian and Andy Stirling (2018), 'Innovation, sustainability and democracy: An analysis of grassroots contributions', *Journal of Self-Governance and Management Economics*, 6 (1), 64–97. doi: 10.22381/JSME6120183.

Smith, Adrian, Mariano Fressoli and Hernán Thomas (2014), 'Grassroots innovation movements: Challenges and contributions', *Journal of Cleaner Production*, 63, 114–24. doi:10.1016/j.jclepro .2012.12.025.

Smith, Adrian, Mariano Fressoli, Dinesh Abrol, Elisa Around and Adrian Ely (2016), *Grassroots Innovation Movements*, New York: Routledge.

Teece, David (1996), 'Firm organization, industrial structure, and technological innovation', *Journal of Economic Behavior & Organization*, 31 (2), 193–224.

UNCTAD (United Nations Conference on Trade and Development) (2017), *New innovation approaches to support the implementation of the sustainable development goals*, Geneva: UNCTAD.

Ustyuzhantseva, Olga V. (2015), 'Institutionalization of grassroots innovation in India', *Current Science*, 108 (8), 1476–82.

Ventresca, Marc and Alex Nicholls (2011), *Rethinking Business Course*, Oxford: Said Sch.

von Hippel, Eric (2005), *Democratizing Innovation*, Cambridge, MA: MIT Press.

von Hippel, Eric (2017), *Free Innovations*, Cambridge MA: MIT Press.

Weightman, Gavin (2015), *Eureka: How Invention Happens*, New Haven: Yale University Press.

Wrigley, Edward Anthony (2004), *Poverty, Progress, and Population*, Cambridge: Cambridge University Press.

Zhang, Liyan and Darshini Mahadevia (2014), 'Translating science and technology policies and programs into grassroots innovations in China', *Journal of Science and Technology Policy Management*, 5 (1), 4–23.

# 14. Frugal innovation: reaching an 'empowered' developing-countries end-user

*Céline Cholez and Pascale Trompette*

## INTRODUCTION

Since the 2000s, the new approaches in development policies and fighting against poverty turn to private partners (companies and NGOs) for the design of specific and small-scale (even individual) products and services for developing countries contexts. The long experience of failures in North/South technological transfers (Rangan et al. 2007; London 2011), and the controversy about the risk of an indebted consumer associated to the initial Bottom of the Pyramid program (Karnani 2007), led to the emergence of design approaches targeting a better inclusion of the end-user at different stages of the design process. Frugal innovation (closed to grassroots innovation) is part of this new design paradigm, which aims at articulating high technology solutions, quality principles and best local practices. Far away from any low quality and low-cost solutions from international mass markets, products and services issued from frugal innovation approach are expected in reaching four criteria: Accessibility, Affordability, Availability and Awareness (Anderson and Markidès 2007) through cost reduction, feature simplification and low resource consumption (Radjou et al. 2012; Sehgal et al. 2010; Zeschky et al. 2011).

The notion of frugality refers to the idea of sobriety in resource consumption, a "good-enough," which would be a necessity in contexts of scarcity and enforced by ecological imperatives. As Bhatti (2012) recalls, the idea is not new: he evokes, for example, the Civilian Clothing 1941 regulation, a series of design patterns and rules elaborated by the British Board of Trade during the Second World War in order to avoid waste in the clothing industry (number of buttons on a shirt, for example). The sociology of consumption and usage has long described and analyzed sober consumption practices guided by economic constraints as well as political protests (Dubuisson-Quellier 2018; Kline 2003; Dobré 2007). Kline has interestingly revealed how rural American consumers had long resisted to many electric appliances offers, favoring traditional rustic solutions addressing many uses with one single device.

Frugal innovation, like other "pro-poor" concepts, mainly aims to provide answers to poverty in developing countries.[1] A challenge over-discussed in the literature is to address the "real needs" of low-income people living in extreme environments. Given the vast literature in innovation studies (particularly the sociology of innovation), about the link between innovation and needs, this proposal questions in many ways. It suggests that needs would pre-exist and that innovation would have to respond to them, invisibilizing the needs' creation process. The "real need answer" challenge also reduces the design process to the designers' steps, as if users were not actors of innovations during uses steps, as if alterations or non-use originated from design mistakes. One may be surprised by these underlying assumptions, which ignores the multiple forms of user presence in the design process from scripts to resistance and diversion through use. Moreover, the literature presents a very reductive conception of the final user.

In this chapter, we investigate how the user, with his skills, is considered in the literature that promotes or criticizes frugal innovation. In line with the sociology of innovation (Oudshoorn and Pinch 2003), we propose to question this new theory of innovation from the way it qualifies its primary target and justification. How are users defined and by whom? Are they conceived as isolated individuals or as a collective? Who speaks for them? How is their integration throughout the design process thought out? Furthermore, to what extent does frugal innovation deployment succeed in impacting people's lives at the base of the pyramid?

This analysis is grounded in a research program on the development of frugal off-grid electrification solutions for deprived rural areas in sub-Saharan Africa. For ten years, we have followed innovative projects lead by MNCs and NGOs through a multi-scaled ethnographic methodology based on, on the one hand, interviews with designers and participation in the design processes and on the other hand, fieldwork in the villages' targets of the innovation experiments. We have also analyzed through fieldwork the growth of a low-cost market of small devices for energy access, made in China and imported by African international traders. This chapter is also based on a state-of-the-art reference article on frugal innovation and other nearby innovations integrating the poor user. Forty references have been systematically analyzed through content analysis.

The chapter is organized as follows. The first section analyses how tackling a poor end-user came central in a frugal innovation approach, considering, on the one hand, the political and organizational context of frugal innovation emergence, and on the other hand, the move in definitions and affiliation along with the use of the concept in the literature. The second section examines the poor user's paradoxical figure, simultaneously destitute and rich. Finally, the last section questions the expected/real impacts of frugal innovation on users' lives and particularly ordinary existing informal economies: "cannibalization" or hard competition.

## HOW (POOR) END-USERS BECAME CENTRAL FOR THE FRUGAL INNOVATION CONCEPT

First coined in 2006 by Renault Chief Executive Carlos Ghosn to describe the competency of Indian engineers in developing low-cost automobiles (Pisoni et al. 2018), the frugal innovation concept could follow a successful trajectory in the same line as similar "blurring"/elastic concepts (resilience, sustainable development or innovation, social innovation (Godin and Gaglio 2019)). In fact, it increasingly shows the same cross-boundary enrolling capacities, gathering different if opposite worlds and the same kind of plasticity in being connected to many other more or less nearby notions. As a boundary object (Star and Griesemer 1989), the frugal innovation concept benefits from an attachment process led by its promoters to different political and organizational contexts, ideas, methodologies, social and thinking movements. As we will see in this section, as large and confusing as the concept appears, the end-user figure becomes predominant as a core convergence point.

### The Political and Organizational Context of Frugal Innovation Emergence: When a Specific South Gets on the Radar for Private Companies

Narrations about emergent contexts of frugal innovation hold a large part in both professional and academic literature on the topic. Among the most evocated ones are the saturation of the western market, the post-2008 crisis austerity age (Bhaduri 2016), and the discovery of under-exploited markets in the global South. They would all require a change in innovation methodologies, based on resource-constrained thinking, to better fit the "needs" of different mass-markets. However, the literature review indirectly suggests other significant changes in international institutions and organizations that could have paved the way for exploring the developing countries' markets by the private sector.

First change: since the 2000s, the new approaches in development policies and fighting against poverty turn to private partners (companies and NGOs) for the design of specific and small-scale (even individual) products and services for developing countries' contexts. It results from a convergence of new principles both in the development and the management fields. On the one hand, the criticism of the Washington Consensus in the 1990s has led to the emergence of new standards based on the centrality of "bottom-up" approaches (against so-called unreliable local States) and the support to civil society and the private sector in fighting against poverty (Stiglitz 2002). Spread within international institutions and western development agencies, the reference works of Amartya Sen popularized the idea that development was a process of human capabilities expansion (Sen 2013) and would arise from the entrepreneurial dynamism of the poor. In this line, aids should consist in first, strengthening the empowerment process "so the poor can become effective market players" (World Bank 2001, cited by Calvès 2009, p. 745), and second supporting a "good" and fair functioning of markets (Sen 2003).

On the other hand, in the same period, scholars in management, mainly from Business School, have joined the arena of research and debates on development issues, suggesting the role of multinational companies in the fight against poverty (Hart and Prahalad 2002). The concept of 'Bottom of the Pyramid' markets, defends a win–win gamble that associates the search for profits and the presumed willingness of poor people in developing countries to consume objects and services and to develop entrepreneurial skills. The delivery of quality and affordable products and services to four billion poor potential consumers would represent some considerable business opportunities for corporations while acting as a driving force for sustaining the economies of the developing countries.

This convergence led to the rise of many programs in development institutions aiming at fostering private initiatives claiming poverty reduction goals. In the electrification domain, one can cite the World Bank affiliates, Lighting Africa, which actively contributes to the design of an off-grid lighting market for developing countries. Lighting Africa promotes the diffusion of many lighting products through different actions such as: lobbying African governments for the tax exemption of solar products, developing expertise in 'BoP market intelligence', acting as a certification and quality assurance agency for products, supplying business development supports and fostering the development of private partnership (MNC with NGO, banks and micro-credit institutions).

Second move: in the same period, organization experts describe a shift in the role of Indian engineering teams in the innovation processes of many international companies. Between 2000 and 2010, the number of MNC R&D centers in India increased from 160 to over 700 (Basant and Mani 2012). Investments would have originated from all over the world and concerned many different sectors (automotive, ICT, pharmaceutical). As indicated by the Indian Senior Vice President of Intuit (a US financial software company) – during a round table published in 2013 (Jha et al. 2013) – an innovation force was already there, in India, when they realized how huge the Indian market could be, with the emergence of the middle class. "These centers are exploring how they can innovate to address the needs of India and India-like markets, which have grown rapidly in the past few years" (Jha et al. 2013, p. 250).

By the way, according to several authors (Bhatti 2012; Bhatti and Ventresca 2012, 2013; Gaglio 2017), frugal innovation can be understood as a challenge opposing western to Indian or Chinese employees, as "doing more with less" could not only talk about the product but also about the innovating process and sound like a guilt-laden message for western engineers. Did not the speech of Ghosn boast Indian engineering ingenuity? More, in many papers (Winterhalter et al. 2017; Neumann et al. 2017) about frugal innovation and its neighbor concepts (Reverse, Grassroots, or Jugaad innovation), the "good understanding" of local needs would be better tackled by local engineering teams. Too sophisticated (Bhatti 2012), western R&D would not have the real capacity to hear the poor customer voice. According to Agnihotri (2015), a "frugal mindset" is required to develop such innovations and to better tackle the culture of the country, MNCs targeting emerging markets should be established in such "low-cost countries."

We had the opportunity to observe such competition while accompanying the French R&D center of an international electric company, in their search for an electrification solution for off-grid rural Africa. Experts in large and middle systems (as micro-grids), the French team developed a small portable device for battery-controlled power, based on ethnographic fieldwork. But the project stopped at the stage of the prototype when the head of the Indian R&D claimed small BoP devices were "their" business. In a way, considering organizational changes and tensions within R&D structuration bring a more nuanced picture of the supposed offensive strategy from MNCs headquarters.

As Bhatti (2012) suggests, frugal innovation also arises from emergent countries' ambitions to engage in the supply and the development of their unserved markets. In this line, an attentive reader of recent papers on the frugal and grassroot approach will note the active engagement of Indian institutions in supporting innovation initiated by local actors (private companies, universities, engineering centers, but also non-expert innovators). As Sharma and Kumar (2019) describe, from the 2000s, the provincial and national Indian states launched programs, funds and institutions to detect grassroot innovators to support them in any innovative issues they encounter, including patenting.

Let us cite the National Innovation Foundation, set up in 2000 by the Department of Science and Technology of the Government of India, whose mission is "to help India become a creative and knowledge-based society by expanding policy and institutional space for grassroots technological innovators" (http://nif.org.in/aboutnif). The NIF was created from the Honey Bee Network, an informal movement, led by Professor Anil Gupta, of the Indian Institute of Management, to scout and diffuse the alternative technologies and traditional knowledge practices of the people "under the radar." With the help of 150 employees and partnership with "various research & development (R&D) and academic institutions, agricultural & veterinary universities and others institutions" (http://nif.org.in/aboutnif), the NIF carries out different activities such as managing a vast database of technological innovations and traditional knowledge practices (more than 300 000) and grassroots innovators (almost 1000), intellectual property management training, business development support and social diffusion of innovation. The NIF is well-known for the organization of events (Annual Innovation Festivals) and for delivering awards to the best grassroot innovator of the year. This program's key objective is to help the innovators enrich their research ideas through interactions with experts and other innovators coming from different parts of the country. Several well-known Indian researchers of the frugal and grassroot innovation field participate in this program.

This investment of the Indian authorities and institutions, for frugal and grassroot innovations, probably explains the high place this country leads in the academic field as well as its reputation as an inspiring example for other countries. Thus, the quantitative study of academic publications on frugal innovation since 1990, led by Pisoni et al. (2018), shows the dominance of Indian researchers (or British researchers who are, for part of them, Indian people working in an English University and with the two nationalities). In the sample, India ranks first as a case studies provider (35%). Would India become the model for frugal engineering? This was suggested by a recent report of the European Commission (Kroll et al. 2016) analyzing the "Foundations, Trends and Relevant Potentials in the Field of Frugal Innovation (for Europe)": "India's efforts to develop frugal innovation have been highlighted as an area of potential learning for Europe" (Pisoni et al. 2018, p.108). Besides, the NIF has recently developed Indo-African cooperation for grassroots innovations with the aims of diffusing Indian innovative ideas and technologies, organizing cultural exchanges, and replicating the Honey Bee model to support African entrepreneurship.

The reality of the Indian public investment in innovation has been discussed (Prathap 2014), so it would be overstated to postulate the birth of a frugal innovation market in India, supported by public initiatives and attractive for international business. However, it is undeniable that the change in R&D organizations and initiatives in development policies, all taken together, contributed to getting "underserved" territories of the South ahead of the radar of market actors and to depicting the concept of frugal innovation in relation to South poverty alleviation.

### The Frugal Concept: From Resource-Constrained Thinking to Communities' Empowerment Goals

Most of the scientists, writing on frugal innovation, converge about the first popularization of the concept in the famous special issue of *The Economist* in 2010. In the different articles, frugal innovation is described as "not just a matter of exploiting cheap labor (through cheap labor helps), it is a matter of redesigning products and processes to cut out unnecessary costs" (*The Economist* April 17, 2010). According to Pisoni et al. (2018), this set of articles established the first generation of publications that associates frugal innovation to the properties of the product or the business model: minimizing the use of material and financial resources, low price, compact design, portable solutions, reuse of existing components, devices or methods, and cutting-edge but uncluttered technology. Far away from any low quality and low-cost solutions from international mass markets, products and services issued from the frugal innovation approach are expected in reaching four criteria: Accessibility, Affordability, Availability and Awareness (Anderson and Markidès 2007), through cost reduction, feature simplification and low resource consumption (Radjou et al. 2012; Sehgal et al. 2010; Zeschky et al. 2011).

However, the definition of the concept quickly moved to integrate moral considerations, as the design process came into the debate. Inspired by Nash (2000),[2] Bhatti (2012) describes frugality as "an ethically conscious choice (…) an intentionally responsive to social and eco-logical conditions of excessive and unfair consumption and production" (p. 14). Inclusivity, co-creation and communities' empowerment appear in many publications. In several research, frugal innovation is discussed regarding a critical reference to the BoP approach, in its first version. In effect, a few years after the first articles of Hart and Prahalad, the BoP paradigms received many critics from the academic arena, mainly about the efficiency of the market as a means of alleviating poverty (Walsh et al. 2005), the over-estimation of the purchas-ing power of the poor (Crabtree 2007; Landrum 2007) and the moral condemnation of the "value-conscious" consumerism philosophy (Karnani 2007) underlying the strategy. First pre-sented as success stories, the experience of an American global company that had developed a broad distribution of some daily sachets of detergent in India led to a significant change in the BoP approaches (Simanis et al. 2008). Academics and practitioners called for a BoP 2.0 stage, with a process based on the involvement of the local producers, and the wish to con-tribute to the development of a local value chain, as well as the introduction of environmental criteria, which reinforced the concept but also made its implementation more constraining.

In the literature about the frugal innovation concept, indispensable attention to local context is highlighted; the BoP target is more frequently designated, and connections to approaches such as Jugaad and grassroots innovations or innovations "below the radar" or again "pro-poor innovations," reinforce a more and more community-oriented definition. In many papers, these concepts appear as almost synonyms, the assimilation creating some confusion. Sharma and Kumar (2019) associate all these terms as grassroots oriented perspectives; Bhaduri (2016) defines frugal innovation within the Juggad movement, and Fressoli et al. (2014) remove the slim boundary hardly established by some authors who related frugal innovation to MNCs and grassroots innovation to communities' initiatives.

It is not a surprise to read in this set of papers, two main "thinking" roots: the Appropriate Technology Movement (ATM) and the Tagore and Gandhi philosophies. Bhaduri (2016), as an example, inscribes the frugal innovation approach in the "spirit" of the work of Stewart (1987)[3] and in the legacy of Schumacher (1973) *Small is beautiful*. He recalls the idea that people in developing countries couldn't remain "inactive recipients of technologies transferred from the Global North" (p. 5). In this work, the researcher calls for a polycentric frugal innova-tion based on a "co-creation" of products all along the R&D process. An influential reference, the ATM is, however, criticized as it never got out of NGO or academic experimentations and got diffused to the private sector. For Sharma and Kumar (2019), the ATM failed as it never reached "to engage the grassroots ingenuity and local communities in the technology devel-opment initiatives" (p. 42). The predominance of the local states and of donor agencies in the different experimentations would have favored mainstream technologies to the detriment of local innovators and poor people's "real needs."

Gandhi and Tagore are also mentioned as inspiring sources for frugal innovation as a process. Both historical figures of the Indian independence would have been the pioneers in building a link between technical developments and pride recovering at the bottom of the society. Only local community-centric approaches promoted by Gandhi and Tagore could reach the basic needs in a distributive and equity lens. Opposite to a solution being imposed from outside, the concept of "swadeshi" refers to

> local self-reliance by incorporating local knowledge and abilities in everyday life. (…) it symbolizes a process of empowerment through self-help. (…) Tagore and Gandhi believed that India could

only regenerate itself, to face the challenges ahead, by seeking out those beliefs, values, knowledge systems, and technologies, which had organically grown from her local communities. (Bhaduri and Kumar 2011, pp. 30 and 31)

The political dimension of such a reference in the Indian current nationalist context sounds remarkable. But beyond this observation that could interestingly be deeply explored, these references provide a political and moral background for a definition of frugal innovation as a process of community inclusion.

For Winterhalter et al. (2017), the incorporation of indigenous knowledge along with value creation is a crucial factor for the success of any frugal innovation projects. Knorringa et al. (2016) put forward two ways of integrating local end-users: directly throughout the design process or through targeting a significant improvement of community living conditions. In the same line, Bortagaray and Ordóñez-Matamoros (2012) identify two complementary paths for inclusive concern: as mentioned before, building the design process as inclusive and/or reaching inclusiveness as a result of the innovation project. The most confident promoters hope for a contribution of frugal innovation to producing a structural change in the forms of participation in the shaping and priority-setting of policies in the technological area (Fressoli et al. 2014). Knorringa et al. (2016) expect frugal innovation "to disrupt existing capital-intensive and top-down forms of innovation, contributing to more inclusive forms of development" (p. 143). Gradually, despite their slight differences, the many articles dedicated to inclusive, frugal, pro-poor, Jugaad and so on innovations converge to the need for incorporating local practices and territorial practices (Gaglio 2017).

This social inclusion perspective redefines the final objective of frugal innovation as critics about these kinds of approaches increase in the academic literature (Dolan and Roll 2013; Elyachar 2012; Meagher 2018). Income generation and capacity building are at the frugal innovation agenda, according to Bhatti (2012). Frugal innovation should not only provide products and services for basic needs but also integrate business models to foster entrepreneurship and allow people to pay for these new offers. "A strong value proposition not only through cost reduction and consequently lower per-unit-prices for the customer but by offering solutions that increase the customers' willingness to pay for them" (Winterhalter et al. 2017, p. 3). Enhance well-being for disenfranchised members of the society through new ideas (Knorringa et al. 2016), initiate a social learning process (Berdegué 2005, p. 15), or foster prosperity (Winterhalter et al. 2017); frugal innovation appears as a "new means" to entailing sustainable economic growth through multi-stakeholder involvement.

The websites of well-known companies of different countries and sizes, involved in the development of PICO solutions for emerging countries give a good overview of how the articulation of innovation can be enacted, through environmental protection and development goals. The d.light homepage opens on the picture of a young girl holding a solar lamp and a text about the vision of the company: "At d.light solar, we envision a brighter future where all people are empowered to enjoy the quality of life that comes with access to reliable and affordable clean energy products" (https://www.dlight.com/). The tabs propose a discovering of their solar products, social impacts ("Empowering communities for a brighter future" and partners (a list mixing big international companies such as Total but also big NGOs such as OXFAM). As Green Planet (https://www.greenlightplanet.com/) and other frugal energy suppliers offer, the company highlights different figures attesting the social benefits of their solutions: the number of reached households and daily users, the numbers of "lives empowered" and the tons of $CO_2$ offset. The French firm Lagazel is probably the more committed in this social trajectory, as it proposes, on its website, to contribute to many social projects in partnership with NGOs to bring light in Africa (see for example the NAFA NAANA program: https://www.lagazel.com/partenariats-solidaires/).

As the boundaries between the different innovation concepts targeting emerging countries are confusing, as the vocabulary of development spread in academic papers as in firms' communication, normativeness about the moral goals of such approaches disseminates. Translated in figures of success (equipped households, $CO_2$ offset), the frugal innovation approach appears as a concrete path for solving insoluble problems, a promise of good efficiency. In line with the X-Innovation movement described by Gaglio et al. (2019) the innovation concept coupled here, with terms associated with the idea of social progress ("inclusive," "frugal," "sustainable") carries an attractive vision of an ideal project, avoiding the trap of the past in the development field (combining market, technology and social goals) and giving an ethic color, a quasi-selfless motivation to the firms' strategy. Frugal innovation and its relatives engage in the positive and moral coloring of innovation through a romanticization of actors' concerns and integration.

## BOTH POOR AND RICH: THE CREATIVE END-USER, TARGET OF FRUGAL INNOVATION

The discourse on frugal innovation is also a discourse about its main target, the poor of emerging countries. This section proposes an overview of the way the "underserved" are perceived both in the academic literature and through the products designed. Many paradoxes or contradictions reveal some ambiguities in the frugal innovation project. A first paradox for scholars in innovation studies lies in the so-called "real needs" the frugal innovation process has to unfold. For Winterhalter et al. (2017), for example, firms have to dedicate means "to focus exclusively on understanding the true customer and market needs to craft proper value propositions" (p. 7). Addressing the real needs is a leitmotiv in the literature on frugal innovation as a stance at the opposite of the western and classic innovation, which is described as top-down (Bhatti 2012) and unable to understand the underserved. Enunciated by experts in innovation, notably some of them being social scientists, such a discourse appears surprising after a rich social history of techniques showing the long collective construction of artifacts (Bijker et al. 1987) and the emergence of needs as a process derived from innovation dynamics. A second paradox, explored in depth in the following pages, relates to the figure of the underserved himself, simultaneously depicted as deprived of anything and yet richer than many R&D organizations.

### A "Naked" (or Deprived) End-User

As we saw before, frugal innovation is associated with the context of "extreme constraints," which is characterized by a profound lack of resources. The notion of "institutional void" is often mentioned as a key concept to describe the absence of reliable public services, the deficiency of infrastructure, the penuriousness of basic facilities, or the weakness of retail and communication chains. Communication, financial services, health care, transportation, energy, water, housing, sanitation and the poor in developing countries are described as wholly deprived. Winterhalter et al. (2017) cite this interesting interview extract where a pharmaceutical company's project leader explains how they tried imagining the Indian rural environment: "We thought: If you have nothing – and I mean literally nothing – what do you need to conduct the analysis? So, we started with a zero-environment in mind: no electricity, no water, and no educated staff (Project Head at Labtech1)" (p. 7).

This poor living condition does not only characterize the environment but also the individual resources held by the end-user or the grassroots innovator. As Meagher (2018) emphasizes, the dynamics of accumulation within informal economies remain outside the frugal innovation promoters' scope. "There is little awareness within the frugal innovation literature that informal economic actors, particularly in Africa, are not necessarily small-scale and not necessarily poor" (Meagher 2018, p. 18). Poverty can also concern abilities. The lack of skills, especially qualified ones, is frequently mentioned, especially regarding maintenance issues. In the line of Hirschman (1958), infrastructure is still considered as fragile in developing countries as long as "maintenance culture" appears underdeveloped. For Bhaduri (2016), the frugal innovation approach would still fail in examining the dynamic learning at the bottom of the pyramid and "the processes through which this body of knowledge sustains itself and grows" (p. 4).

In the energy field, these considerations are widespread and confronted with dissonant local realities. For example, the design team we accompanied was amazed and hardly integrated the fact that Malagasy rural people already had access to electricity even if it was through do-it-yourself solutions. Then it was hard to consider the main demand from potential customers was energy for watching television rather than for educational or health purposes. Taking into account "real needs" could mean addressing less "noble" ambitions in the eyes of corporate social responsibility. Would supplying light to a dark continent (as Africa is often represented in energy access documentation) supposedly deny the fact that lighting is already there? As a consequence, most electrification projects still primarily target community buildings, firms' communication stages, children reading a book, and for most of the products, local maintenance is impossible. In effect, many spare parts are not locally available; the existing skills are not taken into account, so they remain unuseful for the implemented systems. Finally, the design of the lighting products favors a closed technical system where all components are rigidly fixed and associated, preventing people from autonomous forms of do-it-yourself, reparation, remanufacturing, and so on. As the photo-voltaic solution described by Akrich (1987), the proposed products appear unable to integrate the possible diversions and enrichments a more flexible solution as the diesel Genset could allow.

As a symmetry to resource scarcity, the social environment at the bottom of the pyramid, as depicted in the literature promoting frugal innovation, is deprived of any form of complexity. First, regarding many case studies, the design process appears as recurrently focused on two figures quite homogeneous: the individual or the household on the one hand and the community as a whole on the other hand. Grassroots innovation literature is the richer one; recognizing the large family supports local innovators who can benefit from developing their creativity. But even in this approach, social ties are always conceived as positive and empowering; the existence of distinct if not conflicting interests is usually ignored. The community "at the bottom of the pyramid" appears as romanticized without inequalities and power relations (Meagher 2018).

However, a few case studies, not among the most popular, associate failures in frugal innovation projects with ignorance of social complexity. (Fressoli et al. 2014) describe how a water supply project in Brazil, initially based on simple build cement containers, generated a protestation movement: a design change (move to a plastic cistern) initiated by the government, allowed the local political elites to regain power over controlling water, by controlling the distribution and marketing of water cisterns. In the same line, Thomas et al. (2017) describe the solar oven project upheavals in Argentina. Instigated by researchers of the Regional School of Mendoza and based on a long co-design process through broad exchanges with Indigenous households, the project met the opposition of communities' leaders, left out of the innovation process:

> The initial resistance posed by Huarpe leaders against the project shows that researchers had made a narrow interpretation of the complexity of the social problems of the region; the problems of water shortage and advancing desertification could not be addressed separately from conflicts associated with land tenure and ethnic identity. (Thomas et al. 2017, p.190)

This simplified conception of social ties at the bottom of the pyramid is strengthened by an invisibilization of many intermediaries involved in informal economies. For Meagher (2018), frugal innovation approaches the poor through a selection process that would differentiate appropriate from unreliable partners. "The result is that inclusive engagement is highly selective – meaning that some informal actors and innovative ideas are included, while others are excluded" (p. 22). The wood pellets cookstove case analyzed by Iva Peša (2017) gives an excellent example of an over-focusing approach on households' "real needs" that misses the complex socio-economic dynamic of charcoal consumption. The frugal innovative cookstove, developed in co-partnership with "local end-users" and sold by a local company remained backyard as people were still buying charcoal from their neighbors and kinship. A deep analysis shows the charcoal economy is about solidarity. As many people also mentioned during our fieldwork in Madagascar, "charcoal is easy money," anybody can sell something, and anybody can need quick money. People bought the new cookstove, they even tried running it with charcoal, but as it didn't work and as they were persisting in buying charcoal, they returned to their cooking habits.

### Deprived, but Creative: The Bottom of the Pyramid, Realm of Bricolage and Reuse

As we saw before, frugal innovation is often assimilated with other bottom-up trends as grassroots or Jugaad innovation. For Bhaduri (2016), these concepts open new possibilities for integrating the creative improvisations of individuals operating within informal economies. In many papers, researchers remain focused on how people in extreme environments are innovative (Seyfang and Haxeltine 2012; Bhaduri 2016). They share the idea that because of their conditions of deprivation, ordinary people at the bottom of the pyramid have developed "natural," "intuitive" innovating capacities, aligned with frugal innovation goals.

First, they would have long been accustomed to creating solutions with scarce resources, if not pre-existing materials. Reuse, repair, assembling existing objects, or local raw materials are qualified as sustainable practices characterized by a reduced impact on the natural environment. In a way, local innovators would naturally be frugal.

Process in resource-constrained environments is thought to be a potential source of disruptive eco-innovations. These are products and services that are more energy-efficient, use less raw materials, and have a reduced impact on the natural environment (…) Moreover, as already documented in previous studies, the cases reveal that the acute scarcity of resources did not prevent people from being innovative, but rather promoted scarcity-driven innovation based on inventing or attributing new functions to everyday objects (Bhaduri 2016, pp. 14–15).

Second, combining traditional knowledge and improvisation capacities, grassroots innovators are experts in bricolage (de Certeau 1984), which is also drawn as a frugal quality. The extensive literature on bricolage is called to support a description of technical skills enacted in the process of "making do with what is at hand" (Pansera and Sarkar 2016, p. 3): observation of daily life, experimentations through many assemblages. So much aligned with frugal thinking, "bricolage thinking" should be developed in formal team designs involved in such initiatives. "The Poor as Eco-Innovators" defends Pansera and Sarkar.

Another quality of grassroots bricolage would lie in its spontaneous inclusiveness. For Bhatti, this kind of knowledge requires the help of friends and families; thus, involved in the innovation process: the wife of the grassroots innovator and her partners offer situations to improve and testing grounds; relatives supply funding; social networks assume distribution and so on. The description of many grassroots innovators' stories in the academic literature emphasizes how they would be aware of the very practical problems of daily life (Sharma and Kumar 2019). For (Pansera and Sarkar 2016), citing Smith et al. (2014, p. 114),[4] "Grassroots innovators 'seek innovation processes that are socially inclusive towards local communities in terms of the knowledge, processes and outcomes involved'. These are actors with respect to their milieu and their community's specific needs and contexts, which can be hard to grasp by those on the outside" (p. 2). The weak interest, grassroots innovators would show for the business impacts of their innovations (Bhaduri and Kumar 2011; Kumar 2014; Sharma and Kumar 2019) reinforce a picture of selflessness: these people being mainly engaged in finding solutions for their community. As a result, grassroots innovators become endowed with higher developmental abilities. They would have "a tremendous impact not only in terms of serving unmet and ignored consumer needs but also longer-term impacts through enhanced productivity, sustainability, poverty reduction and promoting entrepreneurship" (Pansera and Sarkar 2016, p. 2). Hence, to consider them – and by extension engineers born in developing countries – as better placed to meet the challenges of frugal innovation, it's only a step away. Winterhalter et al. (2017) and Neumann et al. (2017) demonstrate through frugal innovation case study analysis that local innovators and entrepreneurs better address penurious environment requirements and better sense the surrounding context.

Theoretically boasted, these qualities of the targeted ordinary people can constitute a real ordeal in practice. After a few months, the first micro-grid, settled in a small village in Madagascar by the MNC we were accompanying, was completely dismantled, solar panels and batteries sold through informal economic circuits. In another project, an NGO observed, disappointed, the high-quality European batteries had been replaced by low-cost Chinese batteries. Some people mentioned the connection of solar panels with second-hand car batteries, which generates deterioration. In an electrified village, NGO project leaders complained about the collapse of the freezers bought to foster service entrepreneurship: people had discovered they could reduce their energy bill by regularly switching it off, producing more ice by maintaining opened doors. Attractive on paper, bricolage skills are harnessed in the reappropriation process of technologies; they redefine innovations' career, taking them away from designers' ambitions. As Knorringa et al. (2016) note, bottom-up innovations have not been proved to have more potential for generating equitable economic growth than top-down innovations. Ignorance of both the social complexity of surrounding contexts (Meagher 2018) and the dynamics driving innovation adoption (Akrich et al. 2002) leads to a romanticization of the informal economy's potential and critical pitfalls in the field.

## The MNCs Meet the Poor: Cannibalization, Limited Encounter, or Competition?

Like the BoP approach, frugal innovation and similar trends have given rise to many controversies, many of them emanating from the development field. They share taking the issue of the integration of local knowledge and actors seriously, examining it on the ground and all along the design process until the announced change in living conditions.

### Unfair Partnership: MNCs Under Suspicion of Cannibalization

Alongside papers resolutely optimistic about the win–win collaboration that frugal innovation promises, some researchers highlight the unequal power relations between multinationals and poor consumers (Bhaduri 2016; Knorringa et al. 2016). They call for scrutinizing seriously "what does frugal innovation mean," and how the power asymmetry could have negative impacts on informal economies. Referring to earlier controversial approaches as micro-credit, or entrepreneurship empowerment (Elyachar 2012; Dolan and Roll 2013), they remind that market-based solutions for development have sometimes been synonymous with debt and impoverishment. This issue of power inequalities can constitute a dividing line between frugal and grassroots innovations. For example, Sharma and Kumar (2019), promoters of bottom-up innovations, associate frugal innovation with big companies that try through their frugally designed products and services to expand and enhance their market to reach the "unserved" third. In contrast, bottom-up innovations would better target development goals. Questioning "Frugal Innovation and Development: Aides or Adversaries?", Knorringa and his colleagues from the Frugal Innovation Center for Africa challenge all these initiatives to engage with the poor as agents rather than passive consumers (see also Bhaduri 2016).

In her critical paper, recently published in *The European Journal of Development Research*, Meagher (2018) precisely describes four ways for the formal economy to take advantage of the informal economy through so-called pro-poor innovation: copying, free-riding, short-circuiting and shifting risks. Without any real counterpart, big companies would engage in a capture process, allowing overcoming institutional voids: knowledge capture (about native capability, local business practices and consumer behavior) but also capture of the social infrastructure that supports the ordinary life of low-income people (Dolan and Rajak 2016; Dolan and Roll 2013; Elyachar 2012). Another risk for informal economies lies in the potential destabilization of existing trading networks and large informal firms that supply at regional and local scales with low-cost and low-quality products. The distribution of the frugal products would be based on the selection of the "reliable" and unreliable partners to integrate along the supply chain. These business models configuring could be described as a new technique of governance aiming at reformatting "informal economic systems and value chains in ways that privilege the reduction of formal sector costs and while increasing formal sector control" (Meagher 2018, p. 22).

This bleak picture is based on the analysis of empowerment projects led by well-known international companies (Avon, Coca-Cola) or organizations (World Bank). However, ethnography can provide more nuanced observation regarding this capture process. In a study of a project of a social company selling traditional Indian sarees in order to supply a middle-class urban market, Chakrabarti et al. (2018) describe how the initial business model, based on craftswomen empowerment and the eviction of traditional intermediaries was challenged facing the complexity and depth of the relationships within the social networks. Traditional intermediaries were not only women's wholesalers but also their usurers. They also were the only actors able to gather enough sarees to satisfy demands. After a few months, they were re-integrated within the supply chain. We have also depicted the complex equations a world-wide electric company could confront between moral obligations and business opportunities all along the design and development process (Cholez et al. 2012).

The oil company Total also offers an interesting example of the difficulties the formal organization can meet to exploit informal networks. In 2011, the company launched its Awango program, selling solar lamps or small kits to bring 'clean' lighting to poor people in developing countries. After first experimenting in four countries (Cameroon, Kenya, Indonesia and the Republic of Congo), where 125 000 lamps and solar kits were sold through the gas-station network, the sales target of one million lamps was set. Nevertheless, Total met severe difficulties in crossing the last miles. The gas-station network rapidly showed itself to be unsatisfactory because it hardly reached poor rural households. In a second stage, the company built a partnership with a local producer of tobacco, whereby employees could buy lamps at discounted rates and resell in their home villages. It also targeted public actors such as schools and NGOs organizing resale systems for women in villages. As confirmed by the head of the Senegalese program, Total's commercialization network cannot include informal businesses, which are dominant in rural areas for juridical reasons.

### Are Frugal Innovations Invading Developing Countries and Low-Income Everyday Life?

Although poor end-users have been included at the heart of the frugal innovation project, it is challenging to identify the socio-economic impacts of the projects. In the field of energy, for example, a review (Cholez et al. 2017) of 245 articles and reports on off-grid electrification projects (public and private, market and non-market) showed the limitations of assessments (see also Cook 2011): focusing on the number of households connected, not very rigorous methodologies, often established anecdotally.

Likewise, whereas the academic and professional literature on "pro-poor" innovations has been growing since the 2010s, the number of case studies, analyzed (and re-analyzed) across different publications, has been limited. In many papers, the same well-known initiatives are described (The General Electric MAC400, the Mitticool fridge, the Aravind mobile hospitals). Grassroots innovation studies use the NIF database that gathers a lot of quantitative and qualitative data on local innovations. Few studies offer real and in-depth ethnography providing an analysis of both the design stages and the dynamics of dissemination and appropriation. A quick overview of the literature suggests the predominance of Indian cases (Pisoni et al. 2018), Anglo-Saxon countries and a sort of matrix associating large geographical areas and industrial sectors. Thus, innovations in the field of health and agriculture seem to dominate in India, while energy and new information and communication technologies would be privileged in sub-Saharan Africa. To what extent does this distribution not overlap with another map, that of national and international institutional support? We need more studies to tackle this.

In the same line, a study of the competitors engaged in the frugal innovation market remains to be done. In the field of off-grid electrification, MNCs' projects targeting low-income end-users remain supported by their CSR department. In developing countries, both large-scale international companies and startups co-exist and compete to obtain institutional support, NGOs and access to experimental sites. Some local politicians have understood this, such as the mayor of a village in Casamance, who has attracted several electrification projects from different types of actors (an NGO and a local company). In Madagascar, a result is a concentration of projects on a few territories where development actors have long committed, like the very deprived Androy Region. Confronted with very poor households, electrification projects never reach profitability and require continuous institutional and financial support. The experience of the small French company Lagazel reflects how a market for frugal innovation requires strong partnerships with NGOs and development actors (World Bank, micro-financing institutions, development agencies). A long-time associate of the NGO Entrepreneur du Monde, in charge of testing, promoting and distributing the lamps to households, the founder and director of the company devotes a large part of his activity to making its solutions known within the development community through participation in forums and other events bringing together the major players in this sphere. Kogybox, another small French company, only really sells its products through development projects supported by local or international institutions.

Paying close attention to these entrepreneurs as well as to the actors involved in the CSR projects of large companies would also shed light on the proximity of their backgrounds. The founders of energy startups often have in common an ethical or moral trajectory of young engineers who want to break with traditional production/consumption models. Like reverse engineering, frugal innovation, with its developmental and environmental goals, could be in line with a search for a new meaning to technological development.

Finally, in order to identify the impact on end-users' ordinary lives of currents such as frugal innovation, it is essential to consider situations, far from being limited, of resistance or strong competition emanating from targeted social systems and actors in the informal economy. As Fressoli et al. (2014) point out, the encounter between local actors' systems and promoters of pro-poor innovations can produce reciprocal integration but also mobilization regimes leading to direct and indirect contestation. The informal economy can threaten the formal economy in many ways. Harris cites the case of artisans in Nairobi, who destabilize the informal economy through their imitation skills. In the field of energy in sub-Saharan Africa, we have described (Cholez and Trompette 2016) how the multiple initiatives of formal firms and NGOs have to compete with very tinkered with and cheap local solutions and, more recently, with low-cost Chinese solar products imported by international networks of African traders. The latter manages to get Chinese copies of the products designed by the established companies and to distribute them to the most remote areas through their commercial networks. In Senegal, for example, Total has had to align its prices in gas stations with those of the semi-informal market, where a wide variety of products, including solar lamps with USB plugs and even radio and television sets, are available and affordable.

## CONCLUSION

Frugal innovation, like other "pro-poor" concepts, aims to provide answers to poverty in developing countries. This concept is part of a constellation of more or less similar ones (BoP, grassroots innovation, inclusive innovation), all of which place the end-user at the heart of a moral and enchanting innovation. In the literature, the line of the debate runs around the place given to local populations: a simple user, a victim of MNCs, a co-developer, a true innovator. A political debate is taking place in different ways around the question of who is best placed to provide the best and most sustainable answers: MNCs or locals? Who can best address the "real needs" of the poor?

In this chapter, we have tried drawing how this figure of the poor – the local user or inventor – is defined by the promoters of frugal innovation. We have also questioned the meeting between these initiatives and the social systems they wish to improve through their products. Contrary to the angelic discourse found in managerial literature, this encounter can be a source of destabilization and conflict for each of the actors. A large part of the academics (a few of them remain prudent), the institutions and the companies adopt an optimistic and enthusiastic position regarding the concept of frugal innovation. We suggest an alternative approach to dissecting this new proposal for innovation, which is a new contribution to the X-innovation phenomenon (Gaglio et al. 2019): tracing its history, its contexts of emergence, the controversies and questions it raises, the hybridization (of words and worlds) it carries. We suggest following the appropriation and operationalization attempts of actors, the trials and dilemmas they encounter. The theorization of frugal innovation rolled out on an international scale, is a process that requires many more investigations.

A vast field remains open for STS as so many questions remain open: there is a real interest in multiplying empirical analyses of frugal innovation projects, in following the designers (who are they?) in their approaches, in unpacking the usage scripts and the way the design process constructs the partners of the project. Likewise, we must seriously consider the concrete implementation of these projects by dwelling on trial and error, the difficulties encountered and the resistances raised (non-use, circumvention, competition). Following these projects from one end of the process to the other can finally help us identify the real targets of frugal innovation: the poor in developing or emerging countries? Development institutions and their broad development programs? Emerging States looking for quick solutions to accelerate their growth? In her seminal study of a photovoltaic project in Senegal, Madeleine Akrich (1987) already questioned the real recipients of such innovation, ordered by development agencies. Analyzing the 'design and uses' scripts of the solar products experimented, she finally asks: is the photovoltaic system designed for the Senegalese end-users, the state funding agencies, or the solar industry? Her perspective remains incredibly relevant and inspiring when we try to better understand new currents in innovation.

## NOTES

1. Frugal innovation promoters present it as a new turn in the innovation methodologies, addressing any markets lead by sobriety: the poor and the middle-class in developing and emergent markets, but also precarious people as well as activists in favor of sober consumption in western countries. Nevertheless, this intention remains theoretical. Bhatti et al. (2017) and Hossain et al. (2016), observe that the diffusion of frugal innovation is "a blind spot of academic research," few examples always cited being identified and for those having a look at it, a lot of barriers appear: skepticism of western adopters, institutional constraints, organizational curbs in MNCs and so on.
2. Nash, James A. (2000), 'Towards the revival and reform of the subversive virtue: frugality', in A.R. Chapman, R.L. Petersen and B. Smith-Moran (eds), *Consumption, Population and Sustainability: Perspectives from Science and Religion*, Washington, DC: Island Press, pp. 167–190.
3. Stewart, Frances (1987), *Macro-policies for Appropriate Technology in Developing Countries*, Boulder, CO: Westview Press.

4.  See also Smith, Adrian (2005), 'The Alternative Technology Movement: An Analysis of its Framing and Negotiation of Technology', *Research in Human Ecology*, 12, 106–119.

# REFERENCES

Agnihotri, Arpita (2015), 'Low-cost innovation in emerging markets', *Journal of Strategic Marketing*, 23 (5), 399–411.

Akrich, Madeleine (1987), 'Comment décrire les objets techniques?', *Techniques & Culture*, (9), 49–64.

Akrich, Madeleine, Michel Callon, and Bruno Latour (2002), 'The key to success in innovation part II: The art of choosing good spokespersons', *International Journal of Innovation Management*, 6 (2), 207–25.

Anderson, Jamie and Costas Markidès (2007), 'Strategic innovation at the base of the pyramid', *MIT Sloan Management Review*, 49 (1), 83.

Basant, Rakesh and Sunil Mani (2012), 'Foreign R&D centres in India: An analysis of their size, structure and implications', W.P. No. 2012-01-06, Indian Institute of Management, Ahmedabad.

Berdegué, Julio A. (2005), 'Pro-poor innovation systems', Background Paper, IFAD, Rome, pp. 1–42.

Bhaduri, Saradindu (2016), *Frugal Innovation by 'the Small and the Marginal': An Alternative Discourse on Innovation and Development*, Erasmus University Rotterdam.

Bhaduri, Saradindu and Hemant Kumar (2011), 'Extrinsic and intrinsic motivations to innovate: Tracing the motivation of "grassroot" innovators in India', *Mind & Society*, 10 (1), 27–55.

Bhatti, Yasser A. (2012), 'What is frugal, what is innovation? Towards a theory of frugal innovation', Oxford Centre for Entrepreneurship and Innovation (Unpublished).

Bhatti, Yasser A., Matthew Prime, Matthew Harris, Hester Wadge, Julie McQueen, Hannah Patel, Alexander Carter, Gregory Parston and Ara Darzil (2017), 'The search for the holy grail: Frugal innovation in healthcare from low-income or middle-income countries for reverse innovation to developed countries', *BMJ Innovations*, 4 (3), https://dx.doi.org/10.1136/bmjinnov-2016-000186 or https://spiral.imperial.ac.uk/handle/10044/1/50690.

Bhatti, Yasser A. and Marc Ventresca (2012), 'The emerging market for frugal innovation: Fad, fashion, or fit?', *SSRN Electronic Journal*, Available at SSRN 2005983.

Bhatti, Yasser A. and Marc Ventresca (2013), 'How can frugal innovation be conceptualized?', *SSRN Electronic Journal*, Available at SSRN 2203552.

Bijker, Wiebe E., Thomas P. Hughes, and Trevor J. Pinch (1987), *The Social Construction of Technological Systems: New Directions in the Sociology and History of Technology*, Cambridge, MA: MIT Press.

Bortagaray, Isabel and Gonzalo Ordóñez-Matamoros (2012), 'Introduction to the special issue of the review of policy research: Innovation, innovation policy, and social inclusion in developing countries', *Review of Policy Research*, 29 (6), 669–71.

Calvès, Anne-Emmanuèle (2009), 'Empowerment: généalogie d'un concept clé du discours contemporain sur le développement', *Revue Tiers Monde*, (4), 735–49.

Chakrabarti, Ronika, Céline Cholez, Winfred Onyas, Jérôme Queste and Pascale Trompette (2018), 'Changing borders: Constructing boundaries and agencing formality and informality in global networks of trade', 5th Interdisciplinary Market Studies Workshop, June 6–8, 2018, Copenhagen Business School.

Cholez, Céline and Pascale Trompette (2016) 'Africa's mundane solar market: Trading routes, globalization and competition', 4th Interdisciplinary Market Studies Workshop, June 8–10, 2016, University of St Andrews, Edinburgh.

Cholez, Céline, Pascale Trompette and Rhosnie Francius (2017), 'Analyse des activités économiques dans le Sud de Madagascar et en Basse-Casamance au Sénégal. Chaînes logistiques, formation des prix et potentiels d'électrification'. Rapport final dans le cadre du programme PAMELA (FONDEM – AFD): PArtenariat Multi-acteurs pour un accès durable à l'ÉLectricité des Activités économiques: approche croisée à Madagascar et au Sénégal.

Cholez, Céline, Pascale Trompette, Dominique Vinck, Thomas Reverdy (2012), 'Bridging access to electricity through BOP market: Between economic equations and political configurations', *Review of Policy Research*, 29 (6), 713–32.

Cook, Paul (2011), 'Infrastructure, rural electrification and development', *Energy for Sustainable Development* (Special issue on off-grid electrification in developing countries), 15 (3), 304–13.

Crabtree, Andrew (2007), 'Evaluating the "bottom of the pyramid" from a fundamental capabilities perspective', CBDS Working Paper Series, April.

de Certeau, Michel (1984), *The Practice of Everyday Life*, Berkeley. CA: University of California Press.

Dobré, Michelle (2007), 'Consumption: A field for resistance and moral containment', in Edwin Zaccaï (ed.), *Sustainable Consumption, Ecology and Fair Trade*, London: Routledge, pp. 177–91.

Dolan, Catherine and Dinah Rajak (2016), 'Remaking Africa's informal economies: Youth, entrepreneurship and the promise of inclusion at the bottom of the pyramid', *The Journal of Development Studies*, 52 (4), 514–29.

Dolan, Catherine and Kate Roll (2013), 'Capital's new frontier: From "unusable" economies to bottom-of-the-pyramid markets in Africa', *African Studies Review*, 56 (3), 123–46.

Dubuisson-Quellier, Sophie (2018), *La consommation engagée*, Paris: Presses de sciences po.

Elyachar, Julia (2012), 'Next practices: Knowledge, infrastructure, and public goods at the bottom of the pyramid', *Public Culture*, 24 (1), 109–29. doi: 10.1215/08992363-1443583.

Fressoli, Mariano, Elisa Arond, Dinesh Abrol, Adrian Smith, Adrian Ely and Rafael Dias (2014), 'When grassroots innovation movements encounter mainstream institutions: Implications for models of inclusive innovation', *Innovation and Development*, 4 (2), 277–92.

Gaglio, Gérald (2017), '"Innovation fads" as an alternative research topic to pro-innovation bias: The examples of Jugaad and Reverse Innovation', in Benoît Godin and Dominique Vinck (eds), *Critical Studies of Innovation*, Cheltenham, UK and Northampton, MA, USA: Edward Elgar Publishing, Chapter 2, pp. 33–47.

Gaglio, Gérald, Benoît Godin and Sebastian Pfotenhauer (2019), 'X-Innovation: Re-inventing innovation again and again', *Novation: Critical Studies of Innovation*, (1), 1–16.

Godin, Benoît and Gérald Gaglio (2019), 'How does innovation sustain "sustainable innovation"?', in Frank Boons and A. McMeekin (eds), *Handbook of Sustainable Innovation*, Cheltenham, UK and Northampton, MA, USA: Edward Elgar Publishing, Chapter 2, pp. 27–37.

Hart, Stuart and Coimbatore K. Prahalad (2002), 'The fortune at the bottom of the pyramid', *Strategy+Business*, 26, 54–67.

Hirschman, Albert O. (1958), *The Strategy of Economic Development*, New Haven: Yale University Press.

Hossain, Mokter, Henri Simula and Minna Halme (2016), 'Can frugal go global? Diffusion patterns of frugal innovations', *Technology in Society*, 46, 132–9.

Jha, Srivardhini K., R. Rishikesha and T. Krishnan (2013), 'Local innovation: The key to globalisation', *IIMB Management Review*, 25 (4), 249–56.

Karnani, Aneel (2007), 'The mirage of marketing to the bottom of the pyramid', *California Management Review*, 49 (4), 90.

Kline, Ronald (2003), 'Resisting consumer technology in rural America: The telephone and electrification', in Nelly Oudshoorn and Trevor J. Pinch (eds), *How Users Matter: The Co-Construction of Users and Technologies*, Cambridge MA: MIT Press, pp. 51–66.

Knorringa, Peter, Iva Pesa, Andre Leliveld and Cleff Van Beers (2016), 'Frugal innovation and development: Aides or adversaries?', *The European Journal of Development Research*, 28 (2), 143–53.

Kroll, Henning, Madeleine Gabriel, Annette Braun, Emmanuel Muller, Peter Neuhäsler, Esther Schnabl and Andrea Zenker (2016), A conceptual analysis of foundations, trends and relevant potentials in the field of frugal innovation (for Europe): Interim Report for the Project "Study on Frugal Innovation and Reengineering of Traditional Techniques." Publications Office of European Union.

Kumar, Hemant (2014), 'Dynamic networks of grassroots innovators in India', *African Journal of Science, Technology, Innovation and Development*, 6 (3), 193–201.

Landrum, Nancy E. (2007), 'Advancing the "base of the pyramid" debate', *Strategic Management Review*, 1 (1), 1–12.

London, Ted (2011), 'Building better ventures with the base of the pyramid: A roadmap', in Ted London and Stuart Hart (eds), *Next Generation Business Strategies for the Base of the Pyramid:*

*New Approaches for Building Mutual Value*, Upper Saddle River, NJ: Pearson Education, FT Press, pp. 19–44.

Meagher, Kate (2018), 'Cannibalizing the informal economy: Frugal innovation and economic inclusion in Africa', *The European Journal of Development Research*, 30 (1), 17–33.

Neumann, Lukas, Jonas Böhm and Christophe Wecht (2017), 'Knowledge transfer in the context of frugal innovation', in ISPIM Conference Proceedings. The International Society for Professional Innovation Management (ISPIM), pp. 1–11.

Oudshoorn, Nelly E. and Trevor J. Pinch (2003), *How Users Matter: The Co-Construction of Users and Technologies*, Cambridge MA: MIT Press.

Pansera, Mario and Soumodip Sarkar (2016), 'Crafting sustainable development solutions: Frugal innovations of grassroots entrepreneurs', *Sustainability*, 8(1), 51. https://doi.org/10.3390/su8010051.

Peša, Iva (2017), 'Sawdust pellets, micro gasifying cook stoves and charcoal in urban Zambia: Understanding the value chain dynamics of improved cook stove initiatives', *Sustainable Energy Technologies and Assessments*, 22, 171–6.

Pisoni, Alessia, Laura Michelini and Gloria Martignoni (2018), 'Frugal approach to innovation: State of the art and future perspectives', *Journal of Cleaner Production*, 171, 107–26.

Prathap, Gangan (2014), 'The myth of frugal innovation in India', *Current Science*, 374–7.

Radjou, Navi, Jaideep Prabhu and Simone Ahuja (2012), 'Frugal innovation: Lessons from Carlos Ghosn, CEO, Renault-Nissan', *Harvard Business Review Blog Network*, http://blogs. hbr. org/2012/07/frugal-innovationlessons-from/ accessed on December, 3, 2013.

Rangan, V. Kashturi, John A. Quelch, Gustavo Herrero and Brooke Barton (2007), *Business Solutions for the Global Poor: Creating Social and Economic Value*, New York: John Wiley & Sons.

Schumacher, Ernst F. (1973), *Small is Beautiful. Economics as if People Mattered*, New York: Harper & Row.

Sehgal, Vikas, Kevin Dehoff and Ganesh Panneer (2010), 'The importance of frugal engineering', *Strategy+ Business*, 59 (Summer), 1–5.

Sen, Amartya (2003), *Un nouveau modèle de développement économique*, Paris: Odile Jacob.

Sen, A. (2013), 'Development as capability expansion', in J. DeFilippis and S. Saegert (eds), *The Community Development Reader*, London: Routledge, pp. 319–27.

Seyfang, Gill and Alex Haxeltine (2012), *Growing Grassroots Innovations: Exploring the Role of Community-Based Initiatives in Governing Sustainable Energy Transitions*, London: Sage.

Sharma, Gautam and Hemant Kumar (2019), 'Commercialising innovations from the informal economy: The grassroots innovation ecosystem in India', *South Asian Journal of Business Studies*, 8 (1), 40–61.

Simanis, Erik, Stuart Hart and Duncan Duke (2008), 'The base of the pyramid protocol: Beyond "basic needs" business strategies', *Innovations: Technology, Governance, Globalization*, 3 (1), 57–84.

Smith, Adrian, Fressoli, Mariano and Thomas, Hernàn (2014) 'Grassroots innovation movements: Challenges and contributions', *Journal of Cleaner Production*, 63, 114–24.

Star, Suzan L. and Joseph R. Griesemer (1989), 'Institutional ecology, "translations" and boundary objects: Amateurs and professionals in Berkeley's Museum of Vertebrate Zoology 1907–39', *Social Studies of Science*, 19 (3), 387–420.

Stiglitz, Joseph E. (2002), 'Information and the change in the paradigm in economics', *American Economic Review*, 92 (3), 460–501.

Thomas, Hernán, Lucas Becerra and Santiago Garrido (2017), 'Socio-technical dynamics of counter-hegemony and resistance', in Benoît Godin and Dominique Vinck (eds), *Critical Studies of Innovation*, Cheltenham, UK and Northampton, MA, USA: Edward Elgar Publishing.

Walsh, James P., Jeremy C. Kress and Kurt W. Beyerchen (2005), 'Book review essay: Promises and perils at the bottom of the pyramid', *Administrative Science Quarterly*, 50 (3), 473–82.

Winterhalter, Stephan, Marco B. Zeschky, Lukas Neumann and Oliver Gassmann (2017), 'Business models for frugal innovation in emerging markets: The case of the medical device and laboratory equipment industry', *Technovation*, 66, 3–13.

Zeschky, Marco, Bastian Widenmayer and Oliver Gassmann (2011), 'Frugal innovation in emerging markets', *Research-Technology Management*, 54(4), 38–45, doi: 10.5437/08956308X5404007.

# PART V

# SUPPORTING INNOVATION: REFRAMING THE INSTRUMENTS

# 15. X-innovation and international organizations narratives

*Carolina Bagattolli*

## POLITICAL DISCOURSES ON INNOVATION

Since the late 1980s, Policy Analysis has emphasized the importance of incorporating processes of argumentation as analytical dimensions for empirical studies on the grounds that "public policy is made of language" (Majone 1989, p. 1), which means politics is "also about telling good stories" (McBeth 2014, p. xiii), in a constant dispute over problem selection and government strategy (Stone 1989). In their rhetoric political actors select, include and exclude issues from the policy agenda. Indeed, the actual work involved in problem construction is rhetorical and interpretative (Fischer and Forester 1993, p. 6). Hence, policy narratives have received widespread recognition as being highly influential in public policy development (Roe 1994).

Political discourses convey messages and potentially shape behaviors. "What governments say is as important as what governments do" (Dye 2013, p. 66). Policy narratives endow coherence on the public policy process, are able to support or oppose proposals for action. Even in cases where policy arguments bear little direct relationship with the decision-making processes – acting more as rhetorical justifications legitimizing pre-determined courses of action – they express ideology and seek to shape an understanding of a specific issue in particular audiences (Fischer and Forester 1993).

Policy discourse analysis returns an understanding as to what gets privileged and what remains absent in policy discourses regarding the Science, Technology and Innovation (ST&I) arena. ST&I political discourses frequently deploy the authority of scientific discourse to justify their positions and thereby cloaking the politics behind the policy.

Ever since World War II, Science, Technology and (increasingly) Innovation have been heralded as central to economic growth in ST&I policy discourses (Braun 2006; Freeman and Soete 2008; Kallerud 2010; Lundvall and Borras 2004) – with innovation fundamentally understood as technological (Gaglio et al. 2019; Godin 2015). Technological innovation has come to be considered as fundamental to a broad range of economic variables, including international trade, productivity, market share, employment creation, among other benefits and its worth has gone uncontested for decades (Godin and Vinck 2017).

Nevertheless, in more recent decades, innovation narratives have shifted (Gaglio 2017; Godin 2015). The concept of social innovation has gained renewed relevance alongside the proposition and emergence of new concepts – such as open innovation, responsible innovation and green innovation to mention but a few examples. The proliferation of x-innovation concepts represents the latest step in a historical process of enlargement of the innovation concept that contests technological innovation as a hegemonic discourse belonging to the twentieth century (Gaglio et al. 2019; Godin 2019). Innovation hereby becomes a contested and polyvalent entity, imbued with latent transformative contents that may emerge only after the capture of the term per se. In essential terms, these concepts are not merely adopted or rejected but rather 'translated' (Pel 2016).

These x-innovation concepts share two main characteristics. On the one hand, their 'social dimension': an emphasis on social values and participatory governance through public deliberations. On the other hand, a moral imperative regarding the social, ethical and environmental components in such processes. The first theorizations in this field date back to the 1960s. Then commonly associated with a specific area (e.g., technological, organizational, product), later, and from the 1980s–1990s onwards, new innovation concepts were conceived and attaching to – more emphatically than before – adjectives (such as open and ecological, for example), "or attributes" (social, responsibility). Last but certainly not least, it is worth mentioning that these concepts also inherited the ambiguity itself encapsulated within the term innovation (Gaglio 2017; Godin 2019).

These narratives may shape the social imaginary and influence on ST&I policy discourses and agenda and hold relevance to the policy debate. This phenomenon has been considered by Godin (2008, 2015, 2019), Bontems (2014), Pel (2016), Edwards-Schachter and Wallace (2017), Edwards-Schachter (2018), Segercrantz et al. (2017), Gaglio (2017) and Gaglio et al. (2019). However, there still remain only scant efforts to understand the extent to which these narratives have impacted on ST&I policy discourse through primarily considering the narratives of international organizations.

This chapter presents the results of a discursive analysis of the 'x-innovation' concepts (Godin 2019; Gaglio et al. 2019) present in international organization documents. More specifically, we count how frequent these narratives are and, when emerging, unearthing the ways in which x-innovation gets understood and represented as well as identifying whether this appears as a rupture in the core values of technological innovation.

Following this introduction, we organize this chapter into three other sections. In section 2, we contextualize the empirical analysis and the framework applied to deal with the research corpus. The third section displays our central findings, presents the main results, and discusses the trending discourses that emerged from our empirical endeavors. Finally, in section 4, we summarize the main findings and present our final considerations.

## DOCUMENTAL ANALYSIS

International organizations have contributed to developing "increasingly codified conceptions of science" (Drori et al. 2003, p. 6), constructing ST&I narratives (Henriques and Larédo 2013) and influencing various – especially underdeveloped – countries[1] through harmonized ST&I policy orientations and financial resources (Aguiar et al. 2015; Bagattolli et al. 2016; Nupia 2014).

The social positioning of these political actors represents a fundamental factor in the narrative's legitimacy as their (perceived) social importance sustains its discourse among other stakeholders. In this sense, grasping the ongoing discourses of international organizations such as the Organisation for Economic Co-operation and Development (OECD), the European Commission (EC)[2] and the Inter-American Development Bank (IDB) – recognized worldwide as prestigious policy formulation institutions – would enable inferences about future changes in ST&I policies worldwide.

*Table 15.1      The selected international organization documents*

| | |
|---|---|
| **IDB** | ¿Qué Hace el BID en Innovación Ciencia y Tecnología (2016) |
| | [What does IDB do in Innovation, Science and Technology?] |
| | Innovation, Science and Technology Sector Framework Document (2017a) |
| | Políticas Públicas Para la Creatividad y la Innovación: Impulsando la Economía Naranja en América Latina y el Caribe (2017b) |
| | [Public Policies for Creativity and Innovation: Boosting the Orange Economy in Latin America and Caribbean] |
| **EC** | State of the Innovation Union taking Stock 2010-2014 (2014) |
| | State of the Innovation Union taking Stock 2012: Accelerating Change (2013) |
| | Research and innovation: Pushing Boundaries and Improving the Quality of Life (2016) |
| **OECD** | Science, Technology and Industry Outlook (1996) |
| | Science, Technology and Industry Outlook (1998) |
| | Science, Technology and Industry Outlook (2000) |
| | Science, Technology and Industry Outlook (2002) |
| | Science, Technology and Industry Outlook (2004) |
| | Science, Technology and Industry Outlook (2006) |
| | Science, Technology and Industry Outlook (2008) |
| | Science, Technology and Industry Outlook (2010) |
| | Science, Technology and Industry Outlook (2012) |
| | Science, Technology and Industry Outlook (2014) |
| | Science, Technology and Innovation Outlook (2016) |
| | Science, Technology and Innovation Outlook (2018) |

*Source:*    Prepared by the author.

We applied Content Analysis (Bardin 2016) on strategic documents such as reports on the grounds that this document type usually seeks to highlight the strengths and weaknesses in the area and to suggest niches for countries to target their efforts on. The establishment of this research corpus followed the principles of representativeness, homogeneity and pertinence. Given the enormous number of documents of this genre, we recognize this enterprise represents only a first approach and is far from exhaustive. We analyzed eighteen documents published by three international organizations: the IDB, EC and OECD. The OECD and EC play a recognized role in "advising governments on their innovation policies" (OECD 2010, p. 254). Similarly, the IDB serves to provide "advisory services to its clients as well as policy and institutional capacity development."[3] These institutions are the benchmark reference in terms of ST&I policy advocacy. Table 15.1 displays the documental corpus hereby selected.

We engaged an indexing process and developed indicators from textual analysis of the documents selected. Due to the amount of material, we have been using the MAXQDA 2018®[4] software. We initiated the material scanning by searching for the word "innovation" ("innovación" in documents in Spanish). All the mentions (15 168) were then checked and those displaying occurrences of "x-innovation" concepts were categorized and codified in MAXQDA®. Acknowledgments, titles, headers, illustrations, notes, references and annexes were not considered for study. The unit of analysis was the complete sentence (or the entire paragraph if it was inconclusive). After checking the codification, we have interpreted the results. We identified 431 mentions of 27 different "x-innovation" concepts in the corpus, as detailed in Table 15.2, which is very few (less than 3 percent) regarding the 15 168 mentions of "innovation" – denoting the low pervasiveness of these concepts in the corpus.

The first occurrence of an x-innovation concept dates back to the 1960s at OECD's Reports (Godin 2020). In our document corpus the first one (design innovation) appears in 1998 in the

*Table 15.2*     *Frequency of x-innovation concepts present in the corpus*

| Concept | Mentions in The Corpus |
| --- | --- |
| Adaptative Innovation | 1 |
| Agro-Innovation | 2 |
| Branding Innovation | 2 |
| Blue Innovation | 1 |
| Co-Creative \| Co-Creating \|Co-Producing Innovation | 3 |
| Collaborative Innovation | 5 |
| Collective Innovation | 1 |
| Dematerialized Innovation | 1 |
| Design \| Design-Driven \| Design Led Innovation | 8 |
| Digital Innovation | 20 |
| Disruptive And Breakthrough Innovation | 6 |
| Environmental Innovation (Including Eco And Green Innovation) | 166 |
| Financial Innovation | 3 |
| Frugal Innovation | 5 |
| Grassroots Innovation | 10 |
| Inclusive Innovation | 29 |
| Life Innovation | 3 |
| Needs-Driven Innovation | 1 |
| Open Innovation | 63 |
| Responsible Innovation | 15 |
| Social Innovation | 62 |
| Socioeconomic Innovation | 1 |
| Soft Innovation | 3 |
| Sustainable Innovation | 2 |
| User \| User-Driven Innovation \| User-Led Innovation | 18 |
| Total | 431 |

*Source:*     Prepared by the author based on analysis of the documents.

OECD Science, Technology and Industry Outlook (OECD 1998). Furthermore, through to 2006, their incidence in the reports remained infrequent (see Figure 15.1). The mentions rose in occurrence after 2008 before peaking in 2012 (with 162, over one-third of all references). Additionally 2012, 2014 and 2018 registered a greater number of x-innovation concept references than those explicitly mentioning the role of technological innovation for economic purposes.

As Table 15.2 details, among the 27 x-innovation concepts, the majority (19) generate less than ten occurrences, which may express their low level of pervasiveness – at least in this kind of institutional report. Whenever we disregard these, eight concepts account for more than ten occurrences in the documents: responsible innovation, user-driven innovation, digital innovation, eco-innovation, inclusive innovation, social innovation, open innovation and green innovation. Together, they represent almost 90 percent (373 out of 431 mentions) of all x-innovation references (nevertheless, still only 2.4 percent of the total on innovation mentions) in the document corpus and we thus focus our analysis on them in the following section. Figure 15.2 presents a visual representation of the frequency of the main terms.

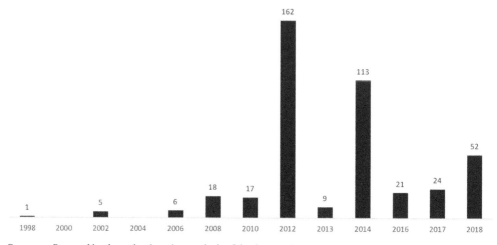

*Source:* Prepared by the author based on analysis of the documents.

*Figure 15.1* *Evolution of x-innovation mentions in the document corpus along time*

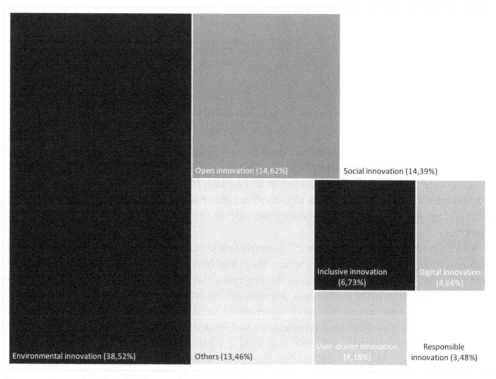

*Source:* Prepared by the author based on analysis of the documents.

*Figure 15.2* *Distribution of the citations of the most frequent x-innovation terms*

## MAKING SENSE OF TRENDING DISCOURSES

As we are here interested in shedding some light on the understanding of the usages of these terms in the ST&I policy narratives of international organizations, this section presents a synthesis of the main findings of the most commonplace x-innovation concepts identified in the corpus; the eight most cited concepts are presented here in order of frequency.

Before presenting our findings for each concept, we would first present some more general details, starting with the evolution of narratives over time. As Table 15.3 sets out, the expansion in mentions of x-innovation concepts is erratic through time. This stems from both their low incidence in the corpus and precludes observations as to whether the entry of a new narrative pushes out another.

Furthermore, this may also raise questions around whether these concepts appear in combination in the narratives. Co-occurrence analysis of the mentions – the intersection of the concepts when two or more concepts are present in the same text segment (in our case, the sentence) – reveals that, although such cases exist, their incidence is quite rare. We further explore below just how the international organization narratives align the concepts in their discourses. For the meantime, we would simply stress that nine of the ten intersections identified involved the open innovation concept and, across the entire document corpus, there is no co-occurrence involving the concept of environmental innovation.

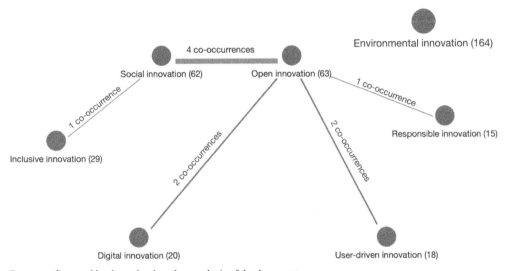

*Source:*   Prepared by the author based on analysis of the documents.

*Figure 15.3*    *Co-occurences of the concepts in the document corpus*

Subsequently, there is relevance in exploring, even if only briefly, the differences in the narratives produced by the international organizations. Figure 15.4 highlights the respective distribution of mentions. As most of the documents (12 out of 17) are by the OECD, to some extent at least, the more significant weighting of this entity in the mentions was expected. Nevertheless, the data also presents some noteworthy trends, among them: (i) the substantial (even though occasional over time – 2012 and 2014) attention attributed by the OECD to environmental innovation; (ii) the emphasis of European Commission documents on social and environmental innovation; and (iii) the more incipient presence of x-innovation concepts in the IDB documents.

Before moving to the narrative analysis, it is still worth mentioning that, considering the book counts on chapters dedicated exclusively either directly to most[5] or to associated concepts,[6] we focus primarily on the narrative. Far from the state-of-the-art, the literature presented here is punctual and deployed strictly to contextualize how the narratives of the international organizations appropriate them. We leave the critiques regarding the limits of this literature to the respective authors.

**Responsible Innovation[7]**

The documents analyzed present responsible innovation in terms of governments "managing the risks and uncertainties around emerging STI developments" (OECD 2016, p. 19). Besides, emerging technologies can even "raise important ethical issues, too" (idem, p. 111) – "…especially in fields (e.g., artificial intelligence [AI], gene editing and neurosciences) where science and technology move faster than legal and ethical rules" (OECD 2018, p. 97).

Accordingly, one of the four trends to "future STI policy practices" is the "growing influence of so-called 'responsible research and innovation,' which places greater emphasis on broader public engagements of policymaking" (OECD 2016, p. 154). In summary, it calls for more open and inclusive governance of "technological change" in which governments should pay considerably more attention to its ethical and societal dimensions. Likewise, the idea is "deliberate and steer the consequences of innovation at an early stage" (OECD 2018, p. 225). According to this mindset, national governments need "to anticipate and assess the potential implications and societal expectations associated with research and innovation" through "responsible research and innovation" policies (RRI) (OECD 2016, p. 154).

The OECD correspondingly treated the design of RRI policies as one of the "policy areas in which governments have been particular active" (OECD 2016, p. 167), "integrated into the general formulation of innovation policy agendas" (idem, p. 189). Horizon 2020 even included responsible innovation as one of its pillars in an alleged attempt to extend this approach across all research activities (OECD 2018).

Even though the earliest theorizations about the x-innovation reach back to the 1960s and the first paper indexed in the Web of Science – probably the most prominent global citation database – including the term Responsible Innovation comes from 1978,[8] the first appearance of the concept in the corpus analyzed is 2016. This constitutes a considerable time lapse.

Briefly, the understanding of responsible innovation prevailing in the literature involves the management of the risks and uncertainties interrelated with ST&I developments, especially new technologies – calling for more open and participatory governance in the ST&I arena. On the one hand, the application of responsible innovation in the corpus reflects the outlook present in the literature on the ethical, legal and social implications of technoscientific developments (Owen et al. 2012). Recognition of the limitations of hegemonic ST&I policy approaches for managing "ethically problematic areas," with an increasing awareness on the innovation impacts on society (idem) – the "irresponsibility" of technological innovation hitherto (Gaglio et al. 2019; Godin 2019) – does feature in the documents even though in a much subtler and vaguer way. However, on the other hand, despite the literature's emphasis on the

*Table 15.3*  Evolution of the most cited concepts through time

| | 2002 | 2004 | 2006 | 2008 | 2010 | 2012 | 2013 | 2014 | 2016 | 2017 | 2018 | Sum |
|---|---|---|---|---|---|---|---|---|---|---|---|---|
| Responsible innovation | | | | | | | | | 11 | | 4 | 15 |
| User driven innovation | | | | 8 | 3 | 2 | | 4 | | | 1 | 18 |
| Digital innovation | | | | | 1 | 1 | | | | 4 | 14 | 20 |
| Environmental innovation | | | | 1 | 2 | 118 | 1 | 40 | 1 | | 1 | 164 |
| Inclusive Innovation | | | | | 1 | 14 | | 9 | 1 | | 4 | 29 |
| Social innovation | | | | | 1 | 5 | 7 | 34 | 2 | 7 | 6 | 62 |
| Open innovation | 5 | | 8 | 5 | | 5 | 1 | 14 | 4 | 10 | 11 | 63 |
| **Sum** | **5** | **0** | **8** | **14** | **8** | **145** | **9** | **101** | **19** | **21** | **41** | **371** |

*Source*:  Prepared by the author based on analysis of the documents.

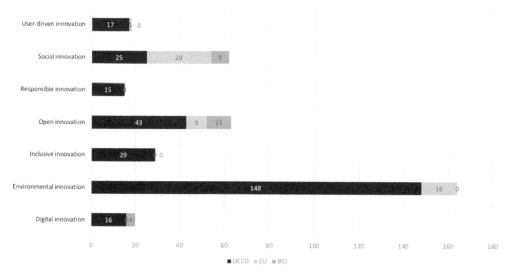

*Source:*    Prepared by the author based on analysis of the documents.

*Figure 15.4    Mentions by international organizations*

importance of a more democratic (participatory) governance of the ST&I system (Karinen and
Guston 2010) – "both setting research agendas and modulating research trajectories towards
socially desirable ends" (Owen et al. 2012, p. 752) – the documents barely elaborate on this
point.

In the reports analyzed, much is said about creating an open and inclusive governance
regime for the ST&I arena even while the emphasis remains on the national government's role.
This policy actor would be responsible for anticipating and assessing the potential implications
of ST&I endeavors. Nevertheless, mutual responsibility among societal actors and innovators
(von Schomberg 2011) then disappears from the narrative. The science for society dimension
is visible in the corpus, even if with a timid presence; but not the science with society facet
(Grinbaum and Groves 2013) – which may reflect how the concept is "being used instrumen-
tally, to smooth the path of innovation in society, and/or to achieve precommitted policies"
(Owen et al. 2012, p. 757).

### User/User-Driven/User-Led Innovation[9]

According to the OECD (2008, p. 102) user-led (e.g., either by consumers or suppliers) innovation ranks among the 'new or alternative' forms of innovation, considered of great importance
to competitiveness. This would hold such significance that supporting non-technological and
user-driver innovation would feature among the main ST&I policy trends in OECD countries
according to its institutional reports (OECD 2008, 2010, 2014, 2018).

Likewise, Norway, Denmark, Finland, Poland and Chile launched initiatives – from occasional policy measures to general programs aimed at improving innovative abilities in companies utilizing 'users' innovation potential' and to diffuse user-driven innovation methods
through the private as well the public sector (OECD 2008, 2010, 2012, 2014). Lastly, more

than emphasizing "the widely acknowledged benefits (and biases) of open or user-led innova-tions," the reports considers user-driven innovation even as a kind of "element of democratic legitimacy to innovation" as this would enable "adjusting trajectories" according [to] potential social concerns (OECD 2018, p. 226).

The first thing worth highlighting is the ambiguity and imprecision with which user-driven innovation emerges in the corpus – which might be a reflection of the literature. Despite the already long trajectory of user innovation studies, the literature is far from consensual and differing as regard conceptualizations, the focus, the findings and the business implications derived from user-driven innovation (Buur and Matthews 2008). The formulations range from "new emphatic awareness of user identities and contexts that designers can bring to compa-nies" to "advanced early-stage customer research." In sum, the concept remains "elusive and subject to debate" (Hyysalo et al. 2016, p. 2). Common to these formulations is the centrality of users in innovation activities as participants, sources of 'information, design and inspira-tion.' Benefits from engagement with users would thus range from improving the information transfers to companies to product enhancements and gains due to a larger likelihood of user acceptance (idem).

Nevertheless, the ambiguity here is not only due to the lack of any consensual definition. The very understanding of participation in user-driven innovation remains hazy: more than participatory governance, this proposes that individual (user) participation represents a source of ideas and information and an asset to value generation. Even though the OECD (2018) considers user-driven innovation as an "element of democratic legitimacy," as social concerns would thus be taken into account by the innovation trajectories, this might be an attempt to capture the term or even an effort to suppress efforts to contest the strategy based on ethical issues. In any case, the document's corpus does not contain any discussion on the ethical issues.

### Digital Innovation

The document's corpus defines digital innovation as "new products and process, enabled by digital technologies or embodied in data and software" "which allow collecting, processing, manipulating, storing and diffusing data automatically, using machines" (OECD 2018, p. 76). Although rather vaguely, the document's corpus presents digital innovation as close to the literature, which basically defines the term as the conception of new products, processes, services, platforms or business models using digital technology (Hargadon and Douglas 2001; Hinings et al. 2018), or alternatively, "the use of digital technology in a wide range of innova-tions" (Nambisan et al. 2019, p. 2).

Discussed within the framework of "Innovation and knowledge as key to productivity growth and economic development," digital innovation is considered as a means to the "capture of rents beyond the traditional dimension of this phenomenon, in a world where the winner gets everything and the runner up almost nothing" (IDB 2017a, p. 11). More than a type of innovation, digital innovation is deemed part of the 'changing nature of innovation.'

Therefore, "(e)xploit the potential of open digital innovation and platforms in areas such as scientific research, business innovation, technology commercialization and talent develop-ment" feature openly among the activities prioritized by the IDB (IDB 2017a, p. 59). Different countries around the world would follow the same path, including Italy that was then encour-aging digital innovation in schools (OECD 2010).

In this scenario, Latin American economies are considered to be at a disadvantage as the region displays "a lack of infrastructure and relatively high costs of [digital innovation] adoption" (IDB 2017a, p. 38). Even so, there are an increasing number of companies "focused on digital innovation, in the form of start-ups" known as "Tecnolatinas" (IDB 2017a, p. 60). The examples highlighted were found in Guadalajara (México), which "deliberately sought to become a hub for digital innovation" and Campinas (Brazil), which "has developed a vigorous IP market and start up ecosystem" (IDB 2017a, p. 36).

Nevertheless, the importance assigned to digital innovation in those reports reaches beyond the business sector given the relevance attributed to the public sector within the scope of efforts to "modernize the public administration, to make it more efficient and transparent, and to improve the quality of services and reduce costs" (OECD 2012, p. 330). Furthermore, and even more curiously, those same reports point out that STI policymakers and stakeholders should consider this a part of the "responsiveness [of innovation policy] to societal demands for inclusiveness and other societal challenges, such as epitomized by the SDGs"[10] (OECD 2018, p. 97).

The potential role of digital innovation in modernizing public administration, making it more transparent and addressing societal demands (such as inclusiveness and other societal challenges) are not present in the literature; not even in Yoo et al. (2010) and their report synthesizing the "Research Workshop on 'Digital Challenges in Innovation Research'" – hosted at Temple University – which brought together a significant group of scholars dedicated to this field from different research fields. This might purport a capture of the innovation term: the uptake of the term in an insufficiently transformative way, compared to their original authors, or perverting its original meaning (Pel 2016).

Still furthermore, according to some authors, among the very characteristics of digital innovation are its rapid development and difficulty to control and predict (Henfridsson et al. 2014; Nylén and Holmström 2015; Yoo et al. 2010). This happens mainly due to its "overall capacity to produce unprompted change, driven by large, varied, and uncoordinated audiences" (Zittrain 2006, p. 1977) as these innovations serve as the basis for creating other new products and services beyond the original design (Nylén and Holmström 2015) – which thus explains the need for responsible innovation. Moreover, the literature also presents digital innovation in relation to another x-innovation concept: open innovation given that "…(t)he advent of digital technologies is making science and innovation more open, collaborative, and global" (Bogers et al. 2018).

## Inclusive Innovation

Inclusive innovation is presented as "consist[ing] in obtaining cheaper (often simplified) versions of existing devices for purchase by lower-income groups" (OECD 2012, p. 124). The central issue involves amplifying and facilitating the consumption of goods although also identifying how this happens by "providing products of lower quality – often cheaper, simplified versions of more sophisticated goods" (OECD 2012, p. 124).

Inclusive innovation also gets termed as 'frugal innovation,' 'grassroots innovation,' and 'innovation by low and middle-income groups': innovations born out of necessity and able to help by improving living standards more than some technical innovations (OECD 2012, p. 126). That means "Inclusive innovation differs in nature (i.e. products and processes targeting lower-income groups) and in its source (i.e. produced by lower-income groups)" (OECD

2012, p. 124). This therefore suggests, "Inclusive innovation is about harnessing science, technology and innovation know-how to address the needs of those at the 'bottom of the pyramid' ... improve their quality of life and reduce social disparities" (OECD 2012, p. 124). This furthermore emphasizes the participatory dimension, "Innovation activities of the poor themselves are also important" (OECD 2014, p. 340). Nevertheless, and according to the same institution, "So far, inclusive innovation has consisted in the adaptation of technologies mainly from developed countries" (OECD 2012, p. 125).

Inclusive innovation is introduced as related to inclusive development, thus reducing the disparities between the richest and poorest in society. Plus, it might "contribute to addressing urgent challenges such as providing access to drinking water, eradicating diseases and reducing hunger," even allowing "the poor to modernize their often 'informal' and low-productivity business" (OECD 2012, p. 15) and helping "the informal sector absorb knowledge and upgrade technology" (OECD 2010, pp. 59–60). In this sense, the Information and Communication Technologies (ICT) may play an important role, consisting of "opportunities to support inclusive innovation by 'democratising innovation' and by extending the circle of individuals and businesses that engage in innovation activities" (OECD 2014, p. 53). More than that, "The potential of ICTs is clear when one looks at the importance of ICT-based products and services among the successes of inclusive innovation initiatives" (idem, p. 149).

The conceptualization of inclusive innovation in these documents broadly resembled that of the literature: innovations, mainly low cost, aiming to benefit the disenfranchised (George et al. 2012; Heeks et al. 2014). Even the dichotomy regarding "which 'aspect' of innovation the marginalized group is to be included in" – only innovation outputs vs. the innovation process – (Heeks et al. 2014, p. 177) finds its way into the international organization's reports.

In terms of the ST&I policy arena, "(t)he discussion of inclusive innovation raises some interesting questions for policy, including issues concerning innovation in the informal sector, the role of grassroots innovation for economic development and the role traditional knowledge can play for economic growth" (OECD 2012, p. 129). That would explain why policy measures designed to stimulate inclusive innovation can be observed in several countries, including China, India, Israel and Malaysia, focusing on "benefiting poor and excluded groups" (OECD 2014).

The reports extend to debate how inclusive innovation is a strategy not only for the informal economy but also for the formal sector. "Established firms are also engaging in more inclusive innovation" (OECD 2018, p. 21). That would be evident, for example, in "practices such as value-based design and standardization," which are becoming "powerful tools for translating and integrating core social values, safeguards and goals into technology development" (OECD 2018, p. 21).

As Figure 15.3 details, co-occurrence analysis indicates an association between inclusive and social innovation. According to the OECD (2016), "new concepts as social innovation, frugal innovation, inclusive innovation [...] are leading to new innovative business models and may contribute to a more inclusive approach to innovation" (OECD 2016, p. 67). At first sight, *inclusive innovation*[11] and *social innovation* might be considered the two terms, among those with greatest frequency in the document corpus, which most contest the idea of technological innovation as a panacea for social problems – the fallout from the assumption of mainstream innovation as an innovation of inequality (Klochikhin 2012; Lazonick and Mazzucato 2013). Although this contestation process is not entirely new (Gaglio et al. 2019; Godin 2019), there have been significant changes in inclusive innovation. Among them stand out the "significant involvement of the private sector ... in the innovation for the poor" and "the development of poor consumers as an accessible mass market" (Heeks et al. 2014, p. 176). Organizations are furthermore engaging in inclusive "innovation activities to connect disenfranchised individuals and communities with opportunities that foster social and economic growth" (George et al. 2012, p. 662). Management studies are paying attention to how low-cost products and business models formulated in the developing world might be transferred to other countries (idem). This highlights the sheer strength of the marketing dimension surrounding this term. Taken to an extreme, this may indeed mean that inclusive innovation is not presented as a counter-concept for technological innovation and is actually more complimentary than antagonistic. This is a perception that also gets shared among the literature and the document corpus.

**Social Innovation**

In keeping with the literature, the document corpus underlines the lack of consensus around the social innovation definition. However, there is recognition that "most tend to emphasise the objective of meeting social goals and, to some extent, the types of actors involved (e.g., not-for-profit, individuals, universities, government agencies, enterprises)" (OECD 2014, p. 148). First, social innovation arises out of attempts to obtain new answers to old social problems "by identifying and delivering new services that improve the quality of life of individuals and communities and by identifying and implementing new labor market integration processes, new competencies, new jobs and new forms of participation that help to improve the position of individuals in the workforce" (OECD 2014, p. 148). Second, this participatory bias stems from "the importance of citizens as key actors of social innovation and the need for broad partnerships for promoting innovation in social policy mechanisms" (European Commission 2013, p. 23). In sum, social innovation may be "defined more by the nature or objectives of innovation than by the characteristics of innovations themselves" (OECD 2014, p. 148).

Social innovation would thus display high social returns but 'low prospects for profits' (OECD 2014, p. 156). Nevertheless, alongside frugal and inclusive innovation, social innovation and social entrepreneurship concepts are flagged as "leading to new innovative business models and may contribute to a more inclusive approach to innovation" (OECD 2016, p. 67) – areas with 'tangible results' and 'highly valued' by IDB clients (IDB 2017a, p. 52).

This view of social innovation as 'essential to the development process' would explain its presence in ST&I policy arenas around the world. Interestingly, this discussion is much more prominent in the European Union in the document corpus than in its OECD counterpart. The European Commission states its support for "social innovation in a number of ways, including though support for the up-scaling of successful projects" (European Commission 2014, p. 61). In this sense, "(s)upport for social innovation against poverty and social inclusion is also provided by the European Platform against poverty and social exclusion" and "a pilot action on networks of incubators for social innovation was launched in 2013 to support two European networks to assess, support and scale up social innovations in Europe" (European Commission 2014, p. 65). A Social Innovation Europe Platform was even launched in 2011,[12] "a virtual hub connecting social innovators and providing an overview of actions throughout Europe" (European Commission 2014, p. 65) – that "attracted nearly 50 000 people in its first 18 months" (European Commission 2013, p. 23). Likewise, a *European Social Innovation*

*Competition*[13] has been organized annually since 2012 to "directly support new solutions and raise awareness about social innovation" (European Commission 2014, p. 65). Apparently, this was not just occasional guidance. In the European Union's current programming period (2014–2020), social innovation is said to be mainstream. "Member States have to programme social innovation-related activities, but they are given the flexibility to target social needs that are particularly relevant to them" (European Commission 2014, p. 65).

Policy actions additionally encompass support for social innovation research, which has been increasing in budget and scope ever since 2011 (European Commission 2013, 2014), including the role of social innovation "in the fight against inequalities," "in the public sector," "in innovative social services" and "for empowering citizens and promoting social change" (European Commission 2014, p. 66). Social innovation was even expressed as relevant to various topics in Horizon 2020, from industrial leadership to addressing societal challenges. While the discussion of social innovation might be much higher profile in the European Commission documents, the initiatives are not restricted to Europe and may be found in different countries all around the world, including Chile, Colombia, China, South Africa, Indonesia and Vietnam (OECD 2010, 2014).

Lastly, it is worth mentioning that the social innovation narrative in our corpus interlinks this with another x-innovation concept: open innovation. This mainly stems from the "open innovation participatory platforms" that provide a means of implementing social innovation programs addressing social challenges such as social inclusion and poverty reduction (IDB 2016 2017a). Open innovation platforms may "provide a foundation for social innovation and the engagement in innovation activities of the population at large, thus contributing to a growing awareness of the impact of innovation across institutions, the public sector and among variety of social groups" (IDB 2017a, p. 58).

The literature regarding social innovation usually defines it as "new ideas that meet unmet needs" (Mulgan et al. 2007, p. 4) and as a "collective social action aiming at social change" (Cajaiba-Santana 2014, p. 43). Furthermore, social innovation represents an old concept with a longer track record than technological innovation (Godin 2019) – the recent reappearance of the term in the literature is 'only a resurrection' (Gaglio et al. 2019). Nevertheless, while the literature tends to tie social innovation to social change, there is less consensus "on what is changed, how it is changed, at what scale it is transformed, and to what end it is changed" (Ney 2014, p. 209). In sum, social innovation "is a term that almost everyone likes, but nobody is quite sure of what it means" (Pol and Ville 2009, p. 891).

Anyhow, the social innovation narrative in the documents corpus does not differ considerably to that found in the literature, for example regarding the defence of enlarging participation in innovation processes – the "democratic dimension."[14] The same happens with the idea that social innovation embraces markets and business strategies. In this way, social innovation is no longer exclusive to the non-profit sector and involving governments, academia and markets. As observed around the concept of inclusive innovation, but perhaps with an even greater emphasis, social innovation propositions do not intend to break with the marketing dimension inherent to the technological innovation concept.

## Open Innovation[15]

In the European Commission's words, "Open innovation means getting more actors involved in the innovation process" (European Commission 2016, p. 4). The claim here is that an increasing amount of business strategy on Research and Development (R&D) is "transitioning from closed, in house research operations, to the use of open innovation," enabling "even dominant and well established companies to access ideas and innovation at the accelerated pace required by the current marketplace" (IDB 2017a, p. 20). Furthermore, "open innovation may offer a faster and less risky route to diversification than internal development" (OECD 2014, p. 73) by "collaborating much more actively than previously with large communities of experts and consumers" (OECD 2018, p. 79).

More than a kind of innovation, open innovation gets heralded by its scholars as a new paradigm, deemed "'open' because there are many ways for ideas to flow into the process, and many ways for it to flow out into the market" (Chesbrough 2006, p. 3). The advance of digital innovations correspondingly plays an essential role in "making science and innovation more open, collaborative, and global," with users (user innovation) centrally positioned in the co-creation process – the core of open innovation (Bogers et al. 2018, p. 6).

Open innovation would enable companies to "bring research results to market more quickly and make it easier for technologies developed in the EU to be commercialised" (European Commission 2016, p. 4). Largely, this relies on "firm's ability to identify and acquire externally produced scientific and technical knowledge, whether in other firms or in universities or government laboratories" (OECD 2002, p. 121). "Openness in science (open science) increases the channels for transferring and diffusing research results while open innovation in business firms creates a division of labor in the sourcing of ideas and their exploitation" (OECD 2012, p. 192). Hence, this establishes the reason "(f)irms, universities and STI actors have clustered around geographical areas, industries or groups of related technologies in order to improve networking and generate more spillovers from open and collaborative innovation" (OECD 2014, p. 43).

This scenario would further explain the restructuring of the innovation process, affecting the positioning of every player. "The globalisation of R&D and the emergence of open innovation platforms are fast redefining how businesses innovate and are leading governments to enhance framework conditions for research and innovation as well as to adapt their specific policies and supporting instruments to the changing nature of innovation" (OECD 2008, p. 58). Although the transition to an open innovation strategy inherently derives from the companies themselves, public policies are identified as central "in facilitating this transition by removing potential barriers" (OECD 2002, p. 121).

Hence, "government support for business innovation is becoming more oriented towards fostering open innovation" (OECD 2006, p. 51). Resulting from this approach, different countries around the world have adopted open innovation platforms, including Colombia, Greece, Norway, Poland, Japan, South Africa (IDB 2017b; OECD 2012, 2014). Open innovation features among the three strategic priorities of the European Union's STI policy (European Commission 2016).

Lastly, we need to comment on the occurrence of several allusions to open innovation as associated with other x-innovation concepts. Aligned with social innovation, it happens mainly through so-called open innovation platforms. This broadly considers the development of social and open innovation platforms as a means of the population in general evolving in terms of the technoscientific development process (IDB 2016, p. 3). Open innovation platforms are considered "a foundation for social innovation and the engagement in innovation activities of the population at large" (IDB 2017a, p. 58). This would provide the means of "turning innovative thinking into the source of practical solutions for the poor and excluded"

(IDB 2017a, p. 16), which "proved to be a powerful instrument to get larger constituencies involved in STI activities and policies" (idem, p. 50).

Albeit coordinated with user innovation, the narrative highlights both "new forms of innovation ... through new methods and greater interaction among innovation actors" (OECD 2014, p. 400) and the "acknowledged benefits" from "pooling external expert knowledge or collective creativity" (OECD 2018, p. 226). In turn, when linked with digital innovation, the corpus expresses "the challenging nature of innovation" (OECD 2018, p. 97), affirming the "the potential of open digital innovation and platforms in areas such as scientific research, business innovation, technology commercialization and talent development" (IDB 2017a, p. 59). Lastly, when associated with responsible innovation, the emphasis of recent years has been "towards more open, co-creative and responsive forms of innovation" (OECD 2018, p. 230). The narratives present in the corpus resemble those in the literature as regards the potential for open innovation to accelerate the development of new products while reducing risks but still getting the results to the market quicker (Tidd 1995). Although there is no mention in the corpus regarding the specific features of two kinds of open innovation; outside-in, inbound open innovation involving opening up the company's innovation processes to external inputs and contributions; and inside-out, an outbound open innovation requiring organizations to allow unused and underutilized ideas to flow outside the company for others to develop (Dahlander and Gann 2010).

However, the idea of open innovation as 'a foundation for social innovation,' as presented in the document corpus, is not entirely developed in the literature consulted. At most, one can encounter statements such as "(i)t is important to consider how we can nurture digital technologies toward a positive social and economic impact" (Bogers et al. 2018, p. 8), without further developments, and that the SDGs can "be harnessed as a further impetus to open innovation" (idem, p. 11). In the end, it is down to the reader to accept whether, besides its role in making firms more productive and competitive, open innovation "makes us better people too" (Bogers et al. 2018, p. 9).

## Environmental Innovations: Eco and Green Innovation

As the literature uses eco-innovation and green innovation as synonymous for environmental innovation (Bernauer et al. 2007), we opted to jointly present the analysis on them. Furthermore, before reporting on the main findings, it is worth highlighting that most mentions (around three quarters) are concentrated into just two years 2012 and 2014 – mostly in the former.

The document corpus presents this as "innovations that reduce the use of natural resources and decrease the release of harmful substances across the entire life cycle" (OECD 2014, p. 146). This "means accelerating not only the rate, but also the direction of innovation towards producing knowledge solutions that address environmental problems" (OECD 2012, p. 92). Thus, this considers environmental innovations as a way to address 'social challenges such as climate change' in ST&I public governance (OECD 2014), fundamental for the economy "to adapt and become more resilient to climate change and more resource efficient, while at the same time remaining competitive" (European Commission 2016, p. 16). Intriguingly, the most cited 'x-innovation' concepts probably reflect the most evasive found in the corpus in a facet that even gains OECD recognition: "the lack of clear definitions of what constitutes green technologies and innovation can hamper benchmarking and policy learning" (OECD 2012, p. 89).

The conceptualization presented in the document corpus resembles that present in the literature. Again, despite the lack of consensus regarding the definition, the literature presents it as "an innovation that improves economic and environmental performance" (Driessen and Hillebrand 2002), attracting "green rents on the market" (Andersen 2002). References of economic variables are not by chance: "there are many factors affecting eco-innovation and only one of those is environmental motivation." Moreover, even assessments of the "improvement of environmental performance" are deemed controversial and an open field (Carrillo-Hermosilla et al. 2009, p. 9). Likewise, "from the social point of view, it does not matter very much if the initial motivation for the uptake of eco-innovation is purely an environmental one" (idem). Far more so, the main goal of environmental innovations involves increasing the firm's "chances of survival and growth" (Laforet 2009, pp. 188–9). Whatever the main determinants for adopting environmental innovations – e.g., regulation, quest for competitiveness – "integrating environmental sustainability issues into business strategy and greening the innovation process are becoming a strategic opportunity for companies" (Albort-Morant et al. 2016, p. 4912).

Delinking economic growth from environmental constraints "requires establishing incentives and institutions that lead to significant green innovations and their widespread adoption and diffusion" (OECD 2012, p. 63). For developing countries, green innovation would even pose "a strong driver for expanding markets and sustainable economic development" (OECD 2012, p. 91). However, green innovation undergoes a 'double externality problem'. On the one hand, the private sector tends to underinvest in research due to "knowledge externalities and the disincentives provided by free riding" alongside uncertain market demand and returns of investment. On the other hand, there is "the negative externalities of climate change and other environmental challenges" (OECD 2012, p. 63). Additionally, beyond the typical market failures associated with any innovation, environmental innovations would experience their own specific failures, "unique to, or more prevalent in, markets for green innovation" (idem).

This would explain why – even if with different policy priorities and support levels – measures to foster environmental innovations have been deployed in a growing number of developed and developing nations. Forty-two countries[16] (OECD 2012, 2014) make mention of policy initiatives to "create critical mass and accelerate the transition to green innovation and technology" in a "crucial impetus for competitive and sustainable economies" (OECD 2012, pp. 62 and 68). Specifically in the European Union, as part of the measures taken to create a "more innovation-friendly business environment in Europe," the "eco-innovation action plan" was launched (European Commission 2014, p. 9) – "a comprehensive set of initiatives to improve the market's uptake of eco-innovation" (European Commission 2014, p. 48). The goal is "greening industry through eco-innovation" (European Commission 2014, p. 146).

Lastly, there is an absent aspect of the literature in these documents but worth highlighting: the relationship between environmental innovation with user and open innovation. On the one hand, user behaviors emerge as a critical dimension for environmental innovations, determining the direction and pace of development as well as the adoption of environmental innovation (Albort-Morant et al. 2016). On the other hand, 'eco-friendly knowledge' sharing and relational learning activities – open innovation principles – also get classified as fundamental here: "a fruitful green innovation process requires collaboration and knowledge exchanges with external stakeholders" (Albort-Morant et al. 2016, p. 4914).

## CONCLUDING REMARKS

Documents such as the reports from international organizations make claims about just how the governance of innovation should take place and towards which priorities. These narratives, from international organizations – renowned both for their major prestige and sheer influence, guiding ST&I policies around the world – could act as frameworks for stakeholders championing modes of ST&I policy governance while also comprising cognitive and normative assets for policy actors. Policy discourses constitute powerful tools in the hands of political actors seeking to foster or hinder particular public policy issues. This thus means understanding these narratives can help in understanding various different characteristics of ST&I policy at the national level, ranging from problem selection to public governance. This also reflects in grasping just what gets prioritized and what falls beyond the scope of policy discourses and therefore making a relevant contribution to policy analysis, especially in the ST&I arena. In saying that, we do not intend to summarize the Science, Technology and Innovation Policy (STIP) process as involving only a single dynamic. We recognize that the STIP process is conditioned, at the national level, through cultural and political traits, local conditions and divergent idiosyncrasies – which underline the policy process is more complicated than the mere emulation of models and practices. More than explain national specificities, we aim to put some light on the power of rhetorical praxis of international organizations to explain potential similarities among STIP national policies.

This chapter reconstructed the x-innovation policy narratives from a few representative international organizations. While not exhaustive, this study is illustrative and would open perspectives regarding the identification of changes in these narratives over time and possible differences between international organizations and national governments or academic papers.

An opening remark focuses on the low pervasiveness of those x-innovation concepts in the international organization documents. While on the one hand, this makes it challenging to extract major conclusions in terms of the impact of these narratives, on the other hand, their marginal position seems to indicate the expansion of these discourses has not yet had any significant impact on the dominant interpretation of innovation prevailing in these international organizations.

Regarding the narratives itself, the main findings reinforce the already mentioned specificities regarding the use of these 'x-innovation' concepts (Gaglio et al. 2019; Godin 2015): (i) claims over enlarging participation in ST&I development, (ii) assertions over their potential for solving social, environmental and ethical issues and, (iii) their ambivalence. As regards the first, all of the concepts analyzed refer to the importance of amplifying individual participation levels. However, when these statements are made, (allegedly) advocating the democratization of ST&I, they appear in vague and imprecise ways. Hardly by accident, participation emerges in greater detail when individuals are identified as the users and the sources of ideas and information (user-driven and open innovation); as audiences that appropriate knowledge and produce unprompted change (digital innovation) or to legitimate and guarantee acceptance of new products and services (eco and green innovation).

Second, the claim about the potential of these kinds of innovation for solving social and ethical issues is present across all of the concepts analyzed even though with different nuances. These extend from the management of the risks and uncertainties regarding ST&I developments (responsible innovation), adjusting their trajectories according to social concerns (user-driven innovation), making public administrations more transparent (digital innovation), promoting better usages of natural resources (environmental innovations) and producing new products and services that improve people's quality of life (social innovation) more cheaply and amplifying the consumption levels of the most disadvantaged (inclusive innovation).

Third, the ambivalence and ambiguity regarding these terms. While the explicit mention of the social and ethical dimensions is a fact in all the analyzed concepts, it is also true that these x-innovation concepts are wedded to economic ideology. Those terms embrace markets and business strategies and serve the interests of firms. The economic connotation of innovation still remains very strong and barely challenged. There is no evident rupture with the core values of technological innovation. On the boundaries, these might just be 'more social into innovation' (Gaglio et al. 2019). Hence, the concepts may claim to be counterhegemonic, but their purposes are not necessarily so.

As the policy narratives of international organizations could act as frameworks for policy-makers, championing governance and helping explain what to hinder and what to prioritize in ST&I policies, it would not be surprising to observe analogous narratives at the national levels. That reflection reaches beyond the scope of this work but further research incorporating documents at national level might turn out especially elucidative.

## NOTES

1. Analysis of the application of "x-innovation" concepts and the role attributed to innovation for (allegedly) counterhegemonic purposes in the STI national policies of Iberoamerican countries is to be found in Bagattolli and Brandão (2019).
2. We may summarize the European Union's institutional set-up as follows: "the EU's broad priorities are set by the European Council, which brings together national and EU-level leaders, directly elected MEPs represent European citizens in the European Parliament, the interests of the EU as a whole are promoted by the European Commission, whose members are appointed by national governments that defend their own country's national interests in the Council of the European Union." https://europa.eu/european-union/about-eu/institutions-bodies_en.
3. https://www.iadb.org/en/about-us/knowledge-generation%2C7916.html.
4. The only exception was the *OECD Science, Technology and Industry Outlook*
5. *Social innovation*, Chapter 6 by Cornelius Schubert; *Responsible innovation*, Chapter 8 by Lucien von Schomberg; *User-centred innovation*, Chapter 9 by Bastien Tavner; *Open innovation*, Chapter 10 by Tiago Brandão.
6. *Sustainable innovation*, Chapter 7 by Frank Boons and Riza Batista; *Grassroots innovation*, Chapter 13 by Fayaz Ahmad Sheikh and Hemant Kumar and *Frugal innovation*, Chapter 14 by Céline Cholez and Pascale Trompette.
7. On this specific topic, see Chapter 8 by von Schomberg in this book.
8. Searching for "responsible innovation," englobing all years and index. https://www.webofknowledge .com/.
9. On that specific topic, see Tavner, Chapter 9 in this book.
10. "The Sustainable Development Goals are a universal call to action to end poverty, protect the planet and improve the lives and prospects of everyone, everywhere. The 17 Goals were adopted by all UN Member States in 2015, as part of the 2030 Agenda for Sustainable Development which set out a 15-year plan to achieve the Goals." They are: (i) no poverty, (ii) zero hunger, (iii) good health and well-being, (iv) quality education, (v) gender equality, (vi) clean water and sanitation, (vii) affordable and clean energy, (viii) decent work and economic growth, (ix) industry, innovation and infrastructure, (x) reduced inequalities, (xi) sustainable cities and communities, (xii) responsible consumption and production, (xiii) climate action, (xiv) life below water, (xv) life on land, (xvi) peace, justice and strong institutions and (xvii) partnerships for the goals. More information can be found at: https://www.un.org/sustainabledevelopment/development-agenda.

11. Also called as 'pro-poor innovation', 'below-the-radar innovation', 'grassroots innovation', 'base-of-the pyramid innovation' and more (Heeks et al. 2014).
12. The Platform (apparently hosted on this website https://webgate.ec.europa.eu/socialinnovat ioneurope/en) is no longer available.
13. Further information about this competition is available at: https://ec.europa.eu/growth/industry/ policy/innovation/social/competition_en.
14. A comprehensive policy narrative analysis, respecting the different views regarding social innovation governance, involving documents from academia, the public sector and citizen sector organizations – highlighting points of conflict and convergence in the different narratives, can be found in Ney (2014).
15. On that specific topic, see Brandão, Chapter 10 in this book.
16. In alphabetical order: Argentina, Australia, Austria, Brazil, Belgium, Canada, Chile, Colombia, Czech Republic, Denmark, Estonia, Finland, France, Germany, Greece, Hungary, Iceland, India, Indonesia, Ireland, Israel, Italy, Japan, Korea, Lithuania, Luxembourg, México, New Zealand, Norway, People's Republic of China, Poland, Portugal, Russian Federation, Slovak Republic, South Africa, Spain, Sweden, Switzerland, the Netherlands, Turkey, United Kingdom, United States.

# REFERENCES

Aguiar, Diego, Francisco Aristimuño and Nicolás Magrini (2015), 'El rol del Banco Interamericano de Desarrollo (BID) en la reconfiguración de las instituciones y políticas de fomento a la ciencia, la tecnología y la innovación de la Argentina (1993–1999)', *CTS – Revista Iberoamericana de Ciencia, Tecnología y Sociedad*, 10 (29), 11–40.

Albort-Morant, Gema, Antonio Leal-Millán, Gabriel Cepeda-Carrión, A Leal-Millan and G Cepeda-Carrion (2016), 'The antecedents of green innovation performance: A model of learning and capabilities', *Journal of Business Research*, 69 (11), 4912–17. https://doi.org/10.1016/j.jbusres.2016 .04.052

Andersen, Maj Munch (2002), 'Organising interfirm learning – as the market begins to turn green', in Theo J. N. M. Bruijn and Arnold Tukker (eds), *Partnership and Leadership – Building Alliances for a Sustainable Future*, Kluwer Academic Publishers, pp. 103–19.

Bagattolli, Carolina and Tiago Brandão (2019), 'Counterhegemonic narratives of innovation: Political discourse analysis of Iberoamerican countries', *NOvation: Critical Studies of Innovation*, 1, 67–105. http://www.novation.inrs.ca/index.php/novation/article/view/6

Bagattolli, Carolina, Tiago Brandao, Amilcar Davyt, Carlos Nupia, Monica Salazar and Mariana Versino (2016), 'Relaciones entre científicos, organismos internacionales y gobiernos en la definición de las Políticas de Ciencia, Tecnología e Innovación en Iberoamérica', in Rosalba Casas, T. S. Pereira and Alexis Mercado (eds), *Políticas de Ciencia, Tecnología e Innovación en Iberoamérica: desafíos en contextos periféricos*, Programa Ibero-Americano de Ciencia y Tecnología para el Desarrollo (CYTED), pp. 189–219.

Bardin, Laurence (2016), *Análise de Conteúdo*, São Paulo: Edições 70.

Bernauer, Thomas, Stephanie Engel, Daniel Kammerer and Jazmin Seijas Nogareda (2007), 'Explaining green innovation: Ten years after Porter's win–win proposition: How to study the effects of regulation on corporate environmental innovation?', *Politische Vierteljahresschrift*, 39, 323–41.

Bogers, Marcel, Henry Chesbrough and Carlos Moedas (2018), 'Open innovation: Research, practices, and policies', *California Management Review*, 60 (2), 5–16. https://doi.org/10.1177/0008125617745086

Bontems, Vincent K. (2014), 'What does innovation stand for? Review of a watchword in research policies', *Journal of Innovation Economics & Management*, 3 (15), 39–57.

Braun, Dietmar (2006), 'The mix of policy rationales in science and technology policy', *Melbourne Journal of Politics*, 31, 8–35.

Buur, Jacob and Ben Matthews (2008), 'Participatory innovation', *International Journal of Innovation Management*, 12 (3), 255–73. https://doi.org/10.1142/S1363919608001996

Cajaiba-Santana, Giovany (2014), 'Social innovation: Moving the field forward. A conceptual framework', *Technological Forecasting and Social Change*, 82, 42–51. https://doi.org/10.1016/j.techfore.2013.05.008

Carrillo-Hermosilla, Javier, Pablo del R González and Totti Könnölä (2009), *Eco-Innovation: When Sustainability and Competitiveness Shake Hands*, London: Palgrave Macmillan.

Chesbrough, Henry (2006), 'Open innovation: A new paradigm for understanding industrial innovation', in Henry Chesbrough, Wim Vanhaverbeke and Joel West (eds), *Open Innovation: Researching a New Paradigm*, Oxford: Oxford University Press, pp. 1–12.

Dahlander, Linus and David M. Gann (2010), 'How open is innovation?', *Research Policy*, 39 (6), 699–709. https://doi.org/10.1016/j.respol.2010.01.013

Driessen, Paul and Bas Hillebrand (2002), 'Adoption and diffusion of green innovations', in Gerard Bartels and Wil Nelissen (eds), *Marketing for Sustainability: Towards Transactional Policy-Making*, the Netherlands: IOS Press Inc, pp. 343–56.

Drori, Gili S., John W. Meyer, Francisco O. Ramirez and E. Schofer (2003), *Science in the Modern World Polity: Institutionalization and Globalization*, Stanford, CA: Stanford University Press.

Dye, Thomas R. (2013), *Understanding Public Policy*, 14th ed., Boston, MA: Pearson.

Edwards-Schachter, Mónica (2018), 'The nature and variety of innovation', *International Journal of Innovation Studies*, 2 (2), 65–79. https://doi.org/10.1016/j.ijis.2018.08.004

Edwards-Schachter, Mónica and Matthew L. Wallace (2017), '"Shaken, but not stirred": Sixty years of defining social innovation', *Technological Forecasting and Social Change*, 119, 64–79. https://doi.org/10.1016/j.techfore.2017.03.012

European Commission (2013), State of the Innovation Union 2012: Accelerating change.

European Commission (2014), State of the Innovation Union: Taking stock 2010–2014. https://doi.org/10.2777/74073

European Commission (2016), Research and innovation: Pushing boundaries and improving the quality of life. https://doi.org/10.2775/903982

Fischer, Frank and John Forester (1993), 'Editors' introduction', in Frank Fischer and John Forester (eds), *The Argumentative Turn in Policy Analysis and Planning*, London: UCL Press, pp. 1–17.

Freeman, Christopher and Luc Soete (2008), *A Economia da Inovação Industrial*, Campinas: Editora UNICAMP.

Gaglio, Gérald (2017), '"Innovation fads" as an alternative research topic to pro-innovation bias: The examples of Jugaad and Reverse Innovation', in Godin and Dominique Vinck (eds), *Critical Studies of Innovation: Alternative Approaches to the Pro-Innovation Bias*, Cheltenham, UK and Northampton, MA, USA: Edward Elgar Publishing, pp. 33–47. https://doi.org/10.4337/9781785367229

Gaglio, Gérald, Benoît Godin and Sebastian Pfotenhauer (2019), 'X-Innovation: Re-inventing innovation again and again', *NOvation: Critical Studies of Innovation*, 1, 1–16. http://www.novation.inrs.ca/index.php/novation/article/view/8/6

George, Gerard, Anita M. McGahan and Jaideep Prabhu (2012), 'Innovation for inclusive growth: Towards a theoretical framework and a research agenda', *Journal of Management Studies*, 49 (4), 661–83. https://doi.org/10.1111/j.1467-6486.2012.01048.x

Godin, Benoît (2008), Innovation: The history of a category. www.csiic.ca

Godin, Benoît (2015), Innovation contested: The idea of innovation over the centuries. Routledge. http://www.csiic.ca/en/the-idea-of-innovation/

Godin, Benoît (2019), *The Invention of Technological Innovation*, Cheltenham, UK and Northampton, MA, USA: Edward Elgar Publishing. https://doi.org/10.4337/9781789903348

Godin, Benoît (2020), *The Idea of Technological Innovation: a Brief Alternative History*, Cheltenham, UK and Northampton, MA, USA: Edward Elgar Publishing.

Godin, Benoît and Dominique Vinck (2017), 'Introduction: innovation – from the forbidden to a cliché', in Benoît Godin and Dominique Vinck (eds), *Critical Studies of Innovation: Alternative Approaches to the Pro-Innovation Bias*, Cheltenham, UK and Northampton, MA, USA: Edward Elgar Publishing, pp. 1–14. https://doi.org/10.4337/9781785367229

Grinbaum, Alexie and Christopher Groves (2013), 'What is "responsible" about responsible innovation? Understanding the ethical issues', in Richard Owen, John Bessant and Maggy Heintz (eds), *Responsible Innovation. Managing the Responsible Emergence of Science and Innovation in Society*, London: John Wiley, pp. 119–42.

Hargadon, Andrew B. and Yellowlees Douglas (2001), 'When innovations meet institutions: Edison and the design of the electric light', *Administrative Science Quarterly*, 46 (3), 476. https://doi.org/10.2307/3094872

Heeks, Richard, Christopher Foster and Yanuar Nugroho (2014), 'New models of inclusive innovation for development', *Innovation and Development*, 4 (2), 175–85. https://doi.org/10.1080/2157930X.2014.928982

Henfridsson, Ola, Lars Mathiassen and Fredrik Svahn (2014), 'Managing technological change in the digital age: The role of architectural frames', *Journal of Information Technology*, 29 (1), 27–43. https://doi.org/10.1057/jit.2013.30

Henriques, Luisa and Philippe Larédo (2013), 'Policy-making in science policy: The "OECD model" unveiled', *Research Policy*, 42 (3), 801–16.

Hinings, Bob, Thomas Gegenhuber and Royston Greenwood (2018), 'Digital innovation and transformation: An institutional perspective', *Information and Organization*, 28 (1), 52–61. https://doi.org/10.1016/j.infoandorg.2018.02.004

Hyysalo, Sampsa, Petteri Repo, Paivi Timonen, Louna Hakkarainen and Eva Heiskanen (2016), 'Diversity and change of user driven innovation modes in companies', *International Journal of Innovation Management*, 20 (2), 1650023. https://doi.org/10.1142/S1363919616500237

IDB (2016), Que hace el BID en innovación, ciencia y tecnología?

IDB (2017a), Innovation, Science and Technology Sector Framework Document. In Inter-American Development Bank (Issue October).

IDB (2017b), Políticas públicas para la creatividad y la innovación: impulsionando la economía naranja en América Latina y el Caribe (José Miguel Benavente and Matteo Grazzi (eds)).

Kallerud, Egil (2010), Goal conflicts and goal alignment in science, technology and innovation policy discourse. EASST 2010 Conference: Practicing Science and Technology, Performing the Social, September 2–4, (1), 1–22.

Karinen, Risto and David H. Guston (2010), 'Towards anticipatory governance. The experience with nanotechnology', in Mario Kaiser, Monica Kurath, Sabrine Maasen and Cristoph Rehmann-Sutter (eds), *Governing Future Technologies. Nanotechnology and the Rise of an Assessment Regime*, Dordrecht: Springer, pp. 217–32.

Klochikhin, Evgeny A. (2012), 'Linking development and innovation: What does technological change bring to the society?', *The European Journal of Development Research*, 24 (1), 41–55. https://doi.org/10.1057/ejdr.2011.20

Laforet, Sylvie (2009), 'Effects of size, market and strategic orientation on innovation in non-high-tech manufacturing SMEs', *European Journal of Marketing*, 43 (1/2), 188–212. https://doi.org/10.1108/03090560910923292

Lazonick, William and Mariana Mazzucato (2013), 'The risk–reward nexus in the innovation–inequality relationship: Who takes the risks? Who gets the rewards?', *Industrial and Corporate Change*, 22 (4), 1093–128. https://doi.org/10.1093/icc/dtt019

Lundvall, Bengt-Åke and Susana Borras (2004), 'Science, technology and innovation policy', in Jan Fagerberg, David Mowery and Richard Nelson (eds), *The Oxford Handbook of Innovation*, Oxford: Oxford University Press, Chapter 22.

Majone, Giandomenico (1989), *Evidence, Argument & Persuasion in the Policy Process*, New Haven, CT: Yale University Press.

McBeth, Mark K. (2014), 'Preface: The Portneuf School of Narrative', in Michael D. Jones, Elizabeth A. Shanahan and Mark K. McBeth (eds), *The Science of Stories: Applications of the Narrative Policy Framework in Public Policy Analysis*, Palgrave Macmillan, pp. xiii–xx. https://doi.org/10.1057/9781137485861

Mulgan, Geoff, Simon Tucker, Ali Rushanara and Ben Sanders (2007), 'Social innovation: What it is, why it matters and how it can be accelerated', *The Young Foundation*. 51 pages. https://youngfoundation.org/wp-content/uploads/2012/10/Social-Innovation-what-it-is-why-it-matters-how-it-can-be-accelerated-March-2007.pdf

Nambisan, Satish, Mike Wright and Maryann Feldman (2019), 'The digital transformation of innovation and entrepreneurship: Progress, challenges and key themes', *Research Policy*, 48, 1–10. https://doi.org/10.1016/j.respol.2019.03.018

Ney, Steven (2014), 'The governance of social innovation: Connecting meso and macro levels of analysis', in Michael D. Jones, Elizabeth A. Shanahan and Mark K. McBeth (eds), *The Science of Stories: Applications of the Narrative Policy Framework in Public Policy Analysis*, London: Palgrave Macmillan, pp. 207–34.

Nupia, Carlos M. (2014), La política científica y tecnológica en Colombia 1968–1991: transferencia y aprendizaje a partir de modelos internacionales, Editorial Universidad de Antioquia.

Nylén, Daniel and Jonnt Holmström (2015), 'Digital innovation strategy: A framework for diagnosing and improving digital product and service innovation', *Business Horizons*, 58 (1), 57–67. https://doi .org/10.1016/j.bushor.2014.09.001

OECD (1996), Science, Technology and Industry Outlook 1996. OECD – Organisation for Economic Co-operation and Development.

OECD (1998), Science, Technology and Industry Outlook 1998. OECD – Organisation for Economic Co-operation and Development.

OECD (2000), Science, Technology and Industry Outlook 2000. OECD – Organisation for Economic Co-operation and Development. https://doi.org/10.1016/s0166-4972(01)00044-x

OECD (2002), Science, Technology and Industry Outlook 2002. OECD – Organisation for Economic Co-operation and Development. https://doi.org/10.1017/CBO9781107415324.004

OECD (2004), Science, Technology and Industry Outlook 2004. OECD – Organisation for Economic Co-operation and Development.

OECD (2006), Science, Technology and Industry Outlook 2006. OECD – Organisation for Economic Co-operation and Development.

OECD (2008), Science, Technology and Industry Outlook 2008. OECD – Organisation for Economic Co-operation and Development. https://doi.org/10.1787/sti_outlook-2008-en

OECD (2010), Science, Technology and Industry Outlook 2010. OECD – Organisation for Economic Co-operation and Development. https://doi.org/10.5860/choice.48-5436

OECD (2012), Science, Technology and Industry Outlook 2012. OECD – Organisation for Economic Co-operation and Development.

OECD (2014), Science, Technology and Industry Outlook 2014. OECD – Organisation for Economic Co-operation and Development.

OECD (2016), Science, Technology and Innovation Outlook 2016. OECD – Organisation for Economic Co-operation and Development.

OECD (2018), Science, Technology and Innovation Outlook 2018: adapting to technological and societal disruption. OECD – Organisation for Economic Co-operation and Development. https://doi.org/ 10.1787/ba79a818-it

Owen, Richard, Phil Macnaghten and Jack Stilgoe (2012), 'Responsible research and innovation: From science in society to science for society, with society', *Science and Public Policy*, 39 (6), 751–60. https://doi.org/10.1093/scipol/scs093

Pel, Bonno (2016), 'Trojan horses in transitions: A dialectical perspective on innovation "capture"', *Journal of Environmental Policy & Planning*, 18 (5), 673–91. https://doi.org/10.1080/1523908X.2015 .1090903

Pol, Eduardo and Simon Ville (2009), 'Social innovation: Buzz word or enduring term?', *The Journal of Socio-Economics*, 38 (6), 878–85. https://doi.org/10.1016/j.socec.2009.02.011

Roe, Emery (1994), *Narrative Policy Analysis: Theory and Practice*, Durham, NC: Duke University Press.

Segercrantz, Beata, Karl-Erik Sveiby and Karin Berglund (2017), 'A discourse analysis of innovation in academic management literature', in Benoît Godin and Dominique Vinck (eds), *Critical Studies of Innovation: Alternative Approaches to the Pro-Innovation Bias*, Cheltenham, UK and Northampton, MA, USA: Edward Elgar Publishing, pp. 276–95. https://doi.org/10.4337/9781785367229

Stone, Deborah A. (1989), 'Causal stories and the formation of policy agendas', *Political Science Quarterly*, 104 (2), 281. https://doi.org/10.2307/2151585

Tidd, Joe (1995), 'Development of novel products through intraorganizational and interorganizational networks: The case of home automation', *Journal of Product Innovation Management*, 12 (4), 307–22. https://doi.org/10.1111/1540-5885.1240307

von Schomberg, René (2011), 'Towards responsible research and innovation in the information and communication technologies and security technologies fields', *SSRN Electronic Journal*, https://doi.org/10.2139/ssrn.2436399

Yoo, Youngjin, Kalle Lyytinen, Richard Boland, Nicholas Berente, James Gaskin, Doug Schutz. (2010), 'The next wave of digital innovation: Opportunities and challenges: A report on the research workshop "Digital Challenges in Innovation Research"', *SSRN Electronic Journal*. https://doi.org/10.2139/ssrn.1622170

Zittrain, Jonathan (2006), 'The generative internet', *Harvard Law Review*, 119, 1975–2040. https://doi.org/10.1145/1435417.143542.

# 16. Transformative innovation policy: a novel approach?

*Markus Grillitsch, Teis Hansen and Stine Madsen*

## INTRODUCTION

The focus and instruments of innovation policy have changed fundamentally over the last decades. Scholars have captured this by proposing various periodisations of innovation policy. For instance, OECD (1963) constitutes one landmark suggesting a shift towards increased interest for economic effect of science through innovation. Lundvall and Borrás (2005) suggest a transition from science, to technology and finally innovation policy. Weber and Rohracher (2012) differentiate innovation policy depending on the failure framework it uses. In a recent attempt, Schot and Steinmueller (2018) argue for three frames for innovation policy based on the linear model, the systemic perspective of innovation and the focus on transformative change.

In this chapter, we critically examine the various periodisations offered in the literature, and consequently debate the novelty of the most recent shift towards transformative innovation policy. A main point put forward in this chapter is that a periodisation of innovation policy in a clear sequence where one frame precedes another, gives insufficient attention to the common elements of these different understandings of innovation policy. Thus, rather than very clear shifts between generations of innovation policy, it might be suggested that innovation policy evolves more gradually, for instance, transformative elements are foregrounded recently under the umbrella of "transformative innovation policy" but can also be identified in previous perspectives on innovation policy.

Even though grand challenges such as global warming and migration flows, as well as technological change such as artificial intelligence and industry 4.0 have fired the debate on transformative innovation policies in recent years, it has a long history – even going back to Schumpeter. Schumpeter (1911) distinguished path-breaking economic activities from such that follow existing paths, thereby foregrounding the role of innovations with transformative character for economic development. We also find a long and substantial debate on lock-ins (Grabher 1993) and policy interventions for structural change (Hassink 2005; Tödtling and Trippl 2004). Following this tradition, and the work on regional path-dependency (Martin and Sunley 2006), more recent studies have discussed mechanisms and sources for new industrial path development (Grillitsch et al. 2018).

In parallel, a new literature has developed focusing on sustainability transitions, which suggests that addressing current grand challenges requires substantial changes in the socio-technical systems that deliver core services such as mobility and nutrition. Due to the urgency of tackling these challenges, incremental innovations are arguably insufficient, and the sustainability transitions literature consequently emphasises the importance of radical innovations (Markard et al. 2012). It is also underlined that transitions require not only innovations in new technologies, but also in social practices, regulations and infrastructures. These prioritisations have important implications for the objectives, instruments and overall design of innovation policy (Grillitsch et al. 2019; Kivimaa and Kern 2016; Weber and Rohracher 2012). In particular, directionality of innovation policy plays a central role because transformative innovation policies build on the explicit aim that change processes should support the achievement of sustainable development outcomes. Consequently, innovation is not viewed as an end, but is instead treated as a means to achieve transformative change. In this view, innovation policy is not merely focused on developing and diffusing new technologies, but also aimed at destabilising and phasing out unsustainable ones (Kivimaa and Kern 2016).

This book chapter unfolds the historical and conceptual roots of transformative innovation policy, compares critically the different strands of literature, and discusses policy implications as well as important questions for future research.

## SCIENCE, TECHNOLOGY AND INNOVATION POLICY – THE TECHNOLOGY-PUSH APPROACH

Policies for promotion of science, technology and innovation take many forms, but here we use the term to refer to a policy approach that implicitly or explicitly builds on a linear understanding of innovation processes, where scientific discoveries precede technology development activities, which eventually lead to the commercialisation of innovations. The development of this policy approach is often attributed to the report "Science: The Endless Frontier" commissioned by the US President Roosevelt and written by Vannevar Bush (1945). However, as examined in great detail by Godin (2017), the intellectual heritage of the linear model of innovation and the associated policy approach is considerably more rich and complicated, going back to anthropologists, rural sociologists and industrialists in the early twentieth century, who all considered the question of how novelty arises, diffuses and changes our societies.

### Model of Innovation

The linear model of innovation describes the development of innovation as a process starting with scientific research, followed by product development, production and finally marketing. A central characteristic of this model, namely the sequencing of the different steps, is also found in early work on the relation between invention and diffusion of cultural practices and traits in anthropology, as well as in writings by rural sociologists on the development and diffusion of agricultural technologies among farmers. However, the first linear model of innovation was conceived by Maurice Holland of the US National Research Council (e.g. Holland 1928), and further theorised by Maclaurin (1947). These contributions explicitly established an understanding of innovation as a sequential process developing from basic research, to applied research, product development, production and commercialisation. This linear conceptualisation of the innovation process has subsequently been reproduced in multiple forms, with significant influence on priority setting in science, technology and innovation policy (Godin 2017).

The linear innovation model has entailed a rather limited focus of science, technology and innovation policy in terms of both actors and instruments. Central types of organisations include universities, research institutes and firms – in particular large firms with own research laboratories – who play different roles with regards to achieving scientific breakthroughs, technology development and commercialisation of new technologies. Correspondingly, policy instruments focus on supporting activities in the various steps of the linear innovation model: provision of public funding for basic research; subsidising applied research directly or indirectly through tax credits; and establishment of intellectual property right regimes, which allow innovators to profit from their research and development (R&D) investments. Implicitly, the rationale behind the introduction of such policy instruments builds on neoclassical economics and the existence of market failures (Laranja et al. 2008; Mowery 1983): firms are disincentivised from investing in knowledge creation due to the difficulties of appropriating returns, leading to sub-optimal levels of investments in R&D from a societal perspective.

Complementary to these instruments, policymakers have since Holland (1928) and Maclaurin (1954) emphasised the importance of gathering statistics and intelligence about levels of investment in science and R&D (Godin 2017). Important manifestations of this are the development of the Frascati Manual (OECD 1963, now in its 7th version) for standardising methods for collection of R&D statistics, as well as the introduction of industrial classifications based on industries' investment levels in R&D. Essentially, the division of industries into categories ranging from high-tech to low-tech has been associated with an clear emphasis in policymaking on the small, R&D-intensive part of the economy, comprising industries such as pharmaceuticals, electronics and aerospace, to the extent that increasing R&D investments has become a goal in itself. As quoted in Hansen and Winther (2011), the European Commission (2008) argues that (p. 11) "[g]iven the weight of high-tech sectors in the overall level of business R&D intensity, a change should include the sectoral composition of the business sector, a move towards a higher share of high-tech companies" and policy should (p. 16) "change the balance of the industrial structure in favour of these research-intensive sectors". Policies for attracting high-skilled labour are also often specifically targeting high-tech industries (Hansen et al. 2014).

### Evaluation and Assessment

The literature abounds with critiques of the linear model of innovation and its associated policy implications, in particular concerning the mismatch between the model's emphasis on science and research as a driver of innovation, and empirical work highlighting the vastly heterogeneous character of innovation processes. Yet, from proponents of a transformative innovation policy approach, a main point of critique has been the assumption that innovation is automatically beneficial for society. Thus, Schot and Steinmueller (2018) suggest that traditional science, technology and innovation policy approaches have been overly focused on innovation as a means of achieving economic growth, thereby marginalising (p. 1561) "broader implications for the environment or human health and welfare".

However, this perspective on the intentions of scholars responsible for early work on innovation appears inaccurate or at least partial. In fact, innovation was directly connected to societal progress not just in the form of increasing possibilities to consume, but also in terms of improving public health, limiting pollution and conserving natural resources. As argued by Giuliani (2018), while some scholars might have subscribed to a simple 'science → innovation → economic growth' logic, social and environmental concerns were not unanimously ignored. To exemplify, improved health and living standards were important arguments to Holland (1931) for supporting innovation; Bush (1945) emphasised multiple socio-economic benefits of science; and Maclaurin researched unemployment and social security measures in other work (Myers and Maclaurin 1942).

Thus, rather than in intentions, the challenge of a linear perspective of innovation in terms of achieving transformative change is arguably found in the appropriateness of the derived policy instruments. Essentially, an emphasis on science-push instruments is adequate for 'solving' technical challenges as those associated with mission-oriented projects such as the Manhattan Project and the Apollo Program. However, addressing grand challenges such as climate change, biodiversity loss and aging societies are fundamentally different in being open-ended and highly complex, not only in technical terms. As pointed out by Mowery et al. (2010), the US government not only funded the Manhattan and Apollo projects, but was also the sole customer, effectively eliminating the need for demand-side policies. Further, the projects addressed technical challenges for which no existing solutions were in use, whereas addressing, for example, climate change, requires replacement of existing technologies, which is associated with significant distributional consequences. Tackling these matters requires a different policy toolbox than the one derived from a linear model of innovation.

## INNOVATION SYSTEM POLICY – THE INSTITUTIONAL APPROACH

Godin (2017) identifies the origins of innovation system policy in seminal work by Jack Morton and Christopher Freeman in the 1960s and 1970s, as well as OECD's effort to push a systemic perspective highlighting the importance of a national innovation policy. The momentum further increased when Bengt-Åke Lundvall became deputy director of the OECD Directorate for Science, Technology and Industry from 1992 to 1995, advocating for the innovation system approach. The innovation system approach differs from the science, technology, innovation (STI) perspective in its underlying theory of innovation, moving from a linear to an interactive model of innovation, in which a large variety of actors collectively engage in conceiving, developing and introducing ideas to the society.

The system approach originated with an appreciation for the many shapes of innovation, which can but do not necessarily entail the commercialisation of new scientific knowledge. Innovation is also viewed as a "counter-concept" to science, stressing that in the process of creating societal benefit many actors contribute in an inclusive process with science representing only one relevant type of actors (Godin 2016). Innovation systems are conceived as open systems, allowing for a variety of perspectives to emerge with a focus on nations (Freeman 1995; Nelson 1993; Lundvall 1992; Edquist 1997), regions (Autio 1998; Cooke 1992; Asheim and Isaksen 1997; Tödtling and Sedlacek 1997), sectors (Malerba 2002; Breschi and Malerba 1997), technologies (Carlsson and Stankiewicz 1991) and functions (Hekkert et al. 2007).

### Model of Innovation

The systemic perspective on innovation opposes the linear model of innovation. Rather than resulting from R&D, innovation processes often start with recognising or believing in a certain market opportunity (demand side), which is then followed by concept development (Kline and Rosenberg 1986; Kline 1985). Research and development comes in to solve technical problems associated with realising the concept and moving into production. The process described by Kline and Rosenberg is characterised by feedback-loops. For instance, results from R&D feed into concept development and opportunity discovery processes. The innovation system approaches therefore rests on a very different understanding of innovation processes, which is non-linear and interactive. Besides firms, universities and research organisations, the importance of a larger variety of actors has been recognised, including customers, intermediary organisations, educational and training facilities and public actors.

The early work on innovation systems in the 1990s highlights the importance of institutions, predominantly understood as set of formal and informal rules. Variations in national institutional configurations were linked to different patterns of innovation processes and outcomes. Closely linked to this literature is the varieties of capitalism concept, differencing in liberal market economies and coordinated market economies (Hall and Soskice 2001; Vitols 2001). It is argued that the institutional configuration of liberal market economies promote radical, breakthrough innovations in science-based sectors while coordinated market economies provide better conditions for incremental innovations based on interactive learning in engineering-based sectors. This also implies a broader understanding of innovation than present in the STI perspective.

Related to that, the innovation system approach provides a more differentiated spatial perspective, which resulted in a number of regional typologies. These typologies showcase systematic patterns that relate to (i) key actors and governance (Asheim and Isaksen 2002; Cooke 1998), (ii) the strengths in radical versus incremental innovations (Cooke 2004), and (iii) regional innovation system failures (Tödtling and Trippl 2005; Isaksen 2001). Essentially, this literature finds that innovation processes and outcomes do not only differ between nations but also between regions. Furthermore, the literature also highlights that innovation comes in many forms and that there is no single best-practice model that would suit all regions. However, innovation policy shall address the particular barriers identified in regional and national contexts.

Appreciating the interactive nature of innovation processes, the involvement of multiple actors in such processes, and the variegated shapes of innovation processes and outcomes at national and regional levels, the innovation system approach broadens the scope for policy intervention considerably as compared to the STI perspective. Policy interventions are to be justified by innovation barriers or system failures. Klein Woolthuis et al. (2005) differentiate in infrastructure, institutional, network and capability failures and link those to a variety of types of actors including demand (consumers, large buyers), companies (small and large firms, start-ups), knowledge institutions (universities, technology institutes) and third parties (banks, intermediaries, sector organisations, employees). Policies related to these various domains need to be coordinated to strengthen innovation systems (Lundvall and Borrás 2005). At the regional level, Tödtling and Trippl (2005) associate regional types to main system failures. Accordingly, metropolitan regions typically suffer from a fragmentation of networks, old industrial regions from lock-in and peripheral regions from organisational thinness. Borrás and Edquist (2013) argue that a mix of policies need to be mobilised in order to address specific failures in regions and nations covering regulatory instruments, economic and financial instruments and soft instruments.

## Evaluation and Assessment

The innovation system approach has been criticised for being static and "concerns structure or institutions rather than time or sequence" (Godin 2017, p. 139). According to this view, the innovation system approach is less concerned with changes over time (a process perspective) than with how different elements of the system play together at any point in time. This invites for a snapshot perspective, which is not suitable for an understanding of system change over time. Hence, historic or long-term studies about how innovation systems have changed over time are rare but would add value to understand the transformative capacity or effects of innovation systems. Yet, this does not necessarily imply that innovation system policies do not promote transformative change because innovation is considered as a key driver of change in the capitalist society (Schumpeter 1911; Shane and Venkataraman 2000). By providing the best possible preconditions for innovation, innovation system policy is about supporting industrial dynamics (Grillitsch and Trippl 2018).

Furthermore, a strand of the innovation system literature has explicitly addressed structural change in old industrial regions (Hassink 2005; Tödtling and Trippl 2004; Coenen et al. 2015b). In essence, this literature investigates the barriers and possible instruments for such regions to overcome lock-ins (Grabher 1993) and transform. Recent literature contributes to a dynamic perspective by emphasising new industrial path development, which is characterised by fundamental changes in actors' capabilities, networks and institutions (Grillitsch et al. 2018; Grillitsch and Asheim 2018; Isaksen and Trippl 2016; Hassink et al. 2019). A virtue of the innovation system approach is that it offers policy implications for a variety of regions and types of innovation, including science-based, radical innovations, which typically emerge in knowledge-intensive urban areas, as well as incremental, engineering based innovations, which are often at the heart of more remote areas.

A fundamental challenge of the innovation system literature, linked to the focus on how elements play together in a system at the expense of a process perspective (Godin 2017), is the lack of theorising and insights on how actors bring about transformative change. The innovation systems literature explains how combinations of elements such as institutions, networks and actor capabilities affect innovation outcomes (top-down explanation) but does not tell us much about how these elements are changed by the actions of individuals, groups of individuals and organisations (bottom-up explanation) (Uyarra et al. 2017; Asheim et al. 2016). As transformations cannot be understood without the micro-processes producing structural change, this is an important shortcoming. This shortcoming is being addressed (but not yet resolved) in emerging literature about agency in regional development (Grillitsch and Sotarauta 2019; Isaksen et al. 2019).

## TRANSFORMATIVE INNOVATION POLICY – THE CHALLENGE-DRIVEN APPROACH

In recent years, numerous scholars have begun to describe yet another shift in the innovation policy domain. This shift marks the emergence of transformative innovation policy (Schot and Steinmueller 2018) also sometimes referred to as challenge-driven innovation policy (Coenen et al. 2015a) or next generation innovation policy (Kuhlmann and Rip 2018). The transformative innovation perspective is often described as a profound break with the STI perspective and the innovation system perspective (Kuhlmann and Rip 2018; Schot and Steinmueller 2018). While scholars are not claiming that transformative innovation policy is fully replacing previous innovation perspectives, the shift to a transformative innovation perspective is emphasised as more than an incremental change in the innovation policy domain (Diercks et al. 2019). We suggest that the novelty of the transformative innovation perspective should not be overstated. In the following sections, we unpack the novel elements of the transformative innovation perspective, but also highlight the ways in which the transformative innovation perspective evolves from previous perspectives. After this we go on to explore the novelty in the implementation of transformative innovation policy.

**Model of Innovation**

We suggest that three aspects of the transformative innovation perspective represent novel elements in innovation policy. First, the aim to transition entire socio-technical systems, second, the emphasis on experimentation and third, the deliberate intention to destabilise unsustainable regimes.

The transformative innovation perspective rests on the aim of transitioning entire socio-technical systems. Socio-technical systems describe the systems that deliver core services for society such as mobility, energy and nutrition (Markard et al. 2012), and transformative innovation policy sets out to address the challenges that contemporary society face with regards to the delivery of these core services. The transformative innovation perspective therefore involves radical innovation across entire systems of both production and consumption, which in turn requires "novel configuration of actors, institutions and practices" (Weber and Rohracher 2012, p. 1037).

Despite a 'system-focus', this contrasts with the innovation system perspective: In the innovation system perspective, the notion 'system' refers to all the factors that influence the generation of innovation. In the framing of transformative innovation policy, 'system' refers to wider socio-technical systems delivering core services for society. Keeping this in mind, transformation challenges illustrated by Weber and Rohracher (2012) concern barriers to change in socio-technical systems, whereas market and system failures concern barriers to innovation performance. Arguably, two of these core challenges for transformative innovation policy – directionality and demand articulation – are qualitatively different compared with previous understandings of innovation policy, while the two remaining challenges – policy coordination and reflexivity – were also recognised before, but require deepened consideration in the transformative perspective.

First, attention to directionality identifies the lack of deliberate attempts to address transformative change through innovation processes. Whereas innovation system policy is broadly oriented towards optimising the structural and institutional environment of the innovation system itself, transformative innovation policy is oriented specifically towards particular societal challenges in socio-technical systems and should proactively steer the direction of innovation towards addressing these. As elaborated below in the point on destabilisation, this has substantial implications for policy formulation.

Second, demand articulation points to a lacking concern for market uptake of innovations and insufficient understanding of user practices and expectations. This reflects the innovation system perspective's main focus on the production side of innovation and the transformative innovation perspective's much greater emphasis on the consumption and demand side of innovation. Although demand and the involvement of actors on the demand side (e.g. users, consumers) is mentioned in both perspectives, their role differs. In the innovation system perspective, the role of users and consumers is mainly to lead and provide inputs into innovation processes, whereas the transformative innovation perspective relies on consumers and users to make innovations part of their everyday practices and life. The goal is to achieve a wide uptake of innovations rather than to focus on a select group of lead users.

Third, policy coordination concerns the difficulty of achieving coherent and timely policy actions across vertical and horizontal policy areas as well as public and private sector institutions. The importance of policy coordination is also emphasised in innovation systems policy (Lundvall and Borrás 2005). However, as transitions between socio-technical systems are complex and connect to multiple policy fields, breaking down policy silos and coordinating across a large number of authorities and organisations are arguably of even greater importance in transformative innovation policy.

Finally, reflexivity regards the need to engage with the long-term and uncertain nature of transformative change. Following from this, emphasis is placed on continuous attention to monitoring, anticipation and evaluation, as well as the establishment of policy platforms bringing together not only governmental representatives but also private sector and civil society actors (Weber and Rohracher 2012). While integrating reflexivity in innovation policy in the form of monitoring, anticipation and evaluation instruments is arguably not reserved for a transformative approach, their importance is likely particularly large in transformative innovation policy: as policy sets a clearer direction for innovation, continuous reflections around the desirability of this direction are needed in government and beyond.

We see this shift in innovation policy articulated in OECD publications. *The System Innovation: Synthesis Report* (OECD 2015, p. 9) clearly articulates this changing perspective on the role of policy by suggesting that

> by and large, most innovation policies aim to foster incremental change; fostering wider system change is a new challenge for innovation policymakers, especially as many of the actions will fall in areas outside the direct remit of research ministries or innovation agencies but where their input, coordination and implementation actions will remain critical.

The emphasis on policy for wider system change is arguably at odds with the OECD's traditional understanding of innovation.

In addition, the role of experimentation has been emphasised in different strands of literature relating to socio-technical transitions. It generally refers to "an inclusive, practice-based and challenge-led initiative, which is designed to promote system innovation through social learning under conditions of uncertainty and ambiguity" (Sengers et al. 2019, p. 153). Thus, the starting point is the uncertainty and complexity of societal challenges, which calls for experimentation that creates a space for learning, reflexivity and failure (Schot and Steinmueller 2018). This highlights the need for testing new technologies, but in a broader sense than previously conceptualised: focus is not only on laboratory tests and demonstration plants, but also on experiments with business models and the use of technologies in practice. Further, experimentation entails specific attention to learning about factors that make emerging technologies successful vis-à-vis established incumbent alternatives. The need for experimentation necessitates a different approach to innovation policy development, compared to STI and innovation system policy: rather than attempting to correct market and innovation system failures in order to 'optimise' innovation processes, transformative innovation policy puts greater emphasis on facilitating experiments, while acknowledging that many experiments will fail. One illustration of this is NESTA's innovation policy approach, including initiatives such as real-world testbeds (Arntzen et al. 2019; Morgan 2018).

The role of experimentation points to an interesting discrepancy in transformative innovation policy between the desire to steer change towards societal challenges and the inherent difficulty of doing so due to the complexity of those same challenges. The extent to which transformative innovation policy can be planned is debatable. A key question in the debate surrounds the role of knowledge and the extent to which scientific evidence can guide policy under uncertainty. Views on this question are somewhat opposing. For example, Weber and Rohracher (2012) suggest that scientific evidence continues to play an important role in establishing visions for change, while Stirling (2008) to a greater extent abandons the idea of evidence based policy advice and instead encourages opening up and democratising the construction and development of knowledge.

Although we suggest that the transformative innovation perspective's aim of transitioning entire socio-technical systems represents something new in innovation policy, it would be wrong to ignore the links to previous perspectives. For example, mission-oriented projects as described in the STI perspective share the challenge-driven nature of transformative innovation policy. Mission-oriented projects also exhibit a sense of directionality and deliberate steering, however, the challenges are of a technical nature, which is unlike the societal challenges addressed by transformative innovation, which are complex beyond the technical. Transformative innovation policy also stresses the importance of collaboration and interaction between different actors across value chains. These interactions and collaborations are believed to be key sites for innovations, which is similar to the innovation systems approach, but quite different from the STI perspective that highlights activities within firms and organisations as key sites of innovations (Bugge et al. 2019).

Beyond the socio-technical transition aspects, a third element sets transformative innovation policy apart from previous perspectives. The transformative innovation perspective departs from the understanding that many of the challenges faced by contemporary society are partly the result of unintended consequences of innovation (Goulet and Vinck 2017; Soete 2013). However, as shown previously in the chapter, scholars in earlier perspectives also considered the complex and sometimes contradictory effects that innovation has on societal progress. What is new in the transformative innovation perspective, however, is the proposed policy action in response to unintended consequences of innovation. Transformative innovation policy implies a deliberate destabilisation of the unsustainable regimes (Turnheim and Geels 2012). Consequently, destabilisation policies are argued to be a necessary component of transformative innovation policy mixes, even if actual policy portfolios are often heavily skewed towards 'creation' rather than 'destruction' policy instruments (Kivimaa and Kern 2016). Further, 'phase-out' or 'exnovation' policies arguably play a central role for transition processes, which for instance is shown in the case study of the German Energiewende (Rogge and Johnstone 2017; David 2017). This focus on destabilisation is a clear difference of transformative innovation policy vis-à-vis previous conceptualisations of innovation policy.

**Evaluation and Assessment**

So far this section has focused on critically examining the novelty of the transformative innovation perspective, but we want to take the discussion beyond conceptual discussions and ask how radically different the transformative innovation perspective is when put into action on the ground? Critical engagement with this question has just begun, but researchers exploring the design and implementation of transformative innovation policy are generating insights that begin to nuance our understanding of transformative innovation policy and its assumed transformative effect.

In a recently published paper, Diercks et al. (2019) conceptualise transformative innovation policy as an emerging policy paradigm. Looking at the policy design of two global initiatives promoting transformative innovation policies, Mission Innovation, an initiative for enhancing innovation in clean energy technologies, and the Global Covenant of Mayors for Climate and Energy, an alliance of cities working towards a just and climate-resilient future, the authors demonstrate that the transformative innovation policy paradigm is characterised by contestation, and that its practical expressions are diverse and even contradictory at times. Both initiatives share a social policy agenda for transformative change, but the respective initiatives differ in their understanding of the innovation process including the perception of relevant actors, activities and modes of innovation. Focusing on singular technological breakthroughs, Mission Innovation represents a narrow understanding of the innovation process that sits rather comfortably with conventional innovation policy, while the Global Covenant of Mayors for Climate and Energy with its more radical aims of reconfiguring urban systems represents a broader understanding of the innovation process that challenges conventional innovation policy.

Drawing on the policy paradigm framework, the authors try to make sense of this diversity. Seeing that the transformative innovation policy paradigm is only in the early stages of emergence it arguably has not yet institutionalised around a set of core ideas and concepts, which in turn makes space for diverse expressions. Diercks et al. (2019) anticipate that the coming years will see a political contest between these diverse paradigmatic expressions before we arrive at a more set understanding of what the transformative innovation policy paradigm entails.

The work by Diercks and colleagues suggests that the transformative effect of transformative innovation policy is not given. Similar conclusions are made by Grillitsch et al. (2019) that set out to generate theoretical and empirical insights by looking at the design and implementation of transformative innovation policies. Grillitsch et al. (2019) analyse two Swedish strategic innovation programmes, which are identified as examples of transformative innovation policy because they explicitly target system-wide transformation and address grand challenges through the coordination of diverse actors, networks and institutions.

The authors find that the design and implementation of the Swedish strategic innovation programmes struggle with some of the same key challenges that transformative innovation policy sets out to address in the first place. For example, conflicting interests between stakeholders in the programmes are not actively dealt with. As a result, the programmes are not steered by a collectively deliberated and well-aligned vision, but rather by a broad programme encompassing various competing agendas. This is an issue because it inhibits the strategic innovation programmes from providing clear objectives and directionality, which are otherwise considered key characteristics of transformative innovation policies. Another example relates to the programmes' intention to promote policy learning and coordination through the involvement of diverse actors. Although the programmes by design involve a broad range of stakeholders, challenges remain in terms of overcoming 'institutional mismatch' between actor communities such as academia and industry. The involvement of industry actors is low during implementation, which in turn weakens the programmes' abilities to promote policy learning and coordination.

While the transformative innovation narrative is attractive in policy, the research looking into transformative policy design and implementation shows that the emerging innovation policy's actual potential for change and its transformative effect is not yet determined and certainly not easily achieved.

## CONCLUDING DISCUSSION

Our account opens up for a more detailed discussion of the development over time of innovation policy. Rather than seeing transformative innovation policy as something that only recently materialised, one can find transformative elements in innovation policy back in time. Thus, the emphasis on innovation as a means for supporting substantial changes in society leading to social progress or greater environmental sustainability is not a recent phenomenon. Rather than discussing entirely different frames of innovation policy, one can observe a gradual change over time where research on innovation policy becomes increasingly centred on the possibilities for innovation policy to deliver transformative change. This does not imply that this perspective was absent until very recently, however, it takes the central position in transformative innovation policy. Furthermore, the attention of transformative innovation policy to deliberately destabilising unsustainable regimes also marks a significant change relative to previous understandings of innovation policy.

However, considering these changes, one may find it quite surprising that work on implementing transformative innovation policies has not progressed to a greater extent. While literature on transformative innovation policy has elaborated considerably on the inability of policy instruments with a foundation in traditional STI or innovation systems understandings to deliver transformative change, it has proven more difficult to give details on how transformative innovation policy instruments should be designed. This is not to argue that no progress has been made (see for example Kivimaa and Kern 2016; Weber and Rohracher 2012; Grillitsch and Hansen 2019). However, the analyses of transformative innovation policy instruments (Diercks et al. 2019; Grillitsch et al. 2019) described earlier underline that even frontrunner countries and alliances are challenged in designing and implementing policy instruments delivering transformative change.

One aspect that sets innovation system policy apart from mission-oriented innovation policy such as the Manhattan or Apollo projects, and recent transformative innovation policy targeting system-wide changes in production and consumption patterns is the lack of "grandeur". Innovation system policy provided a framework for supporting both radical and incremental innovations, responding to the opportunities and addressing the challenges in each context. In that sense, it provided clear recommendations for different types of regions and countries, considering the variegated preconditions they possess. The lack of "grandeur" makes the innovation system approach accessible for all places to make the best out of the respective preconditions. In this vein, Grillitsch and Hansen (2019) discuss the opportunities and barriers, and resulting policy recommendation for green industry development in different types of regions. However, this way of thinking may also work against a transformative innovation policy, in particular if incremental innovations are seen as cementing existing, unsustainable development paths.

Furthermore, the embedding of transformative innovation policy in a capitalist system might need further reflection. Taken all unintended consequences of capitalism aside for a moment, it has proven to be a powerful force of disruptive creation where actors seek new profitable opportunities and take risks to realise them. Maybe, this force is not utilised to the highest possible extent for dealing with grand societal challenges? For instance, by changing incentive structures and creating demand for new products and solutions that contribute to the desired development outcomes, it may be possible to initiate bottom-up experimentation without the need of complex coordination of multiple stakeholders, which is encouraged in many of the current transformative innovation policies. In other words, while experimentation is at the core of transformative innovation policies, such policies may underestimate the degree of (potentially ill-directed) experimentation that exists in modern capitalist societies.

A difference between STI policies, resting on a linear model of innovation, and innovation systems policy and transformative innovation policy, both resting on an interactive model of innovation, is the role of agency. According to STI, the role of agency is relatively clear and positive. Scientific discoveries and thereby scientists are the key agents of change fuelling the innovation process. Innovation systems and transformative innovation policy remain rather silent about the role of agency. Structural elements and global forces are the key explanations for innovation, and the unfortunate situation we are in with grand challenges threatening our societies. Human agency manifests mainly as the unintended consequences of the actions of generations leading to powerful lock-ins that transformative innovation policy is trying to combat. The problem with this lack of theorising and insights on agency are profound because structures are not changing by themselves. It needs agency to change structures and achieve transformative change.

Based on this concluding discussion, we propose three avenues for future research. First, research on transformative innovation policy still has some way to go in providing policymakers with a clear understanding of the types of (mixes of) policy instruments and processes they need to put in place. This also includes an improved understanding of the balance between 'creation' and 'destruction' instruments in policy mixes, and how this might differ between transitions depending on the characteristics and challenges of transition processes.

Second, some insights on change agency are developing now, partially related to the innovation system approach and partially to transformative change (Grillitsch and Sotarauta 2019; Isaksen et al. 2019; Dawley 2014; Simmie 2012; MacKinnon et al. 2019; Jolly et al. 2020; Sjøtun and Njøs 2019). However, it remains unclear what role agency plays in the transformation process and how it may be mobilised through policy.

Finally, we note that while innovations systems research has for decades enjoyed fruitful cross-fertilisation with the field of economic geography, there is still a significant lack of geographical sensitivity in transformative innovation policy. To exemplify, there are very few insights on opportunities for transformative innovation policies at the regional scale (Capasso et al. 2019) and a very limited understanding of how transformative innovation policy can and should take different forms in different geographical contexts (Grillitsch et al. 2019). Thus, in these respects, transformative innovation policy remains considerably under-developed compared to innovation systems policy.

## REFERENCES

Arntzen, Siri, Zach Wilcox, Neil Lee, Catherine Hadfield and Jen Rae (2019), *Testing Innovation in the Real World*, London: NESTA.

Asheim, Bjørn T., Markus Grillitsch and Michaela Trippl (2016), 'Regional innovation systems: past – present – future', in Richard Shearmur, Christoph Carrincazeaux and David Doloreux (eds), *Handbook on the Geographies of Innovation*, Cheltenham, UK and Northampton, MA, USA: Edward Elgar Publishing, pp. 45–62.

Asheim, Bjørn T. and Arne Isaksen (1997), 'Location, agglomeration and innovation: Towards regional innovation systems in Norway?', *European Planning Studies*, 5, 299–330.

Asheim, Bjørn T. and Arne Isaksen (2002), 'Regional innovation systems: The integration of local "sticky" and global "ubiquitous" knowledge', *Journal of Technology Transfer*, 27, 77–86.

Autio, Erkko (1998), 'Evaluation of RTD in regional systems of innovation', *European Planning Studies*, 6, 131–40.

Borrás, Susana and Charles Edquist (2013), 'The choice of innovation policy instruments', *Technological Forecasting and Social Change*, 80, 1513–1522.

Breschi, Stefano and Franco Malerba (1997), 'Sectoral innovation systems: Technological regimes, Schumpeterian dynamics, and spatial boundaries', in Charles Edquist (ed.), *Systems of Innovation: Technologies, Institutions and Organizations*, London and New York: Routledge, pp. 130–56.

Bugge, Markus M., Simon Bolwig, Teis Hansen and Anne Tanner (2019), 'Theoretical perspectives on innovation for waste valorisation in the bioeconomy', in Antje Klitkou, Arne Fevolden and Marco Capasso (eds), *From Waste to Value – Valorisation Pathways for Organic Waste Streams in Circular Bioeconomies*, Abingdon: Routledge, pp. 51–70.

Bush, Vannevar (1945), 'Science: The endless frontier', Washington, United States Government Printing Office.

Capasso, Marco, Teis Hansen, Jonas Heiberg, Antje Klitkou and Markus Steen (2019), 'Green growth – a synthesis of scientific findings', *Technological Forecasting and Social Change*, 146, 390–402.

Carlsson, Bo and Rikard Stankiewicz (1991), 'On the nature, function and composition of technological systems', *Journal of Evolutionary Economics*, 1, 93–118.

Coenen, Lars, Teis Hansen and Josephine V. Rekers (2015a), 'Innovation policy for grand challenges. An economic geography perspective', *Geography Compass*, 9, 483–96.

Coenen, Lars, Jerker Moodysson and Hanna Martin (2015b), 'Path renewal in old industrial regions: Possibilities and limitations for regional innovation policy', *Regional Studies*, 49, 850–65.

Cooke, Philip (1992), 'Regional innovation systems: Competitive regulation in the new Europe', *Geoforum*, 23, 365–82.

Cooke, Philip (1998), 'Introduction. Origins of the concept', in Hans J. Braczyk, Philip Cooke and Martin Heidenreich (eds), *Regional Innovation Systems: The Role of Governances in a Globalized World*, London: UCL Press, pp. 2–27.

Cooke, Philip (2004), 'Integrating global knowledge flows for generative growth in Scotland: Life sciences as a knowledge economy exemplar', in Jon Potter (ed.), *Global Knowledge Flows and Economic Development*, Paris: OECD, pp. 73–96.

David, Martin (2017), 'Moving beyond the heuristic of creative destruction: Targeting exnovation with policy mixes for energy transitions', *Energy Research & Social Science*, 33, 138–46.

Dawley, Stuart (2014), 'Creating new paths? Offshore wind, policy activism, and peripheral region development', *Economic Geography*, 90, 91–112.

Diercks, Gijs, Henrik Larsen and Fred Steward (2019), 'Transformative innovation policy: Addressing variety in an emerging policy paradigm', *Research Policy*, 48, 880–894.

Edquist, Charles (ed.) (1997), *Systems of Innovation: Technologies, Institutions, and Organizations*, London: Printer Publishers/Castell Academic.

European Commission (2008), A more research-intensive and integrated European Research Area. Brussels: European Commission.

Freeman, Christopher (1995), 'The "National System of Innovation" in historical perspective', *Cambridge Journal of Economics*, 19, 5–24.

Giuliani, Elisa (2018), 'Regulating global capitalism amid rampant corporate wrongdoing – reply to "Three frames for innovation policy"', *Research Policy*, 47, 1577–82.

Godin, Benoît (2016), 'Technological innovation: On the origins and development of an inclusive concept', *Technology and Culture*, 57, 527–56.

Godin, Benoît (2017), *Models of Innovation: The History of an Idea*, Cambridge, MA: MIT Press.

Goulet, Frédéric and Dominique Vinck (2017), 'Moving towards innovation through withdrawal: The neglect of the destruction', in Benoît Godin and Dominique Vinck (eds), *Critical Studies of Innovation: Alternative Approaches to the Pro-Innovation Bias*, Cheltenham, UK and Northampton, MA, USA: Edward Elgar Publishing, pp. 97–114.

Grabher, Gernot (1993), 'The weakness of strong ties: The lock-in of regional development in the Ruhr area', in Gernot Grabher (ed.), *The Embedded Firm: On the Socioeconomics of Industrial Networks*, London and New York: Routledge, pp. 255–77.

Grillitsch, Markus and Bjørn T. Asheim (2018), 'Place-based innovation policy for industrial diversification in regions', *European Planning Studies*, 26, 1638–62.

Grillitsch, Markus, Bjørn T. Asheim and Michaela Trippl (2018), 'Unrelated knowledge combinations: The unexplored potential for regional industrial path development', *Cambridge Journal of Regions, Economy and Society*, 11, 257–74.

Grillitsch, Markus and Teis Hansen (2019), 'Green industry development in different types of regions', *European Planning Studies*, 27, 2163–83.

Grillitsch, Markus, Teis Hansen, Lars Coenen, Johan Miörner and Jerker Moodysson (2019), 'Innovation policy for system wide transformation: The case of Strategic Innovation Programs (SIPs) in Sweden', *Research Policy*, 48, 1048–61.

Grillitsch, Markus and Markku Sotarauta (2019), 'Trinity of change agency, regional development paths and opportunity spaces', *Progress in Human Geography*, 1–20.

Grillitsch, Markus and Michaela Trippl (2018), 'Innovation policies and new regional growth paths: A place-based system failure framework', in Jorge Niosi (ed.), *Innovation Systems, Policy and Management*, Cambridge: Cambridge University Press, pp. 329–58.

Hall, Peter A. and David W. Soskice (2001), *Varieties of Capitalism: The Institutional Foundations of Comparative Advantage*, Wiley Online Library.

Hansen, Teis and Lars Winther (2011), 'Innovation, regional development and relations between high- and low-tech industries', *European Urban and Regional Studies*, 18, 321–39.

Hansen, Teis, Lars Winther and Ronnie F. Hansen (2014), 'Human capital in low-tech manufacturing: The geography of the knowledge economy in Denmark', *European Planning Studies*, 22, 1693–710.

Hassink, Robert (2005), 'How to unlock regional economies from path dependency? From learning region to learning cluster', *European Planning Studies*, 13, 521–35.

Hassink, Robert, Ame Isaksen and Michaela Trippl (2019), 'Towards a comprehensive understanding of new regional industrial path development', *Regional Studies*, 53, 1636–45.

Hekkert, Marko P., Roald Suurs, Simona Negro, Stefan Kuhlmann and Ruud Smits (2007), 'Functions of innovation systems: A new approach for analysing technological change', *Technological Forecasting and Social Change*, 74, 413–32.

Holland, Maurice (1928), 'Research, science, and invention', in Frederic Wile (ed.), *A Century of Industrial Progress*, New York: Doubleday, Dran, pp. 312–34.

Holland, Maurice (1931), 'Industrial science – a gilt edge security', *Science*, 74, 279–82.

Isaksen, Ame (2001), 'Building regional innovation systems: Is endogenous industrial development possible in the global economy?', *Canadian Journal of Regional Science*, 14, 101–20.

Isaksen, Ame, Stig-Erik Jakobsen, Rune Njøs and Roger Normann (2019), 'Regional industrial restructuring resulting from individual and system agency', *Innovation: The European Journal of Social Science Research*, 32, 48–65.

Isaksen, Ame and Michaela Trippl (2016), 'Path development in different regional innovation systems', in Mario Parrilli, Rune Fitjar and Andrés Rodríguez-Pose (eds), *Innovation Drivers and Regional Innovation Strategies*, New York and London: Routledge, pp. 66–84.

Jolly, Suyash, Markus Grillitsch and Teis Hansen (2020), 'Agency and actors in regional industrial path development. A framework and longitudinal analysis', *Geoforum*, 111, 176–88.

Kivimaa, Paula and Florian Kern (2016), 'Creative destruction or mere niche support? Innovation policy mixes for sustainability transitions', *Research Policy*, 45, 205–17.

Klein Woolthuis, Rosalinde, Maureen Lankhuizen and Victor Gilsing (2005), 'A system failure framework for innovation policy design', *Technovation*, 25, 609–19.

Kline, Stephen J. (1985), 'Innovation is not a linear process', *Research Management*, 28, 36–45.

Kline, Stephen J. and Nathan Rosenberg (1986), 'An overview of innovation', in Ralph Landau and Nathan Rosenberg (eds), *The Positive Sum Strategy: Harnessing Technology for Economic Growth*, Washington, DC: National Academy Press, pp. 275–306.

Kuhlmann, Stefan and Arie Rip (2018), 'Next-generation innovation policy and grand challenges', *Science and Public Policy*, 45, 448–54.

Laranja, Manuel, Elvira Uyarra and Keiron Flanagan (2008), 'Policies for science, technology and innovation: Translating rationales into regional policies in a multi-level setting', *Research Policy*, 37, 823–35.

Lundvall, Bengt-Åke (1992), *National Systems Of Innovation: Towards a Theory of Innovation and Interactive Learning*, London: Pinter.

Lundvall, Bengt-Åke and Susana Borrás (2005), 'Science, technology, and innovation policy', in Jan Fagerberg, David Mowery and Richard Nelson (eds), *The Oxford Handbook of Innovation*, Oxford: Oxford University Press, pp. 599–631.

Mackinnon, Danny, Stuart Dawley, Andy Pike and Andrew Cumbers (2019), 'Rethinking path creation: A geographical political economy approach', *Economic Geography*, 95, 113–35.

Maclaurin, W. Rupert (1947), 'Federal support for scientific research', *Harvard Business Review*, 25, 385–96.

Maclaurin, W. Rupert (1954), 'Technological progress in some American industries', *The American Economic Review*, 44, 178–89.

Malerba, Franco (2002), 'Sectoral systems of innovation and production', *Research Policy*, 31, 247–64.

Markard, Jochen, Rob Raven and Bernhard Truffer (2012), 'Sustainability transitions: An emerging field of research and its prospects', *Research Policy*, 41, 955–67.

Martin, Ron and Peter Sunley (2006), 'Path dependence and regional economic evolution', *Journal of Economic Geography*, 6, 395–437.

Morgan, Kevin (2018), 'Experimental governance and territorial development', Background paper for an OECD/EC Workshop on 14 December 2018 within the workshop series "Broadening innovation policy: New insights for regions and cities". Paris: OECD.

Mowery, David C. (1983), 'Economic theory and government technology policy', *Policy Sciences*, 16, 27–43.

Mowery, David C., Richard Nelson and Ben Martin (2010), 'Technology policy and global warming: Why new policy models are needed (or why putting new wine in old bottles won't work)', *Research Policy*, 39, 1011–23.

Myers, Charles A. and W. Rupert Maclaurin (1942), 'After unemployment benefits are exhausted', *The Quarterly Journal of Economics*, 56, 231–55.

Nelson, Richard (ed.) (1993), *National Innovation Systems: A Comparative Analysis*, New York: Oxford University Press.

OECD (1963), *The Measurement of Scientific and Technical Activities: Proposed Standard Practice for Surveys of Research and Development*, Paris: OECD.

OECD (2015), *System Innovation: Synthesis Report*, Paris: OECD.

Rogge, Karoline S. and Phil Johnstone (2017), 'Exploring the role of phase-out policies for low-carbon energy transitions: The case of the German Energiewende', *Energy Research & Social Science*, 33, 128–37.

Schot, Johan and W. Edward Steinmueller (2018), 'Three frames for innovation policy: R&D, systems of innovation and transformative change', *Research Policy*, 47, 1554–67.

Schumpeter, Joseph (1911), *Theorie der wirtschaftlichen Entwicklung*, Leipzig: Duncker & Humbolt.

Sengers, Frans, Anna J. Wieczorek and Rob Raven (2019), 'Experimenting for sustainability transitions: A systematic literature review', *Technological Forecasting and Social Change*, 145, 153–64.

Shane, Scott and Sankaran Venkataraman (2000), 'The promise of entrepreneurship as a field of research', *The Academy of Management Review*, 25, 217–26.

Simmie, James (2012), 'Path dependence and new technological path creation in the Danish wind power industry', *European Planning Studies*, 20, 753–72.

Sjøtun, Svein G. and Rune Njøs (2019), 'Green reorientation of clusters and the role of policy: "The normative" and "the neutral" route', *European Planning Studies*, 27, 2411–30.

Soete, Luc (2013), 'Is innovation always good?', in Jan Fagerberg, Ben Martin and Ebsen Andersen (eds), *Innovation Studies Evolution & Future Challenges*, Oxford: Oxford University Press, pp. 134–44.

Stirling, Andy (2008), '"Opening up" and "closing down": Power, participation, and pluralism in the social appraisal of technology', *Science, Technology and Human Values*, 33, 262–94.

Tödtling, Franz and Sabine Sedlacek (1997), 'Regional economic transformation and the innovation system of Styria', *European Planning Studies*, 5, 43–63.

Tödtling, Franz and Michaela Trippl (2004), 'Like phoenix from the ashes? The renewal of clusters in old industrial areas', *Urban Studies*, 41, 1175–95.

Tödtling, Franz and Michaela Trippl (2005), 'One size fits all? Towards a differentiated regional innovation policy approach', *Research Policy*, 34, 1203–19.

Turnheim, Bruno and Frank W. Geels (2012), 'Regime destabilisation as the flipside of energy transitions: Lessons from the history of the British coal industry (1913–1997)', *Energy Policy*, 50, 35–49.

Uyarra, Elvira, Kieron Flanagan, Edurne Magro, James Wilson and Markku Sotarauta (2017), 'Understanding regional innovation policy dynamics: Actors, agency and learning', *Environment and Planning C: Politics and Space*, 35, 559–68.

Vitols, Sigurt (2001), 'Varieties of corporate governance: Comparing Germany and the UK', in Peter A. Hall and David Soskice (eds), *Varieties of Capitalism: The Institutional Foundations of Comparative Advantage*, Oxford; New York: Oxford University Press, pp. 337–60.

Weber, K. Matthias and Harald Rohracher (2012), 'Legitimizing research, technology and innovation policies for transformative change: Combining insights from innovation systems and multi-level perspective in a comprehensive "failures" framework', *Research Policy*, 41, 1037–47.

# 17. Business innovation measurement: history and evolution

*Giulio Perani*

## INTRODUCTION

Measurement is an essential component of scientific enquiry but becomes increasingly relevant in the realm of social sciences in order to move from the pure observation of anecdotal evidence to the identification of "constructs".[1] Innovation, however it is defined and conceptualised, is definitely an abstraction and innovation theorists have been constantly struggling to shore up their constructs by exploring new methods for innovation measurement. This chapter will describe some experiences of providing a quantification of the efforts by business enterprises to implement innovation practices and projects, as well as to achieve a successful innovation output. The economic impact of business innovation will be the focus of the chapter as it has been attracting, compared to competing measurement perspectives (social, technological, etc.) the broadest interest among scholars, practitioners and, not least, policymakers. In this respect, the parallel commitment by the OECD and the European Union, since the 1990s, to define an international standard for business innovation statistics will be described in some detail. On the other hand, the shortcomings of such standardised approach will be identified and proposals for improving or replacing it with alternative approaches will be discussed.

Half a century before the introduction of current innovation statistics, attempts to assess phenomena like technical change (Alford 1929), technological change (Weintraub 1936; Perazich and Field 1940) and, finally, technological innovation (Maclaurin 1950) were already developed, mostly in the USA,[2] focusing on two potential effects of them, those on employment and on productivity.[3] These pioneering experiences paved the way for the diffusion of the standard concepts and definitions currently codified, for instance, in the OECD and Eurostat's Oslo Manual (OM)[4] but failed in defining a consistent measurement framework. They were largely based on case studies[5] (at firm or industry level) or mixed collections of administrative and statistical data[6] and never succeeded in establishing a systematic monitoring of the innovation activities implemented by businesses (with the exception of business R&D survey where the USA took a global leadership already in the 1920s). In the next paragraphs it will be described as to how the European countries managed to fill this gap, becoming a global reference for business innovation statistics.

## MEASURING THE ECONOMIC DIMENSION OF INNOVATION

In the realm of economic statistics, the diffusion of a common international standard to measure the research and development (R&D) efforts by enterprises and institutions has been one of the most remarkable achievements of the 1960s.[7] As a result, economists were able to rely on a second, widely available source of data – to complement the more traditional patenting indicators – to shed light on the processes of technological change influencing the whole economy.

Nonetheless, after just a few years, an increasing awareness, among scholars, of the limitations of R&D surveys and patent databases as statistical sources, led to a number of scientific contributions asking for fresh evidence on the broader process of generation and use of knowledge taking place in enterprises.

The most influential contributions in this stream of literature included: Rosenberg's re-consideration of the role of R&D in the innovation process (Rosenberg 1982), the overcoming of the linear model of innovation (Kline and Rosenberg 1986), Nelson and Winter's evolutionary theory of economic change (Nelson and Winter 1982) as well as Freeman's economics of industrial innovation (Freeman 1974) "which for a long time had a virtual monopoly in presenting the 'state of the art' of knowledge in the field" (Fagerberg et al. 2012, p. 1136; see also, on the same point, Fagerberg and Verspagen 2009).

The perception, at the beginning of this endeavour, was like trying to "measure the unmeasurable" (Brouwer and Kleinknecht 1997). Only two decades after having established the brand new domain of R&D statistics, a new challenge was emerging: that of complementing the existing set of business indicators – mostly based on investments, sales and profits data – with a qualitative assessment of the ability of businesses to compete among them in terms of "novelty".[8] The search for sound concepts and reliable metrics, apparently overwhelming, found an answer from what has been defined as a "Schumpeterian renaissance" (Freeman 2007).

This was largely due to a shift in the economists' focus from large to small firms, paying attention to the environmental conditions that could make firms grow and become profitable (Bolton Committee 1971). An evolutionary approach, like that applied by Joseph Schumpeter,[9] appeared to be the ideal framework to analyse the role of both small and large firms in a dynamic perspective and, even more, to accommodate the needs of conceptualisation and measurement of technological/technical change as a driver of economic development at both macro and micro levels. Within the set of conceptual tools drawn from the extensive – and sometimes "paradoxical" (Freeman 2007) – corpus of Schumpeter's theories in order to lay down the basis for a new economics of innovation, the classification of innovation typologies has to be given a special role. While struggling to design a new methodology to measure "innovation", the adoption of a simplified description of it, as that given by Schumpeter in *The Theory of Economic Development* (2017 [1934]), appeared to be an obvious choice for both analysts and statisticians.[10] By identifying innovation as the commercial or industrial application of something new – a new product, process, or method of production; a new market or source of supply; a new form of commercial, business, or financial organisation – it was possible to focus the measurement activity, either quantitative or qualitative, on specific business functions, looking for any improvement of them as an evidence of innovation. At this stage, the measurement of the innovation output was seen as a priority.

The identification of an innovation process through a few potential outputs (new product, new process, etc.) has proven to be very effective for measurement purposes. An advantage was also the qualification of the firm that developed such innovations as an innovator (i.e. an innovating firm). Thus, it is not surprising that the first experiences to measure innovation were based on collecting data on innovations (mostly, new products): that is the "object" approach.[11]

The best-known database of innovations,[12] which can be taken as an example of such an object approach, was the SPRU Innovations Database that represented

> ... a pioneering attempt to directly identify significant innovative outputs. The Database was established on the commission of the Bolton Committee of Inquiry on Small Firms (Bolton Committee 1971) and was intended to provide evidence on the contribution of small firms to innovation. Subsequently updated, the final database lists 4,378 significant innovations which were introduced into the UK economy between 1945 and 1983. (Tether et al. 1997, p. 20)

This exercise is consistent with the objective of capturing evidence of innovative, non-R&D, activities by small and medium firms.[13]

Indeed, the "counting of innovations" method has been adopted by several studies. Four data collection methods[14] to target specific innovations are usually adopted:[15]

- patent-based counting of innovations;
- citation counts of patents;
- the collection of innovation cases by interviewing experts;
- the counting of innovations from trade journals.

The first two are dependent on a previous patenting activity by the surveyed institutions or firms and are mainly aimed at assessing the success rate of them in transforming inventions into innovations. Of course, non-patented innovations are excluded by definition. The third method is that used for the SPRU database. As highlighted by several authors, the main issue about it "concerns the heterogeneous economic value of the innovations thus identified, since the assembling of the data entails various unexamined biases" (Santarelli and Piergiovanni 1996, p. 691). The fourth method is assumed to have several advantages, compared to the others: it relies on the interest of innovators to make their achievements known by the public; it is relatively not-expensive; considers only marketed innovations (those selected by journal editors as potentially interesting for readers); it prevents any interaction between data providers and analysts.

## THE TRANSITION FROM THE "OBJECT" TO THE "SUBJECT" APPROACH

The needs of minimising both the influence of data collectors on the answers given by data providers and the biases generated by the high level of subjectivity implicit in self-compiled surveys, should have led to excluding the options of asking firms' owners and managers about their new products as a suitable method to measure innovation. Nevertheless, the option to ask inventors/innovators about their activities has often been used by researchers[16] in order to overcome the limitations associated with analyses based on technical literature or administrative data. Biases in the selection of innovation cases are always the highest risk and there is no source of information that would not be influenced by the processes used to collect and process data for research purposes.[17]

It is interesting that most of the studies on measuring innovation output carried out between 1993 and 1996 were stressing the advantages of literature-based or expert-based data collection methods compared to the undertaking of business innovation surveys which were, at that time, still in their infancy.[18] Scholars show different attitudes. For instance, Coombs et al.

(1996) acknowledging that innovation surveys could be a valuable new source to measure innovation output, still argued that the burden associated with large business surveys gives the literature-based method a comparative advantage in this field of study. Conversely, Brouwer and Kleinknecht (1996) emphasised the advantages of innovation surveys compared to other methods, even with reference to the (quantitative) measurement of innovation output. In a few years, the latter opinion became prevalent and innovation surveys were given a key role as data sources for innovation studies.

Two points are still worth a mention: (a) which role has been reserved in innovation surveys, so far, to innovation output measurement and (b) which prospects could be open by the introduction of the concept of "focal innovation" in the OM 2018 (OECD/Eurostat 2018).

The evidence emerged from the wave of experimental innovation surveys carried out (Smith 2005) in several European countries during the 1980s was that the "subject-oriented" nature of such surveys had to prevail on their potential "object-oriented" one. This was part of the assimilation process undertaken in order to involve the National Institutions responsible for official statistics in the new surveys. This process, sponsored by the OECD, was based on the principle that the results from business innovation surveys should have been comparable with any other official business statistics (sales, investments, profits, etc.), thus allowing for the adoption of a single reference population of firms and the availability of a set of consistent indicators.

Moreover, the adoption of a sampling methodology made the answers to open questions (or just to purely qualitative questions) much less useful for researchers compared to quantitative evidence. Survey results had to undergo a double processing: a weighting of the realised sample in order to calculate the quantitative indicators for the whole population and an enumeration of selected answers to extract qualitative information (mostly anecdotal) on the respondents' innovation output. Even when asking about the "number of innovations" implemented in a given period, the information was not suitable for a statistical treatment because of the high heterogeneity among respondents. In short, since the mid-1990s, the object-oriented approach almost disappeared from innovation surveys and, as a result, from innovation studies.

Such evolution cannot be seen as a prescription of the OM as, still in its second edition (1997) the Manual devoted a quite detailed Annex to describe the key features of object-oriented data collection and, more specifically, of literature-based innovation output (LBIO) indicators. In addition, the OM 1997 was allowing for complementing the innovation surveys with object-oriented questions although it clearly stated that:

> ... this approach will never enable statistics to be produced which purport to represent the totality of innovations which occur in a country in a given period. The resulting statistics will only represent a sub-set of the innovations occurring, and analysts will need to avoid drawing conclusions about all innovations. They will however be able to draw conclusions about significant innovations, particularly if they classify them by other characteristics, such as expenditure on the innovation, size of the firm, etc. (OECD and Eurostat 1997, p. 78)

Only with the third edition of the OM (2005) the object-approach, as well as the focalisation of innovation surveys on producing output indicators, were removed from the innovation standard methodological framework:

> From the point of view of current economic development, it is the differential success of firms that shapes economic outcomes and is of policy significance. This favours a subject-based approach, although innovation surveys can combine both approaches by including general questions on the firm

and specific questions on a single innovation. It is the subject, the firm, that is important, and this is the approach has been chosen as the basis for these guidelines. (OECD and Eurostat 2005, p. 21)

The fourth edition of the OM (2018) re-introduced the option to ask, in innovation surveys, questions about the respondents' "focal innovations", also referred to as the "most important innovations". A whole chapter (Chapter 10) has been devoted to deal with the issue in line with the 1997 approach, although in a totally different context. On the one hand, the emphasis on the most important innovations introduces in surveys an issue about the criteria to be used for assessing the relevance of an innovation compared to another (not mentioning multiple innovations merged into a single project or mixed product/process innovations). On the other hand, it could be questioned the possibility itself of singling out a specific "object" in a business environment where phenomena like digitalisation and servitisation make goods and services often indistinguishable.

## THE INVENTION OF 'INNOVATION SURVEYS'

The process that led innovation surveys to adopt, in just a few years, the subject approach as the new standard has been described in detail in several studies, including Godin (2002), Smith (2005) and Arundel and Smith (2013). Differences exist indeed about how this process can be seen: whether as a series of further improvements towards a comprehensive measurement of the innovation phenomenon (with reference to the business sector), or as a more complex trial and error process that needs constant checking and maintenance.

The pilot innovation surveys carried out in countries like Germany, France, the Netherlands and Italy during the 1980s shared, although different in methodology and sample size, a few common features: they were all aimed at operationalising a few key "Schumpeterian" concepts about innovation as a challenge to the statistical evidence supporting the mainstream view in economics of the possibility of a general equilibrium. Thus, it was crucial to emphasise the role of technology in influencing the ability of firms, irrespective of size, to develop new products and processes and improving their competitiveness.

Not surprisingly, the availability of such a set of data, that was originally designed to serve the needs of neo-Schumpeterian economists, was immediately successful and the results of the studies based on it became largely influential among both economists and statisticians. A rather low number of pilot surveys turned to set an international standard in just a few years. Such a standard was based on a few key concepts that have influenced for a long time the interpretation of the innovation-related phenomena and on a relatively fixed structure for innovation surveys. Three examples can be given of such concepts and their evolution: the role of technology, the operationalisation of the novelty criterion and the role of implementation in the definition of innovation.

First, innovation surveys have been designed with a view of innovation as technological improvement of material goods. Of course, manufacturing firms have been the first, obvious targets of innovation surveys. All the 1980s and 1990s pilots mentioned in the literature were considering just industrial firms and only after the publication of the OM 1997 was the service sector included in the scope of innovation surveys. It could be said that – in terms of terminology, examples and even conceptualisation – innovation surveys are not yet carefully designed to accommodate for all the features of innovation in service firms.

Second, the innovation concept has always been qualified in terms of novelty. Unfortunately, the operationalisation of the novelty criterion is still an issue. On the one hand, the "intensity" of novelty is concerned. In principle, any not-irrelevant change in products, processes or business activity could be an innovation but the assessment is left to survey respondents. According to the OM guidelines, respondents should classify their products (as an example) as "unchanged" or "with minor changes", "substantially improved" or "new". The criterion to evaluate novelty has been, over the last decades, based on "characteristics or intended uses" of products or processes (OECD/Eurostat 2005): that is, their "technological contents". In the OM 2018 (p. 46), the emphasis was on "… potential uses, as determined by the characteristics of a product or process compared to alternatives, and by the previous experiences of its provider and intended users". A key issue is, of course, that the timing framework within novelty has to be evaluated: for how long after its "implementation" can a product or a process can be considered new? The reference period of surveys is, usually, three years but a firm in an industry with a high innovation rate may be classified as not very innovative having introduced a new product/service two years ago, compared to a firm in a less dynamic industry which could have a market leadership relying on products introduced more than three years ago. The proposals for a fine tuning of the innovation frequency have never been accepted by the OECD, strenuously defending the OM's "one fits all" approach.

Third, novelty has to be identified at a specific point in time: that of innovation implementation. The OM, consistent across its four editions, argues that the innovation implementation is achieved when a producer deems the output of an innovation project as sufficiently developed to be (potentially) marketed. Indeed, by comparing the definitions given in the second, third and fourth edition of the OM for product innovation, some substantial differences in defining implementation can be pointed out:

OM 1997, p. 9: "A technological product innovation is the implementation/commercialisation of a product with improved performance characteristics such as to deliver objectively new or improved services to the consumer."

OM 2005, pp. 46 and 48: "An innovation is the implementation of a new or significantly improved product (good or service), or process, a new marketing method, or a new organisational method in business practices, workplace organisation or external relations." (…) "A product innovation is the introduction of a good or service that is new or significantly improved with respect to its characteristics or intended uses."

OM 2018, pp. 33 and 34: "A business innovation is a new or improved product or business process (or combination thereof) that differs significantly from the firm's previous products or business processes and that has been introduced on the market or brought into use by the firm." (…) "A product innovation is a new or improved good or service that differs significantly from the firm's previous goods or services and that has been introduced on the market."

The most striking difference is that in OM 1997 and in OM 2005 innovation is the implementation while, in OM 2018, innovation is the output (product or process). This changes dramatically the understanding of innovation, which was previously based on the innovator's activities (subject approach) and now is going back to a pre-OM emphasis on the innovation output (object approach). This change of definition is not without risks: it works well with reference to goods but if we accept that goods are increasingly merged with services and that services are often clustering by generating comprehensive value propositions, it could be questioned that single innovations could be clearly identified and properly compared between them (Perani 2019).

## THE CIS: A SUCCESS STORY

The development of innovation statistics in the last three decades has been based on two pillars: the OM on the one side, and the European Union's (EU) Community Innovation Survey (CIS) on the other side. The manual and the survey have so far been interacting for more than 25 years. Early in the 1990s, European policymakers found it helpful to collectively adopt the brand-new statistical standard proposed by the OECD that was extremely promising as to the delivery of state-of-the-art innovation indicators, easy to understand and largely comparable across countries. Thus, the first CIS (reference year 1992) and the first OM (OECD 1992) were developed in parallel with concepts and definitions discussed in the OECD[19] fora and cost and burden of the survey fell on the EU and the European network of statistical institutions. The CIS 1992 was, nevertheless, a pilot survey that underwent an extensive review (Archibugi et al. 1997) with a new survey planned not before four years later. By challenging a common awareness that several shortcomings of the first survey had to be urgently fixed, the second CIS wave (1996) adopted the same approach of the CIS 1992. It was urgent to adopt a broader concept of innovation – still technology-oriented but now including innovation in services – and a larger scope, surveying consistently manufacturing and service industries (even though with two distinct questionnaires). Such scaling-up of the survey was soon supported by the second OM (1997) that sanctioned the choices made at the operational level. While the impact on policymaking was below the expectations, the CIS found enthusiastic supporters in the academic community, having been the first ever official business survey in Europe whose managers allowed researchers (some of them generously funded by the EU, as well) to access, under some conditions, the survey micro-data. Over the next decade the survey did not introduce substantial changes. The CIS 2000 mostly focused on testing a single questionnaire for both manufacturing and service enterprises and the CIS 2004 was used to pilot a further broadening of the innovation definition (including organisational and marketing changes as potential innovations).[20]

A non-irrelevant step was the introduction in 2004 of a European law[21] making the CIS mandatory for all EU countries, thus preventing any technical discussion on whether the survey had to be redesigned or, for instance, merged with other surveys.[22] In the same perspective of institutionalisation of the CIS, its frequency was increased from four to two years from 2006 onwards.

Such political support for the CIS was a powerful driver to convince other non-European countries to test the survey (of course in addition to those OECD countries, like Canada and Australia, that had already developed their own model of OM survey). By using, as an indicator, the first systematic data collection of innovation data managed by the UNESCO Institute of Statistics (UIS 2012), the diffusion of the CIS model has followed this trend: 12 non-European countries carried out an innovation survey in the second half of the 2000s (8 out of 12 adopting the CIS model). In the next round, 35 countries were able to report innovation data (UIS 2015) and several of them adopted a CIS-like questionnaire in addition to the 33 countries coordinated by Eurostat.[23] Finally, in 2015, four additional countries – including, notably, the US – were able to join the UIS panel (UIS 2017), bringing the total number of countries conducting, more or less on a regular basis, an innovation survey to 72 with, arguably, more than half producing CIS-comparable data.

Other figures have also been used to emphasise how successful the CIS has been. For instance, more than 400 academic published papers in English using CIS data have been found for the period 1994–2011 (Arundel and Smith 2013). On the other hand, it is the impact of the innovation surveys on economic and econometric research that has to be considered. An essential step in this process was the publication of a seminal paper (Crépon et al. 1998) about the use of innovation micro-data for econometric analysis. This paper proposed the use of an econometric model (CDM model from the initials of the authors) to establish an empirical relationship between R&D, innovation and productivity. The combined availability of innovation, mostly CIS, micro-data in many countries and the increasing computational capacity available in universities and research institutions made this paper one of the most cited in the econometrics and economics domain with hundreds of applications and adaptations in more than 40 countries (Lööf et al. 2017). A substantial point is that innovation surveys, designed with the objective of discharging the mainstream economics approach and methods by making available a fresh evidence of a high heterogeneity within the industrial sector and a turbulent dynamics which could be hardly reduced to econometric models, have become a common playing field of a new syncretism where the intellectual barriers between an evolutionary and a general equilibrium approach are mitigated by the sharing of a rich, common data source.

## RE-ASSESSING INNOVATION STATISTICS

As already mentioned, the CIS has been mostly asking a fixed set of questions in a framework that has been constantly adapted to the changes introduced by the OM. The survey and manual are practically intertwined although guided by two different, sometimes conflicting, logics. Timing is an issue because the OM, whose last revision was finalised thirteen years after the previous edition, does provide the survey with updated concepts and definitions with a systematic delay with respect to the expectations of users. The final outcome of this unstable equilibrium has generated some long-standing shortcomings that could undermine the CIS reputation. Two of them are described below.

### Too High Subjectivity of Innovation Statistics

It is a common opinion[24] that the traditional OM surveying approach makes it extremely difficult to evaluate the extent that biases introduced by respondents could affect innovation measurement. The OM 2018 strongly rejects this charge:

> Survey data are self-reported by the respondent. Some potential users of innovation data object to innovation surveys because they believe that self-reports result in subjective results. This criticism confuses self-reporting with subjectivity. Survey respondents are capable of providing an objective response to many factual questions, such as whether their firm implemented a business process innovation or collaborated with a university. These are similar to factual questions from household surveys that are used to determine unemployment rates. Subjective assessments are rarely problematic if they refer to factual behaviours. (OECD and Eurostat 2018, p. 229)

Unfortunately, the most serious concerns affect a key process: the identification of innovations. When extending innovation surveys to non-European countries the results based on the most subjective questions – for instance, product/service innovation – had to be complemented with extensive metadata to be properly interpreted. When presenting the results of its 2015 innovation data collection, the UIS (UIS 2017) avoided any direct comparison of the key innovation indicators between high-income and low-income countries being concerned about the influence of national cultures on survey results. As an example, with reference to product innovation, eight low-income countries, mostly in Africa and Latin America, performed, in their most recent innovation survey, better than Germany, the product innovation leader in the EU (featuring 43.8 per cent of firms with product innovations in the CIS 2014).

An option to address this issue – that of introducing an object-oriented component in the survey – was suggested by Mairesse and Mohnen:

> The Oslo Manual opted for the subject approach, i.e. for collecting data at the firm level, including all its innovation outputs and activities. This implies that we do not have data about particular innovation projects. The object approach in contrast would make the individual innovation the unit of analysis, as is the case for literature-based innovation counts. One important advantage of the subject approach is that the innovation surveys collect comprehensive data at the decision making level of the firm, which is also the level of available accounting and financial data that can be merged with the innovation data for richer analyses and that can be easily related with industry statistics and national accounts. It also naturally covers innovators and non-innovators, generators and users of innovation. In spite of its difficulties, the subject approach is on the whole less demanding than the object approach, which raises specific difficulties to identify, compare and assess individual innovations. The drawback with the subject approach is that it takes as a whole all the innovation projects of a firm, some being highly successful, some less and others not at all. (Mairesse and Mohnen 2010, p. 10)

The issue of finding a balance between a subject and an object approach is the *fil rouge* linking most of the experiences with innovation surveying. Again, the OM 2018[25] does not exclude the option to ask innovators about their projects:[26]

> Respondents can be asked if their firm organises some or all of its work to develop innovations into recognised projects, or they can be asked about a specific innovation project [...]. Information on innovation projects can complement other qualitative and quantitative data on innovation activities. Data on the number of projects for innovation can provide indicators on the variety and diversity of innovation activities. [...] Information on the number of innovation projects is not primarily intended to produce an aggregate figure of the total number of projects for a firm or industry, but rather to derive indicators at the firm level. (OECD and Eurostat 2018, pp. 88–89)

Following Mairesse and Mohnen's reasoning, the issue is not that of collecting information on a "focal" innovation project (similarly to target a focal innovation as envisaged by the OM) but rather to understand if the respondent is a real innovator by considering the whole innovation activity of a firm.[27]

### Poor Measurement of the Innovation Output[28]

The methodological issues preventing a full coverage of the innovation output (unlike of inputs) in business surveys were well known already after the very first waves of innovation surveys.[29] The adoption of the subject approach prevented from asking the innovators about the actual innovations implemented by them; as a result the main indicator of output allowed by the OM (also in its 2018[30] edition) – that is, the share of turnover generated by new products – is very limited in scope and influenced by high subjectivity. Thus, the demand for a reliable indicator of innovation output has not been met, so far, by the OM and whatever the reason would be[31] it makes sense for users to rely on alternative indicators than those produced by innovation surveys. At least, this is the evidence emerging from two exercises regularly carried out by the EU in order to measure the innovation trends for policy purposes: the EU Innovation Scoreboard (EIS) and the EU 2020 Innovation Output Indicator.

The EIS[32] provides policymakers with an annual synthesis of the state of innovation in EU countries based on 27 indicators[33] covering many areas of innovation performance. While it could have been expected as a major role of CIS indicators in this exercise, only six CIS indicators out of 27 are used in the EIS. A strong limitation is, of course, the mismatch between the annual data collection undertaken for the EIS calculation and the availability of fresh CIS results every second year (in addition, with an 18-month time-lag between the end of the survey's reference year and the time results are made available). CIS indicators are used to cover only topics where no alternative data are available (the percentage of small firms with innovations; the non-R&D innovation expenditure; the external collaborations of small firms with innovations; and the key CIS indicators on the sales of new-to-market and new-to-firm product innovations).

When measuring the performance of the EU economy in developing an innovation capacity (Innovation Output Indicator[34]), the EU experts had to exclude any use of CIS indicators. The four areas under monitoring are: (a) Technological innovation as measured by patents; (b) Employment in knowledge-intensive activities as a percentage of total employment; (c) Competitiveness of knowledge-intensive goods and services; and (d) Employment in fast-growing firms of innovative sectors. The CIS indicators have been implicitly discharged as they do not meet the key requirements to be included in this set of high policy relevance indicators. This is a clue that a lack of a clear focus on the measurement of the innovation output is still a drawback for the CIS and related surveys.

## FUTURE PROSPECTS AND ALTERNATIVE VIEWS

The eclectic approach adopted by the OM 2018, with the objective of providing a very broad methodological framework where a range of different data collection exercises could be accommodated[35] (including, quite explicitly, surveys on innovation in the public sector), could have both positive and negative consequences.

In this perspective, the availability by statistical agencies to test new surveying approaches will be crucial. As an example, Eurostat, already before the publication of the new OM, introduced in the CIS 2018 questionnaire a small but substantial change: that of asking about the innovation performance only after having gathered substantial information about the respondent's business strategies and knowledge flows' management. This apparently minor change does affect the nature of the CIS, which has constantly been based on a hierarchical structure where the question "are you an innovator?" was asked at the very beginning of the questionnaire in order to filter out the "non innovators". So, the CIS, for a long time mostly a survey on innovators, could now become a survey about business strategies, including innovation strategies.

*Table 17.1    Alternative approaches to innovation measurement*

|  | Intangible assets | Innovation |
|---|---|---|
| **Macro level** | Growth accounting (macro approach) | Growth-accounting residual (TFP) |
| **Industry level** | Intangible surveys | Innovation surveys |
| **Firm level** | Scoreboards | |

While awaiting the results of the CIS 2018 (at the time of writing they are expected in the second half of 2020), it can be argued that this choice is the result of the strong demand by users for more information about the context where innovation is developed, for instance to understand how much of the overall investment in tangible and intangible assets is devoted to innovation projects. Indeed, the key strategic issue for innovation measurement is the alternative between broadening the scope of the CIS (and other innovation surveys) while keeping a strong focus on innovation, or replacing the CIS with new surveys adopting, by design, an holistic approach to investigate, at a time, business strategies and innovative practices.

In Table 17.1 some options for alternative measurement methods of the innovation capacity or performance at macro, industry or unit level are shown. All of them can be used to replace, partly or totally, the current innovation surveys based on the OM. They are: growth accounting and TFP estimation at a macro-economic level, intangible assets surveys at industry level, and a broad range of self-evaluation exercises to assess the value of own intangible assets or the innovation potential at firm level.

The proposed approaches suggest that the collection of data on the acquisition and management of intangible assets should be a key topic in future innovation surveys. On the one hand, innovation surveys could just extend the scope of questions about innovation costs (R&D, training, fixed investments, etc.) by considering them a share of the total investment in intangible assets. On the other hand, intangible assets surveys have already been tested in several countries, including the UK (Marrano and Haskel 2006; Haskel et al. 2010) and Italy, and in an EU-wide Innobarometer survey (Montresor et al. 2014; Montresor and Vezzani 2016) delivering promising results.

Intangible assets are also the core quantitative evidence used in growth accounting exercises. Some economists have already set out theory and measurement of how intangible investment might capture innovation by processing data on intangibles for the EU, Japan and the US (Corrado et al. 2013; Corrado and Hulten 2014). Such a macro approach features the advantage of addressing innovation as a broad phenomenon that incorporates the effects of the digital transformation, not yet properly assimilated in the OM approach.

A more straightforward approach to a macro accounting for innovation, but not less inspiring, emerged in 2009 when a team of British economists released an original report testing a metric to measure innovation in the UK by estimating total factor productivity (Haskel et al. 2009, p. 12, with a direct reference to Jorgenson's work). Such a process of estimation is described as follows:

> First, the report defines innovation expenditure as spending on new knowledge. Second, the report measures the impact of innovation as the effect of such spending on growth. That is, the view of innovation output as the commercialised outputs of knowledge spend or, more loosely, the commercialisation of ideas. Third, since knowledge can leak across firms (in the way that tangible capital cannot), the report also includes in our innovation index the impact of freely-available knowledge on growth using the growth accounting residual (TFP).

Finally, it should not be forgotten that firms have been constantly developing new and up-to-date methods for measuring their innovation potential, or their innovation performance, irrespective of the efforts by statistical agencies to capture innovation evidence by means of statistical surveys. These experiences can be hardly enumerated but a few studies have attempted to identify the most promising approaches and to build on them as an original approach to innovation measurement. An exercise worth mentioning because of its attempt to include a single framework of different metrics by looking at innovation and investment in intangible assets as two highly correlated phenomena, is the project of Signposts of Innovation of the Conference Board (Hao et al. 2017). The proposed method is based on a grid where different metrics are used to feed with data a set of indicators about the input, the throughput and the output of six areas of business activity ("signposts"). They are: Technology, Digitisation, Environmental and Social Sustainability, Customer Experience and Branding, Internal Innovation Networks, and External Innovation Ecosystems. This is an example of an approach to innovation measurement that sounds very familiar to the business community as they are used to dealing with their management issues by considering all the factors influencing their choices in a combined way.

## CONCLUSIONS

A few points can be taken out of a short description of the evolution of business innovation measurement over the last decades. They refer to three drivers that will probably keep influencing this area again in the coming years and should be properly taken into account.

- Long-standing issues. It is not paradoxical that some issues emerge cyclically in the discussion about innovation measurement. This is the case for the object/subject debate, still unsettled. Also recurring is the question about the role of productivity as a measure of innovation: just re-emerging now after almost one century. This is a key lesson, as the scientific process, also in social sciences, is not linear at all and different tools and concepts have constantly to be adapted to the changing context.
- Flexibility. Although everybody is aware of the advantages of well-established statistical exercises, it has to be acknowledged that innovation measurement can be hardly formalised in a static structure. The availability of different data sources, methods of data collection and indicators should be seen as assets for statisticians dealing with innovation statistics. Innovation is definitely a "fuzzy" concept, describing a phenomenon subject to constant change: the whole panoply of statistical tools has to be used to keep up with this transformation.
- Object focus. Innovation can be described from a range of different points of view but, as far as measurement is concerned, the object has to be carefully identified in order to select the most appropriate metrics. As an example, if the objective is that of measuring the rate of success of an innovation strategy at firm level, to ask for the number of innovations is probably useless as much as it would be asking about a potential "focal innovation". Only quantitative output indicators (including the productivity trend) are essential in order to understand whether an innovation strategy has been effective and successful.

## NOTES

1. "Constructs are invented by researchers to designate conceptual abstractions of phenomena (Kaplan 1964). They are the basic building blocks of theory (Weber 2012). The conceptual meaning of a construct is specified through conceptualisation (Schwab 1980)" (Zhang et al. 2016).
2. A cornerstone of this corpus of studies was the survey carried out by the National Research Project (NRP) of the Works Progress Administration (WPA) in the 1930s on "Reemployment opportunities and Recent Changes in Industrial Techniques".
3. A detailed description of these experiences can be found in Godin (2019).
4. This Manual lays down a set of "Proposed Guidelines for Collecting and Interpreting Technological Innovation Data". Its first edition was published by the OECD in 1992.
5. See Jerome (1932) and (1934), Baker (1933).
6. Godin (2019, p. 138), reports that still early in the 1950s, Simon Kuznets was proposing, as methods for the measurement of technological change: "patents, lags in technology use, censuses of machines (or mechanisation surveys of industries), counting of new (consumer) products and input/output ratios".
7. This achievement was based on the publication of the first OECD Manual on R&D statistics, the "Frascati Manual", that established a common ground to compare R&D data produced in the USA with those from other OECD and non-OECD countries.
8. The issue was already raised by Gilfillan (1935) deeming as unnecessary for inventions being based on some scientific knowledge. Industrialists and scholars brought, by the end of the 1960s, additional evidence to argue for complementing R&D statistics with additional data on (technological) innovation. See Mansfield et al. (1971).
9. The topic is discussed in detail in Witt (1993), Fagerberg (2003).
10. As a matter of fact, economists and statisticians who played a role in developing the field of innovation studies, at least in Europe, identified Schumpeter as the original source of many concepts and terms adopted by them. Actually, Schumpeter was not always explicit on some innovation-related issues and several "Schumpeterian" intuitions have to be credited to other scholars who have been in close contact with him, like Gilfillan (Ballandonne 2017) or Maclaurin (Godin 2008).
11. The key product/process antonymy was, of course, inherited from the initiators of management studies, Taylor (1911) and Gilbreth (1911), whose contributions were thoroughly known by Schumpeter.
12. A review of the exercises aimed at extending the measurement of innovation beyond R&D and patents that have taken place in the second half of the last century can be found in Godin (2002).
13. See also Pavitt et al. (1987).
14. Mentioned, among others, by: Kleinknecht and Bain (1993); Kleinknecht and Reijnen (1993); Coombs et al. (1996); Santarelli and Piergiovanni (1996).
15. The approach of considering "the share of innovative products in a firm's total sales" is not included here because it will be discussed later in the chapter with reference to innovation surveys.
16. Several experiences of innovation output measurement, including the NSF innovation studies in the period 1960s/1970s (Acs and Audretsch 1993), adopt the method to identify a sample of inventors (e.g. patentees) and to ask them about their experience.
17. Technical journals are a case in point where the level of sectoral or technological specialisation, the reputation of peers and the breadth of diffusion affect the quality of the information they can provide researchers with. Another major issue is that of focusing almost exclusively on product innovations neglecting most of the firms' efforts to improve their internal processes (including production processes).
18. Experiences of asking firms about their innovation output have been common even before the launching of innovation surveys, for instance by asking managers to report about sales associated with new products (Brouwer and Kleinknecht 1996).
19. Since 1997, the Oslo Manual is an OECD publication sponsored by the European Union statistical agency (Eurostat). Actually, as the Manual has the ambition to have a global reach, without any specific reference to the CIS, the contribution by Eurostat to the corpus of the Manual has been a minor one, consistently over the last three editions.

20. Extensive information on the CIS survey (including the results of all waves) are available from the Eurostat website (https://ec.europa.eu/).
21. EU Regulation (EC) No 1450/2004 of 13 August 2004 carrying out Decision 1608/2003/EC on the production and development of Community statistics on innovation.
22. Several EU countries argued that the CIS and the R&D survey, both mandatory in the EU, had to be merged in order to reduce both the costs for statistical agencies and the burden on respondents (a description of the Norwegian experience can be found in Wilhelmsen 2012; already the Oslo Manual 1997 – OECD and Eurostat 1997, p. 72 – described the pros and cons of this approach).
23. They include 28 EU member countries, two members of the European Economic Area (Iceland and Norway) and three candidate countries (Montenegro, North Macedonia and Serbia).
24. "Most of the data collected in innovation surveys are qualitative, subjective and censored" (Mairesse and Mohnen 2010, p. 11); "Subjective data: Many of the variables, qualitative and quantitative as well, are of a subjective nature, being largely based on the personal appreciation and judgment of the respondents. One of the most interesting variables and that is relatively well known, the share in total sales due to new products, has, for example, values that tend to be rounded (10%, 15%, 20%…), attesting to its subjective nature and suggesting that perhaps we should treat it as a categorical variable and not make too much out of its continuous variations. What exactly is defined as a new or improved product is not always clear anyway, certainly not to the respondents. There are some examples given in the Oslo Manual, which are themselves more or less debatable and are not always reproduced in the questionnaires. The distinction between "new to the firm" and "new to the market" is also subject to a great deal of subjective judgment. To give a correct answer to this question presupposes a very good knowledge of one's market" (ibidem, p. 12). See also Huovari et al. 2015.
25. "The Oslo Manual, which was designed as a short technical manual back in 1992, has seen its size progressively increasing (+40 percent of words/characters in the 2005 edition compared to the 1997 edition and +100 percent in the 2018 edition compared to 2005). It is now more an original mix between a statistical manual and a treaty on a general theory of innovation than only a collection of technical guidelines for statisticians" (Perani 2019).
26. The definition is given in the same Manual: "An innovation project is a set of activities that are organised and managed for a specific purpose and with their own objectives, resources and expected outcomes. Information on innovation projects can complement other qualitative and quantitative data on innovation activities" (OECD and Eurostat 2018, p. 249)
27. By the way, a breakdown of an innovation strategy, at firm level, by number of "projects" seems more relevant than asking for the number of "innovations", as a project could lead to the development of multiple innovations.
28. Across the four editions of the Oslo Manual, the concept of innovation output is rarely used and always in a generic sense. Nevertheless, it clearly refers to products/effects of the innovation projects reported by innovating firms.
29. Godin (2002).
30. "For output indicators, the share of total sales revenue from product innovations is frequently used. In principle, this type of indicator should also be provided for specific industries because of different rates of product obsolescence" (OECD and Eurostat 2018, p. 227).
31. Godin (2002) identified three potential factors influencing the Oslo Manual approach: (1) the long lasting influence of the "linear model" of innovation could have led to identify only (basic) R&D as the input of the process and the following stages of the linear innovation process as different levels of output; (2) the monopoly of National Statistical Institutes in surveying business innovation could have had an influence on shaping innovation surveys like any other business survey and on opting for the easier option of measuring activities, rather than products and processes; (3) the Oslo Manual definition of innovation itself is complex and difficult to use effectively in a survey (even though only products and processes would be considered as innovations, and more by including organisational and marketing innovations).
32. https://ec.europa.eu/growth/industry/innovation/facts-figures/scoreboards_en.
33. https://ec.europa.eu/docsroom/documents/36282/attachments/1/translations/en/renditions/native.
34. Originated from the EC Communication, *Measuring innovation output in Europe: towards a new indicator* Brussels, 13 September 2013, COM(2013) 624 final. The technical description is avail-

able here: https://rio.jrc.ec.europa.eu/en/stats/innovation-output-indicator and an assessment and scientific analysis in Janger et al. (2017).
35. "Innovation activities occur in all four SNA sectors. Consequently there is a need for a general definition of innovation that is applicable to all institutional units or entities " (OECD and Eurostat 2018, p. 47).

# REFERENCES

Acs, Zoltan J. and David B. Audretsch (1993), 'Analyzing innovation output indicators: the U.S. experience', in Alfred Kleinknecht and Donald Bain (eds), *New Concepts in Innovation Output Measurement*, London: Macmillan, pp. 10–41.

Alford, Leon P. (1929), 'Industry: Part 2 – Technical changes in manufacturing industries', in *Recent Economic Changes in the United States*, Volumes 1 and 2, NBER, pp. 96–166.

Archibugi, Daniele, Patrick Cohendet, Ame Kristensen and Karl-August Schaffer (1997), *Evaluation of the Community Innovation Survey. Report to the European Commission*, Sprint/Eims Report, Luxembourg.

Arundel, Anthony and Keith Smith (2013), 'History of the Community Innovation Survey' in Fred Gault (ed.), *Handbook of Innovation Indicators and Measurement*, Cheltenham, UK and Northampton, MA, USA: Edward Elgar Publishing, pp. 60–87.

Baker, Elizabeth Faulkner (1933), *Displacement of Men by Machines – Effects of Technological Change in Commercial Printing*, New York: Columbia University Press.

Ballandonne, Matthieu (2017), 'On geniuses and heroes: Gilfillan, Schumpeter, and the eugenic approach to inventors and innovators', Paper ESSCA School of Management, April 2017.

Bolton Committee (1971), *Small Firms: Report of the Committee of Inquiry on Small Firms*, HMSO, London.

Brouwer, Erik and Alfred Kleinknecht (1996), 'Firm size, small business presence and sales of innovative products: a micro-econometric analysis', *Small Business Economics*, 8 (3), 189–201.

Brouwer, Erik and Alfred Kleinknecht (1997), 'Measuring the unmeasurable: a country's non-R&D expenditure on product and service innovation', *Research Policy*, 25 (8), 1235–42.

Coombs, Rod, P. Narandren and Albert Richards (1996), 'A literature-based innovation output indicator', *Research Policy*, 25 (3), 403–13.

Corrado, Carol et al. (2013), 'Innovation and intangible investment in Europe, Japan, and the United States', *Oxford Review of Economic Policy*, 29 (2), 261–86.

Corrado, Carol A. and Charles R. Hulten (2014), 'Innovation accounting', in Jorgenson Dale et al. (eds), *Measuring Economic Sustainability and Progress*, Chicago, IL: University of Chicago Press, pp. 595–628.

Crépon, Bruno, Emmanuel Duguet and Jacques Mairesse (1998), 'Research, innovation and productivity: an econometric analysis at the firm level', *Economics of Innovation and New Technology*, 7, 115–58.

Fagerberg, Jan (2003), 'Schumpeter and the revival of evolutionary economics: an appraisal of the literature', *Journal of Evolutionary Economics*, 13 (2), 125–59.

Fagerberg, Jan and Bart Verspagen (2009), 'Innovation studies: the emerging structure of a new scientific field', *Research Policy*, 38 (2), 218–33.

Fagerberg, Jan, Morten Fosaas and Koson Sapprasert (2012), 'Innovation: exploring the knowledge base', *Research Policy*, 41 (7), 1132–53.

Freeman, Christopher (1974), *Economics of Industrial Innovation*, Abingdon: Routledge.

Freeman, Christopher (2007), 'A Schumpeterian renaissance?' in Horst Hanusch and Andreas Pyka (eds), *Elgar Companion to Neo-Schumpeterian Economics*, Cheltenham, UK and Northampton, MA, USA: Edward Elgar Publishing, pp. 130–47.

Gilbreth, Frank Bunker (1911), *Motion Study: A Method for Increasing the Efficiency of the Workman*, New York: David Van Nostrand Company.

Gilfillan, Colum S. (1935), *The Sociology of Invention: An Essay in the Social Causes of Technic Invention and Some of its Social Results; Especially as Demonstrated in the History of the Ship.*

*A Companion Volume to the Same Author's Inventing The Ship*, Chicago, IL: Follett Publishing Company.

Godin, Benoît (2002), 'The rise of innovation surveys: measuring a fuzzy concept', Canadian Science and Innovation Indicators Consortium, Project on the History and Sociology of S&T Statistics, Paper 2002, 16.

Godin, Benoît (2008), 'In the shadow of Schumpeter: W. Rupert Maclaurin and the study of technological innovation', *Minerva*, 46, 343–60.

Godin, Benoît (2019), *The Invention of Technological Innovation: Languages, Discourses and Ideology in Historical Perspective*, Cheltenham, UK and Northampton, MA, USA: Edward Elgar Publishing.

Hao, Xiaohui Janet, Bart Van Ark and Ataman Ozyildirim (2017), 'Signposts of innovation: a review of innovation metrics', The Conference Board Economics Program Working Paper.

Haskel, Jonathan, Tony Clayton, Peter Goodridge, Annarosa Pesole, David Barnett, Gary Chamberlain and Alex Turvey (2009), *Innovation, Knowledge Spending and Productivity Growth in the UK*, Interim report for NESTA, Innovation Index Project, UK.

Haskel, Jonathan, Gaganan Awano, Mark Franklin and Zafeira Kastrinaki (2010), 'Measuring investment in intangible assets in the UK: results from a new survey', *Economic & Labour Market Review*, 4 (7), 66–73.

Huovari, Janne, Mika Nieminen and Olavi Lehtoranta (2015), 'An attempt to measure innovation differently – results of a pilot survey', in Mika Nieminen and Olavi Lehtoranta (eds), *Measuring Broad-Based Innovation*, VTT Technical Research Centre of Finland, Espoo, pp. 81–95.

Janger, Jürgen, Torben Schubert, Petra Andries, Christian Rammer and Machteld Hoskens (2017), 'The EU 2020 innovation indicator: a step forward in measuring innovation outputs and outcomes?', *Research Policy*, 46 (1), 30–42.

Jerome, Harry (1932), 'The measurement of productivity changes and the displacement of labor', *The American Economic Review*, 22 (1), 32–40.

Jerome, Harry (1934), *Mechanization in Industry*, New York: NBER.

Kaplan, Abraham (ed.) (1964), *The Conduct of Inquiry: Methodology for Behavioural Science*, Scranton, PA: Chandler.

Kleinknecht, Alfred and Donald Bain (eds) (1993), *New Concepts in Innovation Output Measurement*, London: Macmillan.

Kleinknecht, Alfred and Jeroen Reijnen (1993), 'Towards literature-based innovation output indicators', *Structural Change and Economic Dynamics*, 4 (1), 199–207.

Kline, Stephen J. and Nathan Rosenberg (1986), 'An overview of innovation', in Ralph Landau and Nathan Rosenberg (eds), *The Positive Sum Strategy: Harnessing Technology for Economic Growth*, Washington, DC: National Research Council, pp. 275–306.

Lööf, Hans, Jacques Mairesse and Pierre Mohnen (2017), 'CDM 20 years after', *Economics of Innovation and New Technology*, 26 (1–2), 1–5.

Maclaurin, W. Rupert (1950), 'The process of technological innovation: the launching of a new scientific industry', *The American Economic Review*, 40 (1), 90–112.

Mairesse, Jacques and Pierre Mohnen (2010), 'Using innovation surveys for econometric analysis', in Bronwyn H. Hall and Nathan Rosenberg (eds), *Handbook of the Economics of Innovation*, New York: Elsevier, pp. 1129–55.

Mansfield, Edwin, John Rapoport, Jerome Schnee, Samuel Wagner and Michael Hamburger (1971), *Research and Innovation in the Modern Corporation*, New York: Norton.

Marrano, Mauro Giorgio and Jonathan Haskel (2006), 'How much does the UK invest in intangible assets?', Working Paper No. 578. Queen Mary, University of London.

Montresor, Sandro, Giulio Perani and Antonio Vezzani (2014), *How do companies 'perceive' their intangibles? New statistical evidence from the Innobarometer 2013*, Institute for Prospective Technological Studies. European Commission

Montresor, Sandro and Antonio Vezzani (2016), 'Intangible investments and innovation propensity: evidence from the Innobarometer 2013', *Industry and Innovation*, 23 (4), 331–52.

Nelson, Richard R. and Sidney G. Winter (1982), *An Evolutionary Theory of Economic Change*, Boston: Harvard University Press.

OECD (1992), *Proposed Guidelines for Collecting and Interpreting Technological Innovation Data: Oslo Manual*, Paris: OECD Publishing.

OECD/EUROSTAT (2005), *Oslo Manual: Guidelines for Collecting and Interpreting Innovation Data*, 3rd edition, Paris: OECD Publishing.

OECD/EUROSTAT (2018), *Oslo Manual: Guidelines for Collecting, Reporting and Using Data on Innovation*, 4th edition, Paris: OECD Publishing.

OECD/EUROSTAT/EU (1997), *Proposed Guidelines for Collecting and Interpreting Technological Innovation Data: Oslo Manual*, Paris: OECD Publishing.

Pavitt, Keith, Michael Robson and Joe Townsend (1987), 'The size distribution of innovating firms in the UK: 1945–1983', *The Journal of Industrial Economics*, 297–316.

Perani, Giulio (2019), 'Business innovation statistics and the evolution of the Oslo Manual', *NOvation: Critical Studies of Innovation*, (1), 36, http://www.novation.inrs.ca/index.php/novation/article/view/5.

Perazich, George and Philip M. Field (1940), *Industrial research and changing technology* (No. 4), WPA, National Research Project.

Rosenberg, Nathan (1982), *Inside the Black Box. Technology and Economics*, Cambridge: Cambridge University Press.

Santarelli, Enrico and Roberta Piergiovanni (1996), 'Analyzing literature-based innovation output indicators: the Italian experience', *Research Policy*, 25 (5), 689–711.

Schumpeter, Joseph A. (2017), *Theory of Economic Development*, London: Routledge (1st edition: 1934).

Schwab, Donald P. (1980), 'Construct validity in organizational behavior', *Research in Organizational Behavior*, 2, 3–43.

Smith, Keith (2005), 'Measuring innovation', in Jan Fagerberg, David C. Mowery and Richard R. Nelson (eds), *The Oxford Handbook of Innovation*, Oxford: Oxford University Press, pp. 148–77.

Taylor, Frederick Winslow (1911), *The Principles of Scientific Management*, New York and London: Harper & Brothers.

Tether, Bruce S., I. J. Smith and Alfred T. Thwaites (1997), 'Smaller enterprises and innovation in the UK: the SPRU Innovations Database revisited', *Research Policy*, 26 (1), 19–32.

UIS (2012), Results of the 2011 UIS Pilot Data Collection of Innovation Statistics, Mimeo.

UIS (2015), Summary Report of the 2015 UIS Innovation Data Collection, Information Paper N. 24.

UIS (2017), Summary Report of the 2017 UIS Innovation Data Collection, Information Paper N. 37.

Weber, Ron (2012), 'Evaluating and developing theories in the information systems discipline', *Journal of the Association for Information Systems*, 13 (1), 1–30.

Weintraub, David (1936), 'Statistical problems confronted in the analysis of the relationship between production, productivity and employment', Read at the American Statistical Association Meeting, Chicago, December 1930, National Research Project on Reemployment Opportunities and Recent Changes in Industrial Techniques, Philadelphia (PA): Works Progress Administration.

Wilhelmsen, Lars (2012), 'The Norwegian innovation survey: combined with the R&D survey versus a separate CIS', Paper presented at the Eurostat CIS Workshop, Luxembourg 2–3 July.

Witt, Ulrich (1993), *Evolutionary Economics*, Cheltenham, UK and Northampton, MA, USA: Edward Elgar Publishing.

Zhang, Meng, Guy Gable and Arun Rai (2016), 'Principles of construct clarity: exploring the usefulness of facet theory in guiding conceptualization', *Australasian Journal of Information Systems*, 20. https://doi.org/10.3127/ajis.v20i0.1123.

# PART VI

# IMMUNE DISCIPLINES AND
# FORGOTTEN THEORIZATIONS

# 18. Religion and innovation: charting the territory
## Boris Rähme

## INTRODUCTION

Judged by the literature in various areas of the study of religion, the intersections between religion and innovation are manifold. Sociologists, anthropologists and historians of religion use the expression 'religious innovation' to refer to phenomena of transformation and change in the religious domain, such as the emergence of new religious movements, departures from orthodoxy and orthopraxy within traditional and institutionalized religions, and the ways in which religion has proved resilient to large-scale modernization and rationalization processes. There are debates over the question of whether religious innovation is merely adaptive or rather creative and, relatedly, about the question of whether the appropriation of innovations in technology (e.g., in the field of information and communication) by religious communities has, in turn, led to religious innovation. Scholars studying the intellectual and conceptual history of innovation argue that religion has had a decisive influence on innovation discourses, with residues of this influence still shaping today's understandings of innovation. Economists conduct research on the impact of religiosity on innovativeness, and social psychologists study the effect of religious beliefs on attitudes towards innovation. Many more items can be added to the list.

Given this variety of claimed intersections between religion and innovation, the purpose of the present chapter is to provide an overview of religion-and-innovation discourses along with conceptual and theoretical considerations that are, or so I hope, useful for approaching the by now extensive research literature. As will emerge from the following sections, the use of the term 'innovation' (and its cognates 'innovative' and 'innovativeness') is far from uniform in debates over religion and innovation. Some authors employ the term generically as an equivalent of the expression 'significant (intentional) change'. Others use it as shorthand for 'significant (intentional) change *for the better*' and thus implicitly convey a positive evaluation of whatever it is they describe as an innovation. Others again use it in ways that draw upon more specific conceptions of technological, commercial or social innovation. The present chapter highlights the varying degrees of normative and evaluative commitment implied by different uses of the term 'innovation' in religion-and-innovation discourses. It thus contributes to answering the question of whether and, if so, to what extent the "pro-innovation bias" widespread in today's innovation discourses (Rogers 1962; Godin and Vinck 2017) has found its way into religion-and-innovation research.

Before turning to these tasks, some remarks on how I will use the word 'religion' are in order. Ideally, research on religion and innovation should employ a concept of religion which is general enough to cover the entire range and diversity of faiths, creeds, practices and worldviews generically referred to by the everyday expression 'religion', including monotheistic, polytheistic, henotheistic and nontheistic traditions as well as their different denominations and sub-denominations. At the same time, the employed concept of religion should be specific enough to enable a distinction between religious and non-religious social phenomena. As is well known, the question of whether such a concept of religion is to be had – or even so much as desirable – has triggered highly controversial debates in the social sciences. Even if somewhat dated, a still useful lens for looking at the debates about how best to understand the term 'religion' is the distinction between substantive and functional definitions (Berger 1974; Bruce 2011a). Functional approaches define religion in light of the "purposes it serves or the needs it meets" (Bruce 2011b, p. 1). They appeal to the functional roles that religious thought and practice play in the lives of individuals and societies. Functional definitions of religion thus rest on a problematic assumption. They assume that only religion is able to discharge the psychological or social functions which are, respectively, claimed to be definitory of religion (Bruce 2011b, p. 1). It is true, presumably, that religion can be conducive to the cohesion of groups and societies, that it can provide sense, purpose and meaningfulness, ethical and moral orientation and guidance. But for any social or psychological function $f$ that religion can fulfil, it is plausible to claim that there can be non-religious functional equivalents of religion which can fulfil $f$ as well. If that is correct, then religion cannot be defined in terms of its functions alone. Purely functional definitions tend to commit their proponents to implausibly counting as religious a series of human practices, belief-systems or communities which – though maybe similar to religion in various respects – are clearly not religious per se (Berger 1974, p. 128). In other words, functional definitions are overly inclusive.

A standard objection to substantive approaches, which define religion with reference to beliefs and practices that are taken to be constitutive of (necessary for) religion, is that they exclude a series of human practices, belief-systems or forms of community which – from the critics' point of view – should be counted as instances of religion. However, contemporary advocates of substantive definitions usually take great care to formulate their accounts in ways that anticipate and pre-empt this objection. Sociologists Stark and Bainbridge (1979, p. 119), for instance, suggest reserving the term 'religion'

> for solutions to questions of ultimate meaning which postulate the existence of a supernatural being, world, or force, and which further postulate that this force is active, that events and conditions here on earth are influenced by the supernatural.

Sociologist Steve Bruce, while disagreeing with almost everything else Stark and Bainbridge have to say about religion (Bruce 1999, 2011c), proposes a similar account. He delineates the phenomenon of religion by appealing to

> beliefs, actions, and institutions based on the [assumed, B.R.] existence of supernatural entities with powers of agency (that is, Gods) or impersonal processes possessed of moral purpose (the Hindu and Buddhist notion of karma, for example) that set the conditions of, or intervene in, human affairs. (Bruce 2011b, p. 1)

What makes these accounts substantive (as opposed to functional) is their reference to assumptions, on the part of religious believers, of the existence of supernatural entities or processes which are relevant to human existence. They take the presence of the assumption, belief or faith in the existence of supernatural entities or processes to be a necessary condition for religion. Taking a leaf out of the books of Bainbridge, Stark and Bruce, for what follows I will use the word 'religion' to refer to socio-cultural contexts composed of practices, beliefs, doctrines, precepts, community bonds, habits, attitudes and institutions which are guided, oriented and structured by commitments to humanly relevant supernatural entities or processes. An immediate consequence of thinking about religion along these lines is that no field of thought or practice that is not guided, oriented or structured by commitments to humanly relevant supernatural entities or processes qualifies as an instance of religion. Much more would have to be said about the question of how to delimit the boundaries of the phenomenon of religion. However, for present purposes the proposed account can serve as a reasonable basis. It is inclusive enough to cover pronounced religious diversity (Simmel 1905, p. 359), including highly individualized forms of religiosity or 'spirituality' (Giordan 2016). At the same time, it is exclusive enough not to commit me to implausible claims to the effect that football fan clubs, environmental activism, nuclear physics, humanitarian NGOs, Kantian or Aristotelian ethics, political parties or the ideological movement of transhumanism are instances of religion (Rähme 2021).

The chapter is structured as follows. The next section introduces and briefly explains a heuristic threefold distinction between aspects or dimensions of the interaction between religion and innovation that have been addressed in the literature: innovation in religion, religion in innovation and religion of innovation.[1] I then give an overview of, and critically discuss, current work on phenomena of innovation within religion (often referred to as 'religious innovation'). The subsequent section addresses examples of how researchers from various disciplines describe and explain the role of religion in innovation processes. Next, I turn to the idea that innovation discourses themselves have become vehicles for religious, quasi-religious or spiritual commitments. The final section draws some conclusions from the preceding considerations, in particular with regard to the idea of a general theory of religion and innovation, and proposes questions for future research.

## INNOVATION IN RELIGION, RELIGION IN INNOVATION, RELIGION OF INNOVATION

At a very general level, and drawing upon the working account of religion proposed above, claims about interactions between the religious social sphere and various forms of innovation can be grouped into three large areas. *Innovation in religion*: contrary to widespread opinion, various religious traditions – such as the different strands of Hinduism, Buddhism, Christianity, Judaism and Islam – have been, and continue to be, subject to innovation from within. Moreover, new forms of religion continue to emerge, and their emergence can be usefully analysed in terms of processes of innovation in the religious sphere (religious innovation). *Religion in innovation*: religious beliefs, motivations, institutions, practices and communities can have a facilitating or inhibiting impact on various, otherwise non-religious, kinds of innovation (technological, social, organizational, sustainable etc.). *Religion of innovation*: innovation itself tends to – or, as some would claim, should – turn into a (secular functional equivalent of) religion. Table 18.1 displays a series of more specific claims about religion and innovation that have been made in the relevant debates.

In suggesting this distinction of three aspects of the interaction between religion and innovation I do not mean to propose a neat and clear-cut categorization. The three aspects may overlap to a greater or lesser extent in any given process of change that is claimed to be an instance of the interaction between religion and innovation. Nonetheless, the proposed distinction is a useful heuristic tool for approaching the relevant research literature.

*Table 18.1*     *Three dimensions of interaction between religion and innovation: Examples*
*of claims made in the literature*

| | Dimensions of interaction | | |
| --- | --- | --- | --- |
| | **Innovation in religion** | **Religion in innovation** | **Religion of innovation** |
| **Examples of claims made in the literature** | Religious traditions continue to be engaged in processes of self-innovation (Williams 1992; Stauning Willert and Molokotos-Liederman 2012; Yerxa 2016a). | The religiosity-level of a country's population is positively correlated with firm-level innovation in that country (Assouad and Parboteeah 2018). | Critical or polemic use of 'religion': innovation can turn (or has turned) into a blind faith, religion or cult (Zax 2013; Ampuja 2016; Godin, Chapter 1 in this book). |
| | Different religious traditions show different degrees of openness to technological innovations (Liu et al. 2018; Recio-Román et al. 2019). | The religiosity-level of a country's population is negatively correlated with country-level innovativeness (Bénabou, Ticchi and Vindigni 2013, 2015). | Affirmative use of 'religion': an attitude of (quasi-) religious or spiritual faith *in* innovation is conducive to firm-level innovation (Buckler and Zien 1996; Pandey 2017). |
| | The emergence of new religious movements (NRMs) can be understood in terms of religious innovation (Dawson 2006; Baffelli, Reader and Staemmler 2011). | Religion can have an impact (positive or negative) on innovation for sustainable development (Narayanan 2013; Tomalin, Haustein and Kidy 2019). | Descriptive use of 'religion': innovation has turned into a secular religion/faith.[2] |
| | The uptake of technological innovations (e.g., in digital technologies) by religious communities and institutions can, in turn, lead to religious innovation (Cowan and Hadden 2008; Campbell 2013). | Religion can play a role in motivating and structuring social innovation and social entrepreneurship (Ataide 2012; Harrison 2016; Samuel Shah 2016). | Belief in the market as the main driver of innovation has turned into religious belief, the market has become God (Cox 2016; Nelson 2019). |

One crucial issue, which concerns religion-and-innovation discourses as much as any other debate over innovation, is the implicit normativity of innovation-speak. The term 'innovation' is today commonly (though not always) used to express a positive attitude towards a given project, process or result $X$. When an author characterizes an $X$ as innovative or an innovation, then this usually implies an attitude of appreciation or endorsement of $X$, that is, an evaluation of $X$ as something good, desirable and useful – as something that improves upon a previous state of affairs.[3] This has not always been so. It was only in the course of the twentieth century that 'innovation' has become a success word – or, as Godin and Vinck (2017, p. 4) put it, "a word of honour".[4] Just as the concept of religion, the concept of innovation is a mutable category whose core meanings and connotations shift diachronically over time and synchronically across cultural, social and political contexts. Such shifts occur both in conceptualizations of what innovation consists in, and in attitudes regarding the goodness or badness of innovation. A project, process or result, then, cannot be said to be innovative *tout court* but only relative to a specific context of reference. To understand a given religion-and-innovation discourse, a good place to start is the characterization of its context of reference, including at least the description of the state of affairs S that constitutes the point of departure of an alleged innovation process, and a reconstruction of the presupposed set of values or goals that determine whether or not a given project, activity or result is to be considered an improvement over S. Essentially the same points are noted by comparative religion scholar Michael A. Williams

(1992, p. 4) with respect to one of the three areas distinguished above, innovation in religion (religious innovation):

> The fact that the very definition of religious innovation is governed by the interpreter's perspective does not render religious innovation meaningless or frivolous as a category, but it does point to the fact that its usefulness is dependent both on a careful establishment of context, as well as on a determination of what perspective(s) is/are in fact germane to a given analysis.

What Williams says about innovation in religion (religious innovation) generalizes seamlessly to the other two areas distinguished above, that is to religion in innovation and religion of innovation. Understanding a given religion-and-innovation discourse with regard to some process or result $X$ requires close attention to a set of interrelated questions: why and with reference to which goals and values is $X$ characterized as an innovation? How is the term 'innovation' used and understood, and is a working definition of the term made explicit? How is the term 'religion' used, and is a working account provided? If so, is it along functionalist or substantivist lines?[5] Is the perspective (insider/outsider, participant/observer) from which $X$ is characterized as an innovation identified? Are normative commitments explicitly acknowledged?

## INNOVATION IN RELIGION (RELIGIOUS INNOVATION)

So far, most academic work on religion and innovation has focused either on processes of change within religious traditions, institutions and communities or on the emergence of new religious movements (formerly called 'cults', see Dawson 2006) – that is, on the dimension of innovation in religion. By now, the term 'religious innovation' is widely used by religion researchers and scholars to refer to such processes of change. Despite this fact, there is a dearth of informative accounts of the conditions under which a process of change within the social sphere of religion qualifies as an instance of religious innovation. Discussions predominantly proceed by way of case studies and examples, and the question of what (if anything) these different cases have in common – apart from being cases of change in religion – is left unaddressed. There are, of course, notable exceptions to which I will turn later in this section. At this point, it is sufficient to notice that authors who use the expression 'religious innovation' usually do not intend it to refer to innovation that is somehow religious or has the quality of being religious. Rather, they use it to refer to processes of innovation in (or of) religion. What is claimed to be innovated are religious practices, organizations and institutions, doctrines, beliefs and communities. Consequently, religious innovation is generally not seen as a specific kind of innovation – as opposed to, for instance, technological, sustainable, or responsible innovation, which are often claimed to be specific kinds of innovation – but rather as a field of innovation, a social sphere in which innovation can or does occur.

As argued in the preceding section, the choice of the term 'innovation' – as opposed to, say, 'transformation' – in studying religious change is significant because it often signals implicit normative commitments. Much work on the interface between religion and innovation is indeed motivated by the goal of disconfirming the commonly held assumption that religion and innovation simply do not match because religion is inherently opposed to change and thus *a fortiori* opposed to change through innovation. A good example of scholarly work that aims to critique this assumption is the volume *Religion and Innovation: Antagonists or Partners?*, edited by the historian Donald A. Yerxa (2016a). In his introduction, Yerxa states that the studies collected in the volume provide evidence for the falsity of generic claims to the effect that religion has "inhibited innovation in human societies", that it "resists change and functions as an agent of tradition and social control" and that "[i]ts relationship to innovation is fundamentally antagonistic" (Yerxa 2016a, p. 1). A second example is a volume edited by historians of religion David N. Hempton and Hugh McLeod (2017):

> In calling our volume "Secularization and Religious Innovation in the North Atlantic World" we start from the assumption that, since the eighteenth century, the threat of secularization has been an inescapable reality for Europeans and North Americans alike, but that churches, devout laypeople, religious intellectuals, and others have responded to the threat in different ways. (McLeod 2017, pp. 2–3)

In what sense, then, do these statements signal implicit normative commitments? Presumably, no one would consider it worthwhile to disconfirm the assumption that religion is inherently opposed to innovation unless they attach positive value both to innovation and to religion, that is, unless they (implicitly) endorse both as, all things being equal, good and desirable. McLeod presents and recommends religious innovation as an appropriate means to avert what he takes to be a threat, the threat of secularization. It is noteworthy that he takes this threat to affect not only the religious portions of the populations of North America and Europe but Europeans and North Americans quite generally. If one considers secularization a threat, one will tend to attach positive normative value to those means of averting this threat which are compatible with one's independently held normative beliefs and values.[6]

## A Process Definition and a Typology

As mentioned above, researchers and scholars of religion who employ the term 'innovation' usually do not pause to give an account of what exactly they intend to convey with this term. If the question is explicitly addressed at all, then answers often remain at a generic level: "Religious innovation is the process by which new ritual traditions, concepts of the divine, and religious institutions are created" (Woolf 2016, p. 2). In most cases, then, religion researchers simply employ the expression 'religious innovation' for their theoretical and empirical purposes. They presuppose that the expression conveys a concept which is sufficiently well-understood to be useful for describing and explaining the introduction and diffusion of novel religious beliefs and practices, novel approaches to the administration of religious institutions, or new forms of religious community.

A significant exception in this regard is Steeve Bélanger and Frédérique Bonenfant's process approach to religious innovation. According to Bélanger and Bonenfant (2016, p. 403), religious innovation is a:

> collective process which, out of a will and / or desire for change in the face of a situation considered as not, or no longer, meeting current needs or aspirations, introduces religious novelty and leads, by negotiation or imposition through a network of communication, to significant, effective and lasting socio-religious change in practices and / or systems of meanings.[7]

This rich account combines several elements which, taken together, suggest a substantive and informative answer to the question of how to delimit the conditions under which a process of change in the social sphere of religion may be considered an instance of religious innovation. First, innovation is described as a process rather than as an outcome or result. Second, the process is taken to be a collective one. Rather than being based on the agency of an individual 'innovator', it is said to involve the agency of a plurality of agents. Third, the account gives prominence to intentionality, will and desire. It thus distinguishes innovation from change that just happens to come about. Fourth, it postulates the perception of unmet or no longer met needs as the motivational basis for religious innovation. Fifth, it stresses the innovation-enabling role of communication networks which, presumably depending on the degree of communicative authority and power of the group that pursues the goal of religious innovation, can be either used to negotiate or to impose religious change. Sixth, a process of change or transformation that meets the preceding requirements is said to qualify as an instance of religious innovation only if it brings about significant, effective and lasting changes in the religious social sphere. Finally, the social sphere of religion is understood in terms of practices and meaning systems.

Bélanger and Bonenfant's reference to significance, effectiveness and longevity, as well as their appeal to needs perceived as unmet, make it very clear that they intend their account of religious innovation to be context-sensitive and relative. In his introduction to the edited volume *Innovation in Religious Traditions. Essays in the Interpretation of Religious Change* (Williams, Cox and Jaffee 1992), which is another one of the very few places in the relevant literature where the idea of religious innovation is not just used but explicitly made the object of discussion, Michael A. Williams (1992, p. 3) makes the following remark about the significance condition on religious innovation: "Defining something as a religious innovation is essentially to define it as 'significant' change – but significant to whom?" He goes on to state that, "by its very nature", the notion of religious innovation is "a relative category, and its analytical usefulness depends precisely upon the recognition of this, and upon the identification of the interpretive perspectives that are appropriate in different analytical situations" (Williams 1992, p. 7). Based on the previous considerations concerning the normativity and context-dependence of innovation-talk, I tend to agree with Williams. The context relativity of Bélanger and Bonenfant's account of religious innovation should be taken to speak in favour of, not against, their account.

A second (or, if we count Williams 1992, a third) exception to the prevalent distaste for explicit accounts of religious innovation can be found in Trine Stauning Willert and Lina Molokotos-Liederman's "attempt to develop an initial typology of innovation in religious traditions" (Stauning Willert and Molokotos-Liederman 2012, p. 9). While introduced in the context of a discussion of Greek Orthodoxy, the proposed typology is intended as a stepping stone for further research on the subject of innovation and religious traditions quite generally (p. 16).[8] As these authors emphasize, the question of whether and, if so, how there can be religious innovation is particularly challenging with regard to religions that have "developed orthodox bodies of belief, custom and practice which are regarded as part of a sacred tradition" (p. 6). Nonetheless, appealing to Williams' (1992) discussion of religious innovation, Stauning Willert and Molokotos-Liederman argue that innovation "is an inherent modality" of Greek Orthodoxy and, more generally, of religions qua social and cultural systems.[9] They distinguish between five types of innovation in religion: "purist", "strategic", "adapting", "unintentional" and "emancipatory" (pp. 9–11).

Purist innovation in a religious tradition is innovation aimed at returning to an allegedly more authentic or original state. As the authors point out, purist innovation in religion often occurs in a concealed form, and its proponents present "their arguments in a traditionalist rhetoric" (p. 4). Strategic innovation is not so much innovation of or in a religious tradition but rather occurs when religious leaders and institutions accept, or even embrace and support, political or economic changes and reforms with the intent of retaining, strengthening or re-establishing their political and cultural standing, where such acceptance does not lead to any substantial change within the religious tradition. Adapting innovation in a religious tradition is innovation motivated by the perceived need or even necessity to react and respond to changes "in the social, political or physical environment" (p. 10) in a way that involves more or less significant changes within the beliefs, practices and institutions of the tradition.[10] Unintentional innovation "is the result of religious practice or thought taking inspiration from external sources without seeing this as a conflict with tradition" (p. 10). In another publication, Stauning Willert suggests that the idea of unintentional innovation depends on the distinction between an insider and an outsider perspective on a religious tradition: "unintentional innovation usually applies to initiatives that are considered so fully integrated within the existing recognized tradition that only an outsider will recognise the initiatives as innovative and as a break from an externally observed tradition" (Stauning Willert 2014, p. 30). Emancipatory innovation within a religious tradition, finally, aims at the creation of spaces of agency for members of the tradition whose capacity for leading an "independent life as active members of society" (p. 12) is significantly constrained by religious precepts. The goal of such innovations, then, is not emancipation from the tradition itself but emancipation from its restrictive forms and interpretations.

By including unintentional religious innovation as a type in their typology, Stauning Willert and Molokotos-Liederman propose an understanding of religious innovation that is incompatible with the one suggested by Bélanger and Bonenfant. However, it might be best to consider the question of intentionality as an unresolved issue in Stauning Willert and Molokotos-Liederman's typology of religious innovation since, a few pages before introducing their typology, they suggest a general characterization of innovation which clearly excludes unintentional innovation: "[W]e take innovation to mean a deliberate and intended form of change or break from the usual way of doing things" (p. 7).

The next section turns to market approaches to the explanation of religious change, developed in what has come to be called the economics of religion. Even though, to my knowledge, no explicit account of religious innovation is to be found in the economics of religion literature, I will argue that from the perspective of the economics of religion it would seem to be quite clear what religious innovation is and requires.

## Economics of Religion: Supply-Side Theories and Religious Product Innovation

'Economics of religion' is the name of a line of research that applies economic concepts, models and theories to the study of religion.[11] Its goal is to explain both large-scale social phenomena related to religion, such as differences in religiosity and church attendance levels between the populations of different countries or shifts in the religious composition of societies, and individual-level religious behaviour, such as conversions or individual commitment levels to practices of worship, in economic terms (Carvalho, Iyer and Rubin 2019): 'market', 'demand', 'supply', 'product', 'competition', 'expected utility' and 'rational choice'. In the words of sociologist Rodney Stark, a protagonist of the field, the approach

> can be characterized as an economic approach because when it analyses religion at the *individual* level, it emphasizes *exchange relations* between humans and the supernatural. At the collective level the economic approach to religion rests on the fundamental concepts of *supply* and *demand*. (Stark 2006, p. 48, emphasis in original)

A religious economy, according to Stark, consists of three elements: "a 'market' of current and potential adherents, one or more organizations ('firms') seeking to attract or retain adherents, and the religious culture ('product') offered by the organization(s)" (Stark 2006, p. 64). As opposed to demand-side theories, which appeal to changes in the religious needs of 'religion consumers' to explain changes in the distribution of religious adherence and activity within societies, or differences between societies, supply-side theories of religion are based on the idea that such dynamics are best explained by appealing to characteristics of the landscape of religion providers (Finke and Iannaccone 1993). On this view, then, religious providers compete among each other in a market environment which, according to Stark, Iannaccone and Finke (1996), is characterized by a relatively stable religious demand among humans.[12] Since demand is assumed to be stable, changes in demand cannot be what explains changes of religious affiliation and participation levels.

For explaining individual-level religious behaviour like (non-)affiliation to religious groups, conversion from one group to another or permanent disaffiliation, economists of religion employ the framework of rational choice theory. It merits emphasis that rational choice theory does not concern the reasonableness of people's religious or spiritual beliefs, commitments and preferences. Economic theories of religion take no stance (or should not take a stance) on the question of whether and, if so, in what sense religious beliefs or commitments themselves can be considered rational, justified or supported by evidence. Rather, the idea driving economic theories of individual religious behaviour is to begin from whatever religious preferences, needs or demands an individual may have at some given time t and then to ask what choice of religious affiliation or non-affiliation it would be rational for S to make at t given S's religious preferences and their weight within S's overall set of preferences and beliefs, including those which have nothing to do with religious questions. A choice or action A of an agent is rational, in this sense of the term, if, and only if, it is in the agent's best interest given whatever it is that they believe and whatever ends or goals they have set themselves. In other words, the choice or action A is rational if, and only if, it maximizes the expected utility or reward. As an application of rational choice theory to the study of religion, supply-side theories thus employ a narrow interpretation of Max Weber's concept of purposive rationality ("Zweckrationalität", Weber [1922] 1978, pp. 24–26) in terms of (purely) instrumental rationality.

My claim that rational choice explanations of religious behaviour (should) take no stance on the rationality of religious beliefs may seem incompatible with what Stark presents as one of the "nine fundamental principles" of the economics of religion: "People are as rational in making their religious choices as in making their secular decisions" (Stark 2006, p. 49). But this incompatibility is merely apparent – even if Stark and other economists of religion would have us believe otherwise (Stark, Iannaccone and Finke 1996). At least two questions regarding the rationality of religious choices may be asked. The first, let us call it the wide-scope question, regards the rationality of a religious choice (of affiliation) against the backdrop of the level of rationality of the choosing person's relevant beliefs and goals. A choice is rational, in this sense, only if it is based on rational, justified beliefs and reasonable goals. The second question, call it narrow-scope, concerns the rationality of a religious choice (of affiliation) independently of the rationality of the choosing person's relevant beliefs and goals – though, of course, not independently of the fact that the person has the beliefs and goals that she does have: given that they believe what they do and that they have the goals that they do, is their choice of religious affiliation rational? A choice can be rational in this narrow sense even if it is based on unjustified and outright irrational beliefs, commitments, goals and preferences, i.e., irrational in the wide sense.

Rational choice approaches to religious behaviour, just as rational choice approaches to non-religious behaviour, respond to the narrow-scope question. And in this regard, Stark may very well be right when he claims that there is no relevant difference in terms of rationality between people's religious and non-religious decision-making and choice behaviour. This is not to say that religious beliefs are always irrational. The debate about the rationality of religious beliefs is open. There are good arguments for the claim that religious beliefs are often acquired and retained in rational ways and there are good arguments for the claim that religious beliefs are mostly acquired and retained in irrational ways (Frances 2015, 2016). The point of the preceding remarks is that rational choice theory is not among the theories that can contribute relevant arguments to the debate over the rationality of religious belief.

Without going any further into the details of the economics of religion, or discussing its explanatory strengths and weaknesses, let me here point out two corollaries of this approach that are particularly relevant to the question of how religion and innovation may be taken to interact. The first corollary is that the theoretical framework of supply-side theories in the economics of religion lends itself to a straightforward appropriation of the idea of technological innovation – with the latter understood along the now standard lines of developing and/or successfully commercializing a product or service which is perceived to be new (Godin 2020): religious innovation occurs when religious 'entrepreneurs' manage to come up with and successfully market a product (in Stark's terms, a religious culture) that either responds to, or triggers new kinds of, demand in significant market niches (Stark and Bainbridge 1979; Finke and Iannaccone 1993; Finke 2004).

The second corollary of the economic approach to religion that merits emphasis in the present context regards the legal idea of religious freedom. Supply-side theories of religion would seem to suggest an interpretation of this idea in terms of a free competitive market for religious and spiritual products. Competition between religious 'firms' or 'suppliers' can be more or less influenced (and distorted) by state and government regulations, for instance through the institution of a state religion or other kinds of unequal legal standings afforded to different religious groups in a country (Grim and Finke 2006). As might be expected, advocates of the supply-side approach emphasize the minimization of government or state regulation of the religious market as a condition conducive to – or even necessary for – religious innovation. The less government regulation of religion there is, the more religious 'entrepreneurship' and innovation are incentivized. In the economics of religion literature, this idea is frequently traced back to Book V of Adam Smith's *The Wealth of Nations* (Smith [1776] 1976), which – according to economist Laurence Iannaccone's interpretation – contains a sustained argument for the claim that "competition would not only generate more religion but also better religion" (Iannaccone 1998, p. 1489, see also p. 1485). According to supply-side theories of religion, then, state and government-level regulation of religion have an inhibiting impact on religious innovation (innovation in religion) and, consequently, on religious diversity and levels of religious activity in a given national context. Iannaccone puts this point as follows: "Theory and data [...] combine to suggest that government regulation of religion tends to reduce individual welfare, stifling religious innovation by restricting choice, and narrowing the range of religious commodities" (Iannaccone 1998, p. 1489).[13]

Some market theorists of religion have further developed the thesis that freedom of the religious market is positively correlated with religious innovation and religious activity into the claim that a free religious market is conducive to technological innovation and, by extension, to the economic competitiveness of societies. Since this issue regards the impact of (freedom of) religion on otherwise non-religious innovation processes – rather than its impact on innovation in religion – I will come back to it in the next section.

## RELIGION IN INNOVATION

Research on the role of religion in innovation presents a variegated and diversified picture. On the one hand, this is due to the ongoing proliferation of ideas, models and frameworks of innovation (innovations of the concept of innovation, if you wish) which then either get accepted and entrenched to a greater or lesser extent or vanish from debates in the manner of fads (Gaglio, Godin and Pfotenhauer 2019).[14] On the other hand, it is due to the complexity of religion as a social phenomenon that encompasses practices, beliefs, doctrines, norms and a variety of societal actors from individual believers and religious leaders to religious communities, church institutions and faith-based national or international organizations. Given the multiple possible combinations of different conceptions of innovation with various aspects of the social phenomenon of religion, it is unsurprising that researchers have found abundant evidence for the effects of religion on, and the involvement of religion in, otherwise non-religious innovation processes. This section takes a brief look at relevant research from economics, development studies, social psychology and management studies.

### Religiosity, Freedom of Religion or Belief and Innovativeness

In recent years, economists have employed quantitative methods to study the effects of religiosity on innovativeness, the latter understood in terms of the number of technological, commercial or business innovations produced by a given population in a given period of time. Whereas the economic study of innovation within religion takes its theoretical inspiration from Adam Smith's *The Wealth of Nations* ([1776] 1976), this branch of the economics of religion, which regards the role of religion in otherwise non-religious innovation processes, is often presented as an empirical exploration of theoretical ideas and arguments developed by Max Weber ([1920] 1930) in his *The Protestant Ethic and the Spirit of Capitalism* (see, e.g., McCleary and Barro 2018, pp. 11 and 45).

In a widely quoted and discussed study, economists Roland Bénabou, Davide Ticchi and Andrea Vindigni present what they take to be evidence for the claim that there is a "robust inverse relationship between religiosity and innovation [...] across both countries and US states" (Bénabou, Ticchi and Vindigni 2013, p. 40). The authors measure religiosity levels by means of self-report answers to religion-related survey questions and innovativeness at country and US-state-levels in terms of (log-)patents per capita.[15] In a subsequent study they investigate "the relationship, at the *individual level*, between religiosity and a broad set of pro- or anti-innovation attitudes" (Bénabou, Ticchi and Vindigni 2015, p. 347, emphasis in original). Again, they find a negative correlation between individual-level religiosity and openness to innovation: "greater religiosity was almost uniformly and very significantly associated to less favorable views of innovation" (Bénabou, Ticchi and Vindigni 2015, p. 350). Liu et al. (2018), studying the impact of religiosity on national creativity (which they take to be closely related to national innovativeness), come to a conclusion largely in line with Bénabou,

Ticchi and Vindigni (2013): "the overall religiosity [of a country's population] has a negative relationship with national creativity". However, they also claim to have found that "different denominations show dissimilar effects on creativity. Protestant[ism] and Catholic[ism] are positively related with national creativity, while Islam is negatively related with national creativity" (Liu et al. 2018, p. 11).[16]

Other studies would seem to contradict Bénabou, Ticchi and Vindigni's claim that religiosity is negatively correlated with innovativeness (Assouad and Parboteeah 2018), as well as the innovation profile assigned by Liu et al. to Protestantism, Catholicism and Islam (Recio-Román, Recio-Menéndez and Román-González 2019). Overall, then, empirical findings concerning the correlations between religiosity and innovativeness point in opposite directions. In part this may be due to diverging definitions and operationalizations of the concepts of religion, religiosity, innovation and innovativeness, as well as to different data sources used in different studies. However, it might also be worthwhile to test the hypothesis that some of this divergence is due to biased approaches to religion and innovation which rest on antecedently held pro- or anti-religious and pro- or anti-innovation attitudes. Very roughly, this hypothesis can be further spelled out as follows: to the extent that positive normative value is attached to innovation, research on the impact of religion on innovation conducted by researchers with pro-religious antecedent attitudes will tend towards finding a positive impact of religion on innovation, and research conducted by researchers with anti-religious antecedent attitudes will show a bias towards negative (inverse) correlations of religion and innovation. I should emphasize that I do not claim this hypothesis to be true. Rather, I suggest that it be submitted to empirical testing.

A different but related strand of recent research focuses on correlations between freedom of religion (or belief) and innovativeness. Heiner Bielefeldt, human rights scholar and former United Nations Special Rapporteur on freedom of religion or belief, canvasses the legal idea of freedom of religion or belief in the following way:

> Freedom of religion or belief demands respect for the identity-shaping existential convictions and concomitant practices of 'all human beings'. As a universal right, it is not confined to recognizing the rights of traditional believers, but equally protects the freedom of dissidents, converts and reconverts, minorities and sub-minorities, schismatic groups and individuals whose positions are considered 'heretic' from the standpoint of certain religious orthodoxies. Moreover, freedom of religion or belief also entitles people not to care about religious issues, to remain indifferent or express critical views. (Bielefeldt 2017, p. 37)

As anticipated in the discussion of supply-side theories in the economics of religion, some scholars and activists who subscribe to an economic approach to religion argue that there is a positive correlation between religious freedom and global economic competitiveness as well as innovativeness at country level (Grim and Finke 2011; Grim, Clark and Snyder 2014). In very broad strokes, the argument goes as follows: "to the extent that governments deny religious freedoms, violent religious persecution and conflict will increase" (Grim and Finke 2011, p. 212, emphasis deleted). But violence, conflict and persecution create environments that are unfavourable to business and, since business is the crucial driver of technological innovation, impede economic competitiveness. The denial of freedom of religion or belief "can drive away local and foreign investment, undermine sustainable development, and disrupt huge sectors of economies" (Grim, Clark and Snyder 2014, p. 4). Therefore, religious freedom is good for business and, by extension, for technological innovation.

This is not the place to evaluate the empirical backing of the premises or the overall cogency of this argument. In the present context it is sufficient to note that the nexus between religion and innovation claimed by these authors is somewhat indirect. It is not the degree of religiosity of citizens or employees per se that is claimed to have an impact on the technological innovativeness of a country or a company but rather the legal institution of freedom of religion (or belief) and the extent to which it is implemented and enforced in a given context. *Prima facie* the claim that freedom of religion is positively correlated with country- and firm-level competitiveness and innovativeness would seem to receive independent support from the studies conducted by Bénabou, Ticchi and Vindigni that have been referred to above. While their main result is that "religiosity is significantly and negatively associated with innovation per capita" Bénabou, Ticchi and Vindigni also claim to have found that "religious freedom [is] positively" correlated with innovativeness (Bénabou, Ticchi and Vindigni 2013, p. 13). Still, answering the question of whether this can count as independent support for Grim, Finke and Snyder's claim that 'religious freedom is good for business' is not as straightforward as it may seem at first glance. Assume for the sake of argument that Bénabou, Ticchi and Vindigni have it right on both counts. Under this assumption, religious freedom can be positively correlated with country-level innovativeness only to the extent that it is inversely (or at least not positively) correlated with country-level religiosity – and this would seem to be incompatible with Grim, Finke and Snyder's version of the 'religious freedom is good for business/innovation' thesis, since the latter authors also subscribe to the claim that religious freedom promotes country-level religiosity (Grim and Finke 2011, pp. 7–8).

If one reads Bénabou, Ticchi and Vindigni's results in the light of Bielefeldt's account of freedom of religion or belief, a conjecture consistent with the claimed negative correlation between religiosity and innovativeness suggests itself: the fact that "freedom of religion or belief also entitles people not to care about religious issues, to remain indifferent or express critical views" (Bielefeldt 2017, p. 37) may be partly responsible for the positive correlation of religious freedom with innovativeness found by Bénabou, Ticchi and Vindigni. Again, without claiming this conjecture to be true I suggest it as worthy of empirical testing in future research. What does seem clear, however, is that Grim, Finke and Snyder's 'religious freedom is good for business and innovation' thesis does not receive support from Bénabou, Ticchi and Vindigni's results, contrary to first appearances.

### Religion, Social Innovation and Innovation for Sustainable Development

Conceptions of innovation that wear their value-ladenness on their sleeves – such as social and responsible innovation, as well as innovation for ecologically, socially and economically sustainable development – bring the issue of normative involvement into full relief (Moulaert et al. 2005, p. 1978). For someone who is committed to an idea of social justice that emphasizes, say, universal human rights and ecological sustainability, the expression 'social innovation' will have a meaning very different from the one assigned to it by people who do not care about social justice or understand it in ways that do not involve human rights or ecological sustainability. Similarly, someone who subscribes to the 17 sustainable development goals (SDGs) set out in the 2030 Agenda for Sustainable Development of the United Nations (2015), and thus thinks that gender equality is an ineliminable ingredient of social, economic and environmental sustainability, will have very different ideas about innovation for sustainable development than a staunch defender of binary biological determinism and the idea that women are inferior to men. At this point it might be objected that white supremacists, sexists and deniers of human-made climate change cannot have coherent ideas about social innovation and sustainable development in the first place. If they use them at all, they 'hijack' these concepts. But is this really so? Instead of insisting on conceptual property, a more promising strategy may be to acknowledge and be upfront about the normative commitments of different conceptualizations of social and sustainable innovation and face the challenge of justifying them – especially today that many of the normative ideals that helped shape and diffuse ideas of social innovation and sustainable development are under assault from nationalist and more or less expressly xenophobic, homophobic, misogynistic and anti-environmentalist political movements capable of winning majorities in democratic elections.

Somewhat surprisingly, work explicitly concerned with the relationship between different religions and social innovation is scarce. The social innovation literature commonly mentions religion generically, if at all, as one of the contextual variables that have to be taken into account in the implementation of projects or policies aimed at promoting social innovation – alongside the compositions of the respective target populations in terms of age, gender and socioeconomic status.[17] Researchers approaching the issue from the religion side emphasize the motivational potential of religious beliefs and worldviews for social entrepreneurship (Ataide 2012), present specific case studies such as "the role of religion in the economic and social outcomes of very poor Dalit micro-entrepreneurs from various faith traditions in Bangalore, India" (Samuel Shah 2016), or point to the general potential of "religious factors to have a positive influence on social innovation" (Harrison 2016, p. 85).

Regarding religion and innovation for sustainable development, most recent work stresses the constructive roles that religious traditions can and do play in SDG-related innovation,[18] in particular through faith-based organizations and initiatives. Providing a list of eight global organizations that promote the agency of religious and faith-based actors in development, Tomalin, Haustein and Kidy (2019, p. 104) even go so far as to speak of a "'turn to religion' by development studies, policy, and practice".[19] At the same time, possibly inhibiting impacts of religion are frequently acknowledged. With an emphasis on gender equality and environmentalism, Yamini Narayanan (2013, p. 132) provides an overview of elements in the teachings and practices of Christianity, Islam, Buddhism and Hinduism that may result in cognitive and ideological opposition to sustainable development. She concludes that "religion clearly can and does have destructive impacts" (Narayanan 2013, p. 134). Having noted various ways in which religions may impede, rather than promote, innovation for sustainability, Narayanan goes on to distinguish three ways in which religion may play a constructive role in innovation for sustainable development: "through [...] important values that religion can offer, through its potential for activism and finally in the more personal realm of self-development" (Narayanan 2013, p. 132).

A similar attention to the ambiguous role of religion can be found in studies that focus on the environmental dimension of sustainability, and more specifically in the growing strand of theological literature that addresses the potential or de facto motivational role of religiously grounded values and beliefs in innovation for environmentally sustainable development.[20] A recurring point of reference for authors working in this area is historian Lynn White's influential 1967 essay "The Historical Roots of Our Ecological Crisis". Emphasizing the extent to which Christian ideas concerning the mastery of nature have shaped western science and technology, White (1967, p. 1207) puts the gist of his line of argument as follows: "Both our present science and our present technology are so tinctured with orthodox Christian arrogance toward nature that no solution for our ecologic crisis can be expected from them alone". It is thus only after acknowledging "that religion is partly responsible for the environmental crisis" that Roger S. Gottlieb (2006b, p. 6), a main contributor to what has come to be called ecotheology or green theology, can proclaim that "world religion has entered into an 'ecological phase' in which environmental concern takes its place alongside more traditional religious focus on sexual morality, ritual, helping the poor, and preaching the word of God" (see also Gottlieb 2006c, p. 22).

## THE RELIGION OF INNOVATION

Authors who use the expression 'religion of innovation' mostly do so to criticize attitudes and behaviours directed at innovation that they deem to be problematic and objectionable. The term 'religion' is supposed to do critical work in these contexts. It is used as shorthand for 'blind faith', 'gullible trust' and 'wishful thinking'. Saying that X is a religion amounts to issuing the judgement that there is something amiss with X. A good example of this line of critique, which presupposes a negatively connotated idea of religion, can be found in a blog post by the philosopher of technology, political theorist and social critic Langdon Winner. Employing the word 'cult' instead of 'religion', and designating the term 'innovation' as "today's central 'god term'", Winner contends that innovation:

> today has become an object of worship in universities, think tanks, corporations, Wall Street broker-age houses, and in the dreams of our social elites. [...] [T]he concept has become for many people the source of their deepest spiritual aspirations and yearnings for transcendence. In fact, it is not an exaggeration to say that [it] has begun to resemble a cult with ecstatic expectations, unquestioning loyalty, rites of veneration, and widely echoing exhortations of groupthink. (Winner 2017)

The lexical field of religion – 'worship', 'cult', 'spirituality', 'transcendence', 'ecstasy', 'veneration' – in this passage serves to generalize a criticism that Winner has levelled as early as 1986 against unjustified and exaggerated expectations regarding the salvific potential of innovations in information and communication technology, that is, against "the almost religious conviction that a widespread adoption of computers and communications systems along with easy access to electronic information will automatically produce a better world for human living" (Winner 1986, p. 105).

Intended as a criticism of innovation discourses in the specific sense just outlined, the claim that innovation has turned into a religion is to be found more often in opinion pieces or the arts and culture sections of newspapers and online blogs than in academic research (e.g., Valéry 1999; Zax 2013). When it does surface in research publications, it is either embedded within a broader context of ideological criticism directed at neoliberalist conceptions of the free market and its beneficial effects on human flourishing (Ampuja 2016) or in critical approaches to the study of innovation that highlight the historical influence of religious contexts on present-day innovation discourses (Godin 2020; Godin, Chapter 1 in this book).

As noted above, Winner's critique of innovation discourses as 'religious' or 'almost religious' operates against the backdrop of a generic concept of religion which is negatively connotated. A rather different line of criticism can be found in the contemporary theological literature. When Protestant theologian Harvey Cox condemns the deification of the market and points out that today's mainstream economic thought "is shaped by a powerful and global system of values and symbols that can best be understood as an *ersatz religion*" (Cox 2016, p. 8, emphasis in original), what is supposed to do the critical work is not so much the term 'religion' but the term 'ersatz'. In other words, it is not religion per se that Cox takes to be problematic but false religion, idolatry. The idea of technological innovation, in particular, is deeply entangled with economic imagery and market ideology. Even if not directly aimed at innovation, Cox's criticism of the deification of the market and the ensuing ersatz religion can therefore be read as a variation on, or at least closely related to, the 'religion of innovation' theme.

So much for critical uses of the expression 'religion of innovation'. In a completely different (affirmative) vein, a growing strand of literature in business and management studies contends that religious and spiritual attitudes towards innovation are actually conducive to (firm-level) innovation processes.[21] An early example is Buckler and Zien's study "The Spirituality of Innovation: Learning from Stories" whose introduction begins as follows:

> Innovation is the heart of value creation in business. New product development is the most important business process and calls for the whole company's energy. In our work we have come to focus on matters of spirit and environment that are critical to providing the context in which innovation can flourish. (Buckler and Zien 1996, p. 392)

In contexts such as this one, the words 'spirituality' and 'religion' are not intended to convey criticism. On the contrary, they serve to contend a positive effect of spiritual or religious attitudes within and with regard to processes of innovation. Such attitudes are considered to be something good and commendable – good, that is, for firms and companies because of their alleged creativity, performance and teamwork enhancing effects. Some of the relevant business and management literature would indeed seem to suggest that innovation should turn into a religion or form of spirituality.

So, what to make of the claim that innovation has turned into a religion? When it is brought forward along the lines of Winner's critique of innovation worship, for instance with regard to the intellectually painful inflation of innovation-speak in contemporary discourses or the tendency to think of innovation as a "a panacea for every socioeconomic problem" (Godin 2015, p. 8), it is difficult not to agree. However, the words 'religion', 'spirituality' and 'cult' in such contexts would seem to be dispensable and unessential. They serve a rhetorical function that may just as well be served by the term 'ideology' and its cognates.

What, then, if the claim that innovation has turned into a religion or a faith is made without evaluative intentions as a purely descriptive statement intended to simply register a fact? Brought forward in this vein, 'innovation is a religion' amounts to saying that innovation satisfies the necessary and jointly sufficient conditions for being a religion – or an adequate number of relevant conditions which, in varying combinations, are sufficient for being a religion – and thus falls within the extension of the general term 'religion'. The plausibility of this descriptive claim depends, among other things, on the meaning assigned to the term 'religion'. If one subscribes to a minimally substantive account and, as suggested in the introduction, takes appeal to some form of supernatural but humanly relevant transcendence to be a necessary condition for religion, then the claim that innovation in general and technological innovation in particular have turned into religions can only be understood as an instance of non-literal use of the word 'religion'. If anything, what the variegated conceptualizations of innovation circulating today have in common is that they refer exclusively to human creativity, communication, collaboration, organization and engineering capabilities. They leave little room for appeals to supernatural transcendence.

Functional accounts of religion, which take appeal to supernatural transcendence to be an optional ingredient of religion, may render the claim that innovation has turned into a religion somewhat more plausible as a descriptive statement. A clear case in point is economist Robert H. Nelson's work on what he calls "economics as religion" (2001) and "economic religion" (2019). Speaking not of innovation but of the closely related idea of (economic) progress, Nelson argues that a "basic role of economists is to serve as the priesthood of a modern secular religion of economic progress that serves *many of the same functions* in contemporary society as earlier Christian and other religions did in their time" (2001, p. xv, emphasis added). A secular religion, according to Nelson (2019, pp. 319–21), is a religion that does neither involve beliefs in nor practices of worship directed at any form of supernatural transcendence. On this view, a system of belief and practice qualifies as a (secular) religion if it fulfils certain functions, such as establishing cohesion in social groups, shaping personal and/or collective identities and goals, providing guidance for action, or offering responses to existential questions. With the idea of supernatural transcendence out of the way, then, Nelson can claim that the "full scope of religion thus includes many 'secular' systems of belief" (2019, p. 321); and he takes this conceptual move, in turn, to clear the way for "extend[ing] the use of traditional theological methods of analysis to secular belief systems such as economics" (2019, p. 320). On Nelson's (2019) view each of the following is a religion: Christianity, Marxism, Judaism, Keynesianism, Buddhism, Neoclassical Economics, Islam, Nordic Social Democracy, Taoism, the Chicago School of Economics, Hinduism, Capitalism, Environmentalism, "and others" (Nelson 2019, p. 320). One may well wonder what is the point of this claim given that Nelson does not intend his 'economics as religion' thesis as a criticism of, but rather as a descriptively adequate statement about economics – as a statement that just registers a fact.

As mentioned in the introduction, functionalism comes at a cost. More precisely, it comes at a theoretical and conceptual cost that, as such, has repercussions in empirical research. It tends towards inflationary religion-spotting and thus runs the risk of depriving the category of religion of its theoretical interest and empirical import. Nelson's list is a clear exemplification of this tendency.

To return to the topic of innovation, I do not deny that research informed by the conceptual tools of theology – religious studies as well as the sociology, psychology and anthropology of religion should be added here – can produce valuable insights, advances of knowledge and understanding. Moreover, heuristically adopting a functionalist perspective on religion can be useful for evincing analogies and intersections between religion and various practices, belief systems and social groups that *prima facie* have nothing to do with religion. Given the pre-eminent role of religion in human cultural history it is to be expected that innovation discourses are suffused with concepts, ideas, themes, topoi and words which have their historical roots in the vocabularies of religious traditions. In this regard, innovation discourses do not differ from discourses about, say, politics, ethics, the environment or the arts. Bringing these intersections to the fore is important. But it is simply a misconception to think that research on religion and innovation can be legitimate only if innovation actually is a (secular) religion, and innovation discourses actually are (secular) religious discourses. To show that X (innovation) resembles, intersects with or bears traces of religion in certain – even many – respects does not require an argument to the conclusion that X (innovation) is a religion.

# CONCLUSION

The fact that the vocabulary of innovation has made it into many areas of religion research and scholarship bears witness to the enormous attraction exerted by the innovation imaginary. This attraction now extends well into academic discourses that would *prima facie* seem to be remote from what are today the most natural habitats of the vocabulary of innovation, economics and business studies. As is evident from the overview presented in this chapter, there is considerable variety regarding the ways in which innovation is conceptualized in different areas of religion-and-innovation research and scholarship. The same holds for the degree of analytical depth with which different authors render their respective interpretations of the term 'innovation' explicit – ranging from detailed definitions at one end of the spectrum to complete neglect of the need for an explication at the other. The threefold distinction between innovation in religion, religion in innovation and religion of innovation introduced in this chapter can serve as a first point of orientation in approaching the constantly growing and diversifying research literature. A second point of orientation is the question of (implicit) normative commitment. How do pro- and con-attitudes towards innovation and towards religion shape the way in which authors write about religion and innovation? Do they introduce biases into religion-and-innovation research? These questions have been touched upon at various points in the preceding pages, but not answered in any useful detail. It is up to future research to provide in-depth reconstructions and case studies of the normative presuppositions of religion-and-innovation discourses.

Considering the prospects of a general theoretical framework for research on religion and innovation, Donald Yerxa notes that there "is simply insufficient, and at times contradictory, evidence for advancing a general theory of religion and innovation" (Yerxa 2016b, p. 6). The material discussed in the preceding sections suggests that Yerxa is right. A general theory of religion and innovation is not in the offing. Yerxa's diagnosis that the available evidence is sometimes contradictory certainly applies to the economic studies briefly reviewed above (Bénabou, Ticchi and Vindigni 2015; Liu et al. 2018), which treat the religiosity of populations as an independent variable and aim to test hypotheses concerning statistically relevant correlations between religiosity and (technological) innovativeness at firm, regional, national and international levels. But even if, counterfactually, there were univocal results in this regard, they would apply only to the relationship between specific operationalizations of religiosity (mostly based on self-report survey responses) and one specific kind of innovativeness (technological, commercial) among the many others that are seriously debated as live options in today's literature.

Alongside the multiplicity and diversity of different conceptualizations of innovation there is religious diversity in beliefs, practices, institutions and forms of community (Harrison 2016, p. 85). Given this twofold diversity, there are good reasons for being sceptical about the prospects of capturing the interactions between religion and innovation in a general theory. However, there are also good reasons for questioning the desirability of a general theory, and even the claim that a general theory can be seriously considered as a reasonable goal for research on religion and innovation. Advances in understanding the intersections of religion and innovation should rather be expected from in-depth and comparative case studies of specific religious groups, communities, beliefs and practices in relation to diverse forms of innovation.

According to sociologist of religion Peter Berger, when it comes to assessing "the role of religion in the affairs of this world", then "there is no alternative to a nuanced case-by-case approach" (Berger 1999, p. 18). Innovation, if anything, is an affair of this world. As long as researchers and scholars of religion make explicit what they intend to refer to by the terms 'religion' and 'innovation', and are upfront about their normative commitments (if any), work on religion and innovation will continue to produce empirically testable and conceptually criticizable hypotheses about past and present transformations of the religious sphere and the involvement of religious actors in various – otherwise non-religious – processes of change and transformation.

## NOTES

1. This threefold distinction has been introduced and developed by researchers of the Center for Religious Studies of Fondazione Bruno Kessler (Trento, Italy) to map interactions between religion and innovation in contemporary societies (Fondazione Bruno Kessler, Center for Religious Studies 2019, pp. 7–10). I would like to thank Marco Ventura and Valeria Fabretti for their permission to use the triangular model in this chapter.
3. In this regard there are close parallels between the notion of innovation and the notion of human enhancement (augmentation of human physical, cognitive, emotional capacities by technological means). See Chadwick 2008, for the relevant distinction between normative and non-normative readings of the term 'enhancement'.
4. For a detailed historical reconstruction of semantic shifts in the normative connotations of the concept of innovation, and the role of religious discourses in bringing these shifts about, see Godin (2015). Summarizing his findings, Godin (2015, p. 8) writes: "The vicissitudes and varieties in the meaning of innovation are a pattern in the concept over the centuries. Negative to the ancient Greeks, the concept of innovation shifted to positive in the Middle Ages. It returned to negative from the Reformation until the nineteenth century, when it gradually acquired a superlative connotation." Also see Godin, Chapter 1 in this book.
5. For the simple reason that they tend to count more human practices as religion than their substantivist opponents, advocates of functionalist approaches to religion are likely to see more interactions between religion and innovation.
6. Given the indexical and context-sensitive grammar of 'innovation', its implicit reference to specific times, places, values, political, economic and cultural contexts, it is no accident that work on the intersection of religion and innovation often comes from historians of religion. Hindsight is needed to decide whether, according to a given set of reference values, a process or result deserves the title of innovation (see Hackett 1991 and the chapters collected in Yerxa (ed.) 2016a, Hempton and McLeod (eds) 2017, and Williams (ed.) 1992).
7. My translation, orig.: "[N]ous proposons la définition suivante de l'innovation religeuse: *Processus collectif qui, par volonté et/ou par désir de changement face à une situation considérée comme ne répondant pas ou plus aux besoins ou aux aspirations actuelles, introduit une nouveauté religieuse et qui conduit, par négociation ou par imposition au moyen d'un réseau de communication, à un changement socioreligieux significatif, effectif et durable des pratiques et/ou du système de significations*" (Bélanger and Bonenfant 2016, p. 403, emphasis in original). See also p. 417, where the authors present a useful schematic depiction of their process approach to religious innovation.
8. Page numbers without further specifications in this and the three subsequent paragraphs refer to Stauning Willert and Molokotos-Liederman 2012.
9. See Williams (1992, pp. 10–11): "[I]t may usually be […] helpful to think of religious innovation as something 'natural' to religious tradition, as a modality of religious tradition itself."
10. For early discussions of adaptive versus creative innovation in religion see Parsons (1962) and Estruch (1972).
11. For overviews of the field see Iannaccone 1998, Iyer 2016, McCleary and Barro 2006, and (from an anthropological perspective) Obadia and Wood 2011.

12. It is important to note that supply-side theories of the religious market are only loosely connected to the analogy of the religious or spiritual supermarket widespread in recent sociology and anthropology of religion. The latter is often used in ways that stress the demand-side of individual religion/spirituality 'consumers' by focusing on the individualization and privatization of religious preferences and the freedom of the individual 'customer' (Aupers and Houtman 2006, for critical discussion Rähme 2018).

13. For a critical assessment of the claim that supply-side theories of religion are supported by empirical data see Bruce 2011c.

14. Gérald Gaglio (2017) draws attention to various "innovation fads" which, while worthy of being considered as research topics, are not (or should not be) considered live options for innovation activism.

15. Bénabou, Ticchi and Vindigni (2013, p. 12) use "two alternative measures of religiosity [...] corresponding respectively to the answers to the World Values Survey (WVS) questions: (i) 'Independently of whether you go to church or not, would you say you are: a religious person, not a religious person, a convinced atheist, don't know', and: (ii) 'Do you believe in God? – Yes, No, Don't Know'." The WVS waves considered are those of 1980, 1990, 1995 and 2000 (see the dataset Barro and McCleary 2003). Patent counts per country are based on statistics provided by the World Intellectual Property Organization (WIPO) for the same years.

16. Liu et al. 2018 base their results on datasets from the 6th wave (2010–2014) of the World Values Survey (Inglehart et al. 2014) and the Global Creativity Index 2015 (Florida, Mellander and King 2015). Their claim that "Islam is negatively related with national creativity" (Liu et al. 2018, p. 11) thus generically refers to present-day Islam (see p. 5). For historically and conceptually more articulated accounts of the relation between Islam and various forms of innovation see the chapters collected in Kamrava (ed.) 2011.

17. See, for instance, the chapters collected in Nicholls, Simon and Gabriel (eds) (2015), and Moulaert et al. (eds) (2013).

18. The pre-2015/16 literature refers to the Millennium Development Goals (MDGs), which are the predecessors of the SDGs stated in Resolution 70/1 of the United Nations (2015).

19. For a critical assessment of the focus "on formalised religious actors engaged in development work" see Jones and Petersen (2011), p. 1297.

20. For a collection of relevant work see Gottlieb (ed.) (2006a). With reference to innovation processes for environmental sustainability in the field of energy transitions, Koehrsen (2015, p. 305) high-lights three potential functions of religion that overlap with the ones pointed out by Narayanan: "(1) campaigning and intermediation in the public sphere; (2) 'materialization' of transitions [through the implementation of concrete projects, B.R.]; and (3) dissemination of values and worldviews that foster environment conscious attitudes and actions".

21. A large number of articles concerned with religion, spirituality and innovation in the context of management can be found the *Journal of Management, Spirituality and Religion* whose first issue was published in 2004 (Biberman and Altman 2004). For an overview see Pandey (2017).

# REFERENCES

Ampuja, Marko (2016), 'The New Spirit of Capitalism, Innovation Fetishism and New Information and Communication Technologies', *Javnost – The Public*, 23 (1), 19–36.

Assouad, Alexander and K. Praveen Parboteeah (2018), 'Religion and Innovation. A Country Institutional Approach', *Journal of Management, Spirituality and Religion*, 15 (1), 20–37.

Ataide, Randy M. (2012), 'The "Porcupine in the Room": Socio-Religious Entrepreneurs and Innovators within the Framework of Social Innovation', in Alex Nicholls and Alex Murdock (eds), *Social Innovation: Blurring Boundaries to Reconfigure Markets*, Basingstoke: Palgrave Macmillan, pp. 178–98.

Aupers, Stef and Dick Houtman (2006), 'Beyond the Spiritual Supermarket: The Social and Public Significance of New Age Spirituality', *Journal of Contemporary Religion*, 21 (2), 201–22.

Baffelli, Erica, Ian Reader and Birgit Staemmler (eds) (2011), *Japanese Religions on the Internet: Innovation, Representation and Authority*, London, New York: Routledge.

Barro, Robert J. and Rachel M. McCleary (2003), *Religion Adherence Data*, https://scholar.harvard.edu/barro/publications/religion-adherence-data (accessed 20 October 2020).

Bélanger, Steeve and Frédérique Bonenfant (2016), 'Pour une approche des processus d'innovation religieuse: Quelques réflexions conceptuelles et théoriques', *Laval théologique et philosophique*, 72 (3), 393–417, doi: 10.7202/1040353ar.

Bénabou, Roland, Davide Ticchi and Andrea Vindigni (2013), *Forbidden Fruits: The Political Economy of Science, Religion, and Growth*, Princeton University, William S. Dietrich II Economic Theory Center Research Paper 065-2014, https://papers.ssrn.com/sol3/papers.cfm?abstract_id=2460787 (accessed 20 October 2020).

Bénabou, Roland, Davide Ticchi and Andrea Vindigni (2015), 'Religion and Innovation', *The American Economic Review*, 105 (5), 346–51.

Berger, Peter L. (1974), 'Some Second Thoughts on Substantive versus Functional Definitions of Religion', *Journal for the Scientific Study of Religion*, 13 (2), 125–33.

Berger, Peter L. (1999), 'The Desecularization of the World: A Global Overview', in Peter L. Berger (ed.), *The Desecularization of the World: Resurgent Religion and World Politics*, Grand Rapids, Michigan: Eerdmans Publishing Company, pp. 1–18.

Biberman, Jerry and Yochanan Altman (2004), 'Welcome to the New Journal of Management, Spirituality and Religion', *Journal of Management, Spirituality & Religion*, 1 (1), 1–6.

Bielefeldt, Heiner (2017), *Il potenziale provocatorio della libertà religiosa / The Provocative Potential of Religious Freedom*, Trento: FBK Press.

Bruce, Steve (1999), *Choice and Religion: A Critique of Rational Choice Theory*, Oxford: Oxford University Press.

Bruce, Steve (2011a), 'Defining Religion: A Practical Response', *International Review of Sociology: Revue Internationale de Sociologie*, 21 (1), 107–20.

Bruce, Steve (2011b), *Secularization. In Defence of an Unfashionable Theory*, Oxford, New York: Oxford University Press.

Bruce, Steve (2011c), 'Secularization and Economic Models of Religious Behavior', in Rachel M. McCleary (ed.), *The Oxford Handbook of the Economics of Religion*, Oxford, New York: Oxford University Press, pp. 289–302.

Buckler, Sheldon A. and Karen Anne Zien (1996), 'The Spirituality of Innovation: Learning from Stories', *Journal of Product Innovation Management*, 13 (5), 391–405.

Campbell, Heidi A. (ed.) (2013), *Digital Religion. Understanding Religious Practice in New Media Worlds*, London, New York: Routledge.

Carvalho, Jean-Paul, Sriya Iyer and Jared Rubin (eds) (2019), *Advances in the Economics of Religion*, Cham, Switzerland: Palgrave Macmillan.

Chadwick, Ruth (2008), 'Therapy, Enhancement and Improvement', in Bert Gordijn and Ruth Chadwick (ed.), *Medical Enhancement and Posthumanity*, Dordrecht, Springer, pp. 25–37.

Cowan, Douglas E. and Jeffrey K. Hadden (2008), 'Virtually Religious: New Religious Movements and the World Wide Web', in James R. Lewis (eds), *The Oxford Handbook of New Religious Movements*, Vol. 1, New York, Oxford: Oxford University Press, pp. 119–40.

Cox, Harvey (2016), *The Market as God*, Cambridge, MA: Harvard University Press.

Dawson, Lorne L. (2006), 'New Religious Movements', in Robert A. Segal (ed.), *The Blackwell Companion to the Study of Religion*, Malden, MA: Blackwell, pp. 369–84.

Estruch, Juan (1972), 'L'innovation religieuse', *Social Compass*, 19 (2), 229–32.

Finke, Roger and Laurence Iannaccone (1993), 'Supply-Side Explanations for Religious Change', *The Annals of the American Academy of Political and Social Sciences*, 527 (May), 27–39.

Finke, Roger (2004), 'Innovative Returns to Tradition: Using Core Teachings as the Foundation for Innovative Accommodation', *Journal for the Scientific Study of Religion*, 43 (1), 19–34.

Florida, Richard, Charlotta Mellander and Karen King (2015), *The Global Creativity Index 2015*, Martin Prosperity Institute, Rotman School of Management, University of Toronto, https://ec.europa.eu/futurium/en/content/global-creativity-index-2015-most-creative-countries (accessed 20 October 2020).

Fondazione Bruno Kessler, Center for Religious Studies (2019), *Religion and Innovation: Calibrating Research Approaches and Suggesting Strategies for a Fruitful Interaction*, https://isr.fbk.eu/wp -content/uploads/2019/03/Position-Paper.pdf (accessed 20 October 2020).

Frances, Bryan (2015), 'The Rationality of Religious Belief', *Think*, 14 (40), 109–17.

Frances, Bryan (2016), 'The Irrationality of Religious Belief', *Think*, 15 (42), 15–33.

Gaglio, Gérald (2017), 'Innovation Fads as an Alternative Research Topic to Pro-Innovation Bias: The Examples of Jugaad and Reverse Innovation', in Benoît Godin and Dominique Vinck (eds), *Critical Studies of Innovation: Alternative Approaches to the Pro-Innovation Bias*, Cheltenham, UK and Northampton, MA, USA: Edward Elgar Publishing, pp. 33–47.

Gaglio, Gérald, Benoît Godin and Sebastian Pfotenhauer (2019), 'X-Innovation: Re-Inventing Innovation Again and Again', *Novation: Critical Studies of Innovation*, 1, 1–16, http://www.novation.inrs.ca/ index.php/novation/article/view/8/6 (accessed 20 October 2020).

Giordan, Giuseppe (2016), 'Spirituality', in David Yamane (ed.), *Handbook of Religion and Society*, Cham, Heidelberg: Springer, pp. 197–216.

Godin, Benoît (2015), *Innovation Contested: The Idea of Innovation Over the Centuries*, London: Routledge.

Godin, Benoît (2020), *The Idea of Technological Innovation*, Cheltenham, UK and Northampton, MA, USA: Edward Elgar Publishing.

Godin, Benoît and Dominique Vinck (2017), 'Innovation – From the Forbidden to a Cliché', in Benoît Godin and Dominique Vinck (eds), *Critical Studies of Innovation. Alternative Approaches to the Pro-Innovation Bias*, Cheltenham, UK and Northampton, MA, USA: Edward Elgar Publishing, pp. 1–14.

Gottlieb, Roger S. (ed.) (2006a), *The Oxford Handbook of Religion and Ecology*, Oxford, New York: Oxford University Press.

Gottlieb, Roger S. (2006b), 'Introduction: Religion and Ecology – What Is the Connection and Why Does It Matter?', in Roger S. Gottlieb (ed.), *The Oxford Handbook of Religion and Ecology*, Oxford, New York: Oxford University Press, pp. 3–19.

Gottlieb, Roger S. (2006c), *A Greener Faith: Religious Environmentalism and Our Planet's Future*, Oxford, New York: Oxford University Press.

Grim, Brian J. and Roger Finke (2006), 'International Religion Indexes: Government Regulation, Government Favoritism, and Social Regulation of Religion', *Interdisciplinary Journal of Research on Religion*, 2 (article 1), http://www.religjournal.com/pdf/ijrr02001.pdf (accessed 20 October 2020).

Grim, Brian J. and Roger Finke (2011), *The Price of Freedom Denied: Religious Persecution and Conflict in the Twenty-First Century*, Cambridge, New York: Cambridge University Press.

Grim, Brian J., Greg Clark and Robert E. Snyder (2014), 'Is Religious Freedom Good for Business? A Conceptual and Empirical Analysis', *Interdisciplinary Journal of Research on Religion*, 10 (article 4), http://www.religjournal.com/pdf/ijrr10004.pdf (accessed 20 October 2020).

Hackett, David G. (1991), *The Rude Hand of Innovation: Religion and Social Order in Albany, New York 1652–1836*, Oxford: Oxford University Press.

Harrison, Peter (2016), 'Religion, Innovation and Secular Modernity', in Donald A. Yerxa (ed.), *Religion and Innovation: Antagonists of Partners*, London, New York: Bloomsbury Academic, pp. 74–86.

Hempton, David N. and Hugh McLeod (eds) (2017), *Secularization and Religious Innovation in the North Atlantic World*, Oxford, New York: Oxford University Press.

Iannaccone, Laurence R. (1998), 'Introduction to the Economics of Religion', *Journal of Economic Literature*, 36, 1465–96.

Inglehart, Ronald F., Christian W. Haerpfer, Alejandro Moreno et al. (eds) (2014), *World Values Survey: Round Six – Country-Pooled Datafile Version*, Madrid: JD Systems Institute, https://www .worldvaluessurvey.org/WVSDocumentationWV6.jsp (accessed 20 October 2020).

Iyer, Sriya (2016), 'The New Economics of Religion', *Journal of Economic Literature*, 54 (2), 395–441.

Jones, Ben and Marie Juul Petersen (2011), 'Instrumental, Narrow, Normative? Reviewing Recent Work on Religion and Development', *Third World Quarterly*, 32 (7), 1291–306.

Kamrava, Mehran (ed.) (2011), *Innovation in Islam: Traditions and Contributions*, Berkeley, CA: University of California Press.

Koehrsen, Jens (2015), 'Does Religion Promote Environmental Sustainability? – Exploring the Role of Religion in Local Energy Transitions', *Social Compass*, 62 (3), 296–310.

Liu, Zhen, Qingke Guo, Peng Sun, Zhao Wang and Rui Wu (2018), 'Does Religion Hinder Creativity? A National Level Study on the Roles of Religiosity and Different Denominations', *Frontiers in Psychology*, 9, article 1912, doi: 10.3389/fpsyg.2018.01912.

McCleary, Rachel M. and Robert J. Barro (2006), 'Religion and Economy', *Journal of Economic Perspectives*, 20 (2), 49–72.

McCleary, Rachel M. and Robert J. Barro (2018), *The Wealth of Religions: The Political Economy of Believing and Belonging*, Princeton: Princeton University Press.

McLeod, Hugh (2017), 'Introduction', in David Hempton and Hugh McLeod (eds), *Secularization and Religious Innovation in the North Atlantic World*, Oxford: Oxford University Press, pp. 1–21.

Moulaert, Frank, Flavia Martinelli, Erik Swyngedouw and Sara González (2005), 'Towards Alternative Model(s) of Innovation', *Urban Studies*, 42 (11), 1969–90.

Moulaert, Frank, Diana MacCallum, Abid Mehmood and Abdelillah Hamdouch (eds) (2013), *The International Handbook on Social Innovation: Collective Action, Social Learning and Transdisciplinary Research*, Cheltenham, UK and Northampton, MA, USA: Edward Elgar Publishing.

Narayanan, Yamini (2013), 'Religion and Sustainable Development: Analysing the Connections', *Sustainable Development*, 21 (2), 131–9.

Nelson, Robert H. (2001), *Economics as Religion: From Samuelson to Chicago and Beyond*, Pennsylvania: Pennsylvania State University Press.

Nelson, Robert H. (2019), 'Economic Religion and the Worship of Progress', *American Journal of Economics and Sociology*, 78 (2), 319–62.

Nicholls, Alex, Julie Simon and Madeleine Gabriel (eds) (2015), *New Frontiers in Social Innovation Research*, Basingstoke: Palgrave Macmillan.

Obadia, Lionel and Donald C. Wood (2011), 'Economics and Religion, Economics in Religion, Economics of Religion: Reopening the Grounds for Anthropology?', in Lionel Obadia and Donald C. Wood (eds), *The Economics of Religion: Anthropological Approaches*, Bingley, UK: Emerald Group Publishing, pp. xiii–xxxvii.

Pandey, Ashish (2017), 'Workplace Spirituality: Themes, Impact and Research Directions', *South Asian Journal of Human Resources Management*, 4 (2), 1–6.

Parsons, Talcott (1962), 'Religion as a Source of Creative Innovation', in John Milton Yinger (ed.), *Religion, Society and the Individual*, New York: Macmillan, pp. 558–63.

Rähme, Boris (2018), 'Digital Religion, the Supermarket and the Commons', *Sociétés*, 139, 73–86.

Rähme, Boris (2021), 'Is Transhumanism a Religion?', in Giulia Isetti, Elisa Innerhofer and Harald Pechlaner (eds), *Religion in the Age of Digitalization: From New Media to Spiritual Machines*, Abingdon, New York: Routledge, pp. 119–34.

Recio-Román, Almudena, Manuel Recio-Menéndez and María Victoria Román-González (2019), 'Religion and Innovation in Europe: Implications for Product Life-Cycle Management', *Religions*, 10 (10) 589, doi: 10.3390/rel10100589.

Rogers, Everett M. (1962), *Diffusion of Innovations*, New York: Free Press.

Samuel Shah, Rebecca (2016), 'Religious Innovation and Economic Empowerment in India: An Empirical Exploration', in Donald A. Yerxa (ed.), *Religion and Innovation: Antagonists of Partners*, London, New York: Bloomsbury Academic, pp. 176–212.

Simmel, Georg (1905), 'A Contribution to the Sociology of Religion', *American Journal of Sociology*, 11 (3), 359–76.

Smith, Adam (1776) [1976], *An Inquiry into the Nature and Causes of the Wealth of Nations*, 2 vols., edited by R. H. Campbell, A. S. Skinner and W. B. Todd, Oxford: Oxford University Press.

Stark, Rodney and William Sims Bainbridge (1979), 'Of Churches, Sects, and Cults: Preliminary Concepts for a Theory of Religious Movements', *Journal for the Scientific Study of Religion*, 18 (2), 117–31.

Stark, Rodney, Laurence R. Iannaccone and Roger Finke (1996), 'Religion, Science, and Rationality', *The American Economic Review*, 86 (2), 433–7.

Stark, Rodney (2006), 'Economics of Religion', in Robert A. Segal (ed.), *The Blackwell Companion to the Study of Religion*, Malden, MA: Blackwell, pp. 47–67.

Stauning Willert, Trine and Lina Molokotos-Liederman (2012), 'How Can We Speak of Innovation in the Greek Orthodox Tradition? Towards a Typology of Innovation in Religion', in Trine Stauning

Willert and Lina Molokotos-Liederman (eds), *Innovation in the Orthodox Christian Tradition? The Question of Change in Greek Orthodox Thought and Practice*, Farnham, Ashgate, pp. 3–17.

Stauning Willert, Trine (2014), *New Voices in Greek Orthodox Thought: Untying the Bond Between Nation and Religion*, Farnham: Ashgate.

Tomalin, Emma, Jörg Haustein and Shabaana Kidy (2019), 'Religion and the Sustainable Development Goals', *The Review of Faith & International Affairs*, 17 (2), 102–18.

United Nations (2015), Transforming Our World: The 2030 Agenda for Sustainable Development, Resolution 70/1, https://www.un.org/en/development/desa/population/migration/generalassembly/docs/globalcompact/A_RES_70_1_E.pdf (accessed 20 October 2020).

Valéry, Nicholas (1999), 'Industry Gets Religion: A Survey of Innovation in Industry', *The Economist*, 20 February: 5–8.

Weber, Max. [1904–1905] [1920] (1930), *The Protestant Ethic and the Spirit of Capitalism*, translated by Talcott Parsons, London: Allen and Unwin.

Weber, Max. [1922] (1978), *Economy and Society: An Outline of Interpretive Sociology*, edited by Guenther Roth and Claus Wittich, Berkeley, CA, London: University of California Press.

White, Lynn (1967), 'The Historic Roots of Our Ecological Crisis', *Science*, 155 (3767), 1203–7.

Williams, Michael A. (1992), 'Religious Innovation: An Introductory Essay', in Michael A. Williams, Collett Cox and Martin S. Jaffee (eds), *Innovation in Religious Traditions: Essays in the Interpretation of Religious Change*, Berlin, New York: Mouton de Gruyter, pp. 1–17.

Williams, Michael A., Collett Cox, Martin S. Jaffee (eds) (1992), *Innovation in Religious Traditions. Essays in the Interpretation of Religious Change*, Berlin, New York: Mouton de Gruyter.

Winner, Langdon (1986), *The Whale and the Reactor. In Search for Limits in an Age of High Technology*, Chicago and London: University of Chicago Press.

Winner, Langdon (2017), The Cult of Innovation: Its Colorful Myths and Rituals, https://www.langdonwinner.com/other-writings/2017/6/12/the-cult-of-innovation-its-colorful-myths-and-rituals (accessed 20 October 2020).

Woolf, Greg (2016), 'Religious Innovation in the Ancient Mediterranean', in John Barton (ed.), *Oxford Research Encyclopedia of Religion*, New York: Oxford University Press, doi: 10.1093/acrefore/9780199340378.013.5.

Yerxa, Donald A. (ed.) (2016a), *Religion and Innovation: Antagonists of Partners*, London, New York: Bloomsbury Academic.

Yerxa, Donald A. (2016b) 'Introduction', in Donald A. Yerxa (ed.), *Religion and Innovation: Antagonists of Partners*, London, New York: Bloomsbury Academic, pp. 1–7.

Zax, David (2013), 'The Religion of Innovation: Enough with Innovation for Innovation's Sake', *MIT Technology Review*, 18 March 2013, https://www.technologyreview.com/s/512621/the-religion-of-innovation/ (accessed 20 October 2020).

# 19. Anthropology of and for innovation

*Ulrich Ufer and Alexandra Hausstein*

## INTRODUCTION

The modern social sciences have routinely conceptualized innovation as a driver of development and progress. It might therefore be expected that anthropology with its core interest in the dynamics of sociocultural change would have a firm grip on innovation in terms of its practices, concepts, discourses and politics. However, any cursory glance at a dictionary of anthropology will confirm the variegated semantic content of the term "innovation" (see Seymour-Smith 2005). On a conceptual level anthropologists link innovation to modernity and change in opposition to tradition, ritualistic repetition, conservatism and equilibrium. On the level of individual agency innovation can be studied as the capacity for cultural creativity, deviance and adaptation. On a structural level innovation may refer to a society's ability to incorporate new elements and recombine existing ones into new configurations. In some anthropological publications innovation still seems to be somewhat of a blind spot. In the recent German language handbook for "Technikanthropologie" the term does not make it into the book's index, nor is it mentioned in the title of the 74 contributions (Heßler and Liggieri 2020). Such findings accord with Benoît Godin's (2017, chapter 1) analysis that anthropology has only developed a belated interest in innovation as an object of research, if at all.

In this chapter we argue that in spite of anthropology's lack of a clear-cut concept of innovation itself, the discipline has made considerable contributions to the conceptual history of innovation, and that it continues to make important future contributions to understanding innovation both as a concept and as a cultural practice. We discuss these interrelated issues by pointing to two general approaches to innovation in anthropology: one being an anthropology *of* innovation, considered as an object of research, the other being an anthropology *for* innovation, considered as an object of engagement.

The first section of this chapter seeks to explain the general neglect of innovation as a concept and analytic category within classical anthropology of the late nineteenth and early twentieth century. Particular attention is paid to the incongruent relationship between the term's prevalent semantic content and anthropology's specific object of research. From this epistemological perspective we also enquire into the potential pitfalls of attributing the term innovation retrospectively to historical practices and conceptualizations of social and cultural change.

Through close analysis of the term innovation in the journal *American Anthropologist* over the first half of the twentieth century the second section seeks to illuminate how innovation was gradually woven into diffusionist and acculturalist conceptions of sociocultural change. Our analysis reveals how the eventual emergence of a cumulative and procedural notion of "cultural innovation" in anthropology developed in close interdisciplinary exchange with sociological conceptualizations of sequential innovation.

In the third section, we discuss how the anthropological notion of cultural innovation was extended to concepts of socially embedded innovation and to ideas about innovating societies after World War II, in contrast to an emerging standard view of technological innovation.

The fourth section discusses how anthropologists from the 1950s onwards when confronted with the near disappearance of their accustomed object of research (namely, "primitive man" and traditional society), turned innovation into an object of anthropological research.

The fifth section examines how innovation turned into a critical object of anthropological engagement, through anthropological interventions into social as well as industrial processes of innovation. This development over the course of the second half of the twentieth century and into the present also raises issues over the normative biases in innovation research.

Finally, we discuss the current Western crisis of innovation as an object for future anthropologic research.

## INNOVATION WITHOUT THE WORD?

Although the term innovation was not absent from early anthropological research, anthropologists before the twentieth century used it only sparingly, in a very general sense, and they developed no clear conceptual definition of the term for their emerging discipline. This is not surprising if one considers the semantic content of innovation around that time (Godin 2016). In one sense, the term was relatively unspecific, having no clear semantic difference from similar terms such as invention, novelty, or change. If anthropologists made use of the term innovation before 1900, it was mostly in this unspecific and general sense.

As well as being a general term innovation was laden with ambivalent values. In 1912, for example, the philosopher Arthur K. Rogers criticized the nineteenth-century arch-conservative Edmund Burke for his exaggerated "fear of innovation," while at the same time expressing himself a good measure of suspicion when he conceded that "[n]o one will deny the danger that lies in a reckless spirit of innovation" (1912, p. 52). Gradually the term came to refer positively to new methods and artefacts in scientific and technological development. In this latter sense, innovation became promoted for its "sound grounding in scientific principles" (Godin 2016, p. 536), thereby relating the term to the acts of individual innovators and setting the path for innovation to become a key concept in modernist and progressivist discourses.

Over the course of anthropology's foundational phase as a discipline in the nineteenth and early twentieth century, however, anthropology was largely concerned with the "representation and interpretation of tradition, convention and persistence" (Welz 2003, p. 257). It therefore conceived of sociocultural change within traditional settings as a reaction to mostly external impacts rather than to internal drives towards change and innovation. This view of traditional society as static and unchanging was summarized by anthropologist Jacques Maquet (1964, p. 50) for whom anthropology was "the study of nonliterate societies and their cultures," defined as "a special discipline for 'primitive,' 'simple,' preindustrial, nonliterate, small-scale societies." Viewed against this backdrop, it is not at all surprising that the semantic content of innovation as an emerging key term for the self-definition of Western modernity, individualism, technological progress, industrialization and capitalist market society simply did not match early anthropology's epistemological aims and was largely incompatible with its emerging vocabulary.

Over the second half of the twentieth century, the term innovation has acquired a widely accepted semantic content that refers in principally to technical inventions and their social diffusion through markets within the settings of profit-oriented cultural practices in capitalist society. For a scientific engagement with the historical usages of the term innovation in anthropology before 1950, as well as for an analysis of retrospective labeling of historical practices as innovative in the modern sense of the term, this poses a number of questions: what did anthropologists precisely refer to, when they used the term innovation? Was the term's historical meaning congruent with its present-day semantic content and can the term innovation retrospectively be applied to cultural practices that might be labeled innovation today, but were not identified as such in the past?

This issue touches upon an early conceptual discussion in historical anthropology concerning the "missing terms" (*les mots qui manquent*). Following Lucien Lévy-Bruhl's (1922) notion of "primitive mentality," Lucien Febvre (1942) argued that a modern term such as atheism (*incroyance*) could not simply be transferred to another period, in this case pre-modern France. Even though Febvre's argument that pre-modern people were incapable of conceptualizing religious unbelief due to a different mental structure has been refuted (Lévi-Strauss 1962; Burke and Hobsbawm 1978) the issue should sensitize researchers to attaching "innovation" to a wide range of practices from the past. It implies that the term's present-day semantic baggage is easily transferrable to other historical settings, even though it reflects a very specific historical conjuncture.

Analysis of historical semantics is therefore relevant as an anthropological object of research in its own terms, especially in terms of its critical engagement with analyses of past anthropological conceptions of sociocultural change as congruent with present-day innovation. Michael O'Brien and Stephen Shennan (2010, p. 4) hold that "[i]nnovation was explicit" in nineteenth-century evolutionist anthropology in so far as anthropologists then, too, were conceptually concerned with "the production of novelties – new ideas, new ways of doing things, and the like." However, a closer look at nineteenth-century evolutionary anthropology reveals a conception of change that is fundamentally different from the present-day semantic content of innovation. Herbert Spencer's 1851 publication *Social Statics* serves to illustrate this incongruity. In his 476 page-long work Spencer only used the term "innovation" twice, both times in the conservative sense of "unsettling effects" on social stability. What is more, he located the evolutionary driving force for change clearly outside human agency, consideration, intentionality, or willful planning. Instead of being a modern manager of innovation, Spencer thought of man as a "servant and interpreter of nature," of the "secret forces," subject to the "gigantic plan" and the "vital principles, ever in action, ever successful" (Spencer 1851, p. 293). Spencer's understanding of man's role in sociocultural change was not only incongruent with, but oppositional to the present-day semantic content of the term innovation. Today, innovation is defined as "change that is human-made and deliberate, as contrasted to that made by God, nature or chance" (Godin 2014a, p. 2). Spencer, however, ultimately perceived the self-acclaimed managers of change in politics and industry as "political schemers" whose clumsy attempts at social engineering were, presumptuous, plainly unnecessary and in the worst case harmful to the natural evolutionary tendency towards perfection, the "mighty movement – towards a complete development and a more unmixed good" (1851, p. 293).

When anthropologists and historians today refer to diverse cultural practices in present and past societies as innovation, this is often more revealing of how academic epistemology is shaped by contemporary policies and funding schemes that seek to promote a culture of innovation, than it is about those past cultural practices to which the term innovation is applied. Thus, historical and archaeological anthropologists as well as historians have repeatedly suggested that innovation was a manifest practice in the early modern and medieval periods (Cipolla 1972; Reith 2000; Davids and De Munck 2014), or even in prehistoric times (Ottaway 2001; O'Brien and Shennan 2010; Pezzarossi 2014; Cooper 2012).

However, motivation and intentionality in the past may differ significantly from today's paradigmatic understanding of innovation as a techno-scientific, progress- and market-oriented driver of change. In this light, we might consider that Johannes Kepler's early seventeenth-century astronomical discoveries sought to reconcile the empirically observed disorder in planetary motion with prevalent holistic cosmology and notions of divine harmony (Gaizauskas 1974, pp. 149–50), or that a big man in a cargo cult society introduces innovative prestige objects with a view to his position in the social structure of kinsmen and client networks (Peace 1979). It is therefore highly questionable, if there has been, in the present-day meaning of the term, an innovation "*avant la lettre*" (Popplow 2019).

In opposition to propositions that in pre-modern times terms such as "invention" or "progress" fulfilled the same semantic function as modern-day "innovation" (Haller 2014) we tend to agree with Godin's analysis (2014a) that more fine-grained attention needs to be paid to the semantic content and discursive functions of the term innovation in different times and places. With a view to present-day discourses and politics of innovation this analysis of the pitfalls of retrospective labeling might be extended to a critical questioning of whether the term innovation in the current hodge-podge of "X-innovations" (Gaglio, Godin and Pfotenhauer 2017) – for example social innovation, disruptive innovation, responsible innovation, and so on – refers to the same semantic content in all its different settings. If not, then we might ask: what are the motivations, strategies and discursive power schemes hidden behind the seemingly uniform term innovation?

## CULTURAL INNOVATION

While the term and concept of innovation was still largely foreign to anthropology over the first decades of the twentieth century, the discipline developed its own vocabulary and concepts to assess and discuss sociocultural change in traditional societies. For this reason it is instructive for our purposes to look in more detail at anthropological usages of the term innovation over the first half of the twentieth century, by taking a closer look at one particular debate on sociocultural change published over a series of papers in the journal *American Anthropologist* between c. 1900 and c. 1950. This debate also illustrates how innovation, both as a term and as a concept, became part of disciplinary boundary demarcations between anthropology and sociology, and how its conceptualizations were influenced by interdisciplinary exchanges.

How did anthropologists' publishing in the *American Anthropologist* over the first half of the twentieth century conceptualize change, and how did they use the term innovation? With his concept of "superorganic" culture, anthropologist Alfred L. Kroeber (1917) raised culture, including notions of cultural dynamics and culture change, to an autonomous level of analysis thus defining American anthropology's field of scientific enquiry and its disciplinary boundaries against neighboring disciplines in the emerging fields of sociology, organic biological evolution, or social and individual psychology (Kroeber 1917). Kroeber suggested a broad vocabulary to describe cultural dynamics on the level of the superorganic including evolution, change, gradual development, progress, imitation, mixture, accumulation, adaptation, assimilation, fashion, invention, or discovery. Notably, his list did not include innovation.

Anthropologists studied cultural change primarily through the emerging key concepts of diffusion and acculturation. Referring to the "diffusion of tales" in the context of native north-American myths anthropologist Franz Boas (1889, 326) introduced the notion of cultural diffusion across space and time to debates in the journal *American Anthropologist* already in the late nineteenth century. The concept of cultural diffusion would absorb terms with related semantic values, like migration or dissemination, and it paved the road for a reified notion of culture as a somehow transferrable essence. The concept of diffusion reflected a particular concern for cultural origins (Linton 1924, 1926) and for retracing centers and lineages of diffusion. Such conceptions of culture would shape anthropological debates in the journal *American Anthropologist* over the first decades of the twentieth century. Even though essentialization of culture has clearly been identified as a conceptual impasse (Friedman 2007a), it has remained influential in recent anthropological studies of cultural globalization in terms of creolization and hybridity. Diffusion was elaborated upon in Boas' (1920) paper 'The Methods of Ethnology' in opposition to evolutionist conceptions of cultural "similarities" and cultural "parallelisms" with independent geographic or temporal origins of cultural development, as observed, for example by Ralph Linton in the settings of almost completely isolated Polynesian island cultures (Linton 1925). Through a memorandum of the U.S. Social Sciences Research Council, the concept of diffusion was then officially superseded in the 1930s by the concept of acculturation, the latter comprising such concepts as diffusion, culture-change, or assimilation, which were all conceived as aspects, or phases of an incremental acculturation process (Redfield, Linton and Herskovits 1936, pp. 150–51; Meggers 1946).

Although the term "innovation" was not completely neglected by anthropologists between 1900 and 1930 it did not play a role in diffusionist or acculturalist conceptualizations of cultural dynamics. Anthropologists often used the term synonymously with "invention" referring to small-scale and rather short-term individual instances of cultural change. Innovation was thus conceived of in terms of a change that had some impact on a stable traditional cultural formation, as a function of an internal invention, or as an external introduction of a new material object, idea, expression or cultural practice – summarily referred to as culture traits – which caused a short-term "decrease in cultural integration" followed by cultural re-integration "as the 'adaptive' culture catches up" (Watson 1953, pp. 138–39). Debates in the journal *American Anthropologist* provide numerous examples of anthropologists' usages of the term innovation as indicating a rather insignificant new material object or practice (Lowie 1942, p. 535), or as referencing short-term and recent events in the incremental processes of diffusion (Lowie 1915, p. 230; Linton 1927, p. 304; Linton 1928, p. 368). However, due to a lack of historical data it was in most cases impossible for anthropologists to assess to what extent and for how long precisely an innovation had had an impact on a given culture. It was equally hard to "reconstruct a picture of the intermediate stages in the creation of the present situation, or ever know the details of the processes whereby native society adjusted itself to some innovations and was dislocated by others" (Culwick and Culwick 1935). The method of dated distributional analysis was one proposition to gain "at least a measure of historic control" on innovation as a process (Hodgen 1945, p. 466).

Procedural or incremental notions of innovation in diffusionist and acculturalist anthropology developed more or less implicitly. Anthropologists considered by what practices new cultural traits diffused geographically and over time within a given culture, or by what means they transgressed cultural boundaries, or how they could have occurred unrelatedly in different cultures, either simultaneously or at different times, or what could be the barriers to their diffusion. In this context, we find the first conceptual link between innovation and cultural change in publications in the *American Anthropologist* in 1932, when anthropologist George P. Murdock introduced the notion of "cultural innovation" as a specifically cumulative and therefore procedural and incremental process (Murdock 1932, p. 202). The term was then taken up in Homer G. Barnett's elaboration of an anthropological procedural theory of culture change involving a classification of different types of innovation (Barnett 1940, 1941, 1942b), and then again a few years later by Albert Heinrich's (1950) notion of "acculturative innovation." Here, however, it is important to note the contrast between, on the one hand, procedural, or incremental notions of innovation as ongoing sociocultural practice, and, on the other hand, parallel conceptions in other disciplines such as economics. Joseph Schumpeter's notion of "creative destruction" (see Godin 2014a, p. 225) for example, defined innovation positively as revolutionary, radical and disruptive.

Anthropological procedural, or incremental, concepts of cumulative "cultural innovation" did not emerge from nowhere. Such concepts were indebted to previous cross-disciplinary studies in anthropology and sociology. In developing his notion of cumulative and procedural cultural innovation Murdock argued against the dominant modernist individualistic conception that cultural innovations spring "full-fledged from the brains of their reputed inventors" (Murdock 1932, p. 206). He criticized this view by drawing both on Kroeber's ethnographic observations and on William F. Ogburn's sociological analyses of "parallel inventions," that is, the simultaneous occurrence of the same innovation from two separate origins. Murdock's principal point that culture is "not only continuous; it is also cumulative" (1932, p. 212) was advanced in reference to the anthropologist Alfred Tozzer who himself had cited sociologist Ogburn on this topic (Tozzer 1928, p. 136). Ogburn himself referred to the ethnologist of religion and philosopher Robert Marret (Ogburn 1922, p. 151; Marett 1920, p. 109). In its final formulation, then, Murdock's anthropological notion of procedural and cumulative "cultural innovation" came agonizingly close to Ogburn's sociological sequential concept of invention, though with a clear focus on culture. Ogburn emphasized that "social heritage in its material aspects grows through inventions and [...] by diffusion" (Ogburn 1922, p. 103 in Godin 2010, p. 296), while Murdock noted that cultural innovation develops "as a synthesis of many previous inventions" (Murdock 1932, p. 206). Overall, mutual influences between anthropologists and sociologists have been constitutive for procedural, or incremental understandings of cultural innovation (Godin 2017, chapter 1), and they have had a lasting influence on innovation studies (Godin 2014a, chapter 12).

Our analysis of usages of the term innovation as they occurred in publications in the journal *American Anthropologist* over the first half of the twentieth century shows that the term's semantic content remained rather equivocal until at least the 1930s. Until then, innovation was used sparingly and almost synonymously with the term invention, referring largely to short-term and rather insignificant steps in the procedural diffusion of culture, for example to the ways in which a new technology, thing, idea, or practice gets introduced to, or acculturated within, a traditional group. Thereafter, some anthropologists specified their usages of innovation by integrating the term into their guiding theories of cultural change. In this endeavor, anthropologists would combine sequential notions of innovation from sociology with their own conceptions of independent and parallel dynamics of innovation, leading to their own understandings of cumulative, culturally and socially embedded practices of innovation. At the same time anthropologists contributed considerably to the development of sequential theories of innovation in sociology through their emphasis on the practices and procedural nature of culture change by borrowing, contact, or exchange.

## SOCIAL INNOVATION

Over the first half of the twentieth century, anthropologists came to understand cultural innovation as a long-term and complex process, depending on interrelations between human practices, their material environment and the human capacity to attribute meaning. This understanding developed in critical contrast to an emerging standard view of technological development and innovation that reigned supreme among economists and engineers, but also among many social and cultural scientists. The standard view of technological innovation concentrated on "the origin of, and the factors responsible for, invention" (Godin 2014b, p. 27). It emphasized necessity as the mother of invention seeing technological function as a necessary answer to material problems, while emphasizing the linear nature of cumulative, technological development, and the genius of great men as the originators of inventions, and presupposed that innovation per se was diffused from its Western centers of origin.

Early critiques of this standard view of technological innovation came from anthropology, sociology and from marginalized positions in economics. Anthropologist Edward Cary Hayes had made an early contribution, notably in the *American Journal of Sociology*, to sequential concepts of cultural innovation as a process over various steps from initial "innovation" over social adaptation by "fashion" to "custom" and "social institution" as continuous and permanent sociocultural manifestations (Hayes 1911, p. 202). For his part the sociologist Ogburn (1926, p. 227) criticized the "hero-worship" of inventors and emphasized the social forces behind innovation, just as the anthropologist Murdock (1932, p. 206) had done when he stated that an "innovation may spread or stagnate, [but] once launched into the stream of culture, it is beyond the power of any individual to control" it. William Ruppert Maclaurin's pioneering, but marginalized studies in economics (see Godin 2017, pp. 65–67) during the 1940s and 1950s also underlined how the standard view of technological innovation stopped short of a complex understanding of the social embedding of innovation by not taking into account contexts and processes. The latter would consider an initial innovative idea, favorable conditions for its concretization, questions of financing for its production as well as the cultural and social adoption and implementation of the innovation, either by creative social reception and widespread behavioral changes, or by placing a new product successfully on the market.

Anthropologist Barnett's 1953 study *Innovation* introduced the term "social innovation" (1953, p. 68) to anthropology. With only one mention in 448 pages, though, the term was far from central to Barnett who employed it in the context of discussing the social changes that a political system might promote among a national population. However, faced with the growing predominance of the standard view of innovation as primarily technological, it became necessary for anthropologists to clearly differentiate between technological innovation and other types of innovation. For Barnett, technological innovation had a limited meaning, which centered on the technical object and referred merely to "a series of inventions," so that "inventors of things" stood in contrast to "social innovators" (Barnett 1953, p. 248). In the sense of diffusing an invented object, then, technological innovation took on a limited meaning in anthropological reasoning close to that of the objectified notion of culture in culturalist discourses of the first half of the century so that technological innovations, that is, inventions, came to refer merely to new technical objects. Anthropological research interest, however, centered on integrative acculturation, adaptation, assimilation and other forms of acceptance by a culture, or on a culture's various forms of exclusiveness and resistance to change. In contrast to object-centered technological innovation anthropologist Clifford Geertz (1959, p. 1006) would define "social innovation" as "consciously motivated [and] explicitly designed," but Robin Horton would also speak of "individual innovators" as "introducer of [...] radical social innovation" (1960, pp. 219 and 223). Against the backdrop of this developing sociocultural conceptualization of innovation in anthropology anthropologist H. Zacharias would ask rhetorically: "No doubt, 'inspiration' of individuals towards innovation exists, [...] even so, unless there is social response to the innovation, will it not remain still-born?" (Zacharias 1953, p. 664).

The French anthropologist André Leroi-Gourhan's integrative and synthesizing work *Gesture and Speech*, published in 1964[1] is an outstanding example for ongoing social and cultural contextualization of innovation in anthropological conceptualizations. While the work only shows one occurrence of the specific term "social innovation," used as a distinction from technical innovation (Leroi-Gourhan 1993, p. 277), it elaborates a historically grounded and socially embedded approach to the dynamic relation between man and the material world. Leroi-Gourhan detected a "technical tendency" (*tendence technique*) in long-term human evolution, which progresses along an "operational sequence" (*chaîne opératoire*). He thereby added to the anthropological (and sociological) critique of inventions as "a unique act of genius whereby an isolated new technique is produced out of thin air" (1993, p. 173). He also demonstrated how technological evolution depends on three decidedly social factors: on a "favourable environment" (*milieu*) (1993, p. 154), on "that uniquely human two-way traffic between the innovative individual and the social community" (1993, p. 228), and on the human collective "operational memory" that allows for "cumulative effects of innovation" (1993, p. 230). From this perspective the driving force in human evolution is "exceptional innovation" (1993, p. 278), that is, the disruption of the established social order. This must be considered in opposition to the processes of merely incremental (technological) "inventions" that proceed slowly and almost unnoticed on the level of everyday adaptations. On this basis Leroi-Gourhan drew implications for a potential politics of innovation. While inventions may originate from the "technician" (1993, p. 168) and incubate among a creative elite (1993, p. 398), for innovation to have wide-ranging and transformative effects it must undergo a final step on the operational chain. Innovation must conclude in the creative reception by the masses, as opposed to the latter's merely passive adoption of inventions. From this insight followed quite practical concerns about potentially interruptive impacts by modern inventions themselves on the long-term evolutionary operational chain, such as television sets, since the "audiovisual language" of mass media threatened to reduce the masses from creative recipients into mere "organs of assimilation" (1993, p. 214).

Over the 1980s and 1990s an emerging social anthropology of technology expanded the analysis of innovation as socio-culturally embedded by concentrating on the systemic relationships that make innovation possible (Pfaffenberger 1988, 1992). Influenced by new insights from science and technology studies, anthropologist Bryan Pfaffenberger's notion of a sociotechnical system addressed the "distinctive technological activity that stems from the linkage of techniques and material culture to the social coordination of labor" (Pfaffenberger 1992, p. 497), thereby not only adding a new layer of critique to the standard view of technological innovation, but also attacking notions of linear developmental progress in the sociocultural framings of innovation. Pfaffenberger conceived of sociotechnical development as proceeding within long-term lifecycles of historical sociotechnical systems which "grow from invention, small-scale innovation, growth and development, and a climax of maximum elaboration and scope, followed by senescence and decay, until the system disappears or is replaced by a competing system" (Pfaffenberger 1992, p. 502). From the anthropological perspective of the sociotechnical system, then, there was no more room for the evolutionist narrative of cumulative and linear technological development. Instead, the sociotechnical system of Western modernity could now be seen as a system with historical origins, maintained by unequal global power relations and with a possible future crisis and end to it. Significantly, a second systemic approach to innovation studies developed only a few years later, albeit without taking into account insights of the social anthropology of technology. Following in the path of OECD policy advisors' discourses for stimulating national economies by technological development (Freeman 1987) the approach of "innovation system research" (Lundvall 2007a) propagated a normative framework to promote innovation as one foundational policy for development (Lundvall 2007b). While this approach lacks any deeper conceptual engagement with innovation beyond defining innovation normatively as a necessity for national development, its timing, programmatic content and notoriety are quite consistent with the concurrent rise of a Western crisis of innovation (see below).

Overall, even though the term social innovation was used sparingly by anthropologists, heightened attention was paid to innovation's social embedding in an effort to contrast the narrow standard view of technological innovation with a more complex picture. However, anthropologists employed the term "social" in social innovation with an ambivalent, two-fold meaning. Understood as the referential object of innovation, it could refer to an innovation of the social as such while in an adjectival sense it could also refer to the non-standard view of technological innovation by pointing to an innovation brought about by social rather than by technological means.

## INNOVATION AS AN OBJECT OF RESEARCH

While anthropologists had made important contributions to the conceptualization of innovation over the first half of the twentieth century they had not typically considered innovation as an object of research. However, this would change over the course of the second half of the twentieth century.

First, this development unfolded against the backdrop of a fundamental disciplinary transformation in which the two accustomed objects of anthropological research, "primitive man" and traditional society, began to wane. Additionally, with the process of decolonization, developing countries experienced progressing cultural and social innovations not only in terms of modernization, development and ideology but also in terms of political and economic practice. Also Western cultures, particularly in Europe, underwent a second wave of modernization and industrialization after World War II, thus allowing comparison of cultural changes between first and second wave modernizations.

Second, anthropology had to come to terms with the fact that innovation had become one of the main paradigms of modernist culture after World War II, both in the West and in developing countries. Its comparative method now allowed for the study of the anthropological "other" not only in terms of contrasts, but also in terms of similarities by, for example, under-standing local agents in (post) colonial modernization as "tribal innovators" (Schapera 1943, 1970). In time this would gradually enhance the discipline's methodological and conceptual self-reflexivity and even establish Western society itself as an appropriate object of anthro-pological research. Thus, in 1952 anthropologist Demitri Shimkin (1952, p. 86) suggested "industrialization" as a suitable research topic for anthropology because its rapid dynamics affected a wide range of cultures, and because its "complexity of content and mechanism" would put the transfer of anthropological methods from traditional to modern societies to the test. In post-World War II Germany, for example, anthropologists promoted industrial folk studies (*industrielle Volkskunde*) as a promising application of the methods of American cultural anthropology to sociocultural transformations of modern (urban) lifestyles (Brepohl 1951, p. 116). And over the next decades a thriving anthropology of industrial work would develop in the U.S. and in Great Britain (Burawoy 1979).

Anthropologists' treatment of modern societies and their analyses of second-wave mod-ernization in the West was informed by the discipline's established concepts of social and cultural change: "For the understanding of personality and culture and of the problems of introducing innovations into isolated communities, there is much to be learned from the older studies of acculturation, early culture contact, and applied anthropology" (Dobyns and Dalton 1971, p. 395). Diffusionist heritages more or less implicitly suggested that Europe and the West were "the world's source of culturally significant innovations" (Blaut 1987, p. 30). However, extensive ethnographies also showed the non-linear process of diffusion of cultural and social innovation in developing countries and required anthropologists to reconsider the cultural barriers and hindrances to diffusion. Social and cultural innovation not only brought diversification of culture, but also the decay of traditional culture. Social innovation not only included improvements in the quality of life, but also brought about critical disruptions of social institutions. From a perspective of cultural skepticism, anthropologists had noted early on that "cultural innovations acquired by acculturation and diffusion" were often the result of unequal power relations that lead to cultural crises on the receiving end (Bidney 1946, p. 541). In this sense, extensive studies on sociocultural change through the impact of modernization and industrialization among traditional groups, such as Ecuadorian Indians, would conclude that in "the so-called backward countries there are advocates of change as such; they see in 'modernization' of their countries an end in itself, and in this vision industrialization is included. Others are not quite so willing to pay the price for innovations" (Salz 1955, p. 6). A study of rural India confirmed that technological innovation might actually work to the "economic detriment" of traditional craftsmen (Opler 1959, p. 129).

Notwithstanding such critiques, anthropology, economic anthropology in particular, turned to the processes of economic innovations in developing countries as an object of research (Dalton 1969, p. 64). Anthropologist George Dalton advanced a theoretical framework based on macro-economic comparison that addressed issues of a traditional culture's receptivity to "technological, economic, and cultural innovations" and aimed at understanding the impacts of innovations on traditional sociocultural settings in terms of "sequential change." For example, "when a group undertakes enlarged production for sale [...] and incorporates other such innovations" (Dalton 1969, p. 69). The economist Joseph Berliner had been suggesting this line of enquiry a few years before by approaching anthropologists for a seemingly fruitful interdisciplinary cooperation. He sought to integrate methods from cultural and social anthro-pology with economic research interests and methods in order to improve prognostics about the successful introduction of "economic innovations" (Berliner 1962). Berliner sought to fuse the notion of endogenous and cyclical dynamics of change from economic theories of business cycles with cultural anthropology's predilections for exogenous and linear change dynamics, as in diffusionism, and with functionalist anthropology's concentration on endogenous statics, as in social equilibrium theory. This allowed him to assess the "capacity for innovation" as a specific cultural property of a given society, which could then be rated according to a list of stereotypical culture-types. Echoing the Marxian notion of a developmental gap between fast moving technological and stunted social innovation, Berliner hoped to promote innovation in developing countries by first deciphering the positive cultural determinants of Western societies, in which research was "consciously designed to change technology, which will cause new changes in the economy, which will cause new changes in technology, and so on" (Berliner 1962, p. 60).

Berliner's paper received a number of immediate comments and replies (Berliner 1962, pp. 61–77). Some anthropologists saw enormous potential in Berliner's suggestions for advancing an applied anthropology that contributed to fruitful conditions for "radical innovations," "useful innovations" and "productive innovations." Others had strong reservations about defining culture-types and declaring some as more open to innovation than others. In a later and more fervent critique, Berliner's whole undertaking of a prognostics of successful innovation became stigmatized as an economist's illegitimate intrusion into the protected realms of culture and society, since "from the point of view of an anthropologist" Berliner's approach reflects "a complete alienation between economic phenomena and their socio-cultural characteristics" (de Oliveira and Berliner 1964, p. 105). But other anthropologists pursued the project of economic anthropology with a decidedly more differentiated perspective on sociocultural dynamics and their implications for innovation. Mary Douglas' work on the Lele in Western Africa (Douglas 1962) is a case in point of conceptualizing change as well as resistance to economic innovation in traditional economies. Studies on primitive economics (Sahlins 1972), or informal economies in developing countries (Hart 1973, 1985) also engaged critically with principles of Western industrial society, like the notion of economic man, or individual rational choice.

## INNOVATION AS AN OBJECT OF (CRITICAL) ENGAGEMENT

Modernization of traditional societies changed anthropological conceptions of traditional culture, and it also changed anthropologists' roles from interpreters and chroniclers of traditional cultures to potential expert advisers on social innovation, managing their pathways into modernity. Thus, while modernization and industrialization became objects of research, they also became objects of practical engagement for consultancy oriented anthropologists. Although the roots of practical, or applied anthropology reach back into the nineteenth century (Barnett 1942a), the practice of giving policy advice on the basis of ethnographic expertise was first suggested by Bronislaw Malinowski (1929) in the late 1920s and then institutionalized both in Britain and the U.S. in the late 1930s and early 1940s. Until the 1960s practical and applied anthropology played an important role in British colonial administrations, as well as American bureaucracy, advancing among the former the sociocultural basis of industrial organization, market developments and the coexistence of peoples in multi-ethnic colonial societies, and contributing among the latter to the development of rural communities and native reservations (Bennett 1996).

However, British practical anthropologists suffered from bad reputations amongst their less policy oriented scientific colleagues. Their work was not only considered to be of low scientific standing (Spencer 2000), but they were considered naïve with regard to the nature of colonial power relations and the strategic interests of colonial governments (Leach 2000, pp. 252–53). They were also often seen to be corrupted by "philanthropic capital" provided by ideologically biased sponsors that financed much of their research (Fisher 1986, p. 7). By contrast, practical anthropology had a much better reputation with public–private funding bodies. National administrations in combination with national industries approached anthropologists to contribute to post-World War II economic and social development, both in the colonies and at home. David Mills (2010) gives a precise account of the institutional and personal relations that developed during the 1950s in the Committee on Anthropology and Industry between leading scientific members of the British Royal Anthropological Institute (RAI) and representatives and lobbyists of British industry and enterprise. The latter offered generous funding schemes for an "application of anthropological knowledge to our industrial society" (Mills 2010, p. 121), suggesting explicitly the idea that British anthropologists should also contribute to increasing productivity. A way forward seemed to lie in the work of the Manchester School around Max Gluckman that followed closely in the footsteps of the ethnographic work of American social psychologist Elton Mayo (1933) who had researched industrial production and organizational behavior at the shop-floor level. However, by the end of the 1950s the British captains of industry and the scientific leaders of the RAI had become somewhat estranged. In spite of generous funding offers anthropologists often had reservations about dedicating their efforts to the national economy's benefit and the shift from fundamental to applied research that such a scheme encouraged. They consequently left such studies and research to psychologists and sociologists who readily stepped into the vacuum that their departure created over the coming decades (see e.g. Burns and Stalker 1961).

The question of anthropology's engagement with (national) development and innovation was posed once again with the onset of Western economic decline from the late 1970s when national economies started to face a general profit crunch due to increasing global competitiveness, geographic shifts of industrial production and global decentralization of capital. In this context, the practice-oriented field of anthropometric anthropology proposed an "industrial anthropology for Europe" as an industry–science cooperation, underlining that Europe "depends much more on the exportation of industrial products than other continents so that the European anthropologist has to adapt industrial products to populations […] to make sure that e.g. clothing, furniture and places of work are adjusted to the proportions of the body" (Jürgens 1976, p. 364). Besides the rather marginal nature of this branch of anthropology, the RAI made a renewed attempt to foster an alliance of mutual benefit between British industry and an applied anthropology for innovation, noting that at "the present time nothing could help British social anthropology more than one or two successful and well-publicized applied projects aimed at assisting the prosperity of British commerce and industry – a widely accepted national goal" (Benthall 1980, p. 2). Outside Britain, too, postindustrial economic transformations reinvigorated the academic projects of industrial anthropology (Moore 1988) and business anthropology (Jordan 2010), which now turned to the study of sociocultural changes under conditions of decline, as opposed to their original interest in the sociocultural contexts of booming post-World War II economies.

Anthropologists held important positions in the UNESCO development programs from the earliest days on (Metraux 1951), and the normative issues raised by anthropological contributions to modernization in developing countries, and at home, encouraged anthropologists to position themselves in discourses on social and cultural innovation. Anthropologist Margaret Mead contributed important publications on the relation between cultural and technological innovation to the UNESCO development program (Mead 1954). Her publication *Continuities in Cultural Evolution* (1966) with replies by several anthropologists (Akhmanova et al. 1966) serves as an illustrative example for the controversies over normative issues in steering, or "manipulating," social and cultural innovation. Based on almost four decades of fieldwork Mead assessed modernization in developing countries with a diffusionist approach to cultural innovation, that is, as a procedural change within evolutionary communicative clusters and networks. In this framework an important role was accorded to the agency of "leaders of innovation" who consciously explored the possibilities for innovation, thereby gaining a measure of control over the direction and acceleration of innovation. At the same time, Mead underlined the social embedding of the innovation process, contrasting "the kinds of innovation and intervention which are possible in the physical sciences with those which are possible in the human sciences, where a much greater participation of society is required to change social forms which are already working after a fashion" (Mead in Akhmanova et al. 1966, p. 67). Positive readings of Mead understood innovation as a means for social advance and improvement, while negative readings tended to understand her work as a cult of innovation bent on disseminating normative Western values throughout the world by manipulative social engineering. This raised "the hoary problems of who is to guard the innovators or direct the directors" (Frantz in Akhmanova et al. 1966, pp. 70–71).

Stimulated by the apparent decoupling of ongoing technological innovation and development from stagnating, or even declining, social development in terms of welfare, effective wages, labor rights and so on, anthropologists came to inquire over the course of the 1970s into the ideological nature of innovation. The British Society for Responsibility in Science united a radical group of leading scientists who initiated the "alternative technology" movement (Dickson 1974) against contributions from academia to the military and arms industries complex. Amongst other sites of publication they spread their central message through the journal *Dialectical Anthropology* warning readers that "innovation is used in a directly political manner to maintain and reinforce the control of a dominant social class" (Dickson 1975, p. 26) and that "technological innovation [...] is only carried out to the extent that it coincides with and maintains the interests of this class" (Dickson 1975, p. 28). Building on this basis, critical applied and interventionist "action anthropology" (Tax 1988, p. 20) sought to contribute even further to the development of alternative applications of technology. The latter should encourage peaceful exchanges through technology as well as social improvements, rather than blindly pursuing technological advance and contributing to the growth of the arms industry.

## CRISIS OF INNOVATION

According to the anthropologist David Graeber's analysis, at least since the late 1970s technological research and development in the West has been turning out a diminishing number of fundamental innovations and has been increasingly focused on fine-tuning and commercializing existing inventions (Graeber 2015, p. 114). Clearly, the rising conjuncture of rapidly growing markets from the 1950s to the 1970s stimulated innovations by offering the greatest possibilities for surplus profit and long-term investments into research and development. During the following declining phase, by contrast, the relatively slow growth of markets has led "to the greatest competitive pressure to innovate to stave off the threat of liquidation" (Clarke 1990, pp. 458–59). Hence, short-term investments into more immediately marketable and profitable innovations would take precedence, thereby shifting innovation increasingly from industrial production into the sectors of finances and services. Today, then, innovation has become an issue of national concern in almost every Western nation state. Edmund Phelps, the 2006 Nobel laureate in economics, has thus identified "rekindling innovation" as the prime strategy to fight stagnation and decline (Phelps 2014). Notably, in his analysis it is not strictly speaking the economy that needs to be fixed, but cultural values that need modification because they function as breaks on innovation and prosperity. Since reflexive "innovation of innovation" has been called "the only way out of the current crisis of innovation in industrial societies" this has also lead to claims from the social and cultural sciences that "disciplines like anthropology need to be involved in the research ventures" (Ruth 2003, pp. 229 and 237). This illustrates quite neatly how innovation as a general concept even outside anthropology has undergone ideological redefinition and culturalization over the past two decades.

Such a re-calibration of attention has put a spotlight on how better to foster the capacity to innovate among both individuals and organizations. As opposed to evolutionary conceptions of naturally advancing innovation this development hinges on a contingent notion of innovation as a process that is hard to plan and to manage and whose successful implementation involves in particular cultural practices and values far beyond the mere introduction of a technological innovation. A lack of innovativeness has thus been attributed to a neglect of the human dimension of work and of the local cultural specificities in business environments, including contexts of social conventions, symbols, codes, practices and patterns of behavior that encompasses discrete personal emotions and power structures. Recognition, security and a sense of belonging among workers has been analyzed as successful factors for creating innovations. In this way informal networks are interpreted as a spur to successful innovation even more so than incentive structures, or technologically optimized working conditions (Bray 2010). Thus, also outside anthropology, innovation is getting increasingly framed as a socio-technical process, rather than being a matter of purely technical concern.

Against this backdrop anthropologists have been increasingly seen as potential aides to innovation processes. The direct engagement of anthropology with innovation policies and practices, its increasing focus on agency in innovation processes and on the experiences of innovation, as well as the discipline's recognition as a contributor to innovation by its methods and expertise is reflected in the role of anthropologists in laboratory studies. The hunger for anthropological expertise in innovation processes made it possible for ethnographic chroniclers and interpreters of innovation processes (Latour and Woolgar 1979; Knorr-Cetina 1999) to turn into potential co-producers of innovation, thus transforming innovation from an object of anthropological research into one of practice-oriented engagement. The use of anthropological methods in the corporate sector has evolved into a tool for investigating both the potential of, and barriers to, industrial or entrepreneurial innovation processes and for enhancing the innovative capacities of companies. For example, ethnographic fieldwork in engineering corporations has contributed to quality improvement (Collins 2003), and ethnographic methods have been applied to better addressing target customers at early stages of the innovation process (Wasson 2000), and to the improvement of marketing strategies (Arnould and Wallendorf 1994). Since the early 1990s, researchers trained in anthropology have found occupations in industry in, for example, the field of applied design anthropology (Gunn, Otto and Smith 2013), which has contributed to computer supported cooperative work (Wasson 2016), and in the Xerox Palo Alto Research Center (Wasson and Squires 2012). At the intersections between academia and entrepreneurship, this trend has led to the frantic branding of new fields of enquiry, such as the real-time ethnography of innovation processes (Hoholm and Araujo 2011), business ethnography (Hara and Shade 2018), or "technoanthropology" (Colobrans et al. 2012) to name only a few. At the same time, anthropological methods have been employed critically in raising consciousness about the implicit social values embedded in innovation processes that are likely to manifest in the eventual product and to impact on innovation processes at "mid-stream" by building capacity for reflexivity and societal responsiveness among laboratory scientists and engineers (Fisher 2007; Fisher and Schuurbiers 2013).

The culturalization of innovation and the parallel innovation of culture under conditions of economic decline have thus added another twist to anthropologists' relations with innovation, namely the "contribution of anthropology to enquiries into how and why innovation can bring competitive advantages to economic actors" (Welz 2003, p. 258). Notably, this quite hands-on practical anthropology of the present stands out for its methodological, rather than its conceptual or self-reflective, nature. One approach to addressing this shortcoming is offered by global systemic anthropology, which has been analyzing the trend from materialism to culturalism in both scientific analysis and in real-word manifestations as an aspect of the decline of Western hegemony (Friedman 2007b). This perspective asks anthropologists to research the cultural contexts and practices of the crisis of innovation, including its discursive representations, as properties of larger global systemic dynamics, such as the global decentralization of capital and the decline of Western hegemony.

## CONCLUSION: ANTHROPOLOGY OF AND FOR INNOVATION

In this chapter we have presented a historical analysis of the variegated usages of innovation in anthropological concepts of, and engagements for, social and cultural change from the late nineteenth century onwards. While innovation over the first half of the twentieth century was often employed in a general and unspecific sense and used interchangeably with the term invention, it still played a role in conceptual anthropological debates about cultural change and contributed, through the process of interdisciplinary exchange, to procedural conceptions of innovation. In response to the emergence of a standard view of technological innovation over the second half of the twentieth century, anthropologists have appropriated cultural and social innovation as differential concepts for their own discipline. Beyond its analytical and descriptive functions the term innovation has gained performative purchase on modernist culture through practical and applied anthropology. In a double sense we have shown the term innovation to be relevant to anthropology as an object of research (the anthropology *of* innovation) and as practice-oriented object of social engagement (the anthropology *for* innovation). With a view to the widespread notion of the Western crisis of innovation and with consideration of the concurrent tendency for the culturalization of innovation (under considerable influence of anthropological methods) we have sought to contribute with this chapter to open up innovation as a relevant topic for further anthropological research.

## NOTE

1. Only selective publications of Leroi-Gourhan's have become available in English, posthumously and with a deferral of three decades or more, which is one explanation why his work has been largely neglected in Anglophone anthropological studies of innovation and technology (see Lemonnier and Pfaffenberger 1989).

## REFERENCES

Akhmanova, Olga, Ernest Beaglehole, Nirmal Bose, Eliot Chapple, Charles Frantz, Roy Grinker, Douglas Haring, et al. (1966), 'Review of Continuities in Cultural Evolution by M. Mead', *Current Anthropology*, 7 (1), 67–82.

Arnould, Eric, and Melanie Wallendorf (1994), 'Market-Oriented Ethnography: Interpretation Building and Marketing Strategy Formulation', *Journal of Marketing Research*, 31 (4): 484–504. https://doi.org/10.2307/3151878.

Barnett, Homer Garner (1940), 'Culture Processes', *American Anthropologist*, 42 (1), 21–48.

Barnett, Homer Garner (1941), 'Personal Conflicts and Cultural Change', *Social Forces*, 20 (2), 160–71. https://doi.org/10.2307/2571335.

Barnett, Homer Garner (1942a), 'Applied Anthropology in 1860', *Applied Anthropology*, 1 (3), 19–32.

Barnett, Homer Garner (1942b), 'Invention and Cultural Change', *American Anthropologist*, 44 (1), 14–30.

Barnett, Homer Garner (1953), *Innovation: The Basis of Cultural Change*, 1st ed., McGraw-Hill Series in Sociology and Anthropology, New York: McGraw-Hill.

Bennett, John (1996), 'Applied and Action Anthropology: Ideological and Conceptual Aspects', *Current Anthropology*, 37 (1), S23–53.

Benthall, Jonathan (1980), 'Industrial Anthropology in Britain', *RAIN*, (40), 2.

Berliner, Joseph (1962), 'The Feet of the Natives Are Large: An Essay on Anthropology by an Economist', *Current Anthropology*, 3 (1), 47–77.

Bidney, David (1946), 'The Concept of Cultural Crisis', *American Anthropologist*, 48 (4), 534–52.

Blaut, James Morris (1987), 'Diffusionism: A Uniformitarian Critique', *Annals of the Association of American Geographers*, 77 (1), 30–47.

Boas, Franz (1889), 'Notes on the Snanaimuq', *American Anthropologist*, 2 (4), 321–28.

Boas, Franz (1920), 'The Methods of Ethnology', *American Anthropologist*, 22 (4), 311–21.

Bray, Zoe (2010), 'Innovation in working practices: an anthropological perspective', *Projectics / Proyectica / Projectique*, 6 (3), 57–68.

Brepohl, Wilhelm (1951), 'Industrielle Volkskunde', *Soziale Welt*, 2 (2), 115–24.

Burawoy, Michael (1979), 'The Anthropology of Industrial Work', *Annual Review of Anthropology*, 8, 231–66.

Burke, Peter and Eric Hobsbawm (1978), 'Reflections on the Historical Revolution in France: The Annales School and British Social History Comments [with Comments and Discussion]', *Review* (Fernand Braudel Center) 1 (3/4), 147–64.

Burns, Tom and George Stalker (1961), *The Management of Innovation*, London: Tavistock Publications.

Cipolla, Carlo (1972), 'The Diffusion of Innovations in Early Modern Europe', *Comparative Studies in Society and History*, 14 (1), 46–52.

Clarke, Simon (1990), 'The Marxist Theory of Overaccumulation and Crisis', *Science & Society*, 54 (4), 442–67.

Collins, Shawn (2003), 'Using Ethnography to Identify Cultural Domains within a Systems Engineering Organization', *Bulletin of Science, Technology & Society*, 23 (4), 246–55. https://doi.org/10.1177/0270467603256091.

Colobrans, Jordi, Artur Serra, Ricard Faura, Carlos Bezos and Iñaki Martin (2012), 'La Tecno-Antropología', *Revista de Antropologá Experimental*, 12 (Monografico), 137–46.

Cooper, Kory (2012), 'Innovation and Prestige among Northern Hunter-Gatherers: Late Prehistoric Native Copper Use in Alaska and Yukon', *American Antiquity*, 77 (3), 565–90.

Culwick, Arthur and Geraldine Culwick (1935), 'Culture Contact on the Fringe of Civilization', *Africa*, 8 (2), 163–70. https://doi.org/10.2307/3180505.

Dalton, George (1969), 'Theoretical Issues in Economic Anthropology', *Current Anthropology*, 10 (1), 63–102.

Davids, Karel and Bert De Munck (eds) (2014), *Innovation and Creativity in Late Medieval and Early Modern European Cities*, Farnham, UK: Ashgate.

de Oliveira, Roberto Cardoso and Joseph Berliner (1964), 'Combinatorial Analysis', *Current Anthropology*, 5 (2), 104–105.

Dickson, David (1974), *Alternative Technology and the Politics of Technical Change (Technosphere)*, Glasgow: Fontana/Collins.

Dickson, David (1975), 'Technology and Social Reality', *Dialectical Anthropology*, 1 (1), 25–41.

Dobyns, Henry and George Dalton (1971), 'On the Economic Anthropology of Postcolonial National Development', *Current Anthropology*, 12 (3), 393–97.

Douglas, Mary (1962), 'The Lele. Resistance to Change', in Paul Bohannan and George Dalton (eds), *Markets in Africa*, African Studies (Northwestern University) 9. Evanston, Illinois: Northwestern University Press, pp. 211–33.

Febvre, Lucien (1942), *Le Problème de l'incroyance Au XVIe Siècle. La Religion de Rabelais*, Paris: Albin Michel. https://www.jstor.org/stable/43012568.

Fisher, Donald (1986), 'Rockefeller Philanthropy: And the Rise of Social Anthropology', *Anthropology Today*, 2 (1), 5–8. https://doi.org/10.2307/3032900.

Fisher, Erik (2007), 'Ethnographic Invention: Probing the Capacity of Laboratory Decisions', *NanoEthics*, 1 (2), 155–65. https://doi.org/10.1007/s11569-007-0016-5.

Fisher, Erik and Daan Schuurbiers (2013), 'Socio-Technical Integration Research: Collaborative Inquiry at the Midstream of Research and Development', in Neelke Doorn, Daan Schuurbiers, Ibo van de Poel and Michael Gorman (eds), *Early Engagement and New Technologies: Opening up the Laboratory*, Philosophy of Engineering and Technology. Dordrecht: Springer Netherlands, pp. 97–110. https://doi.org/10.1007/978-94-007-7844-3_5.

Freeman, Christopher (1987), *Technology, Policy, and Economic Performance: Lessons from Japan*, London; New York: Pinter Publishers.

Friedman, Jonathan (2007a), 'Global Systems, Globalization, and Anthropological Theory', in Ino Rossi (ed.), *Frontiers of Globalization Research: Theoretical and Methodological Approaches*, New York: Springer Science + Business Media, pp. 109–32.

Friedman, Jonathan (2007b), 'Toward a Comparative Study of Hegemonic Decline in Global Systems. The Complexity of Crisis and the Paradoxes of Differentiated Experience', in R. Costanza, L. Graumlich and W. Steffen (eds), *Sustainability or Collapse? An Integrated History and Future of People on Earth*, Dahlem Workshop Reports. Cambridge, MA: MIT Press in cooperation with Dahlem University Press, pp. 95–114.

Gaglio, Gérald, Benoît Godin and Sebastian Pfotenhauer (2017), X-Innovation. Re-Inventing Innovation Again and Again. Montreal: Project on the Intellectual History of Innovation.

Gaizauskas, Barbara (1974), 'The Harmony of the Spheres', *Journal of the Royal Astronomical Society of Canada*, 68 (June), 146–51.

Geertz, Clifford (1959), 'Form and Variation in Balinese Village Structure', *American Anthropologist*, 61 (6), 991–1012.

Godin, Benoît (2010), 'Innovation Without the Word: William F. Ogburn's Contribution to the Study of Technological Innovation', *Minerva*, 48 (3), 277–307.

Godin, Benoît (2014a), *Innovation Contested: The Idea of Innovation over the Centuries*, Routledge Studies in Social and Political Thought. New York: Routledge.

Godin, Benoît (2014b), 'Invention, Diffusion and Linear Models of Innovation: The Contribution of Anthropology to a Conceptual Framework', *Journal of Innovation Economics Management*, 15 (3), 11–37.

Godin, Benoît (2016), 'Technological Innovation: On the Origins and Development of an Inclusive Concept', *Technology and Culture*, 57 (3), 527–56. https://doi.org/10.1353/tech.2016.0070.

Godin, Benoît (2017), *Models of Innovation: The History of an Idea*, Inside Technology, Cambridge, MA: The MIT Press.

Graeber, David (2015), *The Utopia of Rules: On Technology, Stupidity, and the Secret Joys of Bureaucracy*, Brooklyn: Melville House.

Gunn, Wendy, Ton Otto and Rachel Smith (eds) (2013), *Design Anthropology: Theory and Practice*, London ; New York: Bloomsbury.

Haller, Lea (2014), 'Innovation', in Christof Dejung, Monika Dommann and Daniel Speich Chassé (eds), *Auf der Suche nach der Ökonomie. Historische Annäherungen*, Tübingen: Mohr Siebeck, pp. 97–123.

Hara, Yuuki and Lynn Shade (2018), 'International Business Ethnography: Are We Looking for Cultural Differences?', Ethnographic Praxis in Industry Conference Proceedings 2018 (1), 10–22. https://doi.org/10.1111/1559-8918.2018.01193.

Hart, Keith (1973), 'Informal Income Opportunities and Urban Employment in Ghana', *The Journal of Modern African Studies*, 11 (1), 61–89.

Hart, Keith (1985), 'The Informal Economy', *Cambridge Anthropology*, 10 (2), 54–58.

Hayes, Edward Cary (1911), 'The Classification of Social Phenomena', *American Journal of Sociology*, 17 (2), 188–205.

Heinrich, Albert (1950), 'Some Present-Day Acculturative Innovations in a Nonliterate Society', *American Anthropologist*, 52 (2), 235–42.

Heßler, Martina and Kevin Liggieri (eds) (2020), *Technikanthropologie: Handbuch für Wissenschaft und Studium*, Baden-Baden: Nomos.

Hodgen, Margaret (1945), 'Glass and Paper: An Historical Study of Acculturation', *Southwestern Journal of Anthropology*, 1 (4), 466–97.

Hoholm, Thomas and Luis Araujo (2011), 'Studying Innovation Processes in Real-Time: The Promises and Challenges of Ethnography', *Industrial Marketing Management*, Business Networks: Global, Regional and Local The Best from IMP 2010 – Budapest, 40 (6), 933–39. https://doi.org/10.1016/j.indmarman.2011.06.036.

Horton, Robin (1960), 'A Definition of Religion, and Its Uses', *The Journal of the Royal Anthropological Institute of Great Britain and Ireland*, 90 (2), 201–26. https://doi.org/10.2307/2844344.

Jordan, Ann (2010), 'The Importance of Business Anthropology: Its Unique Contributions', *International Journal of Business Anthropology*, 1 (1). https://doi.org/10.33423/ijba.v1i1.1202.

Jürgens, Hans (1976), 'Industrial Anthropology in Europe', Proceedings of the Human Factors Society Annual Meeting. https://journals.sagepub.com/doi/pdf/10.1177/154193127602001604.

Knorr-Cetina, Karin (1999), *Epistemic Cultures: How the Sciences Make Knowledge*, Cambridge, MA: Harvard University Press.

Kroeber, Alfred Louis (1917), 'The Superorganic', *American Anthropologist*, 19 (2), 163–213.

Latour, Bruno and Steve Woolgar (1979), *Laboratory Life: The Social Construction of Scientific Facts*, Beverly Hills: Sage Publications.

Leach, Edmund Ronald (2000), *The Essential Edmund Leach. Volume II: Culture and Human Nature*, Edited by S. Hugh-Jones and J. Laidlaw, New Haven: Yale University Press.

Lemonnier, Pierre and Bryan Pfaffenberger (1989), 'Towards an Anthropology of Technology', *Man*, 24 (3), 526–27.

Leroi-Gourhan, André (1993), *Gesture and Speech*, An October Book, Cambridge, MA, London: MIT Press.

Lévi-Strauss, Claude (1962), *La pensée sauvage*, Paris: Plon.

Lévy-Bruhl, Lucien (1922), *La mentalité primitive*, par L. Lévy-Bruhl. Travaux de l'Année sociologique. Fondateur: Émile Durkheim. Paris: Librairie Félix Alcan.

Linton, Ralph (1924), 'The Origin of the Plains Earth Lodge', *American Anthropologist*, 26 (2), 247–57.

Linton, Ralph (1925), 'Marquesan Culture', *American Anthropologist*, 27 (3), 474–78.

Linton, Ralph (1926), 'The Origin of the Skidi Pawnee Sacrifice to the Morning Star', *American Anthropologist*, 28 (3), 457–66.

Linton, Ralph (1927), 'Report on Work of Field Museum Expedition in Madagascar', *American Anthropologist*, 29 (3), 292–307.

Linton, Ralph (1928), 'Culture Areas in Madagascar', *American Anthropologist*, 30 (3), 363–90.

Lowie, Robert (1915), 'Exogamy and the Classificatory Systems of Relationship', *American Anthropologist*, 17 (2), 223–39.

Lowie, Robert (1942), 'The Transition of Civilizations in Primitive Society', *American Journal of Sociology*, 47 (4), 527–43.

Lundvall, Bengt-Åke (2007a), "Innovation System Research – Where It Came from and Where It Might Go', 2007–01. Globelics Working Paper Series. Globelics – Global Network for Economics of Learning, Innovation, and Competence Building Systems, Aalborg University, Department of Business and Management. https://ideas.repec.org/p/aal/glowps/2007-01.html.

Lundvall, Bengt-Åke (2007b), "National Innovation Systems – Analytical Concept and Development Tool', *Industry and Innovation*, 14 (1), 95–119. https://doi.org/10.1080/13662710601130863.

Malinowski, Bronislaw (1929), 'Practical Anthropology', *Africa*, 2 (1), 22–38. https://doi.org/10.2307/1155162.

Maquet, Jacques (1964), 'Papers in Honor of Melville J. Herskovits Objectivity in Anthropology', *Current Anthropology*, 5 (1), 47–55.

Marett, Robert Ranulph (1920), *Psychology and Folk-Lore*, London: Methuen & co. http://archive.org/details/psychologyfolklo00marerich.

Mayo, Elton (1933), *The Human Problems of an Industrial Civilization*, New York: Macmillan Co.

Mead, Margaret (1954), *Cultural Patterns and Technical Change*, Paris: UNESCO.

Mead, Margaret (1966), *Continuities in Cultural Evolution*, New Haven: Yale University Press.

Meggers, Betty (1946), 'Recent Trends in American Ethnology', *American Anthropologist*, 48 (2), 176–214.

Metraux, Alfred (1951), 'UNESCO and Anthropology', *American Anthropologist*, 53 (2), 294–300.

Mills, David (2010), 'How Not to Apply Anthropological Knowledge. The RAI and Its "Friends"', in David Mills, *Difficult Folk? A Political History of Social Anthropology*, Berghahn Books, pp. 113–28. https://doi.org/10.2307/j.ctv8mdn66.11.

Moore, David (1988), 'Industrial Anthropology: Conditions of Revival', *City & Society*, 2 (1), 5–18. https://doi.org/10.1525/city.1988.2.1.5.

Murdock, George Peter (1932), 'The Science of Culture', *American Anthropologist*, 34 (2), 200–15.

O'Brien, Michael and Steven Shennan (eds) (2010), *Innovation in Cultural Systems: Contributions from Evolutionary Anthropology*, Vienna Series in Theoretical Biology, Cambridge, MA: MIT Press.

Ogburn, William (1922), *Social Change with Respect to Culture and Original Nature*, London: George Allen & Unwin.

Ogburn, William (1926), 'The Great Man versus Social Forces', *Social Forces*, 5 (2), 225–31. https://doi.org/10.2307/3004769.

Opler, Morris (1959), 'Technological Change and Social Organization in a Village of North India', *Anthropological Quarterly*, 32 (3), 127–33. https://doi.org/10.2307/3316898.

Ottaway, Barbara (2001), 'Innovation, Production and Specialization in Early Prehistoric Copper Metallurgy', *European Journal of Archaeology*, 4 (1), 87–112. https://doi.org/10.1177/146195710100400103.

Peace, Adrian (1979), 'Prestige Power and Legitimacy in a Modern Nigerian Town', *Canadian Journal of African Studies / Revue Canadienne Des Études Africaines*, 13 (1/2), 25–51. https://doi.org/10.2307/484637.

Pezzarossi, Heather Law (2014), 'Assembling Indigeneity: Rethinking Innovation, Tradition and Indigenous Materiality in a 19th-Century Native Toolkit', *Journal of Social Archaeology*, 14 (3), 340–60. https://doi.org/10.1177/1469605314536975.

Pfaffenberger, Bryan (1988), 'Fetishised Objects and Humanised Nature: Towards an Anthropology of Technology', *Man*, 23 (2), 236–52. https://doi.org/10.2307/2802804.

Pfaffenberger, Bryan (1992), 'Social Anthropology of Technology', *Annual Review of Anthropology*, 21, 491–516.

Phelps, Edmund (2014), 'Rekindling Innovation', *RSA Journal*, 160 (5558), 36–39.

Popplow, Marcus (2019), 'Die Idee der Innovation – ein historischer Abriss', in Birgit Blättel-Mink, Ingo Schulz-Schaeffer and Arnold Windeler (eds), *Handbuch Innovationsforschung*, Wiesbaden: Springer Fachmedien Wiesbaden, pp. 1–9. https://doi.org/10.1007/978-3-658-17671-6_2-1.

Redfield, Robert, Ralph Linton and Melville Herskovits (1936), 'Memorandum for the Study of Acculturation', *American Anthropologist*, 38 (1), 149–52.

Reith, Reinhold (2000), 'Technische Innovationen im Handwerk der Frühen Neuzeit? Traditionen, Probleme und Perspektiven der Forschung', in Karl Heinrich Kaufhold and Wilfried Reininghaus (eds), *Stadt und Handwerk in Mittelalter und Früher Neuzeit*, Städteforschung, Bd. 54. Köln: Böhlau, pp. 21–60.

Rogers, Arthur (1912), 'Burke's Social Philosophy', *American Journal of Sociology*, 18 (1), 51–76.

Ruth, Klaus (2003), 'Industrial Culture and the Innovation of Innovation: Enginology or Socioneering?', *AI & Society*, 17 (3), 225–40. https://doi.org/10.1007/s00146-003-0278-6.

Sahlins, Marshal David (1972), *Stone Age Economics*, Chicago: Aldine-Atherton.

Salz, Beate (1955), 'The Human Element in Industrialization: A Hypothetical Case Study of Ecuadorean Indians', *Economic Development and Cultural Change*, (1), i–265.

Schapera, Isaac (1943), *Tribal Legislation among the Tswana of the Bechuanaland Protectorate: A Study in the Mechanism of Cultural Change*, Monographs on Social Anthropology 9, London: Lund for London School of Economics and Political Science.

Schapera, Izaak (1970), *Tribal Innovators: Tswana Chiefs and Social Change, 1795–1940*, Monographs on Social Anthropology, No. 43, London: Athlone Press.

Seymour-Smith, Charlotte (2005), *Dictionary of Anthropology*, Digital print, Basingstoke: Palgrave.

Shimkin, Demitri Boris (1952), 'Industrialization, a Challenging Problem for Cultural Anthropology', *Southwestern Journal of Anthropology*, 8 (1), 84–91. https://doi.org/10.1086/soutjanth.8.1.3628556.

Spencer, Herbert (1851), *Social Statics, or, The Conditions Essential to Human Happiness Specified, and the First of Them Developed*, London: John Chapman.

Spencer, Jonathan (2000), 'British Social Anthropology: A Retrospective', *Annual Review of Anthropology*, 29, 1–24.

Tax, Sol (1988), 'Pride and Puzzlement: A Retro-Introspective Record of 60 Years of Anthropology', *Annual Review of Anthropology*, 17, x–21.

Tozzer, Alfred Marston (1928), *Social Origins and Social Continuities*, New York, The Macmillan Company. http://archive.org/details/socialoriginssoc00tozz.

Wasson, Christina (2000), 'Ethnography in the Field of Design', *Human Organization*, 59 (4), 377–88.

Wasson, Christina (2016), 'Design Anthropology', *General Anthropology*, 23 (2), 1–11. https://doi.org/10.1111/gena.12013.

Wasson, Christina and Susan Squires (2012), 'Localizing the Global in Technology Design', in Christina Wasson, Mary Odell Butler and Jaqueline Copeland-Carson (eds), *Applying Anthropology in the Global Village*, Walnut Creek, CA: Left Coast Press, Inc., pp. 251–84.

Watson, James (1953), 'Four Approaches to Cultural Change: A Systematic Assessment', *Social Forces*, 32 (2), 137–45. https://doi.org/10.2307/2573711.

Welz, Gisela (2003), 'The Cultural Swirl: Anthropological Perspectives on Innovation', *Global Networks*, 3 (3), 255–70. https://doi.org/10.1111/1471-0374.00061.

Zacharias, H. (1953), 'Review of Review of Man and His Works. The Science of Cultural Anthropology, by Melville Herskovits', *Anthropos*, 48 (3/4), 662–64.

# 20. Philosophical reflections on the concept of innovation

*Vincent Blok*

## INTRODUCTION

Innovation is often uncritically seen as a good thing (Rogers 1976) and considered as a panacea for all kinds of socio-economic challenges (Blok and Lemmens 2015). At the same time, the concept of innovation itself remains undefined in most policy documents, while its meaning seems to be taken for granted in the scientific literature (Godin 2015). We are familiar for instance with dichotomies like incremental versus disruptive innovation (Christensen 1997) or closed versus open innovation (Chesbrough 2003, 2006), but what does the notion of innovation itself mean? While it is nowadays often understood as the commercialization of technological inventions (von Schomberg and Blok 2018), new developments like the emergence of social, sustainable and responsible innovation make this self-evident conceptualization questionable. If we do not confront this self-evident understanding of innovation, as it is found in economics, innovation studies and business administration, the conceptualization of these new developments remain hemmed in by traditional innovation philosophies.

One would expect that the philosophical tradition can provide guidance in our understanding of innovation, as philosophers traditionally reflect on basic concepts and categories that structure our understanding of the world, such as nature versus technology, humans versus non-humans and so on. It is striking, however, that the notion of innovation is not among the key concepts that traditional philosophers tend to reflect upon (Blok 2020, 2021). One reason might be that philosophers originally tend to *understand* and gain *knowledge* of the world around us, rather than to change or innovate it. Further, the object of philosophical knowledge is traditionally found beyond the world of becoming in which innovations take place, and aims at knowledge about the non-changing or universal categories that underlie this world of becoming, for instance knowledge about the Platonic *ideas* or the Kantian categories. Finally, to the extent that the primary objective of philosophical knowledge is to understand the world, traditional philosophers may have the tendency to *conform* themselves to the world as it is, rather than changing or revolutionizing it. Nonetheless, to the extent that philosophical reflection can contribute to the critical assessment of basic concepts, and innovation is one of the concepts that mark our time, we take a step back in this chapter and philosophically reflect on the notion of innovation. Such a philosophy of innovation enables us to critically reflect on the self-evidence of the techno-economic paradigm of innovation and its applicability on contemporary phenomena like social and sustainable innovation.

In order to open up the phenomenon of innovation, we distinguish between the innovation process and outcome dimension, and between the ontic and ontological dimension of innovation (section 1). The ontic dimension of innovation concerns *beings* like new artefacts, and the ontological dimension concerns the *being* of these beings. These distinctions lead to four characteristics of our understanding of innovation with several implications for the object of innovation and its novelty, as well as for the temporality and human involvement in innovation practices. It will turn out that innovation concerns the ontogenetic process in which primarily a world is constituted – that is, the digital world – in which the invention of new artefacts – that is, blockchain technology – is embedded.[1] This ontogenetic process doesn't follow a chronological temporal order, but is constituted by a temporal iterative process. Human creativity is involved in this ontogenetic process as co-creative capacity (section 2). In section 3, we show the advantage of our conceptualization of innovation and its implications for contemporary and future alternative theories of innovation.

# PHILOSOPHY OF INNOVATION[2]

In this section, we subsequently discuss (a) the innovation process and outcome dimension, and (b) the ontic and ontological dimension of innovation, which leads to four characteristics of the phenomenon of innovation.

## (a)    The Process and Outcome Dimension of Innovation

A first characteristic of innovation can be found if we oppose the innovation process to the outcome of the process. The word innovation has both a substantive meaning, for example the iPhone as an outcome or end-product – and a verbal meaning – the innovation process that results in the iPhone as outcome. On the one hand, this innovation outcome may be considered as something new to the world, for instance the first time Apple introduced the iPhone as an outcome or end-product of the innovation process. On the other hand, the creation of this new end-product will replace predecessors of the smartphone, for instance the dominance of landlines in Western households. Because innovation is not only an outcome but also a process, it is something that can and should be managed. Stage gate models (Cooper 2008) and technology readiness levels of innovation for instance enable the management of this innovation process in such a way that it leads to the best possible outcomes.

In economic thinking, the process of innovation is often understood in terms of a creative destruction; the innovation of the diesel engine in locomotives for instance is not only the creation of a new end-product that can be exchanged on the market, but destructed at the same time the existing industry in steam engines, just like the innovation of the compact disc destructed the industry of cassette tapes and LPs and is now replaced by streaming services (Schumpeter 1943; Blok 2020). But we do not have to refer to Schumpeter's concept of creative destruction to understand the dynamic process of creation and destruction involved in the innovation process. The history of the concept of innovation shows that it always appeared in the context of a dominant "paradigm of orthodoxy, authority and order" (Godin 2015, p. 93). Innovations involve something new that intervenes within the established order. Plato for instance introduces the concept of innovation in the context of the political order. He argues against innovation because it introduces change that threatens the established political order (Plato 1967; Aristotle 1944; Blok 2020). How do we have to understand this dynamic innovation process that creates something new and at the same time threatens to undermine or destruct the existing order?

If we consider the outcome of the innovation process as a concrete *individual* object or artefact, the innovation process itself can be formally conceived as the pre-individual. This reality of the innovation process *before* its individuation in a concrete innovation outcome can be conceptualized as the *ontogenesis* of this outcome. Ontogenesis originally refers to the developmental processes and conditions for the development of an individual organism, from the time of fertilization to its mature form, but can be found in the development of innovations as well. The reality of the ontogenetic *process* of innovation cannot be understood out of its outcome, like incremental and radical innovation, or closed and open innovation, because then the process of innovation is conceptualized based on its outcome, that is, the pre-individual is understood in terms of the individual that comes out of it and not out of this process itself. Precisely this is the problem with many typologies of innovation in the innovation management literature. Although the process of innovation is theorized since the 1920s in disciplines like anthropology and sociology, distinctions like incremental versus radical innovation (Freeman and Soete 1997) or architectural versus modular innovations (Henderson and Clark 1990) characterize different types of innovations but miss its ontogenetic process. On the contrary, the outcomes of the innovation process – concrete individual products or services, its components or the compositions of these components – are taken as point of departure. This focus on the innovation outcome may be explained by what is called the 'culture of things' or material culture: "The origin of this culture goes back to the Renaissance: due to commercial exchanges, exploration and travel, natural and artificial objects have been what is valued in arts, science, and real life" (Godin 2008, p. 21). But if innovation concerns both the process and the outcome of the process, a philosophical reflection can no longer be isolated to outcomes, but must come to terms with the process that is a distinct and integral part of innovation. On the one hand, if we find the point of departure of our reflections in the outcome of the innovation process – that is, the artefact as outcome of innovation – we miss the *operation* that is constituting this innovative outcome, we miss innovation as an ontogenetic process. On the other hand, if we find the point of departure of our reflection in the process of the creation of the innovation outcome that at the same time threatens to destruct the established order, we should no longer think the ontogenetic process out of an *individual* innovation outcome that is created while it destructs a previous *individual* outcome. On the contrary, it should be understood out of the process of creation and destruction itself, that is, at the pre-individual level. With this, we assume a fundamental difference between outcome and process, between individual and *pre-individual*, thereby keeping open the possibility that innovation as process cannot be reduced to innovation as outcome, which is to say that process and outcome are divided by a fundamental difference. We shouldn't take this as an invitation to disregard the innovation outcome – it is highly questionable whether we can understand innovations without taking this outcome into consideration, for instance because these outcomes only account for spatio-temporal differences of their manifestation – but rather as a call to acknowledge both outcome and ontogenetic process as two fundamental aspects of innovation.

This acknowledgement of the ontogenetic process of innovation has certain advantages over established conceptualizations that find their point of departure in either the human creation as input of the innovation process or in the artefact as output of the process. It enables the analysis of general patterns in the ontogenesis of innovation, like the process of *concretization* in which innovations initially form an abstract system of isolated parts that function separately – for instance a cell-phone with a separate screen and key-board – but becomes increasingly integrated and perfected – for instance the integration of screen and key-board in modern smartphones – that cannot be attributed to the subject or object of innovation, but may be associated with the ontogenesis of the innovation process as such (see Simondon 2017).

## (b)    The Ontic and Ontological Dimension of Innovation

At the level of the innovation outcome, a second difference emerges if we consider that innovations involve something *new* that intervenes with the established *order*. If Plato for instance argues against innovation because it introduces change in the political order, he is not interested in the creative destruction of an individual artefact, but in the political order of the world that is threatened by innovation and potentially replaced by a new political order.

The dynamic nature of this creation and destruction of the political order becomes clear if we ask for the measure or unity of the order of the world. The philosophical tradition starting with Plato finds this measure or unity in the ontological characteristic of the being of beings, that is, in the transcendental horizon of the Platonic *idea*. The *idea* is a fixed category or measure, within which the world appears as an ordered whole that makes sense. In light of the *idea* 'human being' for instance, various people appear as human beings and we can understand this variety of humans as human beings. The *idea* human being is itself not a human being, but concerns a given measure, category or value within which the variety of people appears *as* unity. In the philosophical tradition there is a fundamental difference between the *idea*, category or value that establishes the order of beings in the world, and these beings themselves, which can only be perceived and understood in light of the *idea*. What is destructed by innovation, according to Plato, has to be sought at the ontological level of the *idea* as measure for the established political order, and not at the ontic level of things in the world (Blok 2020).

This also becomes clear in Plato's *Republic*. Here, Plato argues that the state should be ruled by the philosopher king, who has the necessary training and education that enables him to intellectually grasp ethical notions such as the *idea* of justice, and has the insights that are required to safeguard the political order. Here the problem with innovation becomes clear. If innovation transgresses the established political *order* of the world, it primarily intervenes at the level of the *ideai*, categories or values within which the world functions as *order*. Innovations are primarily disrupting the existing *ideai*, and consist in the human construction and introduction of new *ideai*. This means that the destructive aspect of innovation does not concern primarily the ontic level of things in the world, but the ontological level of the *ideai*, categories or values that establish and safeguard a world *order*. The idea that innovation primarily intervenes at the ontological level of the *world* order is also confirmed by a later writer on innovation, Francis Bacon, who argues that innovations "have altered the whole face and state of things right across the globe" (cited in Godin 2015, p. 182; Blok 2020).

Also in modern conceptions of innovation, what is at stake is not primarily the innovation of a new artefact from the earliest age to maturity, but the destruction of existing markets and construction of new markets for instance (Schumpeter 1943). What is destructed in the innovation of streaming services is not so much the CD in the literal sense of the word – there are still CDs in the world – but the way value is created and captured via markets in the economic order associated with digital networks like the Internet. What is destructed is not so much an artefact, but the political-economic order that is associated with, for instance, water and the way in which the water mill and the accompanying textile industry was embedded, which in turn gave rise to a new political-economic order associated with steam (railway industry for instance), digitalization and so on.

We see here that innovation operates at two levels of outcomes. The innovation of streaming services concerns first of all the ontogenesis of this service at an ontic level, but secondly the ontogenesis of the political-economic order of the world associated with digital networks at the ontological level. With 'ontology', we do not mean an eternal metaphysical idea, but a temporary category or value that establishes a particular political-economic order of the world – that is, the digital world – that enhances the invention of particular products or services – that is, streaming services – while it makes others obsolete – that is, the LP and CD. To the extent that innovation does not only concern new artefacts but also the structures within which these artefacts appear and are understood as ordered, we think that the distinction between the ontic and the ontological level of thinking may be helpful to understand the phenomenon of innovation. In the current age, the ontological level of innovation concerns the ontogenesis of a world order associated with digital networks, in which the streaming services can emerge, can be applied in various software applications and social media, and can be adopted and used by humans. This distinction between the ontic and ontological level of the innovation outcome provides a new perspective on the nature of innovation. Innovations like the internal combustion engine are innovations at the ontic level of the creation of a new artefact – the first engine for instance – but they involve at the same time the destruction of the economic equilibrium or world *order* associated with a particular set of innovations, in this case innovations associated with the world of steam. Simultaneously, the innovation of the internal combustion engine at an ontic level gives primarily rise to a new world order associated with electricity. The innovation outcome therefore doesn't only concern things in the world, but the *world order* in which these things appear and can be understood (Blok 2020). It is at this ontological level that innovation can be said to change the 'rules of the game' or the 'face of the Earth'. We argue therefore for a dual concept of the innovation outcome: innovation primarily operates at the ontological level of categories that constitute and establish a world, next to its operation at the ontic level within this world where it engenders novel things or innovative outcomes.

This duality doesn't mean that both levels are completely separated. The ontic and ontological levels of innovation turn out to be interconnected and interdependent. On the one hand, the innovation of the internal combustion engine at the ontic level is dependent on a world order associated with electricity. On the other hand, this world of electricity at an ontological level emerges only as world order in case of the innovation of the internal combustion engine that changes the rules of the game and destructs the world of steam. The innovation of the internal combustion engine articulates in a way the condition of its own possibility, namely the creation of the world of electricity as a regime in which this combustion engine can only function in a proper way and conditions further incremental improvements of this artefact; an innovation of the engine that doesn't take the boundary conditions of the world of electricity into account wouldn't make any sense. The innovation of the world of electricity is ontologically first, but not necessarily in the temporal sense of the word.

The interdependency of the innovation outcome at ontic and ontological level already provides good reasons to reject any unilateral focus on either the ontological level of innovation – we can think of a Heideggerian onto-centrism that highlights the importance of the ontological level of the innovation of a world order while neglecting the ontic level of innovations like the Internet, social media and so on – or on the ontic level of innovation – we can think of a post-phenomenological approach that highlights how the innovation of Google glass for instance mediates the world we experience. Roughly speaking, while Heidegger argues in his *Question concerning technology* that this ontological level of technology cannot be found at the level of screws and bolts of an artefact, Verbeek would argue that there is no ontological level beyond the screws and bolts of the artefact (Verbeek 2005). In fact, our reflections on the innovation outcome compel us to rehabilitate the ontic-ontological difference that was

*Table 20.1     Four characteristics of the phenomenon of innovation*

| Innovation Outcome: | Innovation Outcome: |
|---|---|
| artefact (Ontic Level) | world (Ontological Level) |
| Innovation Process: | Innovation Process: |
| ontogenetic process in which an artefact emerges out of predecessors (Ontic Level) | ontogenetic process in which a new world emerges (Ontological Level) |

rejected by post-modernist philosophy – for example, Ihde's idea that there is no Heideggerian 'essence' of technology beyond the many technologies (Ihde 2010) – at least in case we want to reflect on the phenomenon of innovation.

This dual concept of innovation has certain advantages over established dichotomies like radical and incremental innovation. While the distinction between incremental and radical seems to be only a gradual distinction that doesn't make clear what criterion has to be fulfilled for an innovation in order to be called radical, our conceptualization of the ontic and ontological level of innovation helps to operationalize this distinction; while incremental innovation concerns innovations only at an ontic level – a new version of the iPhone – radical innovation concerns innovations both at the ontic and ontological level which involves both the innovation of the steam engine and of the world of steam.

In Table 20.1, we summarize the findings of our reflection in this section. We first distinguished between the process and outcome dimension of innovation and then between the ontic and ontological dimension of innovation. These dimensions provide two axes that enable us to distinguish four characteristics of innovation, namely innovation as innovation outcome at the ontic level, like the innovation of streaming services; innovation as innovation outcome at the ontological level, like the political-economic order of the world – that is, the digital world – associated with digital networks in our current age; innovation as innovation process at the ontic level, like the process in which streaming services evolve out of predecessors (LPs, CDs) and the existing retail market for CDs is destructed and replaced; innovation as innovation process at the ontological level, like the process in which the digital world evolves out the world of petrochemicals and electricity (Blok 2020). A full understanding of the phenomenon of innovation encompasses these four characteristics of innovation.

## CRITICAL ASSESSMENT OF THE NEWNESS, THE TEMPORALITY AND THE HUMAN INVOLVEMENT IN INNOVATION

These four characteristics of the phenomenon of innovation have several implications for our understanding of the object of innovation and the novelty involved, as well as for the temporality and human creativity involved in innovation practices.

In the previous section, we introduced a dual concept of the innovation outcome at the ontic and ontological level while traditionally, innovation is located on the ontic level of new artefacts, ranging from computers to nanotechnological instruments. We broadened our perspective to the innovation of a world order, that is, the innovation of a unifying principle in light of which the world appears as order, in which the innovation of particular artefacts can emerge. Such a unifying principle may be found in a new material resource or element – for example, electricity, steam, digitalization – that articulates for instance the digital world as condition for the possibility of the invention and use of new artefacts like sensing or blockchain technologies that establish the digital world, but also in new concepts and new meanings of concepts like freedom, rationality and so on (Godin 2015, pp. 30–1).

The outcome of the innovation process is traditionally seen as the new. Newness can be seen as one of the key distinguishing factors between technology and innovation (Blok 2020); technology is associated with a type of knowledge which is contrasted with disruptive innovations which are associated with the *un*-known and concerns something new to the world. At the same time, newness and innovation are two separate things. On the one hand, everything we encounter in the world today was once new. On the other hand, the newness of many innovations we encounter in the world can be questioned and rather concern changes or improvements of what is already available in the world. This observation does not necessarily lead to the rejection of newness as distinguishing characteristic of innovation, but makes clear why a further reflection on the newness involved in innovation is needed.

The newness of the innovation outcome is not restricted to something new to the world, like the first combustion engine or the first flat screen TV as we have seen. Already in the literature on innovation management, a distinction is made between innovations that are new to the world and innovations that are only new to the firm. All kinds of gradual differences between categories of new products are acknowledged, ranging from improvement to repositioning and from incremental innovations to disruptive innovations (Christensen 1997). In the literature on invention for instance, innovations are seen as combinations or recombinations of things that already exist, or even a revisit of ancient things (Godin 2019). Also, from a historical perspective, we have to put the newness of the innovation outcome between brackets.

The concept of innovation originates from Ancient Greece, where it is named *kainotomia*. *Kainotomia* means change or the introduction of something new. It comes from *kainon* (new) and *tom* (cut, cutting) and originally meant 'cutting fresh into'. It was originally used in the context of the opening of new mines (Godin 2015, p. 19). Although Plato for instance is opposed to innovation because its newness disrupts the political order and can lead to revolution, the newness of innovation is often understood as *renewal* or *reformation* of the original; *re*-newal stresses newness as *return* to or as a taking back into an original situation (Godin 2015). This shows that the outcome of innovation is in first instance not something completely new to the world without any predecessor, as is sometimes said in case of disruptive innovations like the Internet, the combustion engine etc. The innovation outcome is the product of a historical process of renewal, in which this outcome emerges out of a previous stage and remains embedded in it in its future development. The innovation outcome may therefore consist in a repetition of an original state, or in the transformation of the current state (renovation) or in the renewal in a completely new state.

Such an ambiguous conceptualization of the new in innovation is problematic only if we conceptualize innovation based on its outcome on the ontic level. Then we are looking for a unique characteristic of the innovative product or service as outcome of innovation that didn't exist before and can be protected via patents. Such a unique characteristic will always remain contestable however if we acknowledge that all innovations emerge and evolve out of a previous historical stage and remain embedded in it. How to decide for instance whether the computer is new to the world or the combination of a typewriter and a TV?

Seen from the perspective of our earlier reflections on innovation as process, however, we can conceptualize the new in terms of the process of repetition, renovation, renewal, revolution and so on. The new then concerns the process of repetition, renovation, renewal that characterizes the ontogenetic process from pre-individual to individual, and that results in an innovation outcome that is revolutionary – it overthrows the established world order and is initiates a new beginning of a new world order. This newness of the world concerns a new organizing principle – the digital world – that co-evolves with artefacts that establish this world – sensing technologies, blockchain technologies, artificial intelligence and so on – that could not be expected or anticipated from the worlds that were before; in the world of electricity, innovations like blockchain could not yet be imagined. Innovations concern the absolute new beginning of a world that cannot be expected upfront but that is not detached from history but remain embedded in the temporal dimension of past, present and future. Seen from the perspective of the ontogenetic process of innovation, the newness of the innovation outcome consists in its break with the past (discontinuity) on the one hand, which remains embedded in the history it emerges from (continuity) in its future development on the other. In this respect, we can conceptualize the new involved in innovation in terms of the unexpected and un-known that cannot be calculated upfront, but is an eruption that is born in a world and that constitutes a world.

With this, we receive a further characteristic of the innovation process. The new that is at stake in the process of innovation is characterized by iterability (cf. Derrida 1982), that is, by the paradoxical simultaneity of sameness and otherness. Or framed in terms of the ontogenetic process: to the extent that innovations always remain embedded in the history they emerge from, the 'new' of innovation is always less than itself (pre-individual) and to the extent that the 'new' of innovation always involves a break with this past and is on its way to a possible future, it is always more than itself (post-individual). We see here that the ambiguous notion of newness is less problematic if we conceive it from the perspective of the process of innovation. The new as outcome of innovation is not only *not* a unique characteristic of this outcome but embedded in an iterative process of repetition and renewal. With this, the new is also not created *ex nihilo* or 'out of the blue' but determined by the temporal dimension of innovation it emerges from as ontogenetic process; the new is the individual outcome of the innovation process at the pre-individual level, that is, conditioned by (temporal) iterability.

The idea of the new as outcome of the innovation process at a pre-individual level seems to be at odds with two of the fundamental assumptions in contemporary thinking about innovation: (1) that innovation is not so much the outcome of a historical process, but the product of human *creativity*; (2) that *human* actors at an individual level are the subject of innovation. According to Schumpeter for instance, the innovator is a very special type of disruptive *person*. Also, the OECD and the EU see the human actor (businessmen, entrepreneur) as the primary subject of innovation.[3]

Seen from the perspective of the history of innovation, however, it is not self-evident that the human actor is the subject of innovation in this double sense of the word. In the ancient and medieval reflections on innovation, the human actor is not yet seen as the *subject* of innovation, nor necessarily as the *creator* of innovation (Godin 2015, p. 66). Seen from the Platonic perspective we discussed before, the human construction of a new *idea* should be rejected, first, because the transcendent world of the *ideai* is fixed and eternal and cannot be replaced by new ones according to Plato. Second, human being does not consist in the construction of new *ideai* (we can associate this with the *vita activa*), but should be enabled to grasp the eternal *ideai* to safeguard the political order in light of this *idea* (we can associate this with the *vita contemplativa*) (Blok 2020).

Also in case of Machiavelli for instance, the human actor is not yet seen as the subject of innovation. Not human being is the initiator of the innovation process, but time itself; time is seen as corrupting the established world order that calls for innovation as intervention to withstand this corruption (Blok 2020). A similar idea can be found in the work of Francis Bacon: "Surely every medicine is an innovation; and he that will not apply new remedies, must expect new evils; for time is the greatest innovator; and if time of course alter things to the worse, and wisdom and counsel shall not alter them to the better, what shall be the end?" (cited in Bontems 2014, p. 43). Time itself is seen here as innovator, namely destructing the established world order, that at the same time calls for innovations that withstand this corruption and construct and safeguard a newly established world order.

Seen from this perspective, we can conceptualize the destructive dimension of innovation as entropic aspect of the innovation process, while the creative dimension of innovation can be conceived as a *negentropic* aspect of the innovation process.[4] To be sure, it is not the case that time is entropic, and innovation is negentropic; the ontogenetic process of innovation is both characterized by an entropic aspect *and* a negentropic aspect at the pre-individual level, where the interplay of entropy and negentropy constitutes new individual innovation outcomes. The confrontation between the entropic and negentropic aspect of the innovation process makes clear that the temporal iterability of the innovation process cannot be understood in a chronological way. While the initiation of the innovation outcome could still be understood in a chronological way, namely as embedded in the history it emerges from and the futural state toward which it develops, the entropic and negentropic dimensions of the temporal iterability of the innovation process show that the initiation of the new (innovation outcome) is temporal but can no longer be understood in a chronological way.[5]

We continue to ask for the role of the human actor in the ontogenetic process of creating the 'new'. Although our findings regarding the innovation process indicate a decentralization of the (human) subject, this doesn't imply that the human actor has no role anymore in the innovation process. In ancient times already, innovation is associated with changes by humans, as opposed to changes by God (Godin 2015); the innovator is the one who intervenes in the established political order, who is a violator of boundaries and a *dissenter* (Godin 2015). And although empirical research on the question whether opportunities for new innovations are objective or not is still inconclusive (cf. Kirzner 1997), it is nowadays assumed that opportunities for innovation emerge in the structural interrelation between 'subjective perception' and 'objective realities' (Gregoire et al. 2010; cf. Ploum et al. 2017). Also in the sociological literature on innovation systems, the co-constructive role of the human actor is highlighted (Geels 2005). So although the human actor is not the primary subject of innovation, he is definitely *included* in the innovation process and has a role in it. How do we have to conceptualize the involvement of human being in the innovation process?

Let's turn to the history of innovation for a moment to receive an answer to this question. In the Bible for instance, we encounter innovation in terms of the *metamorphosis* of human existence that is needed according to Paulus, the renewal of the spirit in becoming Christian (Godin 2015). When Paulus for instance calls for a renewal of the spirit to become a true Christian, he doesn't call primarily for a change in our actual behaviour, but for a change of our identity as humans, namely, to become a true Christian. Also in other ancient sources, innovation is associated with 'renewing the soul' (Godin 2015, p. 217). A first sense of the human includedness in the innovation process emerges if we consider that the identity of human being can be the object of innovation.

A second sense of inclusion emerges if we consider such a metamorphosis to the Christian as ontogenesis of this innovation outcome. In this case, the idea of human being as Christian is the outcome of the innovation process. But this implies that human being as Christian is not the subject of innovation. To the extent that human being *as* Christian is an *individual* outcome of the innovation process, the role of human agency in the innovation process has to be conceived at a pre-individual level; the human actor that is involved in the metamorphosis to human being as Christian (individual innovation outcome) is himself not a Christian yet, but underway to a possible futural state in this ontogenetic process (pre-individual). To the extent that the metamorphosis to the Christian only succeeds with a certain human co-creation at the pre-individual level to be constituted at the individual level, human existence is involved in the ontogenetic process of innovation.

This human co-creation in the innovation process is not only at stake in case humans are the object of innovation, that is, in case of a renewal of the soul. In fact, human being is always already involved in the innovations that establish the world *order*. Why? The innovation of the steam engine changes not only the order of things in the world (the emergence of textile industry). To the extent that human being is always already intentionally involved in a meaningful world (Heidegger 2008), the innovation of the steam engine includes at the same time our human responsiveness to the world of steam, as user, adopter, operator, disseminator and so on of this innovation. The innovations that establish a new world order concern the human-world relation as a whole.

Seen from this perspective, human being is not only the co-creator in the ontogenetic process of innovation at the pre-individual ontic level, but also the outcome of this process, namely as the *adopter* and *disseminator* of this innovation within the newly established world order at an individual ontic level, for instance as early adopter (Rogers 1962).[6] This means that human agency occurs at two places in the ontogenetic process of innovation:

(1) Human being can be seen as involved in the ontogenetic process as pre-individual co-creator that contributes to the innovation outcome. In the context of this chapter, we can conceptualize the human co-*creativity* at the pre-individual level only in a negative way: a) human *co*-creativity has to be thought at the pre-individual ontic level, and not at the level of the individual input (human as subject) or the output (human as adopter) level of this process; b) human co-creativity at a pre-individual ontic level contributes to the ontogenetic process at pre-individual ontological level, and has to be understood therefore out of the ontogenesis of the new. This leaves open the question how the co-creativity of human being at the pre-individual level has to be conceived in a positive way.[7]

(2) Human being can be seen as involved in the outcome of the ontogenetic process as adopter of the innovation that establishes a new world order. The new world order is only established thanks to the human adoption and dissemination of this innovation. This contribution doesn't make him the subject of innovation either, because this contribution primarily consists in the *adoption* of the innovation.

This double role of human being in innovation opens a new perspective on the classical idea that innovation is the creation of something new *and* its adoption by the market. While creation as such may still be associated with art, innovation is at stake if the creative act of innovation is accompanied by its adoption. Here, we see these two aspects of innovation represented in the two roles of human being at the individual and the pre-individual level. The question remains how the two roles of human agency as co-creator in the innovation process and as adopter are related to each other.

*Table 20.2    Four characteristics of the phenomenon of innovation and its implications for the novelty, temporality and human involvement in innovation practices*

|  | Ontic level | Ontological level |
|---|---|---|
| Innovation Outcome | Individual innovation outcome at an ontic level: 'new' individual components and compositions of components in individual products or services, like the innovation of streaming services, *and* human being as adopter of these innovation outcomes. | Individual innovation outcome at an ontological level: a 'new' political-economic world order, like the one associated with digital networks, which accompanies the emergence of new products and services within this new world order, like streaming services. |
| Innovation Process | Ontogenetic process at pre-individual ontic level, in which innovation outcomes like steaming services emerge iteratively out of predecessors, in which the human co-creative capacity is involved. | Ontogenetic process at pre-individual ontological level, in which for example the digital world evolves out of the negentropic (creative) and entropic (destructive) dimension of temporal iterability, and in which the human co-creative capacity is involved. |

In Table 20.2, we summarize the findings of our philosophical reflections on the concept of innovation in this section, and its implications for our understanding of the object of innovation and the novelty involved, as well as the temporality and human involvement in innovation.

## CONCLUSION

In this chapter, we challenged the techno-economic paradigm of our understanding of innovation, building on earlier work (Blok 2020, 2021), in order to articulate a philosophy of innovation. We opened up the phenomenon of innovation by distinguishing between the innovation process and outcome dimension and between the ontic and ontological dimension of innovation.

We showed that established theories of innovation in economics and business administration often focus on the innovation outcome, which enables to distinguish between incremental and radical innovation, while they omit particular reflections on the nature of the innovation process itself. Contrary to theories of innovation that conceptualize the process of innovation based on its outcome, for instance the distinction between closed and open innovation, we articulated the ontogenetic process of innovation in this chapter. The first contribution of the philosophy of innovation is the alternative conceptualization of the innovation process as a pre-individual ontogenetic process. It helps alternative theories of innovation to criticize the superficial conceptualization of innovation processes in stage-gate and technology assessment models, and enables us to reflect on the innovation process of creation and destruction as distinct from the innovation outcome. We provided the example of the process of concretization at stake in innovation that cannot be attributed to the input or output of innovation.

We also showed that established theories of innovation often focus on the innovation outcome at an ontic level – for example the iPhone or the steam engine – while they omit to reflect on the world *order* involved – for example the world of steam – that is for instance destructed by the invention of the internal combustion engine and creates for instance the world of electricity. This is strange, as we have seen that economists like Schumpeter already indicated the role of economic waves in innovation, while other disciplines indicated the role of structural socio-technical changes. The second contribution of the philosophy of innovation consists in a philosophical perspective on the role of waves and changes in innovation, and in an alternative conceptualization of the innovation outcome as a combination of an outcome at the ontic and ontological level; the innovation of the steam engine requires the innovation of the world of steam, while it at the same time establishes this world of steam. This dual concept of the innovation outcome helps to criticize the often naïve and superficial conceptualization of the innovation outcome as incremental or radical innovations, and enables alternative theories of innovation to take the innovation outcome at the ontological level into account in further theoretical developments. With this, we also contribute to contemporary debates in philosophy of technology, which focus on either the ontic or ontological level of innovation and technology, while we highlighted the interdependency and interconnectedness of the ontic and ontological level of technology and innovation.

The third contribution of the philosophy of innovation is the provision of a criterion to distinguish between incremental and radical innovation. While this distinction in innovation theory is criticized because it seems to be conventional or at least arbitrary, we propose a clear criterion based on the philosophy of innovation: incremental innovations concern innovations at an ontic level only, while radical innovations concern innovations which involve both the ontic and ontological level.

We then sketched several implications of the philosophy of innovation for our understanding of the object of innovation and the novelty involved, as well as for the temporality and human involvement in innovation practices. If innovation concerns an ontogenetic process in which primarily a world is constituted – that is, the digital world – in which the invention of new artefacts – that is, an Artificial Intelligence application – is embedded, we should consider new developments from this perspective. We should primarily ask for the world order that may be created by these new types of innovation, and how these innovations challenge our current way of living together in the world. The advantage of our philosophy of innovation is that it enables future research to broaden its perspective from the ontic level of the innovation outcome of individual innovations to the ontological level of the political-economic world order that is created and destructed by these new types of innovation.

Although we have seen that the innovation outcome can be seen as something new, the newness involved in innovation is not so much associated with the innovation outcome at an ontic level, as it is here that the newness of new artefacts can always be questioned. On the contrary, it is primarily connected with the new world order that it establishes. This new world order is not the product of a *creation ex nihilo*, but remains embedded in the historical process in which it emerges as *repetition, renewal, reformation, revolution*; it overthrows the established world order and initiates a new beginning of a new world order. In this respect, the new involved in innovation practices concerns an absolute new beginning of a world that cannot be expected upfront (discontinuity), but is not detached from history and remains embedded in the temporal dimension of past, present and future (continuity). This ontogenetic process doesn't follow a chronological temporal order, but is constituted by a temporal iterative process. The fourth contribution of the philosophy of innovation is that it moves beyond the traditional opposition between linear and circular innovation models in innovation management, and highlights the iterability or simultaneity of sameness and otherness involved in innovation. It also moves beyond the traditional idea of human being as creative actor and primary subject of innovation and highlights the human *involvement* in the temporal iterability of this ontogenetic process of historical development itself, in which the established world order is overthrown (entropy) and a new beginning of a new world order is initiated (negentropy). Human creativity is involved in this ontogenetic process as co-creative capacity, namely human identity as an ontological outcome of innovation and at the same time human identity as pre-individual involvement in the process of innovation. The advantage of the philosophy of innovation is that it enables future research to broaden its perspective from the human actor as creator and subject of innovation to the world building capacity of human being, as it is conceptualized in the philosophical tradition (Arendt 1958).

The potential strengths of our philosophy of innovation consists in its ability to move beyond the techno-economic paradigm and to broaden our perspective on the phenomenon of innovation. Although much more dedicated research is needed to analyse its advantages and disadvantages in the context of contemporary developments like social, sustainable and responsible innovation, the philosophical reflections on the concept of innovation we developed in this chapter provide already clear contributions to the further development of our understanding of this challenging phenomenon. It also enables dedicated reflections on the ethics of innovation beyond the techno-economic paradigm, that concentrate on innovation as human ethos (Blok 2018).

## NOTES

1. In the philosophical tradition, *ontogenesis* refers to the process in which not a new being – e.g. a particular computer – emerges, but the being of this new being – e.g. the invention of the first computer, which involved the first computer both at an ontic and an ontological level – emerges.
2. Parts of this section are earlier published in Blok (2021).
3. This also explains the huge amounts of public investments in entrepreneurship education both in the developed and developing world, support of start-ups at technical universities, etc.
4. Negentropy is defined as the opposite of entropy, and means that the innovation process not only destructs the established world order (entropy), but also constructs a new established order (negentropy).
5. The further reflection on the temporal dimension of the ontogenetic process of innovation is beyond the scope of this chapter because the temporality of innovation is not a reinvention of the old (chronological time) and requires a new philosophy of history, in which no longer a dualism between an active human creator and a passive materiality of the Earth is at stake, but both are interdependent and interconnected.
6. This idea philosophically substantiates the findings of Gabriel Tarde, who associates innovation not so much with invention and more with its adoption (cf. Tarde 1903).
7. It is this type of reflections on the nature of human co-creation, that are completely omitted in scientific literature on user-led innovations for instance (cf. von Hippel 2005).

## REFERENCES

Arendt, Hannah (1958), *The Human Condition*, Chicago: Chicago University Press
Aristotle (1944), *Politics*, Cambridge: Harvard University Press
Blok, Vincent, Lemmens, P. (2015), 'The emerging concepts of responsible innovation. Three reasons why it is questionable and calls for a radical transformation of the concept of innovation', in B. Koops, I. Oosterlaken, J. van den Hoven, H. Romijn and T. Swierstra (eds), *Responsible Innovation 2: Concepts, Approaches, and Applications*, Dordrecht: Springer International Publishing, pp. 19–35.
Blok, Vincent (2018), 'Innovation as ethos: Moving beyond CSR and practical wisdom in innovation ethics', in C. Neesham and S. Segal (eds), *Handbook of Philosophy of Management*, Dordrecht: Springer, pp. 1–14.

Blok, Vincent (2020), 'Towards an ontology of innovation: On the new, the political-economic dimension and the intrinsic risks involved in innovation processes', in N. Doorn and D. Michelfelder (eds), *Routledge Handbook of Philosophy of Engineering* (forthcoming), London: Routledge, pp. 273–85.

Blok, Vincent (2021), 'What is innovation? Laying the ground for a philosophy of innovation', *Techné: Research in Philosophy and Technology*, 25 (1), 72–96.

Bontems, Vincent (2014), 'What does Innovation Stand For? Review of a Watchword in Research Policies', *Journal of Innovation Economics and Management*, 3, 39–57.

Chesbrough, Henry (2003), *Open Innovation: The New Imperative for Creating and Profiting from Technology*, Boston: Harvard Business School Press.

Chesbrough, Henry (2006), *Open Innovation: Researching a New Paradigm*, Oxford: Oxford University Press.

Christensen, Clayton (1997), *The Innovator's Dilemma: When New Technologies Cause Great Firms to Fail*, Cambridge: Harvard Business Review Press.

Cooper, R.G. (2008), 'Perspective: The Stage Gate Idea to Launch Process – Update: What's New and Nexgen Systems', *Journal of Product Innovation Management*, 25 (3), 213–32.

Derrida, Jacques (1982), *Margins of Philosophy*, Chicago: University of Chicago Press.

Freeman, Christopher and Luc Soete (1997), *The Economics of Industrial Innovation*, London: Continuum.

Geels, F. (2005), *Technological Transitions and System Innovations: A Co-evolutionary and Socio-technical Analysis*, Cheltenham, UK and Northampton, MA, USA: Edward Elgar Publishing.

Godin, Benoît (2008), Innovation: the history of a category (working paper no. 1).

Godin, Benoît (2015), *Innovation Contested. The Idea of Innovation over the Centuries*, New York: Routledge.

Godin, Benoît (2019), *The Invention of Technological Innovation. Languages, Discourses and Ideology in Historical Perspective*, Cheltenham, UK and Northampton, MA, USA: Edward Elgar Publishing.

Gregoire, D.A., P.S. Barr and D.A. Shepherd (2010), 'Cognitive Processes of Opportunity Recognition: The Role of Structural Alignment', *Organization Science*, 21 (2), 413–31.

Heidegger, Martin (2008), *Being and Time*, New York: Harper Collins Publishers.

Henderson, R.M. and K.B. Clark (1990), 'Architectural Innovation: The Reconfiguration of Existing Product Technologies and the Failure of Established Firms', *Administrative Science Quarterly*, 35, 9–30.

Ihde, D. (2010), *Heidegger's Technologies. Postphenomenological Perspectives*, New York: Fordham.

Kirzner, I.M. (1997), 'Entrepreneurial Discovery and the Competitive Market Process: An Austrian Approach', *Journal of Economic Literature*, 35 (1), 60–85.

Plato (1967), *Plato in Twelve Volumes*, Cambridge: Harvard University Press.

Ploum, L., Vincent Blok, T. Lans, O. Omta (2017), 'Exploring the Relation between Individual Moral Antecedents and Entrepreneurial Opportunity Recognition for Sustainable Development', *Journal of Cleaner Production*, doi: 10.1016/j.jclepro.2017.10.296 (in press).

Rogers, Everett M. (1962), *The Diffusion of Innovations*, New York: Free Press.

Rogers, Everett M. (1976), 'Where are we in the understanding of diffusion of innovations?', in W. Schramm and D. Lerner (eds), *Communication and Change: The Last Ten Years – and the Next*, Honolulu: University Press of Hawaii, pp. 204–22.

Schumpeter, Joseph (1943), *Capitalism, Socialism and Democracy*, London: Routledge.

Simondon, George (2017), *On the Mode of Existence of Technical Objects*, Minneapolis: Univocal Publishing.

Tarde, Gabriel (1903), *Laws of Imitation*, New York: Hold and Company.

Verbeek, P.P. (2005), *What Things Do. Philosophical Reflections on Technology, Agency and Design*, Pennsylvania: Pennsylvania State University Press.

von Hippel, Eric (2005), *Democratizing Innovation*, London: MIT Press

von Schomberg, Lucien and Vincent Blok (2018), 'The Turbulent Age of Innovation. Questioning the Nature of Innovation in Responsible Research & Innovation', *Synthese* (in press).

# PART VII

# THEORIZING THE THEORIES

# 21. Ideology, engine or regime. Styles of critique and theories of innovation

*Brice Laurent*

## INTRODUCTION

The pervasiveness of innovation in the contemporary public discourse manifests itself in the omnipresence of figures like the "start-up creator", the "entrepreneur", or the "investor" in current economic and political debates. But the inevitability of innovation is more and more accompanied by a distinctive atmosphere of skepticism regarding its benefits, and the world-view it encompasses. As the influence of the "tech sector" grows larger, global companies such as Google, Facebook or Amazon are both heralded as seemingly never-ended sources of innovation, and regularly criticized for the ever-growing extension of their power, their disregard of public concerns such as data privacy and the social environment they promote on their grounds. The start-up culture at the heart of current innovation discourses is targeted by journalists, professionals and academics worried about the "collateral damage", as a *Guardian* journalist put it, of disruptive innovation.[1] Eventually, the technological developments on which so many contemporary innovations rely, in fields as diverse as artificial intelligence, renewable energy, or urban mobility, are themselves criticized for their ecological costs and their potential negative consequences in terms of labor structure or social justice.

Scholars, activists and politicians have proposed critical evaluations of the current state of innovation. These critical movements have become integral to the current debates about the objectives and practices of innovation, to the point that they can be said to contribute to the dynamics of innovation itself. "New business models", "inclusive innovation practices" or "new modes of thinking" are presented as answers to the perceived shortcomings of the current innovation environment. Like capitalism, innovation is prone to integrate its own critique to re-invent itself and further extend the perimeter of its actions (Boltanski and Chiapello 2005).

This chapter focuses on the relationships between innovation and critique. A way of ana-lyzing these relationships would be to settle on a specific definition of "innovation", and then examine how it is critiqued and by whom. Another one would be to examine several groups of critics, from anti-technology neo-Luddite movements (Jones 2006) or anti-globalization organizations, to theorists of alternative approaches to innovation (see Godin and Vinck 2017; and Chapters 6 to 14 in this book), and discuss how each of them frames the issues that inno-vation raises. In this chapter, I adopt a perspective that seeks to bring these two approaches together. I contend that producing a critique of innovation is always based on an understanding of what innovation is, and how it should be studied. In other terms, there is no critique of inno-vation without a theory of innovation, be it explicit or not. Accordingly, I want to understand how different styles of critique and different theories of innovation are formulated at the same time. Thus, I start not by defining what innovation is and what it is not, what is "mainstream" or "dominant" and what is "alternative", but by identifying styles of critical approaches to innovation and discussing the (more or less implicit) theories of innovation that these styles of critique propose. An important point to note in these explorations is that this chapter does not restrict who the critics of innovation are. Some of them are scholars, writing in academic or non-academic publications. Others are practitioners of innovation. Yet others are activists and members of social movements. This means that one should clearly identify the identity of the critics, and not take the standpoint of academic critics for granted.

In the following, I start by examining a contemporary style of critique that adopts the religious vocabulary, as it targets the "myths" or the "religion" of innovation. This style of critique sees innovation as an ideology, and contrasts this ideology with facts. I then turn to another style of critique associated with another theory of innovation, which originates from the works of Michel Callon. This approach sees innovation as an engine for critique, and situates the critical position in the midst of innovation processes. The last section examines a third style of critique, which grounds the critique of innovation from an analytical distance from which innovation can be theorized as a situated regime.

## THE CRITIQUE OF A RELIGION OF INNOVATION

### Contesting Myths

A frequent style of critique uses a religious vocabulary to characterize innovation. It is present in the academic literature, but also, and perhaps mostly, in the specialized press and the overall public sphere. Writing in the *New Yorker* in 2014, historian Jill Lepore spoke of the "gospel of innovation" to tell the story of "disruptive innovation" and how it quickly spread in spite of flimsy empirical evidence (Lepore 2014). In 2017, political scientist and STS scholar Langdon Winner wrote on his website about "the cult of innovation, its colorful myths and rituals", and claimed that innovation was "today's central 'god term'", which "has begun to resemble a cult with ecstatic expectations, unquestioning loyalty, rites of veneration, and widely echoing exhortations of groupthink". These academics are joined by a growing number of journalists and public intellectuals who also describe innovation in religious terms. Some speak of "the orthodoxy of unorthodoxy" to describe the unquestioned belief that innovation is the ends and means of all human endeavors (Leary 2018). Already in 1999, the *Economist* published an article entitled "Industry gets Religion", in which "the rhetoric of innovation" was described as "a new theology", and the *MIT Technology Review* talked about the "religion of innovation" in 2013.[2] The history of innovation shows that "innovation" was initially a term that designated a risky, if not utterly dangerous tendency to question the established order of things in the context of institutional religions (Godin 2015, 2019). The current critique of innovation as religion displaces this original trope, as it describes innovation as a new obligatory path for salvation, whether individual or collective. "Salvation", here, has both economic and social undertones, as innovation is expected to both be an engine for economic growth and a source of workable answers to pressing social problems.

The vocabulary of religion can be seen as a convenient figure of speech, which provides telling metaphors and powerful illustrations of what critics of innovation aim to convey. That these metaphors flourish in magazines and blogs is a sign of the appeal of this register and its easy use in public debates. But the vocabulary of religion is also the characteristic of a style of critique that is associated with a particular understanding of what "innovation" is, and why it deserves critical attention. "Innovation", here, designates a mode of action that is based on descriptive and analytical elements, which serve as the basis for prescriptions. "Critique", then, is about the demonstration that the descriptive and analytical elements are mistaken. It is an epistemic critique, that of an "innovation-speak"[3] that functions as a dominant public discourse aiming to provide diagnoses and corresponding solutions. The use of the religious vocabulary by innovation critics is then associated with an approach seeking to "debunk" what is presented as a fact, yet is nothing more than poorly ground belief. As modernist critics of religion sought to confront religious dogma with the reality of historical facts (Poulat 1962), these critics target innovation as an ideology, and contrast it with a discourse meant to be based on facts.

When Jill Lepore speaks about the "gospel of disruption", she targets management scholar Clayton Christiansen and his theory of "disruptive innovation", which is based on empirical examples, and leads into prescriptions for companies to use. Lepore shows that the empirical elements that Christiansen used can easily be contested, which weakens the whole theory he has been defending, and turns it into nothing more than an ideology. The critique contrasting ideology with facts is also visible each time the language of the "myths" of innovation is used, in which case the target is not an individual author or a definite theory (as Christiansen's disruptive innovation), but tacit hypothesis of current innovation discourses. The references to these "myths" are numerous in the academic and non-academic[4] literature. For instance, Mariana Mazzucato argues in *The entrepreneurial state: Debunking public v. private sector myths* that the role of the state is crucial in economic dynamic, whereas the discourse of innovation often makes free markets and the unconstrained interventions of entrepreneurs the sole source of innovation (Mazzucato 2015). She can then argue that if there is a US model and if it is to be replicated, then all its components ought to be included in it. In the online piece I quoted above, Langdon Winner identifies a persistence of a certain technological determinism, which has been contested in a rich body of scholarly work. Pierre-Benoît Joly (2017) discusses the pervasiveness of the linear model, which describes innovation processes in successive phases, in spite of numerous theoretical works and innovation projects that have challenged this model.

## Challenging the Ideology of Innovation

The language of the innovation "myths" points to a critical position that consists in uncovering a discourse by challenging the truth-value of the claims it is implicitly or explicitly based upon. This type of critique can be identified beyond the use of the religious vocabulary, each time scholars or practitioners characterize an overly simple, yet dominant version of innovation that ignores more sophisticated analysis and practices. In a recent volume, Benoît Godin and Dominique Vinck describe an "ideology" that sees innovation as "good a priori" and equates it with a simple vision of radical technological change for market purposes (Godin and Vinck 2017). This simple version of innovation sees it as the articulation between a dynamic of technological change connected to market extension, performed by particularly bright entrepreneurs, for outcomes that are seen as unquestionably positive. This simple version can then be the target of a critique that contests the robustness of the narrative it proposes.

One can track back the challenges to this simple version of innovation to earlier scholarly discussions about the selective gaze that innovation (usually seen as the mechanical effect of radical technological change) pre-supposes. Consider for instance historian David Edgerton's critique of the scholarly focus on innovation. Edgerton speaks of the ways in which "technological futurism has affected our historiography", by which he means that historians and scholars of technology tend to "reproduce the innovation orientation, both in the choice of innovation as subject matter, and in confusing the innovation with the technology-in-use" (Edgerton 1999, p. 128). For Edgerton, innovation implies an empirical and theoretical focus that is simply too selective, as it ignores the reality of technology as it is designed, practiced, and maintained. In the wake of Edgerton's piece, maintenance studies have insisted that the mundane functioning of "technologies in use" is as interesting (if not more) than innovation for anyone wishing to understand technology and its impacts in human lives. This body of work shows that the actual work of scientists and engineers has often little to do with the tenets of innovation-speak (Russel and Vinsel 2019). This set of works is particularly interesting for our reflection on the nature of critique, because it connects an objective of debunking the ideology of innovation using the truth of (social) scientific facts with the problem of exclusion, of individuals and public concerns. The occultation of maintenance has indeed consequences, in terms of public policy priorities, industrial strategy and allocation of costs and benefits in the functioning of technical systems, including at the level of workers' rights.

Thus, the case of maintenance is useful to illustrate a variation on the critique of the ideology of innovation, which contests not only the epistemic quality of the hypothesis on which it is based, but also their consequences in terms of who is included and who is excluded, what is taken into account and what is not. One can connect this perspective to a range of studies that show that who can act as innovators and be recognized as one is deeply skewed in favor of certain people, often at the expense of women and people of color (Cook 2019). Exclusion effects also manifest themselves when considering the "unintended consequences" of innovation, in domains as diverse as digital technologies (Matsumoto and Kawajiri 2012) or the financial sector (Sveiby et al. 2012; see also Sveiby 2017), which are obscured when the discourse of innovation unproblematically hypothesizes that "innovation is always good" (Soete 2013; Sveiby et al. 2012).

## The Religion of Innovation and its Reforms

Phrasing the critique of innovation as a challenge of a quasi-religious discourse is often accompanied by calls for reform. When the *MIT Technology Review* targeted the "religion of innovation", it also had "enough with innovation for innovation's sake" and advised companies to "double down investment in what already works, and take the time to carefully consider new releases".[5] Godin and Vinck complement their critique of the ideology of innovation with a plea for considering "processes like adaptation, withdrawal, learning from failure, alteration of the innovation and unintended consequences (…) as forms of innovation" (Godin and Vinck 2017, p. 3). The critique of the ideology of innovation often fuels, as much as it is fueled by more sophisticated accounts of innovation.

Thus, the reflection about exclusion effects in a version of innovation seen as simultaneously dominant and over-simplifying leads us to consider the possibility of considering alternatives, or, in Joly's terms, of "re-imagining innovation" (Joly 2019). This is precisely what Godin and Vinck ask for as they call for a new research program able to re-open what innovation is about, by including studies of processes that slow down change rather than accelerate it, practices of maintenance or repair, and unintended or unforeseen effects of innovation.

Similarly, practitioners of innovation and innovation scholars often associate the critique of innovation-as-ideology with propositions for reform. They engage in the evaluation of the truth-claims of innovation models and theories, while seeking to refine them. Thus, works in innovation studies have discussed approaches to innovation that are deemed "alternatives", and which the chapters of this volume provide numerous illustration of. After von Hippel's seminal book, numerous works on innovation have discussed the attempts at "democratizing" innovation by opening it up to wider groups of people (von Hippel 2004; Stirling 2008, 2014; see: Hyysalo et al. 2016; Oudshoorn and Pinch 2003). The recent interest for "responsible innovation", which accompanies evolutions in science policy circles, mostly in Europe, can also be interpreted as an alternative approach to innovation, which seeks to integrate into the innovation process the anticipations about potential future effects and possibly affected social groups (Stilgoe et al. 2013). The promoters of these approaches often see them as answers to the shortcomings of mainstream innovation theories and practices. They can be situated within an academic field, innovation studies, which is made of a core dominated by economists, and peripheral approaches that do not always coalesce (Fagerberg and Verspagen 2009; Joly 2017, 2019; Martin 2012), and which often see potential paths for reinvention (Martin 2016). These alternatives have consequences in the policy world, as public institutions embrace calls for "open", "democratic" or "responsible" innovation (see for European examples: Ernst & Young and CEPS 2011).

**Reforms and Critique**

These considerations bring us back to the notion of critique. While the critique of the "religion" or the "myths" targets an overly simple discourse that misrepresents the reality of innovation processes and does not question the means and ends of innovation, the numerous approaches presented as alternatives show that this critique can stem from the actors of innovation and/ or can be integrated in their theories or practices. This is an impetus for further critical works that adopt a similar style as that of the "myths" of innovation, as their purpose is to uncover the extent to which the self-proclaimed alternatives question the basic tenets of the dominant version of innovation. For instance, Joly identifies the persistence of the linear model behind alternative models of innovation (Joly 2017), Edquist laments that "linearity still prevails" (Edquist 2014), and critics of the European program of "responsible research and innova-tion" question the eventual effects on the structural dominance of competitiveness objectives (de Saille 2015). These accounts can be read as further attempts at displaying a kind of "bluff" (Ellul 1988), whereby innovation would only pretend to be reformed. But these works can also lead to further reforms of innovation.

One could then try to differentiate between authors according to the degree to which they are "reformists", possibly in relation to their proximity with the actors of innovation in public bodies or private companies. One could possibly isolate "critics" from "reformers", the former being more interested in challenging the current state of affairs and the latter more concerned with integrating alternative approaches from within. Such an attempt at sorting out reformers and critics is the approach undertaken in a recent collective volume that discusses the various processes through which innovators are actively made (Wisnioski et al. 2019). While there is a clear value in this distinction, it is also striking that the boundary that separates those groups is not always clear. Being a critic of innovation by pointing to its exclusion effects on people leads to calls to integrate these people as potential innovators (Cook 2019). In turn, self-proclaimed "reformists" seeking to include more women and people of color in inno-vation are also critics of current innovation discourses and practices (Sanders and Ashcraft

2019). Pointing to the need to making maintainers might lead others to rethink what it means to be an innovator, for example by introducing an "ethics of care" in the formal training of scientists and engineers (Russel and Vinsel 2019). The porosity of the line between critique and reform is at the heart of the style of critique of innovation that questions the ideology of innovation. One conclusion this can lead to is that this critique is meant to be continuously exercised: as proponents of innovation might be convinced that the simple version relies on myths, one should then question the calls for "democratization", "openness" or "responsibility" and confront them with the reality of innovation practices. This is of course a continuation of a critique meant as an exercise in truth telling. But it also means that one delves ever further into innovation processes, and ultimately participates in refining them.

At this point, one can reflect on the consequences to the problem of critique. If there is a tendency of innovation theory and practice to absorb its own critique (maybe as capitalism itself, see Boltanski and Chiapello 2005), then one can situate the critical activity at the core of innovation processes themselves. This would require elaborating another theory of innovation, as I will discuss in section 2 of this chapter. But one can also envision grounding critique on another type of analytical distance from innovation, then seen as a vehicle for wider transformations. I will discuss this latter approach in section 3 of this chapter.

## CAN INNOVATION BE AN ENGINE FOR CRITIQUE?

### Innovation as a Productive Source of Exclusion

The critique of innovation that I discussed in the previous section constitutes an object of inquiry, "innovation", and evaluates it, possibly before proposing to reform it. But innovation, whether described or practiced, be it "dominant" or "alternative", always has a critical ambition, if only because it is connected to an idea of novelty expected to have an impact, however limited, in the real world. Historical works show that the early use of the term "innovation" pointed to the negative effect of an operation contesting the stability of the collective order (Godin 2015, 2019). This is at the core of Schumpeter's perspective on innovation. Schumpeter makes innovation a central component, if not the main engine, of capitalism. The famous notion of "creative destruction" that he introduced has entered the common vocabulary of innovation scholars, often in a way that directly echoes the idea of "disruption", as a dynamic that reshapes existing orders of things, with unavoidable and potentially negative side effects. But the critical part of creative destruction in Schumpeter's approach to innovation and capitalism is not limited to the regrettable side effects in terms of bankrupting established companies or bringing workers out of their jobs. For Schumpeter, creative destruction is at the core of capitalism, but also plants the seeds for the destruction of capitalism. This is connected to the very success of capitalism, and not, as in Marx's works, because of capitalism inherent contradictions and inevitable instabilities in terms of class relationships. For Schumpeter, innovation requires a permanent flow of radical ideas, which the logic of capitalist growth threatens, as innovation risks "being reduced to routine" (Schumpeter 2003 [1943], p. 132).

Schumpeter's theory of innovation is an invitation to explore at greater length the kind of critique that can be associated with a dynamic theory of innovation. Here, "critical" refers to the ability of innovation to contest existing orders of things. How to theorize this critical ambition? Is it possible to associate it with a theory of innovation, in that it would provide analytical tool to explore the dynamic of innovation and its social consequences? Exploring these questions will force us to discuss what "critique" is about, asking questions such as: what objectives and practices of critical activity? Who conducts it? For the benefit of whom? To discuss these questions, I turn to a theory of innovation originating from Actor-Network Theory (ANT), and particularly the works of Michel Callon. This stream of work has barely been associated to the idea of critique, yet, I will argue, is as much a critical theory as it is a theory of innovation.

Michel Callon is primarily known for his contribution to Science and Technology Studies (STS) in general and ANT in particular. Callon has written extensively about public controversies, democratic practices and lay knowledge, and the making of markets. These threads of analytical works ought to be seen in conjunction with each other. What brings them together is a theory of innovation that focuses on the joint making of technological, economic and social change. In Callon's work, innovation is a dynamic that produces technical and social realities, that re-arranges connections between humans and non-humans, and re-defines them in the process. Callon's theory of innovation is connected to a reflection on markets – not as a ready-made sphere of social activity, but as entities that need to be actively constructed. The dynamic of market construction in Callon's accounts is directly connected to inclusion and exclusion mechanisms, which encompass what economists call "internalization" and "externalization". Markets rely on "framing" processes, which result in the construction of boundaries: between what is integrated in the calculation of costs and prices by buyers and sellers, between who is involved in market exchanges and who is not. But framing is not the end of the story of market evolution. Framing necessarily results in exclusion mechanisms, as potential consequences are not taken into account in the calculation of costs, certain actors are not included in the economic reasoning. For instance, workers hoping to benefit from social welfare will be excluded when markets start to "favor flexibility and mobility" (Callon 2007, p. 156). In Callon's approach, these exclusion effects are the consequences and sources of innovation.

This idea of exclusion is only rarely associated with Callon's works. Yet it is directly present, including in his works that have been read as a plea for deliberative democracy. *Acting in an uncertain world*, which he co-wrote with Pierre Lascoumes and Yannick Barthe (Callon et al. 2009), has been commented in those terms (Pestre 2008; Fuller 2010). Yet all the examples that are analyzed in this book relate to cases of suffering and violence, be they those of patients suffering from rare diseases not taken into account in drug development in pharmaceutical companies, or those of social movements mounting violent opposition against nuclear waste retreatment projects that threaten to forever change the nature of the territory where they live (see also Barthe 2006; Callon and Rabeharisoa 2008). These exclusion effects are important in Callon's theory of innovation, because they are productive. Callon speaks of the "proliferation of the social" in relation to market development, to point to the emergence of social groups engaged in scientific and social inquiries because of the exclusion effects they suffer from. The terms "emergent concerned groups" designate the collectives engaged in these inquiries. Callon describes them in those terms:

> Faced with such difficulties and uncertainties, emergent concerned groups become engaged in investigations and inquiries that sometimes lead them to invest in full-blown research and innovation. They then contribute to the constitution and organization of research collectives, mobilizing not only members of the group but also a wide range of professionals, including researchers and experts. (Callon 2007, p. 146)

Thus, concerned groups are active producers of knowledge. They are innovators themselves in that they propose new arrangements for the production of knowledge and the organization of markets and the political. For instance, patient groups intervene in the production of knowledge and become political actors in doing so (Epstein 1996; Callon and Rabeharisoa 2008). Opponents to nuclear waste disposal projects have forced to re-problematize the issue of nuclear waste in the terms of the reversibility of technical choices (Barthe 2006).

### From Concerns to Critique

The notion of "concern" in the Callonian theory of innovation stems from political theory as much as the sociology of market. It has roots in the American pragmatist tradition, particularly Dewey's theory of the public as exposed in *The public and its problems* (Dewey 2012; see Marres 2007). An important element in this theory of the public is that contrary to common interpretations of Dewey's approach to democratic life, consensus and dialogue are not primary components of what constitutes publics. Instead, publics are affected (and become "concerned groups", in Callon's language) when they suffer. This democratic theory is in that sense a theory of innovation, but also a theory that has a certain degree of violence at its core. This is from these hardly felt concerns that the production of knowledge by concerned groups can occur, and that these concerned groups can organize themselves. In other words, what affects publics is the source of both technical and social production. Here, exclusion is also a positive mechanism in so far as it serves as an engine for new social and technical identities to emerge.

Callon's theory of innovation and its connection with democratic participation have not been unanimously well received. Some have identified a tension between the expression of concerns, often in violent forms, and the expertise in the procedural organization of hybrid fora that some of Callon's works also suggest (Callon et al. 2009), easily integrated in economic projects and oblivious to structural power relationships (Pestre 2008; Fuller 2010). This invites us to discuss the ways in which this theory is a critique, and what it is a critique of. First, it is as much a theory of democratization as it is a theory of innovation. Here, innovation is a destabilization of existing political and economic order, and the outcome of the works of concerned groups. A challenge for innovation and democratization is then to help concerned groups to emerge and voice their concerns. As such, innovation is an engine in "the art of not to be governed that way", to re-use Foucault's famous characterization of critique (Foucault 1997, p. 45). Second, this theory of innovation is about the making of new political subjects. In that regard, it directly echoes one of the central components of the critical approach as proposed by Michel Foucault and Judith Butler, for whom critique ought to be tied with interrogating oneself as a subject (Butler 2002). The notion of "problematization" is used by these authors in that context. It points to an activity that questions the existing state of affairs and simultaneously questions the identity of the subject in the critical activity (Foucault 1984; see: Laurent 2017; Lemke 2011). It also suggests that the role of the analyst is to provide resources (if only analytical) to the concerned groups he or she studies, and thereby to participate in the critical activity in which they are engaged (see: Callon 1999 about his own experience with a patient group organization). Third, the critical theory that derives from Callon's theory of innovation is also related to a collective dimension. "Matters of concern" are politically significant when they are tied to the production of social groups, itself connected to knowledge production (Callon 2007, p. 158; see Latour 2004). This ought to be read as in connection to an understanding of critique that links the production of self and the production of collectives, that ensures "the transition from a fragmentary condition to a collective condition" (Boltanski 2011, p. 42). If we follow Callon's theory of innovation, this task is never ending, as the dynamic of innovation constantly produces overflowing.

A last (and crucial) element of the style of critique that this theory of innovation proposes is that it is *positive*, in that it is connected to the new realities that are spurred by the dynamics of innovation. These new realities are not only the outcomes of the concerned groups engaged in inquiries, exploration, and lay research activities, but also that of the analyst herself, as she studies these concerned groups and contributes to make their contribution visible. Thus, this is a critique that "multiplies the signs of existence", in Foucault's powerful words (Foucault 1994, pp. 106–7, my translation), alongside what the actors themselves are engaged in.

## A Dynamic Process of Critique and Integration

Callon's theory of innovation directs the analytical attention towards the identity of those in charge of innovation. Concerned groups suffer from what is being left out by market dynamics, or from the consequences of what is being done in existing orders of things, and then become innovators themselves in that they re-invent research practices, make new realities emerge, and eventually re-define collective priorities. This process means that "innovation" is both an engine for producing inequalities and suffering, and the activities in which concerned groups engage. In our discussion of Callon's theory of innovation, innovation has appeared as a continuous process, almost synonymous with market development (and it is in that regard that the proximity with Schumpeter is manifest). Here, "innovation" does not point to the same reality as the entity targeted by the critics of the "myth" that we discussed in the first section of this chapter. For the latter group, innovation is a political and economic program, which is contestable because of its weak intellectual foundations and its social consequences. But the two perspectives should not be seen as two distinct intellectual spheres with no contact with each other. One can theorize the critique of the "myth of innovation" in the terms of framing, overflowing, and concerned groups. The questions to ask are then: who are the groups concerned by the dominant versions of innovation? Who suffers from them? What are the counter-propositions that these concerned groups engage in? What can the analyst do to account for them, and participate in their realization? The case of open source provides an illustration at this point, as it can be as a reaction against a model of innovation that is based on property rights and a strict definition of ownership, or, in other words, has pressing framing effects regarding who can and who cannot engage in software development, and earn money out of it. Christopher Kelty's account of the free software movement can be read as the story of a concerned group affected by the closure of software by proprietary rights, and regularly questioning its own means of association in the same time as the content of what it is mobilized about (Kelty 2008).

This dynamic is not limited to the world of software. Among other sectors, agriculture has been a domain where open source has received a growing interest (Chance and Meyer 2017). Another illustration is the somewhat loose domain of the "smart city", where the reference to innovation has been explicitly integrated in urban policies and private companies' strategies in ways that can be analyzed in the terms of a dynamic of framing and overflowing. The initial smart city discourse made technological development the central element of the transformation of the city. Experiments such as Masdar in the United Arab Emirates or Songdo in South

Korea appeared as illustrations of a trend that consisted of creating technologically-optimized cities, usually at the margins of existing metropolis (Evans et al. 2016; Halpern et al. 2013). The problematic relations between a technology-centered city and citizens expected to live in it, the ever-increasing role of private companies, and the soulless nature of projects entirely directed towards technical and economic purposes did not go unnoticed, in the academic world as well as for practitioners. Thus, the initial version of the smart city has spurred counter-reactions, including from the cities themselves. A telling example is that of Medellin, where innovation policies have been meant to be inclusive, as an explicit reaction against a global smart city discourse seen as a threat for the social cohesion of the city (Talvard 2018).

In the previous section, I discussed how the critique of the ideology of innovation led to reforms and spurred a continuation of critical activities related to the actuality of these reforms. The Callonian style of critique implies that the dynamics of framing and overflowing is permanent, yet in this case the critical position is not exterior to innovation but integrated in it. In the Callonian theory of innovation, the critical activities lie in the concerned groups and their interventions. It supposes that concerned groups become themselves innovators, that they participate in yet other processes of framing bound to produce new overflows. It is a positive critique that cannot be separated from the dynamic of innovation itself, and from what the affected actors do about it. It implies forms of intervention on the part of the social scientist that can be described as "interferences" (Law 2010), or "modulation" (Kelty 2008). Is it then possible to envision a critique of innovation that would be extracted from the dynamic of innovation? The next section discusses this point.

# A CRITIQUE OF INNOVATION REGIMES

## Situating a Regime of Intensive Innovation

One can situate the configuration within which the dynamics of framing and overflowing takes place. Callon himself situates the "proliferation of the social" associated with the dynamic of framing and overflowing with a particular regime, that of "intensive innovation". "What is new", he writes, "is not the fact that markets overflow and therefore produce matters of concern; it is the amplitude and frequency of those overflowings" (Callon 2007, p. 151), which results in "a regime favorable to all kinds of overflowing" (Callon 2007, p. 152). The reasons for this evolution are not presented in systematic ways in Callon's writing, but he signals the "singularization" of goods and services, made even more manifest by digital technologies (Callon 2012, 2017). Callon is also clear that this "intensive innovation regime" is only one way of problematizing the relationships between market development and social ordering. One can indeed contrast the singularization regime, characterized by a permanent and intense dynamic of framing, overflowing, and the emergence of concerned groups, with other articulations of innovation processes with political and economic ordering.

An example of such alternative articulations is provided by Callon and Vololona Rabeharisoa, as they comment on the case of a man named Gino, who refused to adopt the expected role of the active member of the concerned group (Callon and Rabeharisoa 2004). In spite of suffering from a genetic disease, he did not engage in collective action as the rest of his family. Callon and Rabeharisoa see in Gino not a reluctant member of a concerned group that would conduct its investigations even when including its most passive members, but as the conveyor of another proposition for political subjectivity. For Gino, what matters is not the re-fabrication of the self in relation to public concerns, and in connection with an active engagement in additional explorations to fight against it. Gino does not engage in the production of knowledge about the disease, nor does he seek to act on it. Rather, he considers that he needs to live with a disease that is part of who he is as a human subject. This suggests that the dynamics of innovation and critique that I described in the previous section mostly through Callon's works should itself be situated, within a regime of "intensive innovation" that might be contrasted with others.

## Innovation Regimes

I call "innovation regime"[6] the stabilized apparatus that associates discourses, practices, regulations, market organizations and social movements in defining the practices and expected objectives of innovation. Innovation regimes are political, economic and social configurations, and the study of these configurations can then provide resources for critique, yet in a different guise than the evaluation of their truth-value (see section 1), or the integration in a permanent dynamic of framing and overflowing (see section 2).

Innovation regimes can be associated with an understanding of innovation as "part of a collectively held imaginary of sociotechnical progress that accompanies a complementary diagnosis of a deficiency in the receiving environment" (Pfotenhauer and Jasanoff 2017, p. 786). In other words, an innovation regime functions within a particular way of defining collective priorities and answering them. It is connected to the organization of public institutions, be they national or international, as they diagnose technological, social or economic deficits that need to be filled thanks to innovation (Pfotenhauer et al. 2019). Understanding the components of innovation regimes also implies that we consider how markets are expected to perform and what their desirable contributions to the common good are. This means that there are close connections between regimes of innovation and the institutions that stabilize them, often through policy instruments (Lascoumes and Le Galès 2007) or market devices (Callon et al. 2007). For instance, the construction, development and extension of instruments such as the linear model of innovation or national innovation systems are tied to international organizations, mostly the OECD (Godin 2006, 2009). A regime of innovation based on "dialogue" and instrumented by various participatory mechanisms emerged in the United Kingdom as a component of Tony Blair's "third way", which radically re-worked the role of the state and the understanding of society (Thorpe 2010). The regime of "responsible innovation" in Europe is part of how the European institutions attempt to re-define the terms of their legitimacy by connecting scientific objectives with the perceived priorities of European publics (de Saille 2015; Laurent 2017). In all these examples, regimes of innovation are situated within national or international institutions. Policy programs are crafted in these institutions, and public priorities are defined, and this results in particular understandings of what innovation is, how it should be conducted and for what purpose. In turn, the formulation of innovation also participates in stabilizing the principles according to which public institutions determine what is legitimate and what is not, what is consistent with the common good and what is not.

The analysis of innovation regimes is not only descriptive, but also tied to a style of critique. Critique can originate from adopting the standpoint of a certain regime to displace another one – pretty much as Gino's passive resistance underlines the fact that being a member of a concerned group can be contrasted with other ways of enacting political subjectivity. Critique can also focus on how devices and instruments that are parts of certain regimes of innovation circulate elsewhere. Thus, authors analyze the circulations of formalized innovation policy instruments outside the arenas where they are produced, and explore the consequences of this circulation. For instance, the National Innovation System (NIS) approach traveled to South America with problematic consequences (Delvenne and Thoreau 2012), and the "MIT model" has given rise to significant variations in non-Western political cultures (Pfotenhauer 2019; Pfotenhauer and Jasanoff 2017). This type of analysis may demonstrate the ways in which certain versions of innovation get sidelined when discussed in unfavorable institutional set-tings. For instance, the World Trade Organization (WTO) is an arena where oppositions about which regulations are scientifically grounded and which are undue trade barriers often reveal contrasted understandings of the relationships between science and society. Detailed studies of the internal functioning of the WTO show that this results in the dominance of "science-based" definitions of innovation, at the expense of approaches that bring together innovation policies and the public management of uncertainty by integrating political concerns in the regulation of technology (Winickoff et al. 2005; Bonneuil and Levidow 2012).

Critique then consists in unearthing the coproduction dynamics between public institutions (and how they imagine the conditions of their legitimacy) and regimes of innovation (Jasanoff 2004, 2005), and discussing both their normative strengths and the stability of their outcomes. This style of critique illuminates the political effects of innovation in that it discusses the political subjects that fit in a given regime of innovation at the expense of others; the public problems deemed relevant in a given regime of innovation at the expense of others. This style of critique is also attentive at the coproduction dynamics themselves, or lack thereof, as inno-vation regimes are tacit propositions for social ordering that might leave some people aside.

### Who Performs a Situated Critique

If innovation is understood as the manifestation of institutionalized practices, in nation-states or international organizations, then critique consists in analyzing innovation discourses and instruments so that the political principles underpinning the functioning of these institutions are made visible. In that, this style of critique differs from the critique of epistemic statements based on the evaluation of their truth-value (see section 1). Innovation, here, is not "debunked", but situated. The analyst is positioned in certain configurations, and/or may circulate from one to another to make their consequences visible. Thus, the analytical position of this last style of critique is neither that of integration within innovation processes (as we encountered in Callon's exploration of framing and overflowing) nor is it a bird's eye view from where the truth about innovation can be unearthed (as in the critique of the ideology of innovation). But the proximity with the style of critique that considers innovation as ideology gets more and more evident as the situatedness of the analytical position is less visible.

Consider for instance Philip Mirowski's *Science-mart*, which exposes the gradual develop-ment of discourses and practices of innovation in the American university, as the introduction of legal instruments such as patents and material transfer agreements made it possible to benefit financially from research results originating from publicly funded research, and go to court over the use of proprietary data (Mirowski 2011). In Mirowski's account, the discourse of innovation, whether it originates from innovation scholars or from research institutions, is not separable from the privatization of public knowledge. Here, innovation plays a part in the extension of neoliberalism through the growing influence of legal devices meant to entangle ever more closely the practices of scientific research and the development of commercial applications. Mirowski documents the transformations of academic science and market, and he shows how these transformations are fueled by private interests. For Mirowski, innovation is situated within a regime stabilized not in national or international institutions, but by a general mode of public reasoning, neoliberalism, which subsumes scientific practices under market objectives.

The tone of Mirowski's work is readily identifiable as "critical", because his target is clear. He discusses how neoliberalism relies on hybrid arrangements mixing scientific research, economic interests and political objectives, thereby resulting in the privatization of knowledge. In doing so, he does not seek to situate this neoliberal regime in an American landscape of science policy, which bears strong differences to European ones (Jasanoff 2005; Laurent 2017; Parthasarathy 2017). Nor does he situate this regime in relation to the particular technical sector, biomedicine, which is the focus of his study, whereas others such as climate science would display different configurations (Porter 2013). These details are not what matters to his approach, and one could argue that not delving into them is a condition for neoliberalism as a whole to be considered a stable enough target.

Mirowksi's style of critique occupies an interesting position at the crossroads of the analysis of innovation regimes (in this case, a global regime of neoliberalism) and the uncovering of what lies behind innovation considered as an ideology (here, corporate and market interests). As such, it can be paralleled with the actions of activist groups also critical of innovation. One of such groups is *Pièces et Main d'Oeuvre* (PMO, "parts and labors"), an activist group based in Grenoble, in South-East France,[7] which sees innovation as a perverse dynamic bound to blur boundaries between activities that would be better kept separated (see e.g. PMO 2012). PMO's critical perspective is close to that of scholars like Mirowski. But while academic critics like Mirowski adopt the epistemological standpoint of the academic scholar, able to see what others cannot thanks to his or her methodological rigor and analytical skills, the position of PMO is different. It is based on what the activists call "critical inquiry". It consists in tracking the connections between the worlds of science, industry and politics. It is deeply skeptical of the value of social science, particularly as it attempts to transform innovation processes from within, under the banners of "participation" or "responsibility", which the group interprets as serving the interest of a new class of self-styled experts, only interested in selling pre-packaged participatory mechanisms. But PMO is not much satisfied with social scientists whose have been keen of distancing themselves with calls for "participatory" or "responsible" innovation. Their position has to do with how the activists envision the social world and their own political position within it. They see a world made of conflicting interests, where everyone competes for promoting his or her particular needs. Thus, a scholar will seek recognition and possibly research contracts, and a civil society group will defend its particular interest. Accordingly, PMO has been careful not to be seen as a civil society group, but as a collection of anonymous "simple citizens". "Simple citizen" is the author of numerous of PMO's papers, and an imaginary figure who is radically separated from all domains of social life.

PMO might be an anecdotal example in the landscape of the critique of innovation, yet it provides a crucial lens for us to understand the connections between the critique of innovation as ideology and the critique of regimes of innovation. PMO's critique sees innovation as the outcome of incestuous relations between domains of social activities – science, economy, politics – which would be better kept separated. Developed through the analysis of the historical evolution of science policy of Grenoble (PMO 2012), it is a critique of a regime of innovation,

*Table 21.1        Three styles of critique of innovation*

| Styles of critique | Contesting the ideology of innovation | Considering innovation as an engine for critique | Analyzing innovation regimes |
|---|---|---|---|
| **Theory of innovation** | Innovation as a program based on descriptive statements and associated prescriptions. | Innovation as the productive outcome of framing/overflowing processes. | Innovation as situated in institutionalized economic, political and social configurations. |
| **Objective of the critique of innovation** | – Debunking myths.<br>– Reforming innovation. | – Integrating overflows.<br>– Re-defining technical and social identities accordingly. | – Making co-production dynamics explicit.<br>– Using comparison to explore alternatives. |
| **Identity and position of the critics** | – Scholars and journalists contesting the truth value of innovation discourses.<br>– Reformists in innovation studies. | – Concerned groups affected by exclusion mechanisms.<br>– Scholars studying/working with these groups. | – Analysts circulating from one situated position to the next. |

but less situated in particular institutional constructs than contestable in general. It can partly be described in the terms introduced in the first section of this chapter in that it seeks to reveal the true nature of innovation, which is, for the critics, a power play pursued for the sake of contestable interests. But contrary to the critics of the religion of innovation, no attempts at reform are undertaken. Rather, the very nature of innovation is considered problematic and it is in that sense that this critique can be considered radical. It is also radical in its means of action, in that it seeks to escape the threat of integration in contestable social dynamics (and of being turned into a reformer) by maintaining an impassable analytical distance. As such, no less than the disappearance of the critical individual behind an anonymous figure is required. By bringing all manifestations of innovation into a single set of contestable practices, this radical version of the critique of innovation also loses the institutional situatedness of innovation regimes, and makes the engaged critical subject disappear behind an imaginary anonymous figure.

## CONCLUSION

In this chapter, we have encountered several perspectives that propose to critique innovation. These explorations have led us to analyze different styles of critique, associated with different theories of innovation (see Table 21.1). These styles of critique all articulate descriptive and normative positions. They all produce statements about what things are and what they ought to be. First, the critique of the ideology of innovation often uses a religious vocabulary to confront "myths" with established facts, and propose reforms. Innovation, here, is a discourse meant to describe and prescribe, which the critic should evaluate according to the validity of the representations of the world it proposes, and the consequences of these representations. Second, innovation can be considered as an engine for critique, if seen as a dynamic process always entangling framing and overflowing. In this perspective, critique is part and parcel of innovation, and performed both by concerned groups and by analysts eager to give voice to them. Third, innovation can be understood as situated in heterogeneous regimes that associate it with more or less explicit propositions for social ordering. Critique then consists in making these associations visible, and possibly using comparison to de-naturalize them.

These three styles of critique are not disconnected with each other. The critique of the ideology of innovation is caught in a flow of integration, reform and further critique that echoes the Callonian dynamics of framing and overflowing. This dynamics is itself situated in a particular regime of innovation, which Callon describes as "intensive". The analysis of innovation regimes can lead to unearth systemic forces such as "neoliberalism", or perhaps even more general dynamics that would fundamentally pervert innovation. If one moves away from its situatedness, then innovation can well be described as an ideology.

But the three styles of critique sketched in this chapter also differ according to the role and position of the analyst. While the first one pre-supposes an analytical distance from which one can compare the ideology with actual social processes, the second relies on an engaged position in the midst of innovation processes. The critique of situated regimes of innovation invites the analyst to experiment with original epistemic and political standpoints, as she circulates from one site to others in order to draw comparisons, is attentive to tensions and controversies, and explores the possibilities of alternatives. The analysis of innovation regimes does not offer definite rules to do so. Instead, it suggests that critics ought to experiment with their analytical distances in order to ensure both meaningful scholarly productions and significant political interventions.

## NOTES

1. https://www.theguardian.com/commentisfree/2017/jun/08/uber-embodies-the-toxicity-of-start-up-culture.
2. *MIT Technology Review*, March 18, 2013.
3. The expression is used by Russel and Vinsel (2019).
4. See for instance 'The myths of innovation' in the *MIT Sloan management review* (https://sloanreview.mit.edu/article/the-5-myths-of-innovation/); or the critique of the 'myths' in order to reveal the true nature of innovation processes by popular authors (e.g. Berkun 2010).
5. *MIT Technology Review*, March 18, 2013.
6. The expression "regimes of innovation" is sparsely used in the STS literature (for an approach that is close to my use of the term see: Barben 2007). Innovation scholars have introduced the expression "innovation regimes" to characterize the "principles, norms and ideology, rules and decision-making procedures forming actors' expectations and actions in terms of the future development of a technology" (Godoe 2000, p. 1034).
7. See Laurent (2017); Meyer (2017).

## REFERENCES

Barben, Daniel (2007), 'Changing regimes of science and politics: Comparative and transnational perspectives for a world in transition', *Science and Public Policy*, 34 (1), 55–69.

Barthe, Yannick (2006), *Le pouvoir d'indécision. La mise en politique des déchets nucléaires*, Paris: Economica.

Berkun, Scott (2010), *The myths of innovation*, Sebastopol, CA: O'Reilly Media.

Boltanski, Luc (2011), *On critique. A sociology of emancipation*, Cambridge: Polity.

Boltanski, Luc and Eve Chiapello (2005), *The new spirit of capitalism*, London and New York: Verso.

Bonneuil, Christophe and Les Levidow (2012), 'How does the World Trade Organization know? The mobilization and staging of scientific expertise in the GMO trade dispute', *Social Studies of Science*, 42 (1), 75–100.

Butler, Judith (2002), 'What is critique? An essay on Foucault's virtue', in David Ingram (ed.), *The political: Readings in continental philosophy*, London: Blackwell, pp. 212–28.

Callon, Michel (1999), 'Ni intellectuel engagé, ni intellectuel dégagé: la double stratégie de l'attachement et du détachement', *Sociologie du travail*, 41 (1), 65–78.

Callon, Michel (2007), 'An essay on the growing contribution of economic markets to the proliferation of the social', *Theory, Culture and Society*, 24 (7-8), 139–163.

Callon, Michel (2012), 'Quel rôle pour les sciences sociales face à l'emprise grandissante du régime de l'innovation intensive?', *Cahiers de recherche sociologique*, 53, 121–65.

Callon, Michel (2017), *L'emprise des marchés: comprendre leur fonctionnement pour pouvoir les changer*, Paris: La Découverte.

Callon, Michel and Vololona Rabeharisoa (2004), 'Gino's lesson on humanity: Genetics, mutual entanglements and the sociologist's role', *Economy and Society*, 33 (1), 1–27.

Callon, Michel and Vololona Rabeharisoa (2008), 'The growing engagement of emergent concerned groups in political and economic life: Lessons from the French association of neuromuscular disease patients', *Science, Technology and Human Values*, 33 (2), 230–61.

Callon, Michel, Pierre Lascoumes and Yannick Barthe (2009), *Acting in an uncertain world*, Cambridge, MA: MIT Press.

Callon, Michel, Yuval Millo and Fabian Muniesa (eds) (2007), *Market devices*, Malden, MA: Blackwell.

Chance, Quentin and Morgan Meyer (2017), 'L'agriculture libre. Les outils agricoles à l'épreuve de l'open source', *Techniques & Culture*, 67 (1), 236–39.

Cook, Lisa (2019), 'The innovation gap in pink and black', in Matthew Wisnioski, Eric Hintz and Marie Stelttler Kleine (eds), *Does America need more innovators?* Cambridge: MIT Press, pp. 221–47.

de Saille, Stevienna (2015), 'Innovating innovation policy: The emergence of "Responsible Research and Innovation"', *Journal of Responsible Innovation*, 2 (2), 152–68.

Delvenne, Pierre and François Thoreau (2012), 'Beyond the "charmed circle" of OECD: New directions for studies of national innovation systems', *Minerva*, 50 (2), 205–19.

Dewey, John (2012), *The public and its problems. An essay in political inquiry*, University Park: Pennsylvania State University Press.

Edgerton, David (1999), 'From innovation to use: Ten eclectic theses on the historiography of technology', *History and Technology, an International Journal*, 16 (2), 111–36.

Edquist, Charles (2014), 'Striving towards a holistic innovation policy in European countries – but linearity still prevails!', *STI Policy Review*, 5 (2), 1–19.

Ellul, Jacques (1988), *Le bluff technologique*, Paris, Hachette.

Epstein, Steve (1996), *Impure science. AIDS, activism and the politics of knowledge*, Berkeley, CA: University of California Press.

Ernst & Young and CEPS (2011), Next generation innovation policy. The future of EU innovation policy to support market growth (https://www.ceps.eu/wp-content/uploads/2011/10/innovation_report.pdf).

Evans, James, Gabriele Schliwa and Katherine Luke (2016), 'The glorious failure of the experimental city: Cautionary tales from Arcosanti and Masdar City', in James Evan, Andrew Karvonen and Rob Raven (eds), *The experimental city*, London: Routledge, pp. 218–35.

Fagerberg, J. and B. Verspagen (2009), 'Innovation studies – the emerging structure of a new scientific field', *Research Policy*, 38, 218–33.

Foucault, Michel (1984), *L'usage des plaisirs*, Paris: Gallimard.

Foucault, Michel (1994), *Dits et Ecrits*, vol. 4, Paris: Gallimard.

Foucault, Michel (1997), 'What is critique?', in Sylvère Lotringer (ed.), *The politics of truth, Semiotext(e)*, pp. 41–81.

Fuller, Steve (2010), 'The new behemoth', *Contemporary Sociology*, 39 (5), 533–6.

Godin, Benoît (2006), 'The linear model of innovation: The historical construction of an analytical framework', *Science, Technology and Human Values*, 31 (6), 639–67.

Godin, Benoît (2009), 'National innovation system: The system approach in historical perspective', *Science, Technology and Human Values*, 34 (4), 476–501.

Godin, Benoît (2015), *Innovation contested: the idea of innovation over the centuries*, London: Routledge.

Godin, Benoît (2019), 'How innovation evolved from a heretical act to a heroic imperative', in Matthew Wisnoski, Eric S. Hintz and Marie Stettler Kleine (eds), *Does America need more innovators?* Cambridge: MIT Press, pp. 141–64.

Godin, Benoît and Dominique Vinck (eds) (2017), *Critical studies of innovation. Alternative approaches to the pro-innovation bias*, Cheltenham, UK and Northampton, MA, USA: Edward Elgar Publishing.

Godoe, Helge (2000), 'Innovation regimes, R&D and radical innovations in telecommunications', *Research Policy*, 29 (9), 1033–46.

Halpern, O., LeCavalier, J., Calvillo, N. and Pietsch, W (2013), 'Test-bed urbanism', *Public Culture*, 25 (2), 272–306.

Hyysalo, Sampsa, Torben Elgaard Jensen and Nelly Oudshoorn (eds) (2016), *The new production of users: Changing innovation collectives and involvement strategies*, London: Routledge.

Jasanoff, Sheila (ed.) (2004), *States of knowledges. The coproduction of science and social order*, London: Routledge.

Jasanoff, Sheila (2005), *Designs on nature. Science and democracy in Europe and the United States*, Princeton: Princeton University Press.

Joly, Pierre-Benoît (2017), 'Beyond the competitiveness framework? Models of innovation revisited', *Journal of Innovation Economics & Management*, 1 (22), 79–96.

Joly, Pierre-Benoît (2019), 'Re-imagining innovation', in Sébastien Lechevallier (ed.), *Innovation beyond technology*, Singapore: Springer, pp. 25–45.

Jones, Steven E (2006), *Against technology: From the Luddites to neo-Luddism*, London: Routledge.

Kelty, Christopher (2008), *Two bits: The cultural significance of free software*, Durham, NC: Duke University Press.

Lascoumes, Pierre and Patrick Le Galès (2007), 'Introduction: Understanding public policy through its instruments – from the nature of instruments to the sociology of public policy instrumentation', *Governance*, 20 (1), 1–21.

Latour, Bruno (2004), 'Why has critique run out of steam? From matters of fact to matters of concern', *Critical Inquiry*, 30 (2), 225–48.

Laurent, Brice (2017), *Democratic experiments. Problematizing nanotechnology and democracy in Europe and the United States*, Cambridge: MIT Press.

Law John (2010), 'The Greer-Bush test: on politics in STS', in Madeleine Akrich et al. (eds), *Débordements: Mélanges offerts à Michel Callon*, Paris: Presses des Mines, pp. 269–81.

Leary, John Pat (2018), 'Innovation and the neoliberal idioms of government', *Boundary*, 2 online: http://www.boundary2.org/2018/08/leary/.

Lemke, Thomas (2011), 'Critique and experience in Foucault', *Theory, Culture & Society*, 28 (4), 26–48.

Lepore, Jill (2014), 'The disruption machine. What the gospel of innovation gets wrong', *The New Yorker*, 23, 30–6.

Marres, Noortje (2007), 'The issue deserves more credit. Pragmatist contributions to the study of public involvement in controversy', *Social Studies of Science*, 37(5), 759-780.

Martin, Ben R (2012), 'The evolution of science policy and innovation studies', *Research Policy*, 41 (7), 1219–39.

Martin, Ben R (2016), 'Twenty challenges for innovation studies', *Science and Public Policy*, 43 (3), 432–50.

Matsumoto, Mitsutaka and Kotaro Kawajiri (2012), 'Information and communication technology as an exporter of $CO_2$ emissions', in Karl-Erik Sveiby, Pernilla Gripenberg and Beata Segercrantz (eds), *Challenging the innovation paradigm*, London, Routledge, p. 229.

Mazzucato, Mariana (2015), *The entrepreneurial state: Debunking public vs. private sector myths*, London: Anthem Press.

Meyer, Morgan (2017), '"Participating means accepting": Debating and contesting synthetic biology', *New Genetics and Society*, 36 (2), 118–36.

Mirowski, Philip (2011), *Science-mart*, Cambridge, MA: Harvard University Press.

Oudshoorn, Nelly and Trevor Pinch (2003), *How users matter: The co-construction of users and technologies*, Cambridge, MA: MIT Press.

Parthasarathy, Shobita (2017), *Patent politics: Life forms, markets, and the public interest in the United States and Europe*, Chicago: Chicago University Press.

Pestre, Dominique (2008), 'Challenges for the democratic management of technoscience: Governance, participation and the political today', *Science as Culture*, 17 (2), 101–19.

Pfotenhauer, Sebastian (2019), 'Building global innovation hub: The MIT model in three start-up universities', in Matthew Wisnioski, Eric Hintz and Marie Stelttler Kleine (eds), *Does America need more innovators?*, Cambridge: MIT Press, pp. 191–220.

Pfotenhauer, Sebastian and Sheila Jasanoff (2017), 'Panacea or diagnosis? Imaginaries of innovation and the "MIT model" in three political cultures', *Social Studies of Science*, 47 (6), 783–810.

Pfotenhauer, Sebastian, Joakim Juhl and Erik Aarden (2019), 'Challenging the "deficit model" of innovation: Framing policy issues under the innovation imperative', *Research Policy*, 48 (4), 895–904.

Pièces et Main d'Oeuvre (PMO) (2012), *Sous le soleil de l'innovation*, Paris: La Fabrique.

Porter, James (2013), 'Science-mart, by Philip Mirowski', *Society+Space*, online (https://societyandspace .org/2013/06/13/science-mart-privatizing-american-science-by-philip-mirowski-reviewed-by-james -porter/).

Poulat, Emile (1962), *Histoire, dogme et critique dans la crise moderniste*, Paris: Casterman.

Russel, Andrew and Lee Vinsel (2019), 'Make maintainers: engineering education and an ethics of care', in Matthew Wisnioski, Eric Hintz and Marie Stelttler Kleine (eds), *Does America need more innovators?*, Cambridge: MIT Press, pp. 249–69.

Sanders, Lucinda and Catherine Ashcraft (2019), 'Confronting the absence of women in technology innovation', in Matthew Wisnioski, Eric Hintz and Marie Stelttler Kleine (eds), *Does America need more innovators?* Cambridge: MIT Press, pp. 323–43.

Schumpeter, Joseph (2003 [1943]), *Capitalism, socialism and democracy*, London: Routledge.

Soete, Luc (2013), 'Is innovation always good?', in Jan Fagerberg, Ben Martin and Ebsen Andersen (eds), *Innovation studies – evolution and future challenges*, Oxford: Oxford University Press, pp. 134–44.

Stilgoe, Jack, Richard Owen and Phil Macnaghten (2013), 'Developing a framework for responsible innovation', *Research Policy*, 42 (9), 1568–80.

Stirling, Andy (2008), '"Opening up" and "closing down" power, participation, and pluralism in the social appraisal of technology', *Science Technology and Human Values*, 33 (2), 262–94.

Stirling, Andy (2014), *Democratising innovation*, SPRU, University of Sussex, http:// sussex.ac.uk/ Users/prfh0/innovation_democracy.pdf.

Sveiby, Karl-Erik (2017), 'Unattended consequences of innovation', in Benoît Godin and Dominique Vinck (eds), *Critical studies of innovation*, Cheltenham, UK and Northampton, MA, USA: Edward Elgar Publishing, pp. 137–56.

Sveiby, Karl-Erik, Pernilla Gripenberg and Beata Segercrantz (eds) (2012), *Challenging the innovation paradigm*, London: Routledge.

Talvard, Félix (2018), 'Can urban "miracles" be engineered in laboratories?', in Claudio Coletta, Leighton Evans, Liam Heaphy and Rob Kitchin (eds), *Creating smart cities*, Routledge, pp. 90–103.

Thorpe, Charles (2010), 'Participation as post-Fordist politics: Demos, new labour, and science policy', *Minerva*, 48 (4), 389–411.

von Hippel, Eric (2004), *Democratizing innovation*, Cambridge MA: MIT Press.

Winickoff, David et al. (2005), 'Adjudicating the GM food wars: Science, risk, and democracy in world trade law', *Yale Journal of International Law*, 30, 81.

Wisnioski, Matthew, Eric Hintz and Marie Stelttler Kleine (eds) (2019), *Does America need more innovators?*, Cambridge: MIT Press.

# 22. Collateral innovation: renewing theory from case-studies

*Gérald Gaglio and Dominique Vinck*

## INTRODUCTION

Theories relating to innovation, whether dominant or alternative, are accompanied by discourses. The messages contained in those discourses are promoted by a whole series of different actors who have an interest in peddling them (political leaders, economists, journalists, but also natural scientists, researchers in the social sciences and humanities, management studies, research policy, etc.). Those discourses on innovation are very often positive, if not enchanting (Godin 2015), and have yet to be entirely decoded.

Concerning social researchers, they produce accounts based on studies, at times of ethnographic inspiration, infused with cases monitored over time and focusing on the actors' experience. This experience is captured within the actors' social and material environments (Knorr-Cetina 1997). The cases themselves serve as units of analysis while the research is inductive. Knowledge about the phenomenon of innovation thus relies greatly on case studies. It homes in on the understanding, description and analysis of innovation journeys, like the pioneering study of the propagation of hybrid corn in American farming communities (Ryan and Gross 1943). In the aftermath and just to mention the most quoted one, Everett Rogers adopted a systematic approach in order to theorize the spread of innovations (Rogers 1962).

It exists criticisms of case studies, which are focused on their ability to lead to theorization. Such monographic approach can be perceived as inadequate when the subject of theorization is the phenomenon of innovation in our societies. Based on a case, a problem is posed, and even exacerbated, in order to underline its importance and suggest its potential for becoming widespread. Although well documented, the studies are based on particular cases whose value may be seen as idiosyncratic only. While they could make it possible to identify questions, formulate hypotheses, develop a case model, and even outline a theory, their broader validity remains to be proven, according to many scholars. Case studies may not readily provide a framework that could be used to put together, model or develop a stylized representation of something that could be generalized (Hammersley, Gomm, and Foster 2000). Their use for such a purpose entails the risk of being trapped inside the "disembodied empiricism" criticized by Charles Wright Mills (1959). To put it differently, the problem posed by case studies, for some scholars, is that these studies could not be considered as strong theoretical contributions in the sense that they do not establish a widely-applicable "grand theory", with real-life concepts inserted into the stories recounted, these being only relative reflections of the indigenous accounts.

Nevertheless, case studies tend to reveal regular characteristics specific to a case and a number of mechanisms, rather than acting as a model they are more often of heuristic worth. Whether we are talking about the spread of stereo-photography analyzed by Howard Becker (1982), or the experimental introduction of a new fish-farming method (Callon 1986), the authors do not present a model applicable to other innovations but simply defend the invention of an analytical stance reflecting the complexity and diversity of innovation dynamics. As a result, case studies are valuable to the research effort, when they help to go beyond singularity. That is the reason why this chapter tries to better understand how case studies could help to contribute to critical discussion about pre-existing theories. Through the analysis of the induction of a specific concept, we would question what has been done in the literature with case studies of innovation pretending to propose new concepts.

In order to do so, we will present two case studies, selected for their convergence, and we will show how we used them to forge the concept of "collaterality" and why we have chosen this angle for its theoretical potential as it qualifies in an original way one aspect of the non-linear process of innovation. We have called "collaterality" or "collateral effects" a series of occurrences and ramifications arising from but different from an initial innovation process, and which thereby multiply it. We have suggested that this original process of innovation is taken in unexpected directions that are always related, in some way or another, to the initial directions. In some cases, these initial directions are erased.

Addressing two case studies, that is, comparing and contrasting them, as suggested by Hammersley, Gomm and Foster (2000), then we have developed a concept as an essential step towards theoretical abstraction. In other words, we have proposed a way to reconcile the case study (based on the actors' accounts) and the theorization of/about innovation. In this way, it was possible to better understand how the cases could have contributed to alternative theoretical development. Comparing and contrasting two cases would be a means of building distance and hence moving towards conceptualization and more wide-spread theorization. The notion of "collaterality" hence fuels this first effort to theorize, which in turn should help to further and even partially challenge innovation theories. In other words, the chapter outlines a means of overcoming the problem of monographic case studies as productions devoid of theoretical objectives.

Consequently, after having recalled three concepts that help to understand the non-linearity of innovation processes and their surprises, like most of the authors do when they propose a new concept, we shall describe two case studies leading to the notion of "collaterality" in order to analyze innovation journeys. The first account focuses on an experiment with emergency telemedicine technology set up between retirement homes and an emergency ambulance centre in the North East of France. The second case outlines a process of exploration and progressive rapprochement between doctors and mechanical engineers. It is set in a hospital centre in the North of France where the aim was to provide new solutions to manage prolapsed organs.

Taking the most salient points from the two case studies, the next section is devoted to a detailed presentation of the concept of "collaterality" and its relevance for reporting on innovation processes.

Finally, we shall conclude this chapter with some thoughts relating to the contribution of case studies and their resulting conceptualization to the development of innovation theories.

# WHAT CONCEPTS HELP UNDERSTAND THE NON-LINEARITY OF INNOVATION AND ITS SURPRISES?

Since several decades, innovation processes have been described as oscillatory and vortexing. The actor network theory, as early as the late 1980s, initiated this mode of analysis, starting with a critique of the diffusionist and linear model put forward by Everett Rogers (1962) (see Akrich et al. 2002). The point is to indicate that innovation trajectories cannot be analyzed through the prism of fate and that there is a risk of reconstructing their logic *a posteriori*. It is then necessary to study innovation processes "in the making". However, concepts were previously proposed to account for the non-linear, surprising and uncertain nature of innovation processes.

With the aim of introducing to the case studies and the emergence of the concept of "collaterality", we will focus on three notions or approaches.

First of all, we mention an analytic tradition sensitive to unexpected effects, which cannot be anticipated, resulting from a series of entangled actions. We might notably refer to the "unanticipated consequences" of bureaucracy, highlighted by Robert K. Merton (1936). He proposed the following definition: "Rigorously speaking, the consequences of purposive are limited to those elements in the resulting situation which are exclusively the outcome of the action, i.e., those elements which would not have occurred had the situation not taken place" (ibid., p. 895). In his functionalist analytical framework, "unanticipated consequences" do not correspond to what was expected and they are "dysfunctional", as K.E. Sveiby (2017) reminds us. Following Merton's steps, "unanticipated consequences" were also pointed to in the methodological individualism of Raymond Boudon (1982). This author repeatedly qualified the social world by referring to the emerging (or perverse) effects generated by aggregated individual behaviours. It stems from the intentional approaches of the actors involved, hence at times creating surprises but targeting wanted effects. Moreover, Boudon primarily describes the unwanted effects that actors are forced to endure since their only footing is formed of individual choices. On principle, this way of thinking could help to apprehend non-linear innovation processes where results are not reduced to what was previously expected but could result from the aggregation of individual similar behaviours.

Second, the concept of "function creep" (Winner 1977) could be also useful. It concerns situations in which users do not follow recommendations for use and use an object for something other than what it was designed to be used for, or in a different manner from the one prescribed. Users invent new uses or unearth existing uses having escaped attention. The advent of the short message service (SMS), when some English teenagers in the 1990s decided to exhume a forgotten feature on their phone to avoid paying the exorbitant cost of calls, is often taken to illustrate this mechanism.

Finally, it seems important to take into account the concept of "overflow", introduced by Michel Callon (1998). Moving away from the economic concept of "externalities", whether negative or positive, that is, the effects induced by economic activity (e.g. toxic fumes given off by a factory or the increasing value of the phone regarding the number of its users), Callon prefers to talk about "overflows" to underline the dialectics at work and in which the effort to "frame" (in the Goffmanian sense) becomes obsolete and must be renewed. The overflows are not necessarily perceived as a problem. According to Callon, they are in fact the norm while any framing is always unfinished and incomplete.

The two later concepts are less canonic than the first in the field of social sciences. Nevertheless, they seem relevant to analyze non-linear processes of innovation. Will these concepts be sufficient to analyze the two cases that will follow?

# FROM CASE STUDIES TO THE CONCEPT OF "COLLATERALITY"

## "Collaterality" as a Response to Failure: The Emergency Telemedicine "Case"

The first case study is that of experimental work carried out between 2012 and 2014. It involves 10 nursing homes in an urban area with about 100,000 inhabitants in the North East of France and the emergency ambulance service of the regional hospital. The experiment consisted in testing a telemedicine "case" (term employed by those involved) containing an electrocardiograph, an oximeter (measuring the rate of oxygen in the blood) and a tensiometer. The "case" made it possible to electronically transfer the readings taken by these devices in the different retirement homes to the emergency centre by mail. The defined objective was to prevent very elderly residents from being sent to the emergency ward when this was deemed unnecessary by the emergency doctors. For example, their electrocardiograph plot (ECG) could be taken by the "case" and electronically transferred to the doctors. This would then allow the doctors to decide whether there was a real danger of infarction or not.

The story behind this technical object had followed various twists and turns. Before landing in the retirement homes, the "case" had travelled across other sectors of activity. Containing only an electrocardiograph, it began its life on the ships of the French merchant navy and was used to check for onboard cardiac events. Today, 500 ships are equipped with the case. The company Maciste Technology, which markets the device, was created in 1996 when an engineer suffering from myocardial infarction was struck by the complexity and amount of wires on the ECG systems in hospitals. Once he had recovered, his meeting with a French merchant navy doctor encouraged him to work on simplifying the device and creating a company, which would later be sold.

Several years later, the experimental work studied here stemmed from an encounter between this evolving technical object and professionals in the health and social service sector. Doctor Van Petegen, who was then in charge of the emergency ambulance centre of the regional hospital concerned, was behind the initiative. Having worked as an emergency doctor for 30 years, he was concerned by the increasing number of elderly persons being brought to the emergency ward. He did not feel that they were receiving suitable medical responses to their problems. He therefore began to think about how to slow down this increasing flow of elderly patients and decided that sending out emergency teams to their homes might be avoided in some circumstances. This would help with the shortage of resources (emergency doctors) and equipment (lack of ambulances). During a congress, he learnt about an experiment allowing fire fighters to electronically transfer medical data from their emergency vehicles. He thought that it might be possible to replicate this system within retirement homes. He talked about this idea with Mrs Parinello, the director of two retirement homes in the area. She had also been wondering about how to send her residents less often to the emergency ward as this exposed them to the cold and a long wait, which meant that they often came back to the home in a worse state than when they left. She agreed with Doctor Van Petegen about how it was important to *"remove"* these very elderly persons from the home as infrequently as possible and leave them instead in the warm and cosy environment of the home. An appointment was quickly set up with the start-up Maciste Technology. The company immediately made two devices available to two associative retirement homes in February 2012. In December 2012, an opportunity to benefit from credit from the Regional Health Agency arose, making it possible to equip eight other retirement homes in the sector and buy the first two "cases". A subsidy of 120,000 euros (the case unit cost was 12,000 euros) was thus granted at the end of the year.

In September 2014, a total of 25 ECGs had been electronically transferred to the emergency centre, that is, an average of 8.3 for the six responding retirement homes. The webcam connected to the "case" took six photos, also electronically transferred to the emergency centre. The number of transmissions was thus quite low given the experimental period duration (over two years) and the number of retirement homes taking part. By way of comparison, roughly 400 calls were made to the centre from retirement homes across the whole area (roughly 60) between April and June 2013. Above all, only the two pilot retirement homes and one other electronically transferred data.

There are many reasons for these limited results, which we shall not go into here (see Gaglio 2018). However, we shall focus on the way in which some actors attempted to quash this upcoming failure, by instilling a vital counter-momentum, and on their efforts to explore other ways to turn the experiment into a success. In fact, the "case" itself was put to one side in order to encourage cooperation between the emergency team and the retirement homes. Here it is important to note that a shared understanding of the clinical and medical state of the residents was required for descriptions given during calls to the centre. Keeping this in mind, and at a time when the technical object was not functional (mid 2012), an informal working group was set up. This included Doctor Van Petegen, the coordinating doctor, and the head nurse of the retirement home directed by Mrs Parinello, together with the head of the geriatric ward in the sector's hospital. The idea was to prepare a questionnaire to describe the clinical state of residents to be used by the care assistants in the retirement homes. The questionnaire was to provide them with support and act as a guide when they called the emergency team. The document produced reflects work on categorization. It contains items such as "*first observations*" and "*vital signs*". Different events are then listed (fall, chest problem, dementia, etc.), together with corresponding questions.

The project leaders were proud of this achievement, which was also well received by the financial backers of the experiment during its assessment. This descriptive questionnaire was referred to as the "*Esperanto*" or "*common language*" document to indicate that the ultimate goal was to help the care assistants "*communicate better*" with the emergency centre's doctors by using terms that would be understandable to a doctor, instead of falling back on the usual "well"/"not well" categories. In short, as far as the initiators were concerned, this document promoted the "*professionalization*" of the care assistants and aimed to "*enhance their skills*". The technical object became relevant, not for itself, and not because it was adapted to emergency situations, but because of the approach that it gave way to, which had nothing to do with using or learning to use the device. This achievement ties in with the desire to set up a more effective[1] form of cooperation, encouraging inter-understanding between the retirement homes and the emergency centre (coordination was based on a call to the centre). Furthermore, this drive for greater cooperation also reflects public discourse about "*opening up*" the health and social service sector.

The project leaders also attempted to stave off failure via the invention of "residential" uses of the device in the retirement homes: a "pre-emergency" use and a "reference" ECG use.

We refer to these uses as "residential" as they are linked to a place, or residence, in other words the retirement homes. These uses were designed primarily for the retirement home residents. The actual uses observed in the three homes using the device followed two directions. The first *"pre-emergency"* use was for *"monitoring"* and *"check-up"* purposes, as opposed to use during a *"vital emergency"*. This pre-emergency use is explained in the words of one of the nurses:

> I believe the first time, I was alone one morning. There was a lady complaining about a pain in her chest and a sense of tightness. To be on the safe side, (it was between 8:30 and 9 AM, and I was alone here with my care assistant colleagues), I did an ECG and transferred the reading to the emergency centre. In fact, I did the ECG and transferred it by mail, then I called them immediately after and said, "I've just sent you a lady's ECG reading", and described the symptoms and explained why I'd done it. And they answered me straightaway saying "We're not sure about the results so we need her to come in, we're sending an ambulance."

The nurse took an initiative. In fact, she did not apply the applicable protocol for use which involved waiting for a doctor's order to examine the resident using the "case". Indeed, this protocol proved to be an obstruction to the device's appropriation. What was important for the nurse was not to avoid having to send the resident to the emergency ward but to feel reassured about her state of health. The nurse felt she had time to get out and use the case as it was difficult to anticipate how the resident's symptoms and complaints might evolve.

The other residential use of the device consisted in doing an ECG in a normal situation, as opposed to doing one to check a resident's condition in an emergency situation. The idea was to be able to compare the reference reading to an emergency reading. This approach was based on anticipation and the comparison of two ECGs performed at different times. By encouraging this use, the promoters of the project hoped to facilitate use of the device in an emergency situation by providing opportunities for the home staff to practice using it. Doctor Van Petegen was also behind this initiative, which he saw as a means of backing up the project, which had been otherwise floundering. This approach was only systematically adopted in three of the retirement homes, but nevertheless represented over 200 examinations. These homes went even further in this planned use: they performed an ECG on each new resident as part of an overall health check carried out under the authority of the coordinating doctor. The "case" was thus presented to the families of potential residents. This ECG then became part of the resident's medical record and acted as a guarantee of the home's medical professionalism, making the home stand out with respect to other rival homes.

Let us now further explore the result of these attempts to overturn the potential failure of the project. We shall start with the *"common language"* questionnaire designed for use by the care assistants to describe the residents' conditions. As underlined earlier, the "case" promoters believed very much in this questionnaire. However, it became an inert document. The care staff did not spontaneously refer to it. Indeed, it was often left lying on the "case", as if it were a second set of instructions. When interviewed, the care assistants said the document was a challenge to their professionalism, when in fact it had been designed to improve their professionalism. They considered its content as a simple reminder of the questions they were already likely to ask. In other words, they disputed the argument that their *"skills needed enhancing"*.

With regard to the "reference" ECGs, these were only applied on a widespread basis in three retirement homes out of ten. The main reason for this lies in a controversy about liability involving the head nurses and coordinating doctors. When invited to perform these exams by Doctor Van Petegen, they voiced reservations that halted the device's widespread use. They argued that if an ECG was performed but not analyzed, the home might be held liable in the event that a resident died from an unidentified condition that could have been identified. In short, they did not feel comfortable about sending these ECGs to an online ether, that is, the information system database where the exam would be stored. The coordinating doctors also objected to being held liable and, to this, two other aspects need to be taken into account. On the one hand, they considered that interpreting these ECGs entailed additional work, when they only spent one day a week in the retirement home. On the other hand, they pointed out that they did not regularly read and interpret ECG plots, which, according to several emergency doctors met during the study, was an obstacle to developing genuine expertise.

## Collaterality as a Component of the Innovation Process

The next case study might be considered among those likely to result in a new form of conceptualization. It did indeed lead to the highlighting of a rarely documented phenomenon as well as a first effort at conceptualization. The concept emerging here may lead to a theoretical development able to challenge existing theories to a small extent.

### A.    From qualifying a problem to seeking a solution

In the case presented (the detailed case study was published by El Maleh and Vinck 2019), the story begins with doctors confronted with a health problem, that of prolapsed organs in women. These doctors turned to researchers in mechanical engineering for help qualifying the problem and seeking an innovative solution. Since the nineteenth century, medicine has relied on science and engineering. This has resulted in biomedical engineering and the creation of new medical specialties (radiology, ultra-sound, medical imaging, electro-cardiology, etc.), and the medical device industry. In the case at hand, the problem is an old one and the solution – the use of pessaries to mechanically support the urethra – has been documented since antiquity (Shah et al. 2006).

Confronted with pathological conditions relating to the pelvic system, in particular prolapsus – descent of the genital and rectal organs affecting a high number of women – a group of researchers, gynaecological doctors and surgeons described the problem as damage to the pelvic tissue, especially the ligaments holding the organs in place, caused by ageing or childbirth. Unable to predict the risk of this happening because of various ill-understood problems, the doctors implemented heavy surgical operations such as the insertion of a surgical mesh sized to hold the organs in place or the replacement of ligaments. However, these treatments were difficult to adapt to individuals and the medical team did not understand why they failed. The project was to design adaptable and customizable prostheses (meshes).

Reporting to a large French university hospital, the team focused its attention on the condition seeking to understand the physiology behind this pelvic descent and eventually decided that the system's behaviour would have to be simulated. They planned to study tissue resistance and, drawing inspiration from scientific literature on the mechanical performance of organs, they designed and performed mechanical testing, but were unable to interpret the results.

### B.    A detour via mechanical engineering

In the literature they read, the doctors discovered that there was a mechanical engineering laboratory specialized in the study of rubber performance. The lab had forged a reputation as a mechanical protocol centre for these elastomers. Since the subject had not been very fully explored, and the tyre industry depended on it, the lab had developed behaviour prediction models. The doctors contacted the lab to obtain an explanation for their results on pelvic system organs.

The mechanical engineers effectively saw the similarities between these organs and rubber elastomers but believed the doctors' results could not be used since their test conditions were not standardized: the doctors' test pieces were of different sizes. Nevertheless, their discussions backed up the idea of modelling the pelvic system as if it was made up of an elastic polymer material. The mechanical engineers hoped that this would lead to a new polymer theory application. Over the course of the discussions, the doctors came to believe that this mechanical modelling was essential to understand the system dynamics. Their cooperation thus aimed to understand disorders affecting pelvic statics in order to find better treatment techniques, notably via suitable prostheses. The ultimate goal of their project was the design of personalized prostheses. To achieve this goal, the doctors agreed to take a detour via the design and development of a sensor and a standardized test protocol. The protocol covered tissue sampling (taken from young cadavers, even though dead tissue is different from living tissue), the conditions for dissecting and storing the tissue (freezing), and then performing the traction and compression measurements, taking into account tissue orientation, testing speed (owing to the tissue's visco-elastic behaviour), and the temperature at which the tests were carried out. The tests were to provide reliable data, making it possible to develop a static model of the pelvic system and then simulate its mechanical behaviour (rather than physiological or histological). The simulations thus became a source of knowledge to help design the solutions. Different studies were performed jointly, leading notably to the numerical simulation of a child being born.

Once the protocol was stabilized, the tests performed provided data. The data was then interpreted according to two theories, one relating to the mechanical behaviour of macromolecular polymers with non-linear and viscoelastic behaviour and the other to histological and biochemical aspects. Using the data from the tests and the literature on polymer laws, they built a numerical and 3D geometrical model of the tissue's behaviour. The model was validated by the doctors as being capable of adequately simulating hyper-elastic pathological tissue as well as tissue ageing.

Both the engineers and the doctors used MRI images to identify the parts of the pelvic system and geometrically rebuild each organ making it up. Once the segmentation work had been validated by the doctors, they were able to cut around the organs to define "fixed" areas and "mobile" areas. This made it possible to model the system then analyze it using topological optimization models. Using the functional model, it was then possible to study organ dynamics, without having to refer to their histology. After having studied the movement of the bones, they added the ligaments to the model so that the bladder and uterus movements could be simulated. The end result was that they had produced a functional anatomical model of the pelvic system.

## C.    The by-products of the innovation process become starting points for new projects

The functional anatomical model of the pelvic system developed by the mechanical engineers as a means of producing knowledge held the doctors' attention. They began to imagine a different use for the model, which had not been anticipated: that of a tool for training doctors.

The model could thus enable students to learn how the system worked in an original and interactive way. It could be used both to raise awareness and train doctors. It thus became an educational tool, contributing to the digitalization of medical culture by formalizing medical and mechanical knowledge differently and opening up new routes for learning that did away with the use of cadavers. The next challenge was to transform the scientific culture of doctors while improving their ability to diagnose and prescribe treatments.

At this stage in the collaboration, the engineers and doctors had modelled the pelvic system and improved understanding of the pathologies affecting it, but had not yet come up with personalized treatments for prolapsed organs. The static modelling could be used to link the pressures exerted to organ movements but these pressures would have to be known. Hence a new research problem emerged. This led to the development of a dynamic model of the pelvic system, which resulted in yet another unforeseen innovation in the form of a diagnosis tool. Although the solution initially sought had not been found, several "by-products" had been uncovered along the way: use of the pelvic system models as educational tools (for awareness-raising and training), and use of the dynamic model as a diagnosis tool. These by-products became separate innovative projects. So, from one innovative project a bundle of three innovative projects emerged, all stemming from the intermediary results of the first process and targeting new end goals.

According to the mechanical engineers, for a mechanical system to be perfectly modelled the force and displacement fields had to be determined together with the geometry. As the geometry had been set by the static model, they designed a probe – yet another by-product of the process – to measure the pressures inside the vagina, taking into account the effects of the device on the pelvic system measured, along with the risks of interference with the sensors from the MRI magnetic field and ergonomics. They designed a protocol for pairing the pressure measurements with the displacements observed on the MRI images and developed an electronic and computer system to retrieve and process the signal and data. The invention of this sensor constituted the starting point for a fourth innovative project transferred to industry to create a measurement tool. This allowed other researchers to produce new data about the pelvic system, together with its dynamics and pathologies. The data was published hence contributing to the digitalization and modelling of the human body. An experimental database on tissue was also set up and became an additional resource for research and teaching.

Thus, one of the three by-products of the initial project was able to support research in the biomedical field while the other two influenced the demand for therapeutic solutions. At this point, it must not be forgotten that the initial project itself aimed to produce an offer of treatment solutions.

This account of innovation reminds us that simply inventing a solution is not enough to turn that solution into a widespread innovation. The users (doctors) seized on the intermediary products to meet other needs (e.g. awareness-raising and training of doctors with the development of educational models). These educational spin-off products participated in the innovation process, albeit unintentionally, by creating apparently favourable conditions for the receipt of the therapeutic innovation. In other words, the educational and therapeutic paths followed only seemed to be different since the solution to the patients' problem necessitated the invention of a therapeutic technological offer and the creation of a demand for these products. This demand came from the doctors, who had become aware of the problem, and were capable of understanding the interest of the solution, deciding on the right diagnosis and defining the right treatment strategies. From this point of view, the deflected educational use of the modelling work contributed to the innovation dynamics by acting on demand. The intermediary and spin-off products had an influence on the dynamics by acting on the supply of solutions stemming from the research as well as on demand by raising awareness of this pathology and its solution among the medical public. The various by-products thus fashioned a potential market.

## THE CONCEPT OF "COLLATERALITY": EXPLANATION AND RELEVANCE

Those two case studies lead us to the concept of "collaterality". This concept, that we have to characterize, is not radically new. It has a back history, a background, starting with the concepts and approaches that we presented briefly in the first part of this chapter, which refer to close modes of analysis. Introducing a different term underlines that there is something different within it. But let's return to the former concepts and approach and explain what is different with "collaterality".

For the "unanticipated consequences", while the social world according to Boudon is the result of aggregated individual behaviours in a situation of inter-dependence, our theoretical framework instead suggests an interactional setting up of arrangements leading to some unexpected effects. Furthermore, we also take into account the material aspects of the action, whereas Boudon concentrates exclusively on the rationality of the actors, whether this is axiological or cognitive. At last, to go back to Merton's conceptualization, "the terminology (i.e. "unanticipated consequences") is laden with negative connotations and it has had no impact on innovation literature" (Sveiby 2017, p. 138), which both (the negative connotation and the lack of impact on innovation literature) constitute a problem, according to us.

Second, although it draws inspiration from the notion of creep function, the notion of collateral effects does not need to be tied to the technical object whose acquisition triggers the innovation process. This is what happened with the telemedicine "case". Its use led to the drawing up of a questionnaire for care assistants to describe the symptoms of retirement home residents in an emergency situation. The technical object itself was no longer the centre of attention. Conversely, creep functions never lose sight of the original device, made available to a group or a population. The notion of collateral effects is different for another reason since creep functions contribute to the success of an innovation, even if this innovation is different from the one planned. Here, the inventiveness of users comes into play as they struggle against designers and suppliers. Collateral effects do not necessarily result in success. Again, this is illustrated with the telemedicine "case" since the descriptive questionnaire is not adopted on a broad scale, just like the "reference" ECG. As a result, collateral effects should be seen rather as new occurrences, with their own difficulties and their own paths strewn with as many obstacles as that of the initial project; changing the use of something does not necessarily stabilize an innovation. Detours leading to unexpected collateral effects are not necessarily creep functions, but rather explorations of alternative routes for reaching an initial end goal. Whatever emerges from these detours cannot be foreseen, as with the pelvic system treatment case described.

Third, whereas the concept of "overflow" is close to that of "collaterality", we do not think of this last concept in terms of framings with unintentional overflows. The actors may move away from the initial framing (an objective and means), but they do this intentionally, either to explore other possibilities or to find a solution allowing them to return to the initial frame.

While collateral effects might be considered as overflows, they concern either "intentional overflows" or unexpected effects produced during intentional and theoretically provisional detours. From a heuristic point of view, the notion of "overflow" has the drawback of losing sight of the direction followed by the actors.

To go further, the presentation of our two case studies stemming from the medical field point to adjacent empirical occurrences: crossings, branch trails and unexpected developments. These move away from the linear and eschatological vision of innovation processes which was a long time dominant in research, even this has not been the case for several decades. The question to be raised here is whether a new concept is necessary to characterize these peculiar occurrences. After all, hypothetically speaking, the same events could have happened else-where in a comparable manner. Our answer to this question is "yes", which is why we propose the concept of "collaterality" or "collateral effects" since the peculiar nature of the phenome-non identified is that the different branches of the original innovation process (with its targets, envisaged solutions, etc.) could not have occurred without the problem underlying the initial project being re-raised (Callon 1986, refers to "re-problematization"). This re-problemati-zation produces effects that exceed the directions and strategy initially set up (attribution of resources, planning, etc.). In this way, the effects are "collateral" in that they follow on from a first displacement and first set of events (intermediary results, meeting of unforeseen obstacles, change of actors, etc.), hence broadening the initial scope. Metaphorically, we have drawn inspiration from the notion of "collateral damage" used in times of war, that is, the fact that civilians are unintentionally, if not inevitably, affected by attacks although they are not the initial targets. However, our notion does not take on the negative connotations associated with the term "collateral damage". The meaning we place behind "collateral effects" reflects neither a positive nor a negative judgement. The concept's vocation is to report on a phenomenon and its complexity, rather than to offer a Manichean view of the subject studied.

Examining the history behind the concept of "collateral effects" in the field of innovation has enabled us to characterize it. Let us now explore the notion in relation to the cases studied.

First, with the telemedicine "case", collaterality arises from the solutions to be found faced with the lack of use of the device in emergency situations. Routine use is put in place (the "reference" ECG), along with the standardization (the questionnaire to describe symptoms) of a known interaction between the actors, that is, the call to the emergency centre from the retirement home. With the pelvic system treatment, collaterality stems from the detour taken to produce an intermediary result allowing the initial objective to be pursued. In both cases, the collateral effects concern the uses and/or resources to be put in place and matched to the initial project taking into account the difficulties encountered in the pursuit of the initial objectives. The collateral effects relate to a series of actions peppered with twists, new hopes and cog-nitive investments to counter potential failure (case n° 1) or to achieve a first short-term goal (case n° 2). The actors react to things and explore new paths to further a temporarily fruitless project, even if this means changing it, which is something that has been well documented in the literature on innovation. However, what can be seen in the second case is not a response to potential failure (the chopping off of financing in the first case), but to identified difficulties. In the second case, the collateral effects are produced through an iterative process of discovery and progressive learning, resulting in unprecedented collaboration between the doctors and the mechanical engineering researchers. The actors involved in the project discover realities and ideas for solutions that they had not thought about. They take a number of detours to guarantee a solid base for inventing a solution. However, as the project moves forward they also produce models and measuring instruments whose use goes beyond the intermediary steps of the initial innovative project. In the end, the models and knowledge generated prove to be more useful for training future doctors than inventing a treatment solution, at least in the immediate present.

Second, the collateral effects take the form of transitional products (the transition being towards a final desired result that remains the same), or by-products of the original innovation process. The focal point for analysis lies in the degree of autonomy gained along the new paths taken in relation to the path initially imagined. This autonomy appears systematically, but varies according to the case. For example, in the first case, the "reference" ECG is designed as a means for the care staff to become more familiar with the use of the "case" so that it can be used in emergency situations, notably at night. The potential for collaterality takes on a life of its own and is transformed into the systematic performance of an ECG upon the arrival of new residents in three of the ten pilot homes. In the second case, the autonomy attained is even greater, leading to the creation of a start-up in the medical and medical implant textile field. The spin-off and intermediary products (e.g. the training of young doctors or modelling of the pelvic system) go so far as to partly replace the initial objective, that is, predicting prolapsus in women and defining the appropriate treatment. This is why the "by-product" category to qualify the collateral effects is not entirely satisfactory. On the one hand, studying these products is interesting in itself, if not even more stimulating than the initial project. On the other hand, the innovation dynamics results in productions and associated attempts to produce something that take on a life of their own and follow other innovation paths. This is especially apparent in the second case where the branch paths taken contribute to the digitalization of the human body and provide the foundation for new scientific, educational, therapeutic and industrial developments. The actors are swept up in the dynamics of cooperation, which throws up new knowledge in its wake. This new dynamic movement takes over from the initially targeted objective. When it comes to characterizing the concept of "collaterality", we suggest that these detours and associated creations fashion the sociotechnical environment (in other words the "context", Akrich et al. 2002), create the need and prepare the demand, in such a way that the environment readies itself to receive the initially planned innovation. Collateral effects provide fertile ground for future successes and initiatives coming under the same generic designation as the initial project, which may still be in its infancy. For example, the "case" project made telemedicine more socialized via the dialogue initiated between the homes and the hospital: at the end of the investigation, several nursing homes had set aside a room in their building for telemedicine, without yet equipping it but in view of providing teleconsultations in dermatology.

Third, the collateral effects of an innovation process generate a series of movements creating new potentialities. These movements may relate to professionality as demonstrated through the innovation project to predict and treat organ prolapsus where the doctors and mechanical engineers broaden their range of skills and, at least partly, redefine their professional identity. Or it may be movement with respect to the initial "problematization", as demonstrated in the telemedicine "case" story where unnecessary costly transport of residents from their nursing homes to the emergency centre is avoided. Indeed, the collaterality of the process notably leads towards the challenge of improving shared understanding between the emergency centre doctors and the home care assistants when they communicate over the phone (using the "Esperanto" questionnaire). This mechanism is well-known in actor-network sociology and is subsumed by the pivotal concept of translation (Callon 1986). Yet, here, there is a nuance since several innovation paths cohabit, and may even overlap, while at the same time being triggered by an originating initiative to which they remain attached to varying degrees. To a certain extent, the first innovation process unfolds and its ramifications are studied. These may lead off in directions quite remote from the original epicentre and create new resources and opportunities.

## CONCLUDING THOUGHTS: CASES, CONCEPTS AND DISCUSSION ABOUT EXISTING THEORIES

### The Ability of a New Concept to Challenge Innovation Theories

Both case studies and their cross examination allowed us to develop a concept, which we pretend to be relatively different from those pre-existing. As other authors could have done, we now need to assess the worth of this notion beyond the two cases in which it is grounded. In other words, how does it contribute to theorization of the innovation phenomenon?

Literature on innovation has for a long time been based on a linear conception of the process. The process is portrayed as sequential, starting with the production of scientific knowledge used to develop, industrialize and sell applications, which are then adopted and used. This top-down model, which assumes a high level of scientific autonomy, as well as a firm separation between science and its application, has been disputed (Edgerton 2004; Godin 2006). This model has also been used to reject many authors who have demonstrated through case studies or statistical surveys that the process is not linear but zigzagging, involving trips back and forth between research, development, industrialization, commercialization and use. Several theoretical developments have thus appeared over the course of time. A first model suggests that the process is pulled by the market, which implies that demand works its way up the chain, from the user to science via market research and design offices, before moving back down in compliance with the basic supply model. A second model places either an individual actor (e.g. a Schumpeterian entrepreneur), or an organized group at the heart of the process. This actor moves back and forth from upstream to downstream until the right combination is found, the challenge being to pair science and technology with the market. A third model introduces retroactive loops between different links in the chain, either from one link to the next or by skipping intermediary links. Furthermore, the "open innovation" model as opposed to the "closed innovation" model (Chesbrough 2006), consists in seeking ideas to develop differentiating products outside of the R&D department of a large firm and allowing innovation processes of different origins to decant rather than making a quick decision inside the firm about what needs pursuing and what should be abandoned.

Some theorizations nevertheless break up the linearity by highlighting the displacements and transformations of ideas, intentions, demands, projects and knowledge, which in turn introduce branching, slippage, shifts and reorientations. Combined with the previous model, this less linear conception notably results in a swirl pattern conception of the process. This process does not evolve along a single axis and the axis itself can be twisted and even broken as the innovation progresses, thus leading to something very different from what was imagined.

All these models nevertheless share the fact that thinking gravitates around a unique object. However, some authors have shown that innovations combine often different components that synergize with each other (innovation bundles) or are articulated (e.g. architectural innovation or innovation through subtraction). These situations assume that different innovations and innovation axes interfere with each other and that the interactions between the innovation axes must be added to the back and forth movements and retroactive loops specific to each. These developments in the conceptualization of innovation and the modelling of its processes are backed up by case studies and/or feed into new case studies that facilitate reporting on the complexity of the processes.

Our notion of collaterality developed from the case studies presented in this chapter contributes to the conceptualization efforts listed above. It does not fundamentally question the existing theories. Rather, the notion tends to reinforce theories by pinpointing the importance of non-linearity, actors' interactions, displacements and reorientations, and the emergence of unexpected elements, without considering innovation dynamics as a mere coincidence. The notion completes the non-linear theory of innovation processes by underlining the role of surprising results generated by voluntary detours. These lead either towards new innovative projects taking on a life of their own in relation to the initial project, but not necessarily calling it into question, or towards complementary innovations preparing the way for the success of the initially planned innovation by modifying it and/or readying its receiving environment (e.g. by making the demand originate from this environment).

The concept of collaterality also invites us to defend the paradigm of exploration to describe and analyze innovation processes. It encourages us to go beyond a whole series of useful but paralyzing dichotomies: invention and innovation, supply and demand, intermediary results and final result. It allows us to discuss and move beyond the dichotomy between failure and success, and perhaps even the principle of symmetry on which this dichotomy is based (Gaglio 2018). Finally, it invites us to explore all of the productions and transformations, whether expected or unexpected, and the interdependence arising between them. In this respect, the notion also reinforces theories that refuse to portray innovation as focused on a unique object or a unique process and instead combine a variety of elements that overlap and interfere with each other to varying degrees. It allows us to highlight the importance of by-products and intermediary results as well as the paths that take on their own life and create new arrangements – including in the receiving context – for the initially planned innovation, hence drawing out innovative properties hitherto unimagined.

The concept of collaterality was not conceived with a normative outlook, or with a view to suggesting new best practices, but as a means of characterizing the mechanisms at work in the way innovations are deployed. Nevertheless, the notion does help us to take a step back from the dominant theories and hence allows for a critical approach. The fact that the dominant theories act as a reference for thinking about innovation management policies and methods implies that an alternative theoretical contribution inevitably lends it critical weight with respect to these policies and methods. When an alternative concept is used to promote a political or managerial alternative, it becomes normative and ideological. Furthermore, although the development of this notion did not aim to produce an instrument of criticism, in reality it can be seen as such since it questions the policies and methods in place. It also becomes normative in that it implicitly suggests that other management policies and methods might be conceived. Indeed, such policies and methods might be more in line with the detours, unexpected occurrences, by-products and potentially autonomous sub-projects, or with the intentional exploration and co-building of supply and demand, the inventions and uses, the innovative object and the actors involved, and so on. To put it in other words, and unlike the "unanticipated consequences" concept, collaterality is a positive approach insofar as that surprises or detours are not bad news and do not constitute a problem. In a way, it is the logical flow of things and it could lead to new opportunities. The perspective of collaterality is then "positive", analytically speaking, without venerating innovation.

**The Ability of Case Studies to Support Alternative Conceptual Developments**

Case studies are regularly proposed in literature. They document processes that have not always been conceptualized, for example with respect to the regulation mechanisms and policies interfering with the dynamics of innovation, or negative externalities and the unequal distribution of the spin-offs from innovations, diversion and bypassing strategies, and so on. This wealth of case studies questions existing theorizations and potentially contributes to new conceptualizations. The case studies presented in this chapter also support these efforts to conceptualize and discuss existing theories. They underline a rarely documented phenomenon and outline a concept able to lead to a theoretical development and potentially renew existing theories, at least to a small extent. Such case studies contribute to incremental rather than disruptive theoretical development.

Prior to any conceptual development, case studies first and above all produce accounts. Their narration forms the empirical basis for an analysis taking into account the temporality and complexity of the phenomena studied. As long as they are not overly stylized and, above all, not fashioned by a theoretical framework that would limit their analysis, these accounts provide information, making the discussion of existing theories and models possible and paving the way for new concepts. Thus, the account of the second case described in this chapter teaches us that during the innovation process the interesting results produced are sometimes collateral effects. These were unplanned but take on value because they are used by the actors to generate new innovations and/or consolidate the initial project.

This question of case studies and their associated narratives concerns the entire field of STS (Science and Technology Studies), including suggested modelling of the dynamics at work (e.g. credibility cycles, mode 1 and mode 2, S&T regime theory or cycles in the promise economy). Although concepts are sometimes assimilated with theories – this is the case with the triple helix theory (Leydesdorff and Etzkowitz 1996) – theories about innovation seem to be somewhat lacking. On the one hand, ANT (Actor-Network Theory) presents more as a means of describing innovation dynamics (with the help of methodological principles and conceptual tools) than an explanatory theory (Latour 2005). On the other hand, much research focuses on the production of knowledge and research government modes, scientific and technical discoveries, and the role of users in conception, without even mentioning innovation, whereas this research really concerns techno-scientific transformation dynamics. It exploits existing theoretical developments – such as the absence of a break between science and society or the refusal of technological determinism (Oudshoorn and Pinch 2003) – but does not claim to be theoretical. As the Durkheimian sociologist François Simiand puts it, these researchers expose themselves to critics accusing them of accumulating facts without theory.

Finally, we suggest it would be better to talk about contributing to the theoretical development rather than to the theory of innovation. First, this corresponds to the idea of "middle-range theory" (Merton 1949), in other words the priority given to conceptual, intermediary and situated production, which fits in with the study of innovation processes. Second, producing concepts implies contributing to the theoretical effort on innovation not by submitting a new theory but, more marginally, by discussing and extending existing developments. It is a question of being part of a cumulative framework not of starting from scratch and seeking originality at all costs. Third, this chapter reminds us (beyond the concept of collaterality itself) that extracted from case studies strange or new facts help to conceptualize and then contribute to theoretical discussions.

## NOTE

1.  This is also reflected in the fact that the care staff of the retirement homes participating in the project visited the emergency centre three times.

## BIBLIOGRAPHY

Akrich, Madeleine, Michel Callon, Bruno Latour and Adrian Monaghan (2002), 'The key success in innovation. Part 1: The art of interessment', *International Journal of Innovation Management*, 6 (2), 187–206.
Becker, Howard S. (1982), *Art Worlds*, Berkeley, CA: University of California Press.
Boudon, Raymond (1982), *The Unintended Consequences of Social Action*, London: MacMillan Press.
Callon, Michel (1986), 'Some elements of a sociology of translation: Domestication of the scallops and the fishermen of St Brieuc Bay', in John Law (ed.), *Power, Action and Belief: A New Sociology of Knowledge?*, London: Routledge, pp. 196–223.
Callon, Michel (1998), 'An essay on framing and overflowing: Economic externalities revisited by sociology', in Michel Callon (ed.), *The Laws of the Markets*, Oxford: Blackwell, pp. 244–69.
Chesbrough, Henry (2006), *Open Innovation: Researching a New Paradigm*, Oxford: Oxford University Press.
Edgerton, David (2004), '"The linear model" did not exist: Reflections on the history and historiography of science and research in industry in the twentieth century', in Karl Grandin and Nina Wormbs (eds), *The Science–Industry Nexus: History, Policy, Implications*, New York: Watson, pp. 31–57.
El Maleh, Cédric and Dominique Vinck (2019), 'Innovation collatérale. Lorsque les sous-produits de l'invention préparent la demande en innovations dans le diagnostic et le traitement du système pelvien', *Technologie et Innovation*, 4 (4), https://doi.org/10.21494/ISTE.OP.2019.0412.
Gaglio, Gérald (2018), *Du neuf avec des vieux ? télémédecine d'urgence et innovation en contexte gériatrique*, Toulouse: Presses Universitaires du Midi, collection Socio-Logiques.
Godin, Benoît (2006), 'The linear model of innovation: The historical construction of an analytical framework', *Science Technology & Human Values*, 31 (6), 639–67.
Godin, Benoît (2015), *Innovation Contested: The Idea of Innovation Over the Centuries*, London: Routledge.
Hamel, Jacques (1998), 'Défense et illustration de la méthode des études de cas en sociologie et en anthropologie. Quelques notes et rappels', *Cahiers internationaux de sociologie*, 104, 121–38.
Hammersley, Martyn, Roger Gomm and Peter Foster (2000), 'Case study and theory', in Roger Gomm, Martyn Hammersley and Peter Foster (eds), *Case Study Method. Key Issues, Key Texts*, London: Sage Publications, pp. 234–58.
Knorr-Cetina, Karine (1997), 'Sociality with objects: Social relations in postsocial knowledge societies', *Theory, Culture & Society*, 14 (4), 1–30.
Latour, Bruno (2005), *Reassembling the Social. An Introduction to Actor-Network Theory*, Oxford: Oxford University Press.
Leydesdorff, Loet and Henry Etzkowitz (1996), 'Emergence of a triple helix of university-industry–government relations', *Science and Public Policy*, XXIII, 279–86.
Merton, Robert K. (1936), 'The unanticipated consequences of purposive social action', *American Sociological Review*, 1 (6), 894–904.
Merton, Robert K. (1949), *Social Theory and Social Structure*, New York: The Free Press.
Mills, Charles W. (1959), *The Social Imagination*, Oxford: Oxford University Press.

Oudtshoorn, Nelly and Trevor Pinch (2003), *How Users Matter: The Co-Construction of Users and Technologies*, Cambridge MA: MIT Press.

Rogers, Everett (1962) (1st edition), *Diffusion of Innovations*, New York: Free Press.

Ryan, Brycw and Neal Gross (1943), 'The diffusion of hybrid seed corn in two Iowa communities', *Rural Sociology*, 8 (1), 15–24.

Shah, Sheetle M., Abdul H. Sultan and Ranee Thakar (2006), 'The history and evolution of pessaries for pelvic organ prolapse', *International Urogynecology Journal*, 17 (2), 170–75. https://doi.org/10.1007/s00192-005-1313-6.

Sveiby, Karl-Erik (2017), 'Unattended consequences of innovation', in Benoît Godin and Dominique Vinck (eds), *Critical Studies of Innovation: Alternative Approaches to the Pro-Innovation Bias*, Cheltenham, UK and Northampton, MA, USA: Edward Elgar Publishing, pp. 137–55.

Winner, Langdon (1977), *Autonomous Technology: Technics-out-of-Control as a Theme for Political Thought*, Cambridge: MIT Press.

# Conclusion to the *Handbook on Alternative Theories of Innovation*

*Gérald Gaglio, Dominique Vinck and Benoît Godin*

Now that we have reached the end of this Handbook, let us return to its underlying ambition. Our goal was to analyze and assess the approaches put forward over at least the past twenty years and whose aim has been to rethink the understanding of the phenomenon of innovation, in particular by suggesting new avenues for reflection. An associated and equally important objective was to investigate whether new innovation theories were in the making or gaining ground (without these necessarily being promoted as such by the authors), in relation to the lessons learnt and established theories, whether widespread or not, presented by Benoît Godin (Chapter 3). As part of this dual challenge to pinpoint and decode such approaches, it was suggested that the authors' viewpoint should be historical (by reporting on the genesis of notions and their developments), but also conceptual and critical. This was so that the type or area of the innovation dealt with, or the academic discipline used as the authors' starting point and on which their innovation thinking is based, could be thoroughly explored. In short, the authors of the Handbook were provided with a mission statement. This appeared to be essential in order to identify what is potentially original or alternative, according to our terminology, and theoretically fertile in the approaches developed by focusing mainly on the shaping and circulation of ideas in a given academic field. The idea was therefore to contribute to feedback on theoretical production in a given area and not to produce more empirical work based on detailed in-the-field investigations.

The Handbook therefore deals with a good many of the notions, approaches and disciplines conceptualizing innovation, from this recommended triple perspective, without managing to cover the full array identified initially. The main reason for this is the incredible profusion of concepts and notions accompanying innovation, often in adjectival form, which strive to characterize it afresh or rather to "locate" it in a given area. By drawing inspiration from the idea of X-innovation (Gaglio, Godin, Pftotenhauer, 2019), the contribution of Mónica Edwards-Schachter (Chapter 5) gives a perfect account of this thick jungle of ideas, painting a rich and almost exhaustive panorama. It would not have been possible to find contributors for each of these X-innovations. A knowledgeable reader will also notice the absence of disciplines such as management and psychology in the first part of the Handbook. Concerning management, notably in the field of organizational theory, there is a patent link with innovation (Lam, 2005), especially in terms of the organizational forms most able to help with innovation (in the sense of producing novelties). With respect to psychology, the themes of creativity and ideation (with the methods to back them up) are often preferred to that of innovation, probably due to the focus on invention rather than on innovation processes as a whole. In spite of a few exceptions here and there, our discussions with authors from these disciplines were not conclusive, possibly because our perception of their contribution was too specific – there is manifestly very little critical questioning on innovation in these disciplines – or because, at closer range, the work and skills of the authors contacted proved to be further removed from our expectations than planned. Discussions also stalled with authors who appeared too interested in a promotional (and self-promotional) approach with respect to a supposedly innovative concept linked to innovation. The idea behind this Handbook is not to provide a showcase where authors can exhibit the type of innovation they recommend, like in a supermarket, but to query theoretical production in a given area. Furthermore, beyond this concept marketing, the small misadventures we experienced during our enrolment of authors revealed a more fundamental and thorny issue, which seemed difficult to avoid altogether (but we will let our readers be the judge of this): how can any form of normativity be neutralized when it comes to innovation and, even more importantly, alternatives to the dominant ways in which it is addressed? As to whether normativity is a genuine problem to be evacuated without remorse or assumed and dealt with through a theme-based approach in an effort to drive new theoretical developments, this remains an open question. After all, the social sciences have the peculiar characteristic of being founded on values, as argued by Max Weber (1978 [1922]). In fact, theoretical conceptions are at work in action, either explicitly or inexplicitly, especially when it comes to narrating the innovation addressed by, among others, Carolina Bagattolli (Chapter 15), the innovation policies analyzed by Markus Grillitsch, Stine Madsen and Teis Hansen (Chapter 16), or the innovation measuring instruments discussed by Giulio Perani (Chapter 17).

Nevertheless, in spite of these gaps in terms of disciplines covered and the knotty question of normative and promotional approaches, this Handbook provides several major contemporary starting points to think about innovation. The chapters contained within its pages are concise and adopt the three suggested viewpoints (historical, critical and analytical) to varying degrees. They also report on how some disciplines relate to the theme of innovation, as in Chapter 4 by Irwin Feller with respect to economics, or problematize a discipline's lack of interest in innovation as well as the possible paths to be followed to get a better grasp on this research subject. This is the case of Ulrich Ufer and Alexandra Haustein with respect to anthropology (Chapter 19) and Vincent Blok regarding philosophy (Chapter 20). Quite different approaches offer food for thought by looking at the relationship between innovation and religion, which is what Benoît Godin (Chapter 1) and Boris Rähme (Chapter 18) explore, or the role of the imaginary (Chapter 2 by Harro van Lente), offering new theories and innovations. The bibliographic information provided will allow those readers with a specific interest in a subject to explore it further.

What can be said about the initial ambition to explore the emergence of alternative innovation theories? First, this Handbook serves as a reminder, if one were necessary, of what the alternatives actually oppose, or could oppose, of what they stand against or could stand against:

(a) an approach to innovation focused exclusively on economic profit, propped up by its market dimension alone in the commercial and utilitarian sense;

(b) an assimilation between innovation and industrial and technological innovation, to the detriment, if not exclusion, of other dimensions including social innovation, as addressed by Cornelius Schubert (Chapter 6);

(c) an acritical approach considering innovation as a good in itself: Brice Laurent (Chapter 21) shows how innovation can be at once a "target of criticism" and a "driver of criticism", which leads him to distinguish between different types of criticism;

(d)     an obligatory reference to Everett Rogers' linear diffusion model to think about the phenomenon of innovation and the conditions for its concrete deployment given that this model is partly outmoded, as argued by Gérald Gaglio and Dominique Vinck (Chapter 22) with the notion of "collaterality", or Daryl Cressman who, in his own way, deconstructs the concept of disruptive innovation (Chapter 11).

The initial ambition was to scrutinize allegedly new, contemporary and successful approaches or theorizations of innovation, not to promote new theorizations. To put it another way, the aim was to examine existing approaches and theories of innovation. However, and in spite of the considerable amount of writing about innovation, there are no theories of innovation appearing on the horizon. This raises questions, which we shall now attempt to shed light on. Thus, it has to be said that none of the starting points addressed leads us to a conceptual architecture on which a theory with the potential to be generally applied and explain a vast array of phenomena can be founded. Such an observation is undeniable. Moreover, it is reassuring and can be seen positively as a form of resistance to the all-encompassing excesses of the "grand theory" (Mills, 1959) of which Talcott Parsons remains the principal embodiment. Yet, the authors of this Handbook do not evoke this argument. Instead, based on the associated literature, some build local "middle range" theories (Merton, 1949) that help to describe and categorize parts of reality by digging small inroads. Indeed, these are not so much imposing theories but rather angles of attack on innovation. Thus, Lucien von Schomberg addresses the moral dimension (Chapter 8), while Frank Boons and Riza Batista (Chapter 7), Fayaz Ahmad Sheikh and Hemant Kumar (Chapter 13) and Céline Cholez and Pascale Trompette (Chapter 14) turn their attention to the environmental dimension. Peter Swann concentrates on common innovation (Chapter 12), Bastien Tavner on user-centred innovation (Chapter 9) and Tiago Brandão on open innovation (Chapter 10), to develop an innovation perspective that might be qualified as "democratic" since this innovation does not have and is not limited by expert brains monopolizing operations. In sum, these dimensions act as firewalls against the dominant views of innovation. They bring to light realities or ways of doing things that have been undervalued or sometimes eclipsed.

    Finally, we also feel that there is a risk of creating subfields, devoid of any interactions and suffering from mutual blindness. Many close links can be drawn between innovation thinking (frugal innovation, sustainable innovation, grass-roots innovation), without these necessarily leading to debates. By at least bringing together similar themes within a single work this Handbook is an invitation to start a dialogue. The authors sometimes defend the specific aspects of an approach or notion without necessarily lending a critical eye to other approaches (see notably Chapter 12 on "Common Innovation" by Peter Swann). They do not want to argue with other authors. Yet, it would be better to dig much deeper and embark on a constructive confrontation to prevent a proliferation of sub-communities cut off from each other in spite of the obvious bridges that could be built. There seems to be some considerable overlapping and general convergence, which could indeed lead to the unfolding of alternative innovation theories. What is able to subsume highly localized knowledge production spaces? Unfortunately, this proliferation of diverse approaches reluctant to dialogue with each other stems from the pressure weighing down on the academic environment with its push for originality and distinction. The existence of specialized publication places and the race for publication, exhorting authors to impose their own brand or choose their own (restricted) discussion forum, are not conducive, in our view, to adopting the right way forward.

Another factor of fragmentation is the relative discretion of traditional academic disciplines among the alternative perspectives proposed (see Parts III and IV in this Handbook). In fact, the field or specific subject prevails without claiming to be rooted in or originate from a specific discipline. These disciplinary origins and roots can be guessed at, are at times suggested but never structure the arguments developed by the authors. Consequently, references to disciplinary debates beyond the specific subject discussed or innovation theme are rare or entirely absent. The causes and consequences of the erosion of disciplinary grounding behind the work presented in this Handbook can be questioned. This erosion must largely stem from the internationalization of research and publication spaces, with their increasingly narrowed focus on specific themes. To a certain extent, this is something to be happy about with the parallel emergence of inter and even cross-disciplinary studies focused on reciprocally fertile research fields: Science and Technology Studies, in spite of their diversity, are a good example. This field would not have developed if the figureheads had not freed themselves of their disciplinary shackles with their proclivity to immobilize and exclude. On the other hand, disciplinary roots can act as supports for producing theories and contributing to a cumulative theoretical production process. Their epistemology is firmly grounded, they are manifestly able to administer proof and be sure of their facts when faced with the question, "what is the scale and target of the learning and knowledge produced?" Is this necessarily the case with "studies"? And to what type of studies do the contributions in this Handbook belong? Are they part of "innovation studies" (which could thus be likened to a large church with multiple chapels instead of being, as is the case now, a subfield characterized by a specific approach)? Or are there as many "studies" as there are themes derived from innovation? As indicated above, the risk here would be to extend ignorance and build even more fences than at the time (still recent and partly current) when disciplines were the cornerstone of conceptual and theoretical evolution. Consequently, would it be better to rely upon the "studies" operating mode or be more conventional and base reasoning on disciplines when working on alternative innovation theories? Extensive arguments would have to be developed to answer this question but the tendency is to think that the two are not incompatible.

We must also heed the invitation of Irwin Feller in this Handbook (Chapter 4). According to this economist, bridges need to be built between disciplines. If we were to develop this idea further, thinking must be based on disciplinary foundations and reasoning modes so that gaps able to be filled by other disciplines can be filled. Despite the wishful thinking behind this idea (even though we fully adhere to it!), and as underlined by the author, this supposes above all that the last thing we should do is confine ourselves to a caricature of the discipline with which we want to enter into dialogue. According to Feller, this is often what economics suffers from, in spite of its diversity.

An alternative has been proposed by the actor-network theory (ANT) – sometimes referred to as "sociology of innovation" – which both in theory and practice encourages a clearly interdisciplinary and "interfiled" transversality. However, this "theory" is not a theory in the sense of structured concepts and explanatory hypotheses. It above all strives to be an infra-language, offering purposefully weak concepts – entity, relationship or network, for example – which above all help to free it from overly numbing theories or conceptual dualities in order to drive the investigation further. Theoretical production is therefore local and closely linked to specific empirical investigations with hardly any theoretical development whether of the grand or middle range variety. One – a conceptual infra-language – does not cancel out the other – the structuring of grand or middle-range theories – but in practice ANT has not (yet) led to the production of alternative theory.

This discussion brings us at last to the conditions for the possible emergence of alternative theories relating to innovation, which could quite simply take the form of a consolidation or generalization of the already structuring approaches developed in this Handbook.

At least three paths seem to open up before us. The first is to more closely target the type of theories or concepts we wish to produce: are they analytical, do they have a normative side (and to what extent), or do they simply aim to be descriptive? In short, on what *episteme* do we wish to found the conceptualization and theorization we are proposing? Regarding the idea of targeting specific theoretical developments, the identification of structuring or cross-cutting themes upon which to meditate (ethical, environmental and democratic innovation in particular), as already noted, is another path likely to engender more general theorization.

Thus, by picking out the blind spots in current scholarly productions, it is possible to identify lines of questioning that can then be used to put forward conceptual developments. This is the case of the question of incentives, addressed by Markus Grillitsch, Stine Madsen and Teis Hansen (Chapter 16) with respect to innovation policies but which theoretically also makes sense for all of the works exposed in this Handbook whether with respect to the opening of the innovation process, its democratization, or its connections with ethical and societal questions, even though this question is not explicitly dealt with in these different approaches. The thing that drives actors to innovate could also be dealt with as a cross-cutting theme and lay the foundation for the development of a theory. The same applies to the question of agency, including materiality and infrastructures, or indeed to that of politics in its various accepted forms (Latour, 2007), not only as a procedure for governance and the establishing of sovereignty, but also as a public affair, involving controversy and the constitution of a public – in the sense of John Dewey – or creating worldly or cosmopolitical upheaval as new, notably technological entities are introduced and imposed, or in the sense of the routine management of problems whose political dimension implies work on genealogy in the sense of Michel Foucault. Another theme emerging from the different contributions, which is rarely explained and could benefit from cross-cutting theoretical developments, is that of uncertainty and collective exploration. While these aspects are very much present in the various contributors' approaches, they are rarely explicitly addressed as a theme. Finally, it would also be useful to explore the themes that are particularly absent but whose importance in other areas linked to innovation is growing. This is the case of maintenance and repair studies, which report on what happens once the innovation has taken place hence providing the first positive feedback from an adopter or user. Here the idea is not so much to study the diffusion but rather the effective inscription of the novelty within collectives and the development of efficient routines to ensure the novelty's performance. Maintenance, repairs, continuous improvements, adjustments, and so on are undervalued dimensions in the conceptualization of innovation which, at the end of the day, always gives priority to design over use and disruption over routine. Research work focuses more on the question of origin – where the type of innovation addressed has "come" from – rather than what it "becomes" once the innovation has been established.

The second path is to quite simply dare to conceptualize and theorize more. This would not be just for the beauty of it or to derive pleasure (often fleeting) from the novelty but for the risk (in the positive sense of the term) of pushing knowledge even further, helping other authors to think, laying out hypotheses, contradicting these, and so on. In short, conceptualization and theorization are always heuristic and inspiring since they help to describe the world and its phenomena. They provide insight into the real functioning mechanisms of our societies, beyond statements of principle and opinions steered by interests. Ultimately, as several chapters in this Handbook underline in specific sections – especially the contributions dealing with alternative innovation approaches or types in Parts III and IV – conceptualizing and theorizing involve summarizing the research work of other authors and seeing what brings them together, what differentiates them, and how public discourse and institutional watchwords are relayed in scholarly circles (and vice versa). An objective of future research would be to focus more on the way concepts circulate and shift by extending the investigation, that is, by asking whether (and how) experimentation linked to a specific type of innovation returns to a concept and helps it to evolve: Chapter 14 by Céline Cholez and Pascale Trompette on frugal innovation attempts to do just this, as does Chapter 9 by Bastien Tavner on user-centred innovation.

The last path suggested is to go beyond the mission entrusted to the authors in order to decipher possible alternatives in terms of thinking about innovation. In this respect, we should notably refer to Chapter 5 by Mónica Edwards-Schachter in which she proposes to group together indigenous concepts, which leads to typologies of typologies stemming from X-innovation. The method, proposed by Frank Boons and Riza Batista (Chapter 7) or Tiago Brandão (Chapter 10) based on bibliometrics and data analysis also appears to be a fertile approach for picking out phenomena and pinning down what they might imply.

## BIBLIOGRAPHY

Gaglio, Gérald, Godin, Benoît and Pfotenhauer, Sebastian, 2019, X-Innovation: Re-Inventing Innovation Again and Again, *NOvation*, First Issue, Introduction of the special issue, pp. 1–17, file:///C:/Users/ggaglio/AppData/Local/Temp/8-Article%20Text-47-4-10-20200226.pdf.

Lam, Alice, 2005, Organizational innovation, in Fagerberg, Jan, Mowery, David C. and Nelson, Richard R. (eds), *The Oxford Handbook of Innovation*, Oxford, Oxford University Press, pp. 115–47.

Latour, Bruno, 2007, Turning Around Politics: A Note on Gerard de Vries' Paper. *Social Studies of Science*, 37 (5), 811–820. https://doi.org/10.1177/0306312707081222.

Merton, Robert King, 1949, *Social Theory and Social Structure*, New York, The Free Press.

Mills, Charles Wright, 1959, *The Sociological Imagination*, New York, Oxford University Press.

Weber, Max, 1978 [1922], *Economy and Society, an Outline of Interpretive Sociology*, Berkeley, CA, University of California Press.

# Index